装备科技译著出版基金

# Intelligent Materials and Structures
## 2nd Edition

# 智能材料与结构

## （第2版）

［以色列］海姆·阿布拉莫维奇（Haim Abramovich）著

汪刘应　葛超群　刘顾　王龙　等译

许可俊　赵雪琦　审校

国防工业出版社

·北京·

著作权合同登记　图字：01-2023-2902 号

图书在版编目（CIP）数据

智能材料与结构：第 2 版/（以）海姆·阿布拉莫维奇著；汪刘应等译．—北京：国防工业出版社，2024.3
书名原文：Intelligent Materials and Structures（2nd Edition）
ISBN 978-7-118-13037-9

Ⅰ．①智…　Ⅱ．①海…②汪…　Ⅲ．①智能材料　Ⅳ．①TB381

中国国家版本馆 CIP 数据核字（2024）第 045138 号

Abramovich, Haim. Intelligent Materials and Structures © Walter de Gruyter GmbH Berlin Boston. All rights reserved. This work may not be translated or copied in whole or part without the written permission of the publisher（Walter De Gruyter GmbH, Genthiner Straβe 13, 10785 Berlin, Germany）.

本书简体中文版由 Walter De Gruyter GmbH 授权国防工业出版社独家出版发行。版权所有，侵权必究。

※

国防工业出版社出版发行
（北京市海淀区紫竹院南路 23 号　邮政编码 100048）
雅迪云印（天津）科技有限公司印刷
新华书店经售

*

开本 710×1000　1/16　插页 5　印张 27¼　字数 494 千字
2024 年 3 月第 1 版第 1 次印刷　印数 1—1500 册　定价 188.00 元

（本书如有印装错误，我社负责调换）

国防书店：(010) 88540777　　书店传真：(010) 88540776
发行业务：(010) 88540717　　发行传真：(010) 88540762

# 译 者 序

智能材料与结构是能够对环境行为进行感知、响应并优化自身功能的一种集传感、控制和驱动于一体的材料结构系统，因其在高科技前沿领域的广泛应用前景而备受国内外学者关注。智能材料与结构的研究和发展涉及多学科交叉与多技术融合，与材料、信息、生命科学等前沿学科以及光电技术、信息技术、仿生技术、材料设计与合成技术等多种高新技术领域密切相关，具有较高的热度和较强的前沿性。当前，国内智能材料与结构的研究尚处于快速发展阶段，但在智能材料与结构的基础理论和设计技术、多功能化与集成化等方面仍需要进行大量探索性研究，这也是译者翻译本书的主要原因。

本书回顾了当前智能材料与结构的最新进展，采用独立的章节对层压复合材料、压电材料、形状记忆合金、电流变和磁流变体、电致伸缩材料和磁致伸缩材料等典型智能材料与结构的理论和应用进行了详细介绍，运用大量方程和图表重点介绍了智能材料与结构研究中涉及的材料和结构基础理论、物理模型和应用性能。书中包含大量的理论分析和应用实例，为读者掌握智能材料与结构的基础理论和设计方法提供了指导。

全书共 10 章。第 1 章概述压电材料、形状记忆合金、电流变和磁流变体、电致伸缩和磁致伸缩材料等典型智能材料与结构；第 2 章介绍层压复合材料；第 3 章介绍压电材料；第 4 章介绍形状记忆合金；第 5 章介绍电流变体和磁流变体；第 6 章介绍磁致伸缩材料和电致伸缩材料；第 7 章介绍智能材料在航空航天、医学等领域中的应用以及压电发动机研究的新动态；第 8 章介绍基于智能材料的能量收集；第 9 章介绍光纤；第 10 章介绍压电纤维复合材料、声能采集、基于形状记忆合金和压电传感器的能量收集等。

本书第 1 章由刘顾翻译；第 2 章由赵雪琦翻译；第 3、4 章由葛超群翻译；第 5~7 章由王龙翻译；第 8 章由许可俊翻译；第 9、10 章由王伟超翻译。全书由汪刘应统稿，许可俊、赵雪琦、王伟超、王滨等校稿。此外，在本书的翻译过程中，得到了智剑实验室（火箭军工程大学）各位领导和同事的支持与帮助，以及复旦大学吴仁兵教授和中国兵器工业第五九研究所张天才研究员的指导，在此一并表示感谢。感谢陕西省"特支计划"科技创新领军人才项目、陕西高校青年创新团队给予的支持。特别感谢国防工业出版社对本书翻译与出版给予的热心指导，感谢装备科技译著出版基金给予的资助。

  在本书的翻译过程中，译者力求忠实于原著，但由于水平和经验有限，翻译疏漏和不妥之处在所难免，敬请读者不吝批评指正，以便日后继续完善，不胜感激！

译者<br>2023 年 8 月

# 前　　言

《智能材料与结构》第 1 版于 2016 年出版，旨在利用相对较新的多功能材料传播先进的工程理念。本书编写的目的是让学生和学者更好地理解智能/机敏结构这一新出现的跨学科主题。本书可作为希望研究智能材料与结构的研究生的入门书籍，在这里他们能够获得必要的物理和数学工具来理解这个新工程领域的各种复杂问题。

本书以作者在过去 5 年里发表的关于智能结构创新研究为基础，更新了第 1 版的内容。

本书包含了第 1 版的 8 章内容，即关于压电材料、形状记忆合金（Shape Memory Alloys，SMA）、电流变和磁流变体、磁致伸缩和电致伸缩材料等主要智能材料与结构的详细介绍章节；层压复合材料包括经典层压理论和一阶剪切变形理论；压电及其本构方程和各种梁、板模型；SMA 和超弹性；电流变体和磁流变体；磁致伸缩材料和电致伸缩材料；智能材料在航空航天、医药领域等结构和器件中的应用以及压电发动机的新面貌。第 8 章介绍了使用基于压电和电磁的装置进行能量收集的基本方程及其相关文献。

本书增加了两个新章节：第 9 章介绍了光纤，第 10 章介绍了其他主题，如配备形状记忆合金丝增强板材的抗弯性能、压电纤维复合材料、声能采集、使用形状记忆合金的能量收集和基于压电传感器的道路交通能量收集。

除了这些新章节，第 3 章还增加了关于压电耦合系数定义的附录，第 8 章增加了两个小节，涉及振动激励下压电双晶片电源和具有增强频率带宽的压电收集器。

希望本书能够以更广泛的视角介绍各种智能结构，使读者更加熟悉智能材料与结构这一新的工程主题。

作者希望感谢他的妻子 Dorit 以及他的孩子 Chen、Oz、Shir 和 Or，感谢他们给予的支持、理解、爱和奉献。他们在整个写作期间的持续支持使本书得以出版。

海姆·阿布拉莫维奇
以色列 海法，2021 年 4 月

# 目 录

**第1章 智能材料与结构概述** ·································· 1
1.1 智能材料的类型 ········································· 1
1.2 压电材料、形状记忆合金、电流变体和磁流变体以及
磁致伸缩和电致伸缩材料综述 ························· 19
   1.2.1 压电材料 ········································ 19
   1.2.2 形状记忆合金 ···································· 21
   1.2.3 电流变体和磁流变体 ······························ 23
   1.2.4 磁致伸缩和电致伸缩材料 ·························· 23
1.3 智能材料的典型特性 ····································· 24
1.4 智能材料与系统的应用 ··································· 34
参考文献 ··················································· 40

**第2章 层压复合材料** ········································ 83
2.1 经典层压理论 ··········································· 83
   2.1.1 简介 ············································ 83
   2.1.2 位移 ············································ 84
   2.1.3 应变 ············································ 84
   2.1.4 应力 ············································ 85
   2.1.5 正交各向异性材料 ································ 85
   2.1.6 单向复合材料 ···································· 87
   2.1.7 单层的特性 ······································ 88
   2.1.8 应力和应变的转换 ································ 89
2.2 一阶剪切变形理论模型 ··································· 96
2.3 附录一 ················································· 101
   2.3.1 附录 A ·········································· 101
参考文献 ··················································· 102

**第3章 压电材料** ············································ 109
3.1 本构方程 ··············································· 109

VII

  3.1.1 介电材料分类 ······ 110
  3.1.2 重要的介电参数 ······ 111
 3.2 销力梁模型 ······ 127
 3.3 均匀应变梁模型 ······ 139
 3.4 伯努利-欧拉梁模型 ······ 147
 3.5 一阶剪切变形（Timoshenko 型）梁模型 ······ 157
  3.5.1 静态和动态情况的解 ······ 163
  3.5.2 由 $E_{11}^0$ 产生的轴向诱导力计算 ······ 163
  3.5.3 小横向振动的运动方程的解 ······ 166
  3.5.4 静态情况的解 ······ 169
  3.5.5 诱导轴向力的实验验证 ······ 173
 3.6 压电片复合板 ······ 175
 3.7 附录二 ······ 190
  3.7.1 附录 A ······ 190
  3.7.2 附录 B ······ 192
  3.7.3 附录 C：表 3.4 中所示的常数 ······ 192
  3.7.4 附录 D：表 3.5 中所示的常数 ······ 193
  3.7.5 附录 E ······ 194
 参考文献 ······ 194

## 第 4 章　形状记忆合金 ······ 198
 4.1 形状记忆合金的基本性能 ······ 198
  4.1.1 形状记忆效应 ······ 201
  4.1.2 超弹性 ······ 203
  4.1.3 应用 ······ 204
 4.2 本构方程 ······ 210
  4.2.1 形状记忆合金材料的一维本构方程 ······ 211
  4.2.2 电流加热的形状记忆合金材料 ······ 216
 4.3 文献中的形状记忆合金模型 ······ 218
 参考文献 ······ 224

## 第 5 章　电流变体与磁流变体 ······ 227
 5.1 电流变体与磁流变体的基本特性 ······ 227
 5.2 电流变体和磁流变体建模 ······ 239
 5.3 电流变体和磁流变体的阻尼 ······ 248

5.4 附录三 ·258
  5.4.1 附录A ·258
  5.4.2 附录B ·260
  5.4.3 附录C ·260
参考文献 ·261

# 第6章 磁致伸缩与电致伸缩材料 ·268
6.1 磁致伸缩材料的特性 ·268
6.2 磁致伸缩材料的本构方程 ·275
6.3 电致伸缩材料的特性 ·281
6.4 电致伸缩材料的本构方程 ·285
6.5 附录四 ·292
  6.5.1 附录A ·292
  6.5.2 附录B ·294
参考文献 ·300

# 第7章 智能材料在结构中的应用 ·303
7.1 航空航天领域 ·303
7.2 医疗领域 ·321
7.3 压电发动机 ·327
  7.3.1 线性压电发动机 ·329
  7.3.2 旋转压电发动机 ·330
  7.3.3 超声压电发动机的特性和分类 ·331
  7.3.4 超声压电发动机的工作原理 ·337
参考文献 ·338

# 第8章 基于智能材料的能量收集 ·342
8.1 压电能量收集 ·342
8.2 电磁能量收集 ·354
8.3 振动激励下的双晶片电源 ·360
8.4 具有增强频率带宽的压电收集器 ·365
  8.4.1 引言 ·365
  8.4.2 运动方程及其解的推导 ·366
  8.4.3 数值验证 ·371
  8.4.4 实验验证 ·372
  8.4.5 3个与2个双晶片系统 ·378

|     8.4.6 结论和建议 ········· 380
|  参考文献 ········· 380

## 第9章 光纤简介 ········· 386
### 9.1 几何光学：基本概念 ········· 386
|     9.1.1 光的反射 ········· 386
|     9.1.2 斯涅耳定律-折射定律 ········· 387
|     9.1.3 临界角和全内反射 ········· 387
|     9.1.4 光纤的数值孔径和接收角 ········· 388
|     9.1.5 干涉 ········· 390
|     9.1.6 光的衍射光栅 ········· 391
### 9.2 光纤的基本特性 ········· 391
### 9.3 光纤-历史视角 ········· 395
### 9.4 光纤布拉格光栅 ········· 397
### 9.5 用作应变传感器的光纤布拉格光栅 ········· 400
### 9.6 光纤布拉格光栅询问器 ········· 401
参考文献 ········· 401

## 第10章 其他主题 ········· 404
### 10.1 形状记忆合金丝增强板材的抗弯性能 ········· 404
|     10.1.1 形状记忆合金特性用于驱动技术：受限恢复现象 ········· 404
|     10.1.2 实验结果 ········· 406
### 10.2 压电纤维复合材料 ········· 412
|     10.2.1 PZT 压电纤维 ········· 412
|     10.2.2 PVDF 共聚物纤维 ········· 416
### 10.3 声能采集 ········· 417
|     10.3.1 声学基础 ········· 417
|     10.3.2 声功率增强 ········· 418
### 10.4 使用形状记忆合金转化能量 ········· 418
|     10.4.1 磁性形状记忆合金 ········· 419
|     10.4.2 形状记忆合金和磁性形状记忆合金的能量转化 ········· 419
### 10.5 利用压电传感器进行道路交通能量收集 ········· 420
参考文献 ········· 422

# 第 1 章 智能材料与结构概述

## 1.1 智能材料的类型

本章内容主要是向读者介绍智能结构的基本知识。虽然科研人员已对智能结构中的智能材料进行了数十年的研究,但自 2000 年以来关于智能材料的文章在各种期刊上发表数量急剧增加,并几乎在所有工程领域得到应用。

智能结构(在航空航天领域也称为智慧结构)通常属于更大研究领域的小子集,如图 1.1 所示[1-2]。

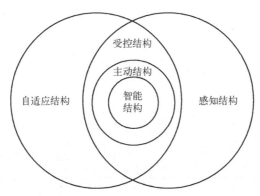

图 1.1 智能结构是主动结构和受控结构的子集[1]

Wada 等将致动器遍布整体的结构定义为自适应结构,如具有铰接式前缘和后缘控制面的传统飞机的机翼[1]。相应地,将传感器遍布整体的结构称为感知结构。这些结构具有可以检测位移应变的传感器,从而能够对力学特性、电磁特性、温度或损伤的显现或累积进行监测。层叠结构包含致动器和传感器或隐式包含用于连接致动器和传感器的闭环控制系统,该结构也可称为受控结构。主动结构是受控结构的子集,主动结构与受控结构的区别在于,主动结构中高度分布的致动器具有结构功能和系统承重功能[2]。在该层级系统中,智能结构只是主动结构的一个子集,它们具有高度分布式的致动器和传感器系统,以及分布式控制功能和计算能力。因此，智能结构是一种能够通过集成各种元件（如传感器、致动器、电源、信号处理器和通信网络）以能预测和期望的方式感知环境并对其做出反应的结构。除了可以承载机械载荷，智能结构还可以减轻振动、降低噪声、监测自身状况和环境、自动执行精确对准以及根据指令改变其形状或力学特性（图1.2）。智能结构经常出现在文献中的另一种定义是：智能结构是包含多功能部件的系统，可以执行传感、控制和驱动，是生物体的初级类似物。例如参考文献[3]所述，类比人类和动物的仿生系统可以看出，对于任何智能材料，以下机制都是必不可少的：

（1）传感器：感知外部刺激的传感工具（如感知热梯度的皮肤或感知光信号的眼睛）。

（2）控制器：通过通信网络将感测到的信号传输到决策机制（如人类和动物的神经系统）和决策装置（如人脑）。

（3）执行装置：材料固有的或与其外部耦合的（如人或动物的肌肉硬化以抵抗外部载荷引起的变形）具有致动器功能的驱动装置。

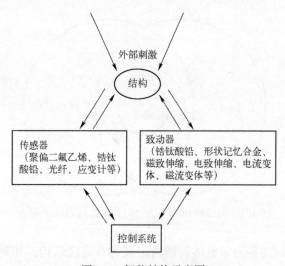

图1.2 智能结构示意图

所有这些装置都需要在实时应用中处于主动状态，以便材料在最佳时间段内做出智能响应。

智能材料用于构建这些智能结构以执行传感和驱动功能。因此，智能/机敏材料是经过设计的具有一种或多种特性的材料，这些特性可以通过外部刺激（如压力、温度、湿度、pH值、电场、磁场等）以可控方式发生显著变化。

智能/机敏结构的另一有趣之处在于其跨学科特性，通常涉及航空航天或机械工程（或任何其他应用工程领域）、电气工程、计算机工程、材料科学、应用物理学和系统工程等多个工程学科领域。因此，该领域吸引着来自不同工程和科研领域的研究人员发表了大量的研究成果。尽管这些研究人员会在各自领域相关的期刊上发表此类工作，但有3本专业期刊几乎可以涵盖新型智能结构领域。

（1）《智能材料与结构》(Smart Materials and Structures)[①] 是一本多学科期刊，致力于推动智能材料、系统和结构（包括智能系统、传感和驱动、自适应结构和主动控制）的技术进步与应用。

（2）《智能材料系统与结构学报》(Journal of Intelligent Material Systems and Structures)[②]（JIMSS）是一本发表高质量原创研究的国际同行评审期刊，旨在发表关于智能材料系统和结构的研究，包括智能结构、智能材料、主动材料、自适应结构和自适应材料的实验结论或理论工作。

（3）《智能结构与系统》(Smart Structures and Systems)[③] 旨在为智能结构和系统领域的研究人员提供相关科研成果的出版渠道。该期刊出版的研究主题包括传感器/致动器（材料/装置/信息学/网络）、结构健康监测与控制以及诊断和预测。

除了专门的期刊，开源期刊《执行器》(Actuators) 和《传感器与材料》(Sensors and Materials)[④] 中也有许多与智能结构相关主题的文章。另外，还有很多关于智能结构及应用的书籍[4-21]汇集了多年来的研究成果，如 Ghandi 和 Thompson 的早期著作就主要涵盖了诸如振动、动力学和健康监控等方面[6-14]的智能结构，并以解决覆盖该主题所有方面[15-20]的智能结构的内容而收尾[4]。此外，还有机械电子学[⑤][13]、自适应电子学[⑥][21]、德国模式的自适应结构相关的书籍。其中，两本较为适合作为本科和研究生教学书籍：一本由 Leo 编著[12]，另一本由 Chopra 和 Sirohi 合著[20]。其中，Chopra 和 Sirohi 合著书籍的内容更翔实、结构更合理，且参考资料全面。

压电材料是可用在智能结构中的主要候选材料。压电性也称为压电效应，其定义为某些材料（如石英、陶瓷和罗谢尔盐）在受到机械应力或振动时产生电压的能力，或在遇到电压时发生伸长或振动的能力，如文献 [22]

---

① 由 IOP Publishing（英国）出版，网址为 http://iopscience.iop.org/09641726。
② 由 Sage Journals 出版，网址为 http://jim.sagepub.com/。
③ 由 Techno-Press 出版，网址为 http://techno-press.org/journal=sss&subpage=5#。
④ 由 MDPI 出版，网址为 www.mdpi.com。
⑤ 机械电子学是一个多学科的工程领域，包括机械工程、电气工程、电信工程、控制工程和计算机工程的结合。
⑥ 自适应电子学（全称为自适应结构技术），是一种用于结构系统优化创新的新截面技术。

中所述，皮埃尔·居里（Pierre Curie）和雅克·居里（Jacques Curie）于1880年发现了压电特性，该名称源自希腊语中 piezein（压电力）一词的直接翻译[23]。

通过与热释电晶体中的温度感应电荷进行类比，居里兄弟观察到了某些晶体在机械压力下的带电现象，包括电气石、石英、黄玉、蔗糖和罗谢尔盐。在后续的参考文献[24]中，居里兄弟证实了压电的第2个特性，即逆效应。从那时起，一个多世纪以来，压电性一词一直被广泛用于描述材料产生电位移的能力，用字母 $D$ 表示，它与施加的机械应力（用希腊字母 $\sigma$ 表示）成正比（图1.3（a））。按照这个定义，若应力从拉伸变为压缩，则出现在电极上的电荷符号会反转。由热力学观点可知，所有压电材料同样会受到逆压电效应的影响（图1.3（b）），即在施加的电场下会发生材料变形。同样，若电场 $E$ 的方向反转，则应变 $S$（伸长或收缩）的符号将变为相反。当剪切机械应力或应变与电荷线性耦合时，剪切压电效应（图1.3（c））也可能存在。

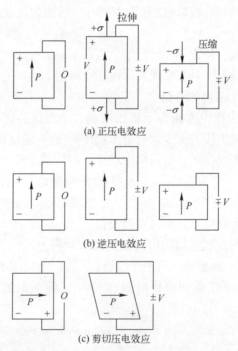

图1.3 正压电效应、逆压电效应和剪切压电效应示意图

要了解压电物理性能，必须深入了解该效应的晶体学原理（文献[25-26]中 Voigt 在1910年的开创性工作和 Jaffe 等在1971年的工作）。表1.1列

出了主要压电材料的物理性质及其对称性[22]。本书不包含压电晶体学部分，感兴趣的读者可阅读文献［26］以及其他文章。应该注意到，压电陶瓷的发明[26]使压电材料在许多工业应用领域中表现出惊人的性能，产生了数十亿美元的产业，具有广泛的应用和用途。

表1.1 主要压电材料的物理性质及其对称性

| 参数 | 材料 | | | | | | |
| --- | --- | --- | --- | --- | --- | --- | --- |
| | 石英 | 钛酸钡 | 钛酸铅：Sm | PZT$_5$H | LF$_4$T | PZN-8%PT [001] | PZN-8%PT [111] |
| 对称性 | 32 | 4mm | 4mm | 3m/4mm | 2mm/4mm | 3m/4mm | 3m/4mm |
| $d_{33}/(pC/N)$ | 2.3 | 190 | 65 | 593 | 410 | 2500 | 84 |
| $d_{31}/(pC/N)$ | 0.09 | 0.38 | 0 | −274 | −154 | −1400 | −20 |
| $\varepsilon_{33}^T/\varepsilon_0$ | 5 | 1700 | 175 | 3400 | 2300 | 7000 | 1000 |
| $T_c/℃$ | — | 120 | 355 | 193 | 253 | 160 | 160 |

锆钛酸铅（PZT）是最广泛使用的压电材料[27]，表现出正压电效应（在有应变时产生电压）和逆压电效应（置于电场中时会产生应变）。因此，PZT陶瓷既可用作传感器，也可用作致动器。压电陶瓷本质上是多晶的，在原始状态下并不表现出任何压电特性。为了在PZT陶瓷中产生压电效应，这些材料在2kV/mm（该值为平均值）的高直流电场中，在居里温度下发生极化，单元格的极轴平行于所施加磁场而排列，进而实现材料的永久极化。陶瓷极化过程中的另一个重要现象，是由于陶瓷电畴的重新定向而导致陶瓷体产生的永久机械变形。纵向、横向和剪切效应之间应加以区分。纵向效应是极化方向的主动应变并与电场平行；横向效应是由于产生的泊松平面应变造成的；剪切应变的方向平行于极化方向并垂直于电场。从应用的角度来看，通常只使用纵向和横向效应而忽略剪切效应。需注意的是，PZT陶瓷的最大应变相对较小（0.12%~0.18%），受到饱和效应和去极化的限制[28]，且对非常小的信号表现出2%的滞后，在额定电压时可达10%~15%[28]。PZT的密度通常为7.6g/cm$^3$[29]。

从结构的角度来看，由于极化过程，PZT不仅电各向异性，而且机械各向异性。PZT的纵向杨氏模量（弹性模量）通常为50~70GPa，而在横向上的值为35~49GPa（更多有关信息，请参阅文献［30］）。

压电效应还取决于另一个限制因素——环境温度。在低于260K时，它会随着温度的降低以约0.4%/K降低[31]，其上限温度受居里温度限制。在居里温度（介于 150~350℃）的 70% 以下，PZT 致动器能可靠地驱动[30]。

温度高于居里温度会导致去极化。高机械应力也会使 PZT 陶瓷去极化，因此通常习惯于将施加的应力限制在其机械压缩载荷极限（200~300MPa）的 20%~30%[28]。PZT 陶瓷的另一个重要问题是，当施加与极化方向相反的电压且大于其矫顽力电压时，可能会发生去极化①。

压电陶瓷可以制成薄片（贴片）、管、短棒、圆盘、纤维/条纹等形式，也可以堆叠形成离散的压电堆叠驱动器（图 1.4）。

(a) 压电式堆叠　　　(b) 管、棒、盘、板和堆叠

(c) 1~3 纤维复合材料　　(d) 超细纤维复合材料贴片　　(e) 锆钛酸铅纤维

图 1.4　PZT 陶瓷的典型形式
（资料来源：美国宇航局智能材料公司（(a)、(c)~(e)）；PI 陶瓷（b））

聚偏二氟乙烯（PVDF）、三氟乙烯（TrFE）和四氟乙烯（TFE）的共聚物为半晶态氟高聚物[31]，是另一种基于聚合物生产压电材料的原材料（图 1.5）。与 PZT 一样，PVDF 可用作传感器和致动器，并且也显示出纵向、横向和剪切应变三种效应，但在技术层面通常只涉及横向效应。PVDF 和 PZT 之间的显著差异与其机电材料特性有关，即 PVDF 在电场方向上会收缩而非 PZT 表现出的伸长；但在平面上，PVDF 发生伸长而 PZT 则表现为收缩。比较压电常数（$d_{31}, d_{32}, d_{33}$）的值，PZT 的压电应变常数比 PVDF 大 10~20 倍[31]。PVDF 在 100kHz 以下产生 0.1% 的工作应变，与 PZT 在同一数量级，但需要施加更大的电场，即 10~20kV/mm。PVDF 无定形区域的玻璃化转变温度（约 -40℃）决

---

①　矫顽力是铁电材料承受外部电场而不去极化的能力。

定了聚合物的力学性能,而微晶的熔化温度(约180℃)决定了其温度的上限。但是,压电效应受到相对较低的居里温度(约100℃)的限制,PVDF只可在60~80℃的温度下可靠地使用。电极化可通过使用50~80kV/mm量级的电场来实现。单轴拉伸时,聚合物片材平面内的电性能和力学性能是高度各向异性的,双轴拉伸时是各向同性的。在平面方向上PVDF的杨氏应变常数则更大。与PZT类似,必须注意避免施加过大的电压、机械应力或温度,以防止PVDF的去极化。PVDF相对于PZT的优势在于其密度低(约1.47g/cm$^3$)和加工灵活,其坚韧、易于制造成大面积材料,而且可以切割并形成复杂的形状。PVDF的主要缺点是低刚度,显著降低了其作为结构材料的优越性,使其只能应用于对力学性能要求比较低的场景中。

(a) 大幅PVDF薄膜贴片

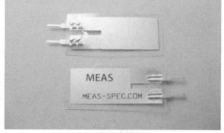
(b) 带接头的PVDF贴片

图1.5 典型的PVDF样式

铁电和陶瓷(包括压电)材料(如PZT和PVDF)的另一个重要特性是其热释电效应①,即由于环境温度的变化(包括加热和冷却)而在其电极上产生电压。温度的变化导致晶体结构内原子的轻微移动,从而产生自发极化。需注意的是,温度是随时间变化的,热电系数定义公式为

$$p_i\left[\frac{c}{m^2K}\right]=\frac{\partial P_{s,i}}{\partial T}, \quad i=1,2,3 \tag{1.1}$$

式中:$p_i$为热电矢量;$P$为极化矢量;$T$为温度。忽略介电损耗[32],热释电方程可以写为[32]

$$\dot{D}=\varepsilon\dot{E}+p\dot{T} \tag{1.2}$$

$$\dot{S}=p\dot{E}+\frac{c\dot{T}}{T} \tag{1.3}$$

---

① 热释电(Pyroelectricity)来自希腊语pyr(意为火)和electricity这两个词。

式中：$D$、$E$、$T$ 和 $S$ 分别为电感应、电场、温度和熵；$\varepsilon$、$p$ 和 $c$ 分别为介电常数、热电系数和热容。需注意，$(\dot{\ })$ 表示对时间求导。

基于热释电效应，文献 [33-41] 设计了红外传感器和采集装置来捕获与热时间相关的能量，并将其转化为电能。使用热释电效应的典型应用如图 1.6 所示。

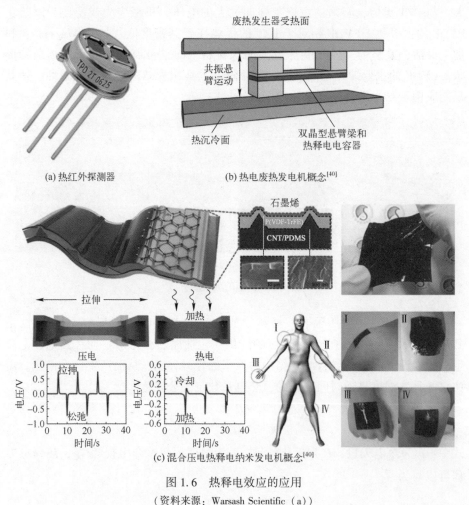

图 1.6 热释电效应的应用

(资料来源：Warsash Scientific (a))

另一类智能材料是电致伸缩材料①[42-51]。电致伸缩材料具有与外加电压（或外加电场）的平方成正比的诱导应变，并在拉伸时产生电流。这些材料主

---

① 电致伸缩效应在外加电场的影响下引起材料的尺寸变化。

要用于精密控制系统,如工程中的振动控制和声学调节系统、楼面系统中的减振和建筑施工中的动态加载。

电致伸缩是所有绝缘体或电介质中的基本机电现象[42]。电场/极化诱导的应变($S_{ij}$)可描述为与电场($E_i$)或极化($P_i$)的平方成正比,并由以下两个方程表示:

$$S_{ij} = Q_{ijkl} P_k P_l \tag{1.4}$$

$$S_{ij} = M_{ijkl} E_k E_l \tag{1.5}$$

式中:$Q_{ijkl}$ 和 $M_{ijkl}$ 为电致伸缩系数。由于四阶张量的性质,电致伸缩可以在所有晶体对称性中观察到[44]。电致伸缩材料不像压电材料那样极化,而且比压电材料的迟滞率更低。但它们在本质上更具电容性,因此需要更大的驱动电流。同时,因为它们对施加的电压/电场的平方产生响应,所以它们也是高度非线性的。基于弛豫铁电体的电致伸缩陶瓷表现出与压电材料相当的应变,已经在许多商业系统中得到应用[51]。最著名的电致伸缩材料是铌镁酸铅-钛酸铅(PMN-PT)① 和锆钛酸铅镧(PLZT)。这些材料的杨氏模量约为700GPa,非常脆,具有快速响应时间和低磁滞回线,适用于高达50kHz的频率,并且只能用作致动器。图1.7所示为典型的电致伸缩材料。

磁致伸缩与上述电致伸缩特性类似,是材料受到电磁场[52]作用时改变其长度的特性,也称为焦耳效应[53]。第二个磁致伸缩特性称为维拉里效应,会导致材料在受到外力变形时产生电磁场(会引起磁通密度的变化)。第三个磁致伸缩效应称为巴雷特效应[54],是指材料的体积随着磁场的作用而变化。因此,磁致伸缩材料可用于传感和驱动。铽镝铁合金(Terfenol-D)② 是一种市售的磁致伸缩材料,在室温下具有"超"磁致伸缩特性,可用作传感器和致动器。铽是较稀有的稀土材料之一,因此价格非常昂贵。Terfenol-D可以承受较大的应变(2%),并且杨氏模量为200GPa,还可以产生大至kN级的驱动力并且具有较低的迟滞性。图1.8所示为一些典型的磁致伸缩材料及其用途。

另一类智能材料是电活性聚合物(EAP)(图1.9(a)③) 和磁活性聚合

---

① 它们被称为具有高相对介电常数(20000~35000)和高电致伸缩系数的弛豫铁电体。
② Terfenol-D 是一种分子式为 $Tb_x Dy_{1-x} Fe_2 (x \approx 0.3)$ 的合金,是一种磁致伸缩材料。最初由美国海军军械实验室于20世纪70年代开发。20世纪80年代,在美国海军资助的项目下,艾姆斯实验室开发了有效制造该材料的技术。它以铽、铁(Fe)、海军军械实验室(Naval Ordnance Laboratory, NOL)命名,而D来自镝。
③ 摘自 www.hizook.com/blog/2009/12/28/electroactivepolymers-eap-artificial-musclesepam-robot-applications#。

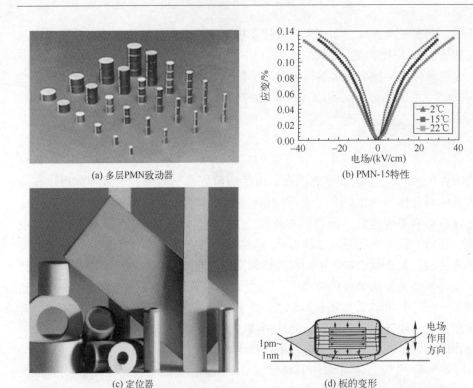

图 1.7 电致伸缩材料

(资料来源：AOAXinetics (a)、TRS Technologies (b)、Piezomechanik GmbH (c) 和 MURATA (d))

物（MAP）（图 1.9（b））。电活性聚合物[55]是一种在电场刺激下会发生尺寸或形状变化的聚合物，这种材料最常见的应用是在致动器和传感器中。电活性聚合物的典型特性是它在承受较大力的同时也会发生较大的形变。

　　大多数基于智能材料的致动器由陶瓷压电材料组成。虽然这些材料能够承受很大的力，但它们的形变通常只有百分之几。20 世纪 90 年代末，已经证明一些电活性聚合物可以表现出高达 32% 的应变，远远超过任何陶瓷致动器[55]。电活性聚合物最常见的应用之一是在机器人领域用于开发人造肌肉，因此电活性聚合物通常被称为人造肌肉。

　　磁活性聚合物是基于聚合物的复合材料，响应磁场作用可产生大的形变或可调力学性能。虽然这类材料很多，但大多数是软聚合物基体与磁性颗粒填料的复合材料。磁活性聚合物中的多个物理场相互作用为它们提供了两个非常显著的特征。首先，它们响应磁场的作用产生可变的力学特性（如刚度）。其次，它们的形状和体积可能会在磁场中发生显著变化。这两个特征都可以通过

# 第1章 智能材料与结构概述

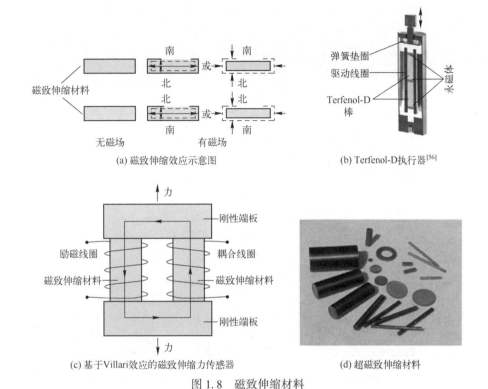

图 1.8 磁致伸缩材料

(资料来源:John Fuchs:http://www.ctgclean.com/techblog/2011/12/ultrasonicstransducers-magnetostrictive-effect/(2011年12月30日发布)(a);甘肃天兴稀土功能材料有限公司(d))

设计复合材料的微观结构来调整。磁活性聚合物的潜在应用包括传感器、致动器、生物医学和增强现实(Augmented Reality,AR)。

应该注意到,通过将铁粒子嵌入聚合物基质中形成一种固体弹性体。与磁流变体(Magnetorheological Fluid,MRF)不同(参见本章后面的讨论),电活性聚合物或磁活性聚合物中的粒子运动非常有限。施加磁场导致了磁活性聚合物硬化。

另一类智能材料是形状记忆合金(Shape Memory Alloy,SMA),这是一类独特的金属合金,当它们被加热到一定温度以上时,可以恢复表观永久应变,并呈现出具有大应变的超弹性效应(Pseudoelasticity Effect,PE)[58-61]。形状记忆合金有两个稳定相:高温相称为奥氏体,具有立方晶体结构;低温相称为马氏体,具有单斜晶体结构。此外,马氏体状态可以是孪晶和退孪晶两种形式之一,如图1.10所示。在加热/冷却时这两个相之间发生的相变是形状记忆效应(Shape Memory Effect,SME)和PE等形状记忆合金独特性质的基础。

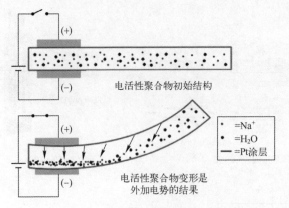

(a) 电活性聚合物激活示意图

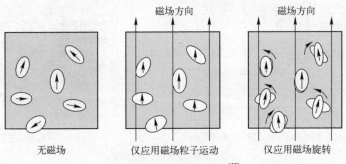

(b) 磁活性聚合物激活示意图[57]

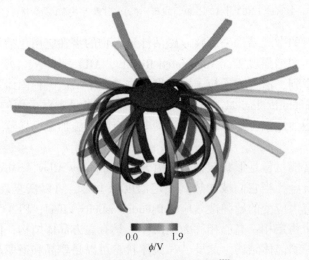

(c) 基于电活性聚合物的海星抓手[58]

图 1.9　电活性聚合物和磁活性聚合物

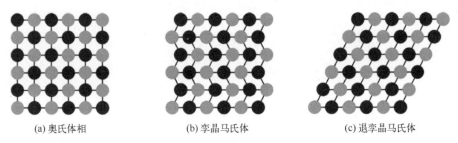

(a) 奥氏体相　　　　　(b) 孪晶马氏体　　　　　(c) 退孪晶马氏体

图 1.10　奥氏体相、孪晶马氏体和退孪晶马氏体的示意图

可以通过加热、机械载荷或温度和机械载荷的组合来实现相之间的转变。在没有施加载荷的情况下冷却时，材料从奥氏体转变为孪晶马氏体，而没有宏观的形状变化。加热马氏体相的材料，则发生逆相转变，产生奥氏体相（图 1.11）。

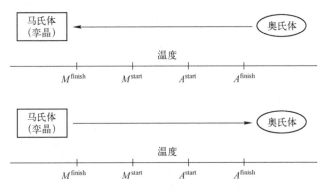

图 1.11　形状记忆合金——没有机械载荷时温度诱导的相变

图 1.11 显示了 4 个特征温度：马氏体起始温度（$M^{start}$），它是材料开始从奥氏体转变为马氏体的温度；马氏体结束温度（$M^{finish}$），在该温度下转变完成，材料完全处于马氏体相；奥氏体开始温度（$A^{start}$），逆转变（奥氏体到马氏体）开始温度；奥氏体结束温度（$A^{finish}$），在该温度下完成逆相转变，材料为奥氏体相。

若在孪晶马氏体状态（低温下）对材料施加机械载荷，则可以获得退孪晶马氏体。当载荷被释放时，材料保持变形。若将材料加热到高于 $A^{finish}$ 的温度，将导致反相转变（马氏体到奥氏体）并导致完全的形状恢复（即 SME），如图 1.12 所示。

当在奥氏体相中施加机械载荷并且冷却材料时，相变将导致退孪晶马氏体的产生。因此，将观察到非常大的应变（5%～8%），重新加热材料将导致形

13

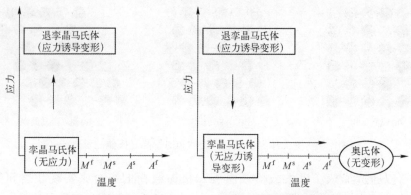

图 1.12　形状记忆合金——应力诱导的相变和形状恢复

状完全恢复。上述加载路径如图 1.13 所示①。需注意，这种情况下的转变温度很大程度上取决于所施加载荷的大小，施加的载荷值越高，转变温度值越高。通常假定施加的机械载荷与转变温度之间存在线性关系，如图 1.13 所示。

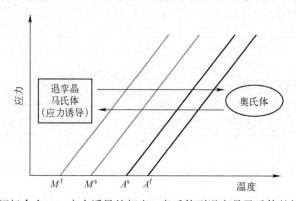

图 1.13　形状记忆合金——应力诱导的相变：奥氏体到退孪晶马氏体的转变和反相转变

图 1.14 所示为典型形状记忆合金材料在不同温度下的应力-应变曲线②。注意超弹性效应（c）和记忆效应（b）之间的差异。

最著名的形状记忆合金材料之一是镍钛诺合金（NITINOL，有时也称为镍钛合金），它是一种由镍和钛组成的金属合金。镍钛诺一词源自其成分和发现地点：镍钛-美国海军军械实验室，其特性如表 1.2 所示[61]。镍钛诺合金可提供可靠、先进的致动器和弹簧，在医疗领域（图 1.15）以及机械、民用和航空领域有着广泛的用途。

---

① 来自 smart. tamu. edu/overview/overview. html。
② 来自 http://heim. ifi. uio. no/~mes/inf1400/COOL/RobotProsjekt/Flexinol/ShapeMemoryAlloys. htm。

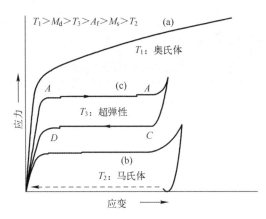

图 1.14　形状记忆合金——不同温度下相对于相变的典型应力-应变曲线：
(a) 奥氏体，(b) 马氏体和 (c) 超弹性行为

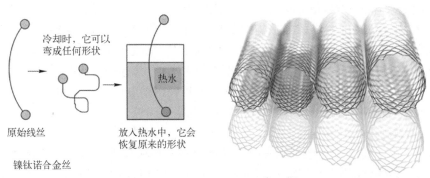

(a) 简单实验　　　　　　　(b) Protage@GPS™自扩镍钛诺合金胆道支架系统

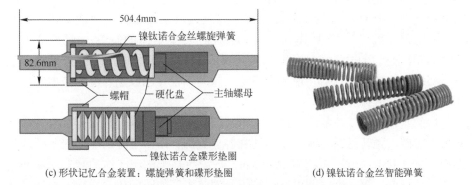

(c) 形状记忆合金装置：螺旋弹簧和碟形垫圈　　(d) 镍钛诺合金丝智能弹簧

图 1.15　镍钛诺合金应用和用途

（资料来源：www.talkingelectronics.com/projects/Nitinol(a)；M & JCo.Ltd,www.twmt.url.tw(b)；
www.neesrcr.gatech.edu(c)；www.preproom.org(d)）

最后要介绍的智能材料是电流变（Electrorheological，ER）体和磁流变（Magnetorheological，MR）体。

表1.2 镍钛诺合金的力学性能和电性能

| 性　　能 | 马氏体相 | 奥氏体相 |
|---|---|---|
| 密度/(g/cm$^3$) | 6.45 | 6.45 |
| 杨氏模量/GPa | 28~40 | 75~83 |
| 屈服应力/MPa | 70~140 | 195~690 |
| 泊松比 | 0.33 | 0.33 |
| 热膨胀系数/10$^{-6}$ | 6.6 | 11 |
| 电阻/(10$^{-6}$Ω·cm) | 76 | 82 |
| 热导率/(W/(cm·K)) | 0.086 | 0.18 |

电流变体是一类可以通过施加电场来控制其流变特性的材料。典型的电流变体由介电粒子在低介电常数液体中的胶体分散体组成。当施加电场时，电流变体的有效黏度急剧增加，超过一定的临界值时，电流变体可能会变成凝胶型固体，其剪切应力会随着电场的进一步加强而增加。从结构的角度来看，已观察到电流变体中的介电粒子倾向于形成跨越装置两个电极的链（图1.16（a）和（b）），随着外加电场的增加，这些链会聚集形成粗柱，导致电流变体的表观黏度急剧增加。Winslow于1947年获得关于电流变体的专利[62]并于1950年发表文章[63]。电流变体在专利中称为功能流体，屈服应力可以通过施加的电压改变。其特点是流体特性可以根据基础油的种类和特性以及分散在流体中的颗粒大小和密度而改变[64-65]。目前，已有大量研究致力于提高电流变体的性能和稳定性，并将其应用于机械部件，如悬架、减振器、发动机支架、离合器、制动器和阀门（图1.16（d））。激活后电流变体的行为类似于宾汉模型①，其屈服点由电场强度决定。达到屈服点后，流体的剪切行为等同于液体，即增量剪切应力与剪切速率成正比，这与不存在屈服点、应力与剪切应变成正比的牛顿流体相反。

虽然基于电流变体的装置已经上市，但两个主要问题仍然阻碍着电流变体的重大突破：颗粒悬浮液倾向于沉淀在容器底部，阻碍了施加电场引起的链的形成；空气的击穿电压大约为3kV/mm，与激活基于电流变体的装置所需的电场接近。目前，第1个问题可通过粒子及液体密度的匹配或使用纳米粒子来解

---

① 宾汉塑料（以Eugene Cook Bingham（1878—1945年）的名字命名）是一种黏性塑性材料，在低应力下表现为刚体，但在高应力下则表现为黏性流体[66]。

# 第1章 智能材料与结构概述

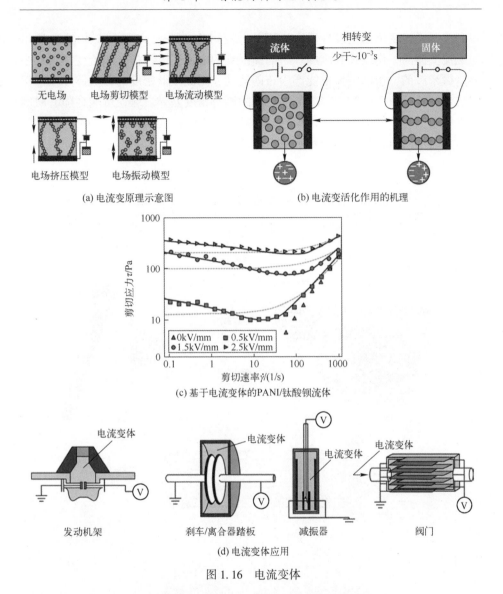

图1.16 电流变体

决。第2个问题可通过对电流变体装置进行绝缘处理并防止其与空气直接接触来解决。

与电流变体类似，磁流变体是一种智能流变流体，会在施加磁场时改变其表观黏度。当受到磁场作用时，流体的表观黏度会大大增加，直至变成黏弹性固体。流体屈服应力可以通过改变磁场强度实现非常精确的控制，如使用一个电磁铁能够控制由磁流变体产生而到达指定接收器的力。文献 [67-70] 中可找到关于磁流变体的物理性质和应用的广泛讨论。图 1.17（a）和（b）所示为磁流变体的工作模式和应用。与前面提到的电流变体一样，磁流变体颗粒主要是微米级的，密度太大不能产生布朗运动，因此在低密度的载流体中无法保持悬浮，并且由于颗粒与其载流体之间的固有密度差异，颗粒可能会随着时间的推移而沉降。如前所述，该问题的处理方式与电流变体相同，或可以通过保持恒定磁场来防止悬浮颗粒及其载流体的沉降。

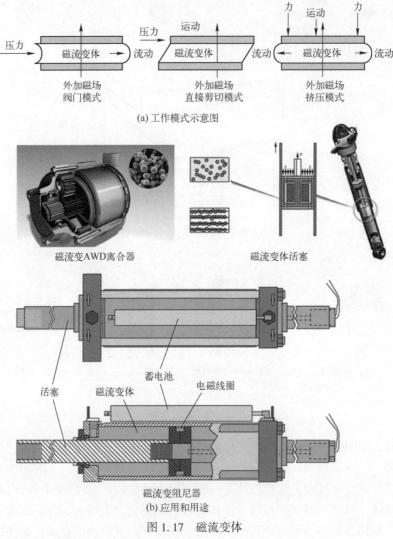

图 1.17　磁流变体

（资料来源：scmero.ulb.ac.be/project（a）；www.atzonline.com、www.autoserviceprofessional.com、NEES 项目数据库（b））

## 1.2 压电材料、形状记忆合金、电流变体和磁流变体以及磁致伸缩和电致伸缩材料综述

接下来的扩展综述①旨在介绍本书中所涉及内容的典型研究。其中，文献[71-337]涉及压电材料及其用途，文献[338-443]则主要介绍形状记忆合金的相关研究，文献[444-599]侧重于对电流变体和磁流变体的研究做出概述，文献[600-665]则涉及磁致伸缩和电致伸缩材料及其应用。

### 1.2.1 压电材料

压电材料在科学和工程方面一直以来都是国内外研究的热点，所提供的系列参考资料为读者提供了参考文献已发表内容中的重要数据。在本书的第3章，额外的参考资料则进一步阐明了压电材料相关内容。

当前压电材料相关综述将从PVDF压电膜开始，对其特征及初步应用进行了讨论，详见文献[71-78]。文献[79-80]概述了PVDF在梁和板的振动控制中的应用，文献[81-82]则讨论了其在太空领域的应用，文献[83-86]介绍了PVDF作为传感器的应用，而文献[87-88]概述了PZT/PVDF复合物阵列在振动控制领域的研究。文献[84]研究了PVDF换能器在Lamb波测量中的应用，PVDF在制造业相关的概念和性能的改性方法在文献[89-105]做出了讨论和介绍，文献[108]描述了其在采集环境能量方面的应用，其中微米带状PVDF可从呼吸中采集到$1\sim2mJ$能量。此外，文献[109]报道了在传统PVDF薄膜顶部，多孔表面的应用可以进一步提升采集能量。

毫无疑问，各种梁、板和壳理论对带有压电层与贴片的梁的静、动态行为研究是所有智能材料出版物中数量最多的。文献[110-124]中涉及的相关书籍和综述对该领域进行了很好的介绍，首先，文献[110-111]是两本关于压电材料的基础书籍，而后是对铁电陶瓷最新技术水平的相关综述，详见文献[112-116]。接下来，文献[117-119]对适用于压电层压板的理论进行了综述，而文献[120-124]对压电材料的多种用途，如能量采集器，或作为致动器以抑制振动和噪声进行了概述。文献[125-174]涉及以下主题：配备压电连续层和/或贴片的梁状结构的静态和动态性能；压电材料作为分流器或闭环控制来抑制振动；如何利用主动约束层减振装置来进行压电层压梁和增强减振

---

① 参考文献按时间顺序出现在文献中，并大致分为与其包含的内容相关的部分。

能力的控制。主动约束层减震装置借助有源元件（通常为压电层）来加强无源约束层（通常是黏弹性层），从而增强阻尼过程的能量耗散。参考文献包括：描述横梁压电驱动的模型（通常应用伯努利-欧拉理论）；利用层压复合板的圆柱形弯曲将二维层面的问题简化为适用于梁状结构的一维问题；先进的机械理论，如 Timoshenko 的理论和复合压电层压梁、带有压电贴片的夹层梁的耦合分层分析，以及各种以服务梁的主动减震的控制定律的应用。上述大部分参考文献都提供了理论/数值计算，其中一些参考文献还附有必要的实验数据以提升研究中所提出各种模型的应用性能。另一重要主题是带有压电贴片的直升机旋翼的性能，旨在诱导应变使旋翼叶片变形以提供根据飞行条件改变旋翼形状的控制能力，相关的典型研究在文献 [175-179] 中做出了介绍。文献 [180-193] 涵盖了压电材料的传感器应用，以探测各种冲击以关闭控制回路或检测层压复合结构中的分层现象。这些文献给出了理论和实验结果，并讨论了所用理论的准确度。文献 [194-220] 强调了通过应用压电层、双压电晶片层或叠层以驱动各种构件的研究，这些研究包括自感应型致动器、能够提供大应变的增强型致动器、致动器功率特性的计算、压电材料基致动器的实验和理论计算，及其在各种结构（如梁、板和翼）中的性能。需要指出的是，上述研究统一以 $d_{31}$ 作为应变系数进行驱动和传感，而较大应变系数 $d_{33}$ 则应用于叠层结构。在描述压电材料的第 3 章内容中，可应用剪切应变系数 $d_{15}$ 来获得驱动和传感，相关内容在文献 [221-231] 中做出了强调，基于剪切式传感器或拉压式传感器的梁和板的封闭解，结合剪切力共同增强结构的减震效果或能量的收集。以上所有引用的研究均提供了解析/数值计算。

在文献 [232-250] 中，介绍了压电材料的另一个应用情景。分析和数值研究表明，在特定状态下，压电层或贴片的使用无论在静态和动态情况下均可以增强类梁结构和/或板结构的弯曲刚度。应用于传感器和致动器的压电材料，通过应用足够的驱动来防止结构的损失，以控制结构的弯曲载荷，甚至提高结构的稳定裕度。同时，通过提高结构的刚度，其固有频率也随之改变。结构在受到拉应力时（由于压电诱导应变），固有频率增加；相反，当梁结构上施加为压应力时，该值减小。这些研究提供了分析/数值预测，有些文章中还包含实验结果。文献 [251-255] 通过圣维南（Saint-Venant）原理[①]来解决弹性压

---

[①] 圣维南原理（为了纪念法国弹性理论家 Adhémar Jean Claude Barré de Saint-Venant 而命名）指出："如果作用在弹性体表面一小部分上的力被另一个作用在弹性体表面上同一部分的静态等效力系统代替，这种载荷的重新分布就会在局部产生显著的应力变化，但对远处的应力影响可以忽略不计，这里远处是指相对受力变化部分表面的线性尺寸来说比较远的位置。"（S. Timoshenko 和 J. N Goodier，《弹性理论》（*Theory of Elasticity*），第 2 版，1934 年，第 33 页，McGraw-Hill。）

电问题并获得复合压电梁的弹性解。

在文献［256-291］介绍的大量研究中，阐述了具有压电贴片作为传感器和/或致动器的板和壳的解决方案。这些研究利用 Kirchhoff-Love 和其他诸如"一阶剪切变形"等先进理论对各类型压电板和壳进行了分析，实现层叠复合材料层结构中压电部分的传感和驱动，其中有源压电层实现了对静态和动态行为的闭环控制。论文对预测位移、模态电压、结构的形状控制和自适应结构的电气特性形式进行了数值计算。

另一广泛发表的研究课题是利用压电材料从环境振动和其他寄生能源中获取能量，并将其转化为电能以提供当地用户的电能需求。文献［292-317］代表了过去 15 年内发表的大量论文中的典型文献。首先是利用配备与环境频率匹配的压电层/贴片的振动悬臂梁的方式从振动中提取机械能，并通过电路将其转换为电能存储在电池或其他储能器件（如电容器或超级电容器）中。大量研究表明，收集能量（或功率）的多少可以进行数值预测，并以精心设计的测试支撑理论预测结果，以加大电力输出的先进电路的发展。众多先进设计包括改进装置方法以产生更高电能供应潜在用户。

在对 PZT（锆钛酸铅）型传统陶瓷钙钛矿型[①]压电材料进行研究的同时，还有许多研究用于制造和预测由铌酸锂单晶、合成石英或其他材料制备的单晶换能器性能。尽管其中一些材料在本质上是电致伸缩性质的，如 PNM-PT，也因为这些文献中大多材料的特性涉及与锆钛酸铅的比较，所以这里提供了相关参考文献。相较于锆钛酸铅元件，这些压电材料表现出优异的压电特性，对温度不敏感并且表现出非常高的电/机械能量转换系数。因此，这些材料在多种应用方面表现出巨大的吸引力。文献［318-330］中开发了多种模型用来预测单晶材料构建的装置的输出，它们用于能量采集器及其控制。最后，大量研究都讨论了压电材料作为传感器的结构健康监测。文献［331-337］中列出了一些典型的示例，所涉及的研究是利用压电贴片以感知结构的状态，并在结构开始失去完整性时提前发出警告，最终发展为监控结构（如桥梁）内各种关键部分的有用工具。这些研究涉及传感器的集成及其与监控结构状态的通用电气系统的连接。

## 1.2.2 形状记忆合金

对形状记忆合金的研究一直都在进行，并且已经涉及科学与工程的各个方

---

① 钙钛矿（Perovskite，以俄罗斯矿物学家 Lev Perovski 的名字命名），是一种由钛酸钙组成的钙钛氧化物矿物，化学式为 $CaTiO_3$。

面。文献［338-443］提供了已发表的形状记忆合金相关的数据，同时，本书的第4章内容还包含了更多的参考资料以进一步阐述形状记忆合金相关的问题。

既然被命名为形状记忆合金，其具有两个主要特征：一是形状记忆效应，它描述了合金在特定温度下发生变形的能力，然后通过将其加热到高于其转变温度以恢复其原始形状；二是PE或称为超弹性，其发生在高于合金转变的温度下，无须加热即可恢复未变形的形状。这种特性表现出巨大的弹性应变，是普通金属的10~30倍，恢复时没有任何塑性变形。

为了理解形状记忆材料复杂的行为，读者可参阅文献［338-358］所列出的书籍和综述。其中，文献［338-354］涵盖了形状记忆合金的基础知识，同时包括在非医学和生物医学领域的各种应用示例。文献［355-357］对形状记忆合金方面的最新技术，包括建模和基于形状记忆合金的应用做出了介绍，随后的一些参考文献将介绍高级应用。文献［359-366］介绍了通过插入形状记忆合金线来提高结构的稳定性，从而提高结构的刚度、主动振动控制和形状控制。生物医学是形状记忆合金的主要应用领域之一，由于镍钛诺合金[①]这种形状记忆材料具有与人体的兼容性，故可将支架或其他装置植入人体，从而执行与康复相关的高等任务。镍钛诺合金在生物医学领域的应用可参阅文献［367-377］。

已发表的大量参考文献对形状记忆合金引入杆、柱、轴和制动器进行了研究，可参阅代表性文献［378-394］。模型构建以及对所建模型的实验验证，使工程师能够设计和制造带有形状记忆合金线的结构。文献［395-399］中概述了将形状记忆合金引入更先进的结构（如板和壳）中，构建出基本模型并对这些结构的静态、动态和屈曲行为进行了表征。

很多研究致力于对从马氏体相到奥氏体相，再到马氏体相的转变进行建模的复杂问题，以便能够正确预测形状记忆合金在PE和记忆效应轨迹中的表现。文献［400-428］中引用了上述部分文献，包括高级数值工具，如使用Preisach模型[②][410]来捕获形状记忆合金的滞后曲线。

众所周知，形状记忆合金线在从一个相到另一个相的转变过程中可以承受非常大的力，因此在文献［429-434］中对它们在变形机翼中的应用进行了探究和表征，这些只是涉及该应用的一小部分文献。与形状记忆合金相关的其他研究，如新型形状记忆合金系统、控制问题、疲劳和循环载荷，甚至传感器的

---

① 镍钛诺合金(NITINOL)＝镍钛-海军军械实验室，Ni＝镍，Ti＝钛。
② Preisach, F., Über die Magnetische Nachwirkung, Zeitschrift für Physik, Vol. 94, 1935：277-302.

应用，在文献［435-443］中有详细介绍。

## 1.2.3　电流变体和磁流变体

电流变体和磁流变体是科学与工程方面另外两种深入研究的智能材料。文献［444-599］中报道了涉及电流变体和磁流变体的物理性质以及将其应用于阻尼器的数据。本书的第 5 章包含更多涵盖电流变体和磁流变体研究内容的参考资料。

电流变体和磁流变体包含悬浮在流体中的颗粒，施加电场或磁场将改变其流变行为，从而呈现出具有流体/固体状材料行为的一类新流体。对这些流体的特性已有深入的研究，并发表了很多相关研究工作，此外，电流变体和磁流变体知识的相关书籍与综述也有发表[444-451]。在提供的文献［447-451］列表中包括两本书，涵盖了流体相关的众多科学和技术方面内容，并附有许多综述和最新的研究成果[442-444]，旨在提供设计并构建基于这两种流体装置所需的技巧和经验。

许多发表的研究工作中涉及了电流变体现象（如文献［453-509］），且涵盖了电流变体的各个方面，包括建模、通过改变流体特性以合理地进行器件设计、智能流体的实际应用。文献［510-537］介绍了电流变体和磁流变体的应用，对两种流体的各种性能进行了比较，研究了它们的流变特性，评估了各种开发模型的性能，探索了它们的结构应用。

文献［538-599］介绍了对磁流变体的研究、采用各种数值方法实现滞后曲线的建模分析（包括 Preisach 模型[564]）、作为高级阻尼器的性能评估及其在实际结构（如直升机旋翼叶片的滞后阻尼器或汽车工业的制动器）中的应用。

## 1.2.4　磁致伸缩和电致伸缩材料

磁致伸缩材料和电致伸缩材料是本书要介绍的另一类智能材料。文献［600-665］对磁致伸缩和电致伸缩材料的众多研究做出了介绍，并涵盖了科学与工程的电磁和磁学方面的应用。和其他智能材料一样，本书后面有专门的章节（第 6 章）进行论述，以进一步完善磁致伸缩和电致伸缩材料的相关知识。

磁致伸缩是铁磁材料在磁化过程中改变其形状或尺寸的特性。同样，所有非导体（电介质）都具有在电场作用下改变形状的电致伸缩特性。这种磁能或电能实现动能的转化方式可被用来构建致动器和传感器。随着新型材料的产生，如超磁致伸缩致动器 TERFENOL-D[①] 和其他磁致伸缩材料，以及各种电致伸缩材料，如 PMN（含铅、镁、铌酸盐）、PLZT（含铅、镧、锆酸盐、钛酸盐）或 PMN-PT（含铅、镁、铌酸铅、钛酸盐），引发了对这些新材料的研究，并已有众多相关文献发表。

综述型文献［600-602］中，一篇涉及电致伸缩和磁致伸缩[607]，另外两篇进行了磁致伸缩现象的综述[601-602]。涉及磁致伸缩材料的文献数量非常多，文献［603-636］中仅概述了一些典型示例。学者们报道了对磁致伸缩现象的建模、TERFENOL-D 和 GALFENOL[②]/ALFENOL[③] 特性的研究、磁致伸缩材料的传感和驱动、应力影响以及评估磁致伸缩材料的各种装置的使用实验。对基于电致伸缩装置的研究更受限制，发表的文献也相对较少，文献［637-648］进行了 PMN 建模，对具有电致伸缩致动器的板状和壳状结构进行了评估，描述了压电与电致伸缩材料的特性，并进行了响应性能的评估。最后，文献［649-665］概述了电磁和磁现象在振动抑制、振动控制、磁约束层中的用途，以及在结构和能量采集中产生增强的阻尼效应。

## 1.3 智能材料的典型特性

本节旨在向读者展示各种称为智能或机敏材料的机械和电性能方面的数据。首先，介绍德国 PI 公司[④]给出的压电材料的典型特性。该公司是压电材料和器件的知名公司之一。类似的产品也可以在诸如 Morgan[⑤]、CeramTec AG[⑥]、APC 国际有限公司[⑦]、Noliac[⑧]、TRS 陶瓷公司[⑨]和全球其他众多公司中找到。表 1.3 列出了软质 PZT 材料性能，表 1.4 列出了硬质 PZT 材料性质，表 1.5 列出了由 PI 公司发布的无铅材料的特性。

---

① TERFENOL-D：TER=铽，FE=铁，NOL=海军军械实验室（现为海军水面作战中心卡德洛克分部（NSWC-CD）），D=镝。
② GALFENOL：GAL=镓，FE=铁，NOL=海军军械实验室（现为海军水面作战中心卡德洛克分部（NSWC-CD））。
③ ALFENOL：AL=铝，FE=铁，NOL=海军军械实验室（现为海军水面作战中心卡德洛克分部（NSWC-CD））。
④ www.physikinstrumente.com。
⑤ www.morgantechnicalceramics.com。
⑥ www.ceramtec.com/piezo-applications。
⑦ www.americanpiezo.com。
⑧ www.noliac.com。
⑨ www.trsceramics.com。

表 1.3 软质 PZT 材料性能（资料来源：PI 公司）

| 性　　能 | | 符号/单位 | PIC151 | PIC255/PIC252* | PIC155 | PIC153 | PIC152 |
|---|---|---|---|---|---|---|---|
| 物理和介电性能 | | | | | | | |
| 密度 | | $\rho/(\text{g/cm}^3)$ | 7.80 | 7.80 | 7.80 | 7.60 | 7.70 |
| 居里温度 | | $T_c/℃$ | 250 | 350 | 345 | 185 | 340 |
| 相对介电常数 | | $\varepsilon_{33}^T = \varepsilon_0$ ① | 2400 | 1750 | 1450 | 4200 | 1350 |
| | | $\varepsilon_{11}^T = \varepsilon_0$ ② | 1980 | 1650 | 1400 | | |
| 介质损耗因子 | | $\tan\delta/10^{-3}$ | 29 | 20 | 20 | 30 | 15 |
| 机电性能 | | | | | | | |
| 耦合系数 | | $k_p$ | 0.62 | 0.62 | 0.62 | 0.62 | 0.48 |
| | | $k_t$ | 0.53 | 0.47 | 0.48 | | |
| | | $k_{31}$ | 0.38 | 0.35 | 0.35 | | |
| | | $k_{33}$ | 0.69 | 0.69 | 0.69 | | 0.58 |
| | | $k_{15}$ | | 0.66 | | | |
| 压电应变常数 | | $d_{31}/(10^{-12}\text{C/N})$ | -210 | -180 | -165 | | |
| | | $d_{33}/(10^{-12}\text{C/N})$ | 500 | 400 | 360 | 600 | 300 |
| | | $d_{15}/(10^{-12}\text{C/N})$ | | 550 | | | |
| 压电电压系数 | | $g_{31}/(10^{-3}\text{V·m/N})$ | -11.5 | -11.3 | 12.9 | | |
| | | $g_{33}/(10^{-3}\text{V·m/N})$ | 22 | 25 | 27 | 16 | 25 |
| 声力学性能 | | | | | | | |
| 频率系数 | | $N_p/(\text{Hz·m})$ | 1950 | 2000 | 1960 | 1960 | 2250 |
| | | $N_1/(\text{Hz·m})$ | 1500 | 1420 | 1500 | | |
| | | $N_3/(\text{Hz·m})$ | 1750 | | 1780 | | |
| | | $N_t/(\text{Hz·m})$ | 1950 | 2000 | 1990 | 1960 | 1920 |
| 弹性柔度系数 | | $S_{11}^E/(10^{-12}\text{m}^2/\text{N})$ | 15.0 | 20.7 | 19.7 | | |
| | | $S_{33}^E/(10^{-12}\text{m}^2/\text{N})$ | 19.0 | | 11.1 | | |
| 弹性刚度系数 | | $c_{33}^D/(10^{10}\text{N/m}^2)$ | 10.0 | | 11.1 | | |
| 力学品质因数 | | $Q_m$ | 100 | 80 | 80 | 50 | 100 |
| 耐热性 | | | | | | | |
| $\varepsilon_{33}^T$ 温度系数③ | | $TK\varepsilon_{33}^T/(10^{-3}/\text{K})$ | 6 | 4 | 6 | 5 | 2 |
| 时间稳定性（参数每 10 年的相对变化，%） | | | | | | | |
| 相对介电常数 | | $C_\varepsilon$ | | -1.0 | -2.0 | | |
| 耦合系数 | | $C_k$ | | -1.0 | -2.0 | | |

\* 用于多层胶带技术的材料，根据要求提供系数矩阵。

① 平行于极化方向。

② 垂直于极化方向。

③ 在 -20~+125 的范围内。

表1.4 硬质PZT材料性能（资料来源：PI公司）

| 性能 | 符号单位 | PIC181 | PIC18[①] | PIC144[①] | PIC241 | PIC300 | PIC300 |
|---|---|---|---|---|---|---|---|
| 物理和介电性能 | | | | | | | |
| 密度 | $\rho/(\text{g/cm}^3)$ | 7.80 | 7.75 | 7.95 | 7.80 | 7.80 | 5.50 |
| 居里温度 | $T_c/\text{℃}$ | 330 | 295 | 320 | 270 | 370 | 150 |
| 相对介电常数 | $\varepsilon_{33}^T/\varepsilon_0$ [①] | 1200 | 1015 | 1250 | 1500 | 1550 | 950 |
| | $\varepsilon_{11}^T/\varepsilon_0$ | 1500 | 1250 | | 1500 | 1550 | 950 |
| 介质损耗因子 | $\tan\delta/10^{-3}$ | 3 | 5 | 4 | 5 | 3 | 15 |
| 机电性能 | | | | | | | |
| 耦合系数 | $k_p$ | 0.56 | 0.55 | 0.60 | 0.50 | 0.48 | 0.30 |
| | $k_t$ | 0.46 | 0.44 | 0.48 | 0.46 | 0.43 | 0.42 |
| | $k_{31}$ | 0.32 | 0.30 | 0.30 | 0.32 | 0.25 | 0.18 |
| | $k_{33}$ | 0.66 | 0.62 | 0.66 | 0.64 | 0.46 | |
| | $k_{15}$ | 0.63 | 0.65 | | 0.63 | 0.32 | |
| 压电应变常数 | $d_{31}/(10^{-12}\text{C/N})$ | −120 | −100 | −110 | −130 | −80 | −50 |
| | $d_{33}/(10^{-12}\text{C/N})$ | 265 | 219 | 265 | 290 | 155 | 120 |
| | $d_{15}/(10^{-12}\text{C/N})$ | 475 | 418 | | 265 | 155 | |
| 压电电压常数 | $g_{31}/(10^{-3}\text{V}\cdot\text{m/N})$ | −11.2 | −11.1 | −10.1 | −9.8 | −9.5 | |
| | $g_{33}/(10^{-3}\text{V}\cdot\text{m/N})$ | 25 | 24.4 | 25 | 21 | 16 | −11.9 |
| 声力学性能 | | | | | | | |
| 频率系数 | $N_p/(\text{Hz}\cdot\text{m})$ | 2270 | 2195 | 2180 | 2190 | 2350 | 3150 |
| | $N_1/(\text{Hz}\cdot\text{m})$ | 1640 | 1590 | 1590 | 1590 | 1700 | 2300 |
| | $N_3/(\text{Hz}\cdot\text{m})$ | 2010 | 1930 | | 1550 | 1700 | 2500 |
| | $N_t/(\text{Hz}\cdot\text{m})$ | 2110 | 2035 | 2020 | 2140 | 2100 | |
| 弹性柔度系数 | $S_{11}^E/(10^{-12}\text{m}^2/\text{N})$ | 11.8 | 12.7 | 12.4 | 12.6 | 11.1 | |
| | $S_{33}^E/(10^{-12}\text{m}^2/\text{N})$ | 14.2 | 14.0 | 15.5 | 14.3 | 11.8 | |
| 弹性刚度系数 | $C_{33}^D/(Z10^{10}\text{N/m}^2)$ | 16.6 | 14.8 | 15.2 | 13.8 | 16.4 | |
| 力学品质因数 | $Q_m$ | 2000 | 400 | 1000 | 400 | 1400 | 250 |
| 温度稳定性 | | | | | | | |
| $\varepsilon_{33}^{T\,(3)}$温度系数 | $TK\varepsilon_{33}^T/(10^{-3}/\text{K})$ | 3 | 5 | | 2 | | |
| 时间稳定性（参数每10年的相对变化，%） | | | | | | | |
| 相对介电常数 | $C_\varepsilon$ | | | −4.0 | | | −5.0 |
| 耦合度 | $C_k$ | | | −2.0 | | | −8.0 |

① 初步数据，可能会发生变化。

# 第1章 智能材料与结构概述

表1.5 无铅材料的特性（资料来源：PI公司）

| 性 能 | 符号/单位 | PIC50* | PIC700** |
|---|---|---|---|
| 物理和介电性能 | | | |
| 密度 | $\rho/(g/cm^3)$ | 4.70 | 5.6 |
| 居里温度 | $T_c/℃$ | >500 | 200① |
| 相对介电常数 | $\varepsilon_{33}^T/\varepsilon_0$① | 60 | 700 |
| | $\varepsilon_{11}^T/\varepsilon_0$ | 85 | 1650 |
| 介质损耗因子 | $\tan\delta(10^{-3})$ | <1 | 30 |
| 机电性能 | | | |
| 耦合系数 | $k_p$ | | 0.15 |
| | $k_t$ | | 0.40 |
| | $k_{31}$ | | |
| | $k_{33}$ | | |
| | $k_{15}$ | | |
| 压电应变常数 | $d_{31}/(10^{-12}C/N)$ | | |
| | $d_{33}/(10^{-12}C/N)$ | 40 | 120 |
| | $d_{15}/(10^{-12}C/N)$ | 80 | |

\* 晶体材料。
\*\* 初步数据，可能会发生变化。
① 最高工作温度。

按PI网站上所写，其提供的所有PZT材料都具备以下属性：
比热容：WK约为350J/(kg·K)；
比热导率：WL约为1.1W/(m·K)；
泊松比（横向收缩）：$\nu=0.34$。
热膨胀系数：
$\alpha_3=(4\sim6)\times10^{-6}(1/K)$（近似值，极化方向，短路）；
$\alpha_1=(4\sim8)\times10^{-6}(1/K)$（大约垂直于极化方向，短路）。
静压强度：>600MPa。
数据是根据En 50324-2标准几何尺寸测试件确定的典型值。所有数据均为环境温度(23±2)℃条件下极化24~48h后测得①。

另一种广泛使用的聚合物薄膜压电材料为PVDF，如文献［666］所述，PVDF具有以下特性：

---

① 若需要某个材料的完整系数矩阵，应向PI Ceramic (info@piceramic.de) 申请。

(1) 柔软性（可应用于非水平表面）。
(2) 高机械强度。
(3) 尺寸稳定性。
(4) 薄膜平面内均衡的压电作用。
(5) 90℃以下高且稳定的压电系数，不随时间而变化。
(6) PVDF 特有的化学惰性。
(7) 缠绕在滚筒上可在较长的长度内连续极化。
(8) 厚度介于 0.9~1mm。
(9) 声阻抗接近水的声阻抗，具有平坦的响应曲线。

PVDF（图 1.18 和图 1.19）特性列于表 1.6~表 1.8 中，数据来源于文献[666]。

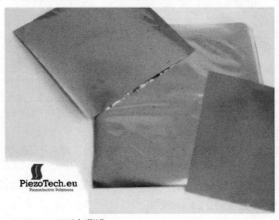

PVDF-TrFE压电薄膜

图 1.18 PIÉZOTECH S. A. S. 公司（www.piezotech.fr）的 PVDF-TrFE 压电薄膜

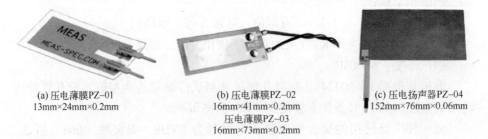

(a) 压电薄膜PZ-01
13mm×24mm×0.2mm

(b) 压电薄膜PZ-02
16mm×41mm×0.2mm
压电薄膜PZ-03
16mm×73mm×0.2mm

(c) 压电扬声器PZ-04
152mm×76mm×0.06mm

图 1.19 图像科学仪器（Images Scientific Instruments）公司（http://www.imagesco.com/）的各种 PVDF 压电薄膜传感器

表 1.6~表 1.8 使用了文献 [666-672] 中提供的数据,以供选择合适的材料类型并理解其目的、驱动或传感特性。

表 1.6 PVDF 压电膜材料双向拉伸膜的特性
(资料来源:PIÉZOTECH S.A.S.,www.piezotech.fr)

| 性　能 | 厚度 9μm (±5%) | 厚度 25μm (±5%) | 厚度 40μm (±5%) |
|---|---|---|---|
| 23℃下压电/热电性能 | | | |
| $d_{33}$/(pC/N) | 16±20% | 15±20% | 15±20% |
| $d_{31}$/(pC/N) | 6±20% | 6±20% | 6±20% |
| $d_{32}$/(pC/N) | 6±20% | 1±20% | 6±20% |
| $g_{33}$(1Hz)/(Vm/N) | 0.15±20% | 0.14±20% | 0.14±20% |
| $p_3$/(μC/(m²K))① | −20±25% | −25±25% | −19±25% |
| 23℃下介电性能 | | | |
| $\varepsilon_r$(0.1kHz) | 11.5±10% | 11.5±10% | 11.5±10% |
| $\varepsilon_r$(1kHz) | 11.5±10% | 11.5±10% | 11.5±10% |
| $\varepsilon_r$(10kHz) | 11.0±10% | 11.0±10% | 11.0±10% |
| tan$\delta$(0.1kHz) | 0.010±10% | 0.010±10% | 0.010±10% |
| tan$\delta$(1kHz) | 0.015±10% | 0.015±10% | 0.015±10% |
| tan$\delta$(10kHz) | 0.035±10% | 0.035±10% | 0.035±10% |
| 直流击穿电压/V | 750±30% | 760±30% | 540±30% |
| 23℃下力学性能 | | | |
| 杨氏模量/MPa② | 2500±20% | 3200±20% | 2500±20% |
| 杨氏模量/MPa③ | 2500±20% | 3200±20% | 2500±20% |
| 断裂拉伸应力/MPa② | 175±15% | 240±15% | 170±15% |
| 断裂拉伸应力/MPa③ | 190±15% | 60±15% | 190±15% |
| 断裂伸长率/%② | 50±30% | 20±30% | 50±30% |
| 断裂伸长率/%③ | 50±30% | 5±30% | 50±30% |
| 23℃下热学性能 | | | |
| 熔点/℃ | 175±5% | 175±5% | 175±5% |
| 横向 | 90~100 | 90~100 | 90~100 |

① 热电系数。
② 纵向。
③ 横向。

表 1.7　P(VDF-TrFE)共聚物 75/25 薄膜的特性
（资料来源：PIÉZOTECH S. A. S.，www.piezotech.fr）

| 性　能 | 厚度 12μm（±5%） | 厚度 25μm（±5%） | 厚度 50μm（±5%） | 厚度 110μm（±5%） |
| --- | --- | --- | --- | --- |
| 23℃下压电/热电性能 | | | | |
| $d_{33}/(pC/N)$ | 16±20% | 15±20% | 15±20% | 15±20% |
| $d_{31}/(pC/N)$ | 6±20% | 6±20% | 6±20% | 6±20% |
| $d_{32}/(pC/N)$ | 6±20% | 6±20% | 6±20% | 6±20% |
| $g_{33}(1Hz)/(Vm/N)$ | 0.15±20% | 0.18±20% | 0.18±20% | 0.18±20% |
| $p_3/(\mu C/(m^2 K))$[①] | -20±25% | -19±25% | -19±25% | -19±25% |
| 23℃下介电性能 | | | | |
| $\varepsilon_r(0.1kHz)$ | 9.4±10% | 9.6±10% | 9.6±10% | 9.6±10% |
| $\varepsilon_r(1kHz)$ | 9.3±10% | 9.4±10% | 9.4±10% | 9.4±10% |
| $\varepsilon_r(10kHz)$ | 9.1±10% | 9.2±10% | 9.2±10% | 9.2±10% |
| $\tan\delta(0.1kHz)$ | 0.014±10% | 0.015±10% | 0.015±10% | 0.015±10% |
| $\tan\delta(1kHz)$ | 0.014±10% | 0.016±10% | 0.016±10% | 0.016±10% |
| $\tan\delta(10kHz)$ | 0.028±10% | 0.032±10% | 0.032±10% | 0.032±10% |
| 直流击穿电压/V | 575±30% | 395±30% | | |
| 23℃下力学性能 | | | | |
| 杨氏模量/MPa[②] | 950±20% | 1000±20% | 1000±20% | 1000±20% |
| 杨氏模量/MPa[③] | 1500±20% | 1200±20% | 1200±20% | 1200±20% |
| 断裂拉伸应力/MPa[②] | 90±15% | 60±15% | 40±15% | 40±15% |
| 断裂拉伸应力/MPa[③] | 30±15% | 20±15% | 30±15% | 30±15% |
| 断裂伸长率/%[②] | 150±30% | 300±30% | 400±30% | 400±30% |
| 断裂伸长率/%[③] | 30±30% | 300±30% | 450±30% | 450±30% |
| 23℃下热学性能 | | | | |
| 熔点/℃ | 150±5% | 150±5% | 150±5% | 150±5% |
| 居里温度/℃ | 135±5% | 135±5% | 135±5% | 135±5% |
| 横向 | 90~100 | 90~100 | 90~100 | 90~100 |

① 热电系数。
② 纵向。
③ 横向。

表1.8 P(VDF-TrFE)共聚物70/30薄膜的特性
(资料来源:PIÉZOTECHS.A.S.,www.piezotech.fr,见图1.18)

| 性　能 | 厚度20μm (±5%) | 厚度25μm (±5%) | 厚度40μm (±5%) | 厚度6μm (±5%)[①] |
|---|---|---|---|---|
| 23℃下压电/热电性能 | | | | |
| $d_{33}/(pC/N)$ | −20±20% | −20±20% | −20±20% | −19±20% |
| $d_{31}/(pC/N)$ | 6±20% | 6±20% | 6±20% | 6±20% |
| $d_{32}/(pC/N)$ | 6±20% | 6±20% | 6±20% | 6±20% |
| $g_{33}(1Hz)/(Vm/N)$ | 0.15±20% | 0.2±20% | 0.2±20% | 0.2±20% |
| $p_3/(\mu C/(m^2 K))$[②] | | | | −20±25% |
| 23℃下介电性能 | | | | |
| $\varepsilon_r(10Hz\sim1kHz)$ | 8±10% | 8±10% | 8±10% | 9.4±10% |
| $\tan\delta(1kHz)$ | 0.016±10% | 0.016±10% | 0.016±10% | 0.016±10% |
| 23℃下力学性能 | | | | |
| 杨氏模量/MPa | 1000±20% | 1000±20% | 1000±20% | 950±20% |
| 断裂拉伸应力/MPa | 60±15% | 60±15% | 60±15% | 30±15% |
| 断裂伸长率/% | 60±30% | 60±30% | 60±30% | 30±30% |
| 23℃下热学性能 | | | | |
| 熔点/℃ | 156±5% | 156±5% | 156±5% | 150±5% |
| 居里温度/℃ | 112±5% | 112±5% | 112±5% | 135±5% |
| 最大温度/℃ | 95 | 95 | 95 | 120 |

① P(VDF-TrFE)共聚物77/23薄膜。
② 热电系数。

表1.9比较了标准PVDF薄膜和标准PZT。显然,虽然PVDF与PZT相比重量轻且具有较低的杨氏模量和较低的压电电荷系数,但它具有较大的电压系数,使其适用于传感应用。

表1.9 标准PVDF薄膜和标准PZT之间的比较

| 性　能 | 符号/单位 | PVDF | PZT |
|---|---|---|---|
| 压电应变系数 | $d_{31}/[(C/N)$或$(m/V)]$ | 22×10⁻¹² | −175×10⁻¹² |
| | $d_{33}/[(C/N)$或$(m/V)]$ | −30×10⁻¹² | 400×10⁻¹² |
| 压电电压系数 | $g_{31}/(Vm/N)$ | 216×10⁻³ | −11×10⁻³ |
| | $g_{33}/(Vm/N)$ | −330×10⁻³ | 25×10⁻³ |

续表

| 性　　能 | 符号/单位 | PVDF | PZT |
|---|---|---|---|
| 耦合度 | $k_{31}$ | 0.14 | 0.34 |
| 相对介电常数 | $\varepsilon_r = \varepsilon_{33}^T/\varepsilon_0$ | 12 | 1700 |
| 最大工作温度 | ℃ | 80 | 150 |
| 密度 | $\rho/(kg/m^3)$ | 1780 | 7600 |
| 杨氏模量 | $T$/GPa | 2 | 71 |
| 声阻抗 | $Z/(10^6 kg/(m^2 \cdot s))$ | 2.7 | 30 |

表 1.10 所示为用于驱动的各种材料的比较，列出了各种相关属性，以评估不同类型致动器的性能。

表 1.10　基于智能材料的致动器之间的比较[667]

| 性　　能 | PZT G-1195 | PVDF | PMN* | TERFENOL-D | 镍钛诺合金 |
|---|---|---|---|---|---|
| 驱动机构类型 | 压电陶瓷 | 压电薄膜 | 电致伸缩 | 磁致伸缩 | 形状记忆合金 |
| 最大应变/$10^{-6}$ | 1000 | 700 | 1000 | 2000 | 20000 |
| 杨氏模量 | 70 | 2 | ~120 | 43[①]~107.5[②] | 28~40[③]和83[④] |
| 频带宽度 | 高 | 高 | 高 | 中 | 低 |
| 响应时间 | μs | μs | μs | μs | s |
| 应变电压特性 | 一阶线性 | 一阶线性 | 非线性 | 非线性 | 非线性 |

① 不饱和磁态。
② 饱和磁态。
③ 马氏体相。
④ 奥氏体相。
* PMN = 铌镁酸铅。

PZTG-1195 的特性如下：

$\rho = 7.5 g/cm^3$, $c_{11}^E = c_{22}^E = 148 GPa$, $c_{33}^E = 131 GPa$, $c_{12}^E = 76.2 GPa$

$c_{13}^E = c_{23}^E = 74.2 GPa$, $c_{44}^E = c_{55}^E = 25.4 GPa$, $c_{66}^E = 35.9 GPa$, $e_{31} = -2.1 C/m^2$

$e_{33} = 9.5 C/m^2$, $e_{15} = 9.2 C/m^2$

式中：

$$\begin{aligned} &d_{31} = e_{31}S_{11}^E + e_{31}S_{12}^E + e_{33}S_{13}^E, \quad d_{33} = e_{31}S_{13}^E + e_{31}S_{13}^E + e_{33}S_{33}^E \\ &d_{15} = e_{15}S_{44}^E, \quad e_{15} = d_{15}c_{44}^E, \quad e_{33} = d_{31}c_{13}^E + d_{31}c_{13}^E + d_{33}c_{33}^E \\ &e_{31} = d_{31}c_{11}^E + d_{31}c_{12}^E + d_{33}c_{13}^E \end{aligned} \quad (1.6)$$

Shahinpoor 等描述了其他类型的致动器，包括离子交换聚合物-贵金属复合材料（Ion-exchange Polymer-noble metal Composite，IPMC）和电活性陶瓷

（Electroactive Ceramic，EAC）[668]。它们与形状记忆合金致动器的特性比较如表1.11所示。

表1.11 基于IPMC、SMA和EAC的致动器性能比较[668]

| 性　　能 | IPMC | EAC | 形状记忆合金 |
|---|---|---|---|
| 驱动位移 | >10% | (0.1~0.3)% | <8% |
| 力/MPa | 10~30 | 30~40 | ~700 |
| 响应时间 | 从μs到s | 从μs到s | 从s到min |
| 密度/(g/cm$^3$) | 1~2.5 | 6~8 | 5~6 |
| 驱动电压/V | 4~7 | 50~800 | 6~10 |
| 能量功耗 | W | W | W |
| 断裂韧性 | 韧性，弹性 | 脆性 | 弹性 |

随着 Terfenol-D 等巨磁致伸缩材料（Giant Magnetostrictive Material，GMM）和 NiMnGa 合金等磁性形状记忆材料（Magnetic Shape Memory Material，MSM）的开发，基于这两种材料和类似材料构建了新的致动器。文献［671］对有望用作致动器的材料进行了研究，将它们的特性进行总结得到了表1.12，包括 Terfenol-DGMM、MSN、复合 GMM（由磁致伸缩合金颗粒与电绝缘黏合剂如聚合物制成）、Galfenol（由铁和镓制成的磁致伸缩合金-FeGa）、PZT-4（一种硬质压电陶瓷材料）和用于致动器的软 PZT 多层技术（Multilayer Technignue for Actuator，MLA）。

表1.12 中显示的应变为峰值，由在 Cedrat Technologies 进行的实验中获得①。

表1.12 GMM与其他相关驱动材料的比较[671]

| 性　　能 | Terfenol-D GMM | MSM | GMM 复合物 | Galfenol | 硬质PZT PZT-4 | 软质PZT MLA |
|---|---|---|---|---|---|---|
| 最大静态应变/ppm① | 1800 | 50000 | 1000 | 320 | 600 | 1250 |
| 耦合度/% | 70 | 75 | 35 | 40 | 67 | 65 |
| 杨氏模量/GPa | 25 | 7 | 20 | 45 | 60 | 40 |
| 最大预应力/MPa | 50 | 1 | 30 | 80 | 50 | 40 |
| 最大动态应变（共振）/ppm | 4000 | 140② | 3000 | 3500② | 1600 | 2000 |

① ppm=百万分比。
② MSN 和 Galfenol 动态应变使用了应力极限进行计算。

---

① CENELEC 规范委员会 BTTF63-2 先进技术陶瓷（WGⅡ：NG13），www.cedrat-technologies.com。

前文讨论了使用智能材料作为致动器。下面将描述智能结构的第 2 个角色，即传感器。由于压电陶瓷和压电薄膜能够制造成用作传感器所需的任何尺寸，因此它们是该角色的理想选择。它们与其他传感器的比较如表 1.13 所示。

表 1.13　各种传感器的比较

| 传感器类型 | 激活的方式或性质 | 敏感度 | 定位/mm | 频带宽度 |
| --- | --- | --- | --- | --- |
| 电阻计 | 用 10V 激活 | 30（V/$\varepsilon$） | 0.2038 | 0Hz～声频 |
| 半导体应变片 | 用 10V 激活 | 1000（V/$\varepsilon$） | 0.7620 | 0Hz～声频 |
| 光学纤维 | 1.016mm 干涉仪测量长度 | $10^6$（(°)/$\varepsilon$） | 1.0160 | 0Hz～声频 |
| 介电薄膜 | 厚度为 0.0254mm | $10^4$（V/$\varepsilon$） | <1.0160 | 0.1Hz～GHz |
| 压电陶瓷 | 厚度为 0.0254mm | $2\times10^4$（V/$\varepsilon$） | <1.0160 | 0.1Hz～GHz |

## 1.4　智能材料与系统的应用

本小节将介绍智能材料与系统的应用方面在进行与已发表的工作。所选的内容代表了许多大学、研究机构和行业正在活跃进行的一些研究项目。

第 1 个主题涉及变形体飞行器（近期典型的工作见文献［673-677］）。该主题已被许多研究者和研究课题广泛报道（参见欧洲计划 SARISTU[①]-智能飞机结构和相应的美国 NASA 变形机翼计划[②]）。尽管"变形"一词没有量化定义，但可将变形飞机定义为一种适应性强、随时间变化，且几何形状的变化会影响其空气动力学性能的飞行器（图 1.20）。根据任务改变机翼轮廓的想法来自对鸟类翅膀的观察，鸟类的翅膀会根据起飞、着陆、潜水和巡航等飞行阶段来改变其横截面形状。机翼的变形通过形状记忆合金、Terfenol、锆钛酸铅等致动器来实现，这可将其主题与智能材料联系起来。变形机翼的显著优势在于活动部件少、重量轻、改进的空气动力学、燃料消耗低以及更少的机翼布线。但是，变形方法存在一些缺点或风险：如何克服由于附加驱动系统而导致的重量增加、高材料成本、维修成本、所用材料的耐用性以及在系统故障的情况下如何运行。虽然这种解决方案限制了机动性和效率，所产生的设计对许多飞行状态而言也不是最佳情况，但可展开襟翼的机制为当前自适应机翼几何形状提供了标准。文献［673］给出了详细的最新综述，总结了变形飞机的最新技

---

① 网址：www.saristu.eu。
② 网址：www.nasa.gov。

术、重点介绍了固定翼和旋转翼结构、形状变化概念以及有源系统。作者在文献中指出，"虽然已经综合了许多概念，但很少有人进行机翼风洞测试，更少有人进行飞行测试的相关研究"。在文献［678］中可以查阅到有关机翼变形的充气式解决方案的参考。Joo 等在论文中讨论了复合皮肤问题及其领域存有的技术挑战[674]。为了获得具有较低的平面内刚度和较高的平面外刚度的材料，他们提出了一种两阶段的设计方案，并基于该方案过程展示了一种二维皮肤设计，通过快速原型技术制造了模型并且进行了相应测试。其他学者[675]建议使用弹性基体复合材料来制备可发生被动变形的皮肤，其包括弹性体-纤维复合材料基表面层，并由柔性蜂窝结构作为支撑，且每个结构平面内具有接近于零的泊松比。完整的原型变形皮肤可以表现出 100% 的单轴延伸和 100% 的表面积增加。同时，在高达 9.58kPa 的压力下，可以在各种面积变化水平下保持小于 2.5mm 的平面外偏转。

荷兰代尔夫特理工大学团队提出了一种结合优化设计方法的通用气动弹性变形机翼分析[676]。该方法可用于预测空气动力学性能、载荷分布、气动弹性变形以及所需的致动力和力矩以及相应的致动能量。此外，该程序在理论上可应用于折叠式弯曲机翼性能的预测。Lesieutre 等提出了一种二维柔性蜂窝桁架结构以替代给定机翼内部的固定结构[677]。这种桁架结构能够改变翼展、展弦比和机翼面积。作者在研究中表明，对于重达 100 英磅（1 英磅 = 0.454kg）的飞行器，他们提出的变形结构能将跨度减少 74%，同时还能保持与传统飞机相似的重量分数（包括致动器重量），该值为 12.2%。将该设计放大在大型飞机上使用发现，其可实现跨度的大幅减小。然而，结构和致动器的重量分数需要使结构变形，并限制了二维蜂窝桁架结构大规模使用时的可行性。最后，Pecora 等报道了他们在意大利公司（Alenia Aeronautica S. p. A.）资助下进行的一项用于可变弧度后缘的新型变形结构的研究[679]。他们基于 CS-25 类典型民用区域运输飞机的全尺寸机翼确定了参考几何形状，然后用可塑性筋骨材料为基础的变形后缘替代了传统的襟翼组件。所提出的这一架构可以产生很高的变形能力，同时保持良好的负载承受能力。这种变形后的机翼后缘结构被构建成一系列相互连通的变形翼肋，该结构通过翼肋内嵌入的基于形状记忆合金的原始致动器来驱动。Previtali 等提出了另一种可行的方法，开发了一种能够同时承受弯曲和剪切载荷，并允许平面内大幅度拉伸的皮肤[680]。上述方案的实现基于双波纹的双壁结构，该双壁结构具有高弯曲刚度，同时能实现平面内 20% 拉伸（图 1.20）。通过实验数据对数值结果进行了验证，显示出应用于这一新领域的可行性。

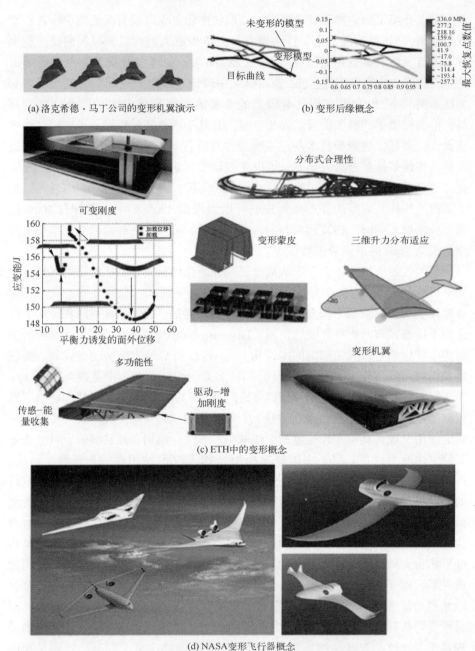

图 1.20 各种变形飞机概念

(资料来源：www.newscientist.com/news/news/jsp?id=as99994484 （a）；Sergio Ricci 教授-DAST PoliMI （b）；ETH Zurich-DMAVT-复合材料与自适应结构实验室 （c）；www.nasa.gov （d））

第 2 个主题是利用智能材料收集能量,其目的是收集寄生环境中的振动能量并将其转化为电能为各种装置供电,使其自给自足。典型装置如图 1.21 所示,其中配备压电层的悬臂梁由于环境激励而振动。为了降低梁的固有频率,尖端加装了质量块以使固有频率与激励频率相匹配。发生共振时,悬臂梁将以其最大振幅振动而产生电压,电压收集并存储到电容器、超级电容器、电池等储能装置中。由于振动形式是谐振,所以需要用到二极管桥(图 1.21(b)),以防止电压返回到压电层。

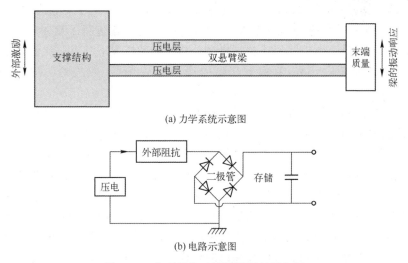

(a) 力学系统示意图

(b) 电路示意图

图 1.21 典型的基于悬臂梁配备压电层

已有大量论文致力于多方面能量采集系统和方法的实现,最近相关的典型工作将做出介绍[681-693]。Tadesse 等提出了一种由电磁和压电采集系统构成的创新混合机制,由长度为 125mm 的铝悬臂锥形梁配备 3 对单晶压电板组成①[678]。将固定电感线圈连接到悬臂梁的尖端,使得悬臂梁尖端的质量块与线圈同心,当悬臂梁振动时,磁铁进出线圈,根据法拉第定律将产生电压。连接到尖端的线圈既用作附加质量(从而降低悬臂梁的固有频率),又用作电磁收集器。实验结果表明,在加速度为 35$g$、频率为 20Hz(悬臂梁的第一固有频率)时,同时以电磁和压电两种混合机制可获得的功率为 0.25W,而仅使用压电晶体时该值为 0.25mW。将频率增加到 100Hz 发生非共振,获得的功率显著下降,混合系统获得功率仅为 0.025W。Liao 和 Sodano 鉴于能量采集效率没

---

① 压电板由 Pb(Zn$_{1/3}$Nb$_{2/3}$)O$_3$-PbTiO$_3$(PZN-PT)单晶($d_{31}$模式)制成[694]。

有公认的定义，故将每个周期振动梁的应变能与输出功率的比值定义为系统能量收集效率[682]。该定义类似于材料的损耗因子，仿真研究表明，最大效率通常出现在阻抗匹配处。然而，对于具有高机电耦合的材料，最大功率则在接近开路和闭路谐振时产生，此时的效率较低。Sun 等配备了两个单晶板（其夹紧边界与文献 [681] 中使用的材料相同，自由尖端为钢质块体）的铝悬臂梁进行数值和实验研究[683]。结果表明，在外部电阻负载为 80kΩ 时所得功率最大，约为 0.6mW。Liu 等展示了另一种通过开关式电子装置来控制压电装置上相对于机械输入的电压和/或电荷，优化能量转换进而解决电子端能量收集的难题[684]。他们提出的新方法通过多层 PVDF 聚合物装置进行了展示①，与优化的基于二极管整流器电路相比，在机械位移相同的情况下，有源能量采集方法收集的能量增加了 5 倍，在有效机械应变为 1.37% 时，收集到了约 22mW 的能量。

来自美国弗吉尼亚理工大学的 Erturk 等[685]提出了一种新型梁结构，以替代传统悬臂梁用于能量收集。这一新型 L 形梁具有两个集中质量块，一个在水平梁的末端，另一个沿着垂直梁，压电材料沿着上述两梁而成 L 形。调控 L 形梁两端的固有频率至接近，将产生更宽频带的能量采集系统。通过数值方法研究了 L 形压电能量采集器配置在无人机中作为起落架的情景，在与已发表实验中呈现的以弯曲梁同目的配置测试结果相比，同样显示出良好的结果。来自法国里昂 INSALGEF 的 Guyomar 团队详细介绍了利用压电材料收集能量并将其转化为电能的方式。他们通过增加压电材料中存在的热释电效应来扩展压电材料的采集能力，从而建立了从温度变化中采集能量的方式，其中的温度变化能量是一种常见的寄生能源。

Farinholt 等对 PVDF 进行了研究，并与离子导电聚合物换能器进行了比较[687]。在该研究中，他们假设能量收集器为轴向加载，并将分析模型与实验结果进行比较，研究表明，厚度为 64μm 的 PVDF 收集的功率为 $4.494 \times 10^{-4}$ mW，而离子导电聚合物换能器收集的功率为 $3.15 \times 10^{-6}$ mW。其中，PVDF 的尖端质量为 1208g（第 1 固有频率为 138Hz），而第 2 个换能器为 346g（第一固有频率仅为 70Hz）。Olivier 和 Priya 提出了一种基于 for-bar 电磁装置的采集器，分析/数值计算和实验结果表明，在频率约为 150Hz 时所得平均功率为 17.9mW[688]。

Aldraihem 和 Baz 提出了一种新型能量收集器[689]。分析/数值研究中，他们在压电元件和移动底座之间放置了弹簧质量系统，从而放大了压电元件所承

---

① 来自 Measurement Specialties 公司。PVDF 膜的厚度为 2mm。

受的应变,最终实现了收集器电力输出的放大。结果表明,与没有动态放大的普通能量收集器相比,新型能量收集装置可将功率放大 20 倍。

Zoric 等基于粒子群优化算法的模糊优化策略,通过在配备压电贴片的薄壁复合梁上应用最优振动控制系统来解决能量收集的难题[690]。Tang 等报道了一项全面的实验研究,结果表明,在各种振动模式下,磁铁的使用可以达到改进压电材料的能量收集器功能的目的[691]。换言之,磁铁的引入诱导产生的多种非线性使得振动能量收集器的性能显著提高。

Elvin N. G. 和 Elvin A. A. 对柔性压电收集器中引入大挠度的影响进行了研究,并对静态、自由振动和受迫振动情况下的实验结果与数值预测进行了比较[692]。结果表明,与常见的小挠度线性采集器相比,大挠度能量收集装置往往会改变谐振频率并增加阻尼,最终引发输出电压显著降低。当振动梁的尖端偏转超过梁长度的 35% 时,这些非线性动态效应会变得更为显著。

Mikoshiba 等尝试利用谐振器晶格系统来提升收集的能量,所研究的基本单元器件由装有弹簧的磁铁组成,该磁铁封装在配有铜线圈的带盖的聚甲基丙烯酸甲酯管里,形成既可用作谐振器又可用作线性发生器的单元结构[693]。研究结果表明,单元器件通过 $1\Omega$ 负载电阻可产生 36mW 的连续有效功率。Green 等将环境振动源视为随机振动,通常也称为"白噪声"[695]。他们将此假设应用于两个振动产生源:人类步行运动和悬索桥的振动,并指出这些非线性引入能量采集器而实现的性能提升幅度与其所受到的环境激励类型有关。Xiong 和 Oyadiji 使用分布式参数机电模型进行了对电阻负载压电收集器的功率输出的预测[696]。结果表明,对于部分覆盖有压电层的收敛和发散型锥形悬臂与矩形悬臂梁,能够产生比全覆盖压电层的传统矩形悬臂梁更高的机电耦合系数。此外,在振动能量收集器的不同位置合理地添加额外质量可能会产生更大的功率密度。Anton 等从不同的方向进行了研究,并在研究中发现无铅聚合物基驻极体(一种新型压电泡沫材料)展现出压电特性,并对其低功率能量的产生进行了探究[697]。在介绍压电驻极体泡沫的制造和操作后,对各种试样进行了机械测试并获得相应的力学性能。结果显示,所得杨氏模量在 0.5~1GPa,拉伸强度在 35~70MPa,$d_{33}$ 系数在 10Hz~1kHz 的范围内相对恒定,约为 175pC/N。在 60Hz 和 ±73μm 的位移下对预拉伸的样品(尺寸为 15.2cm×15.2cm)进行谐波激发,产生的能量传递到 1mF 存储电容器的平均功率为 6.0mW,电容器在 30min 内充电至 4.67V,表明压电驻极体泡沫具有为小型电子元件供电的能力。Abramovich 等提出了另一种使用压电材料不借助振动模式采集能量的方法,该方法是在磁盘的厚度方向上施加循环载荷[312-313]。该项实验进行了对施加压缩载荷的影响探究,结果显示,当施加 8 个锆钛酸铅盘(每个叠放在另一个的顶部)时,在 40MPa 压力下所获得的功率为 0.43W。

另一篇由 Stoppa 和 Chiolerio 撰写的综述聚焦于纺织行业新趋势的概述,所述电子纺织品具有电子和相互交织的功能,从而呈现出其他现有电子制造技术无法实现的物理灵活性和特有尺寸[698]。在可穿戴电子产品的展望中,未来电子系统将成为日常服装不可或缺的一部分,因此,这类电子装置必须满足有关可穿戴性的特殊要求。可穿戴系统的特点是能够自动识别穿着者的活动、行为状态及其周围的情景,并以这些信息来调整系统的配置和功能。他们的综述侧重于智能纺织品领域的最新进展,描述各种可用材料及其制造工艺,并在灵活性、人体工程学、低功耗、集成性和自主性之间进行着重权衡时,客观评价其展示出的优、缺点。

Yu 和 Leckey 报道了智能材料的其他方面的应用[699]。他们应用压电传感器阵列、定量裂纹检测和兰姆波聚焦阵列算法,并利用三维弹性动力学有限积分技术研究了薄壁板上的兰姆波传播以及波与裂纹损伤的相互作用。Annamdas 和 Radhika 发表了一份有关监测方面的重要进展,以及与机电阻抗的金属和非金属结构健康监测相关的有线、无线和能量采集方法的最新综述[700]。Spaggiari 等在报道中模拟并测试了由形状记忆合金制成的 Negator 弹簧,这种弹簧由带状形状记忆合金制成并以给定的曲率缠绕,每个线圈都紧紧地缠绕在其内部的相邻线圈上[701]。Negator 弹簧在机械上具有两点优势:一是带状组件解旋过程中具有恒力-位移特性,二是它安装在转鼓上且具有很长的长度。作者报道了由形状记忆合金制成的 Negator 弹簧的分析/数值模型,实验结果证实了对于这种几何形状的适用性。Ghaffari 等研究了另一种智能材料——磁流变体[702]。他们综述了当前用于磁流变体的各种模型和模拟方法,包括连续介质模型和离散相模型两种通用方法。此外,对使用磁共振流体的各种流变和结构模型以及连续统方法和离散方法的计算进行了总结。Meisel 等将增材制造方法与形状记忆合金的优势相结合,生产带有嵌入式形状记忆合金驱动线的 3D 打印部件,并详细介绍了使用紫外光固化成型的 3D 打印的制造过程,重点展示了初步实验结果[703]。

# 参 考 文 献

[1] Wada, B. K., Fanson, J. L. and Crawley, E. F., Adaptive structures, Journal of Intelligent Material Systems and Structures 1, April 1990, 157-174.

[2] Lazarus, K. B. and Napolitano, K. L., Smart structures, An overview, AD-A274 147, WL-TR-93-3101, Flight Dynamics Directorate Wright Laboratory Air Force, Material Command Wright-Patterson Air Force

Base, Ohio 45433-7552, September 1993, 60 pp.
[3] Iyer, S. S. and Haddad, Y. M., Intelligent materials-An overview, International Journal of Pressure Vessels and Piping 58, 1994, 335-344.
[4] Gandhi, M. V. and Thompson, B. S., Smart Materials and Structures, 1st edn, London, New York, Chapman & Hall, 1992, 309.
[5] Culshaw, B., Smart Structures and Materials, Artech House, 1996, 207.
[6] Preumont, A., Vibration Control of Active Structures, Dordrecht, the Netherlands, Kluwer Academic Publishers, 1997, 259.
[7] Adeli, H. and Saleh, M., Control, Optimization, and Smart Structures: High-Performance Bridges and Buildings of the Future, John Wiley & Sons, May 3 1999, 265.
[8] Srinivasan, A. V. and McFarland, D. M., Smart Structures: Analysis and Design, 1st edn, Cambridge University Press, 2001, 223.
[9] Watanabe, K. and Ziegler, F., Dynamics of Advanced Materials and Smart Structures, Springer Science & Business Media, July 31 2003, 469.
[10] Wadhawan, V. K., Blurring the distinction between the living and nonliving, Monographs on Physics and Chemistry of Materials 65, Oxford University Press, 18 October 2007, 368.
[11] Bandyopadhyay, B., Manjunath, T. C. and Umapathy, M., Modeling, Control and Implementation of Smart Structures: A FEM-State Space Approach, Springer, April 22 2007, 258.
[12] Leo, D. J., Engineering Analysis of Smart Material Systems, John Wiley and Sons, Inc., 2007, 556.
[13] Cetikunt, S., Mechatronics, John Wiley and Sons, Inc, 2007, 615.
[14] Giurgiutiu, V., Structural Health Monitoring: With Piezoelectric Wafer Active Sensors, 1st edn, Elsevier, Academic Press, 2008, 747.
[15] Gaudenzi, P., Smart Structures: Physical behavior, Mathematical Modeling and Applications, John Wiley and Sons, Inc, 2009, 194.
[16] Pryia, S. and Inman, D. J., (eds.), Energy Harvesting Technologies, Springer Science & Business Media, 2009, 517.
[17] Carrera, E., Brischetto, S. and Nali, P., Plates and Shells for Smart Structures: Classical and Advanced Theories for Modeling and Analysis, John Wiley & Sons Ltd., September 2011, 322.
[18] Holnicki-Szulc, J. and Rodellar, J., Smart Structures: Requirements and Potential Applications in Mechanical and Civil Engineering, Springer Science & Business Media, December 6 2012, 391.
[19] Vepa, R., Dynamics of Smart Structures, 1st edn, Wiley, April 2010, 410.
[20] Chopra, I. and Sirohi, J., Smart Structures Theory, New York, Cambridge University Press, 2013, 905.
[21] Janocha, H., (ed.), Adaptronics and Smart Structures: Basics, Materials, Design, and Applications, Springer Science & Business Media, November 11 2013, 438.
[22] Kholkin, A. L., Pertsev, N. A. and Goltsev, A. V., Piezoelectricity and Crystal Symmetry, Ch. 2 from Piezoelectric and Acoustic Materials for Transducer Applications, Safari, A. and Akdogan, E. K., (eds.), Springer, 2008, 482.
[23] Curie, P. and Curie, J., Developpement par pression, de l'electricite polaire dans les cristaux hemiedres a faces inclines, Comptes Rendus (France) 91, 1880, 294-295.
[24] Curie, P. and Curie, J., Contractions et dilatations produites par des tensions electriques dans les cristaux

hemiedres a faces inclines, Comptes Rendus (France) 93, 1881, 1137-1140.
[25] Voigt, W., Lerbuch der Kristallphysik, Leipzig-Berlin, Teubner, 1910.
[26] Jaffe, B., Cook, W. R. and Jaffe, H., Piezoelectric Ceramics, New York, Academic Press, 1971.
[27] Monner, H. P., Smart materials for active noise and vibration reduction, Novem-Noise and Vibration: Emerging Methods, Saint-Raphaël, France, 18-21 April 2005.
[28] www.piceramic.com/pdf/PIC_Tutorial.pdf.
[29] Giurgiutiu, V., Pomirleanu, R. and Rogers, C. A., Energy-Based Comparison of Solid-State Actuators, University of South Carolina, Report # USC-ME-LAMSS-2000-102, March 1, 2000.
[30] http://www.piceramic.com/pdf/material.pdf.
[31] Kawai, H., The Piezoelectricity of Poly (vinylidene fluoride), The Japan Society of Applied Physics 8, 1969, 975.
[32] Damjanovic, D., Ferroelectric, dielectric and piezoelectric properties of ferroelectric thin films and ceramics, Reports on Progress in Physics 61, 1998, 1267-1324.
[33] García, C., Arrazola, D., Aragó, C. and Gonzalo, J. A., Comparison of symmetric and asymmetric energy conversion cycles in PZT plates, Ferroelectrics Letters Section 28 (1-2), 2000, 43-47.
[34] Navid, A. and Pilon, L., Pyroelectric energy harvesting using Olsen cycles in purified and porous poly (vinylidene fluoride-trifluoroethylene) [P (VDF-TrFE)] thin films, IOP Publishing, Smart Materials and Structures 20, 2011, 025012, 18 August 2010, 9. doi: 10.1088/09641726/20/2/025012.
[35] Xie, J., Mane, P. P., Green, C. W., Mossi, K. M. and Leang, K. K., Energy harvesting by pyroelectric effect using PZT, Proceedings of SMASIS08 ASME Conference on Smart Materials, Adaptive Structures and Intelligent Systems, October 28-30, 2008, Ellicott City, Maryland, USA.
[36] Sebald, G., Guyomar, D. and Agbossou, A., On thermoelectric and pyroelectric energy harvesting, Smart Materials and Structures 18, 2009, 125006.
[37] Xie, J., Experimental and numerical investigation on pyroelectric energy scavenging, 2009, VCU Theses and Dissertations, Paper 2041.
[38] Lee, F., Experimental and analytical studies on pyroelectric waste heat energy conversion, Master of Science Thesis in Mechanical Engineering, 2012, University of California, Los Angeles, USA, 138 pp.
[39] Erturn, U., Green, C., Richeson, M. L. and Mossi, K., Experimental analysis of radiation heat-based energy harvesting through pyroelectricity, Journal of Intelligent Material Systems and Structures 25 (September), 2014, 1838-1849.
[40] Bowen, C. R., Taylor, J., LeBoulbar, E., Zabek, D., Chauhan, A. andVaish, R., Pyroelectric materials and devices for energy harvesting applications, Energy and Environmental Sciences 7, 2014, 3836-3856.
[41] Navid, A., Lynch, C. S. and Pilon, L., Synthesis and characterization of commercial, purified, and Porous Vinylidene Fluoride-Trifluoroethylene P (VDF-TrFE) thin films, Smart Materials and Structures 19, 2010, 055006 (1-13).
[42] Li, F., Jin, L., Xu, Z. and Zhang, S., Electrostrictive effect in ferroelectrics: An alternative approach to improve piezoelectricity, Applied Physics Reviews 1, 2014, 011103, AIP Publishing LLC. doi: http://dx.doi.org/10.1063/1.4861260.
[43] Surowiak, Z., Fesenko, E. G. and Skulski, R., Dielectric and electrostrictive properties of ferroelectric

relaxors, Archives of Acoustics 24 (3), 1999, 391-399.

[44] Newnham, R. E., Properties of Materials: Anisotropy, Symmetry, Structure, OUP, Oxford, November 11, 2004, 390 pp.

[45] Uchino, K., Nomura, S., Cross, L. E., Jang, S. J. and Newnham, R. E., Electrostrictive effect in lead magnesium niobate single crystals, Journal of Applied Physics 51, 1980, 1142-1145.

[46] Uchino, K., Nomura, S., Cross, L. E., Newnham, R. E. and Jang, S. J., Review Electrostrictive effect in perovskites and its transducer applications, Journal of Materials Science 16, 1981, 569-578.

[47] Cross, L. E., Jang, S. J., Newnham, R. E., Nomura, S. and Uchino, K., Large electrostrictive effects in relaxor ferroelectrics, Ferroelectrics 23, 1980, 187-191.

[48] Nomura, S., Kuwata, J., Jang, S. J., Cross, L. E. and Newnham, R. E., Electrostriction in Pb($Zn_{1/3}Nb_{2/3}$)$O_3$, Material Research Bulletin 14, 1979, 769-774.

[49] Jang, S. J., Uchino, K., Nomura, S. and Cross, L. E., Electrostrictive behavior of lead magnesium niobate based ceramic dielectrics, Ferroelectrics 27, 1980, 31-34.

[50] Uchino, K., Cross, L. E., Newnham, R. E. and Nomura, S., Electrostrictive effects in non-polar perovskites, Phase Transitions 1, 1980, 333-342.

[51] Zhang, Q. M., Bharti, V. and Zhao, X., Giant electrostriction and relaxor ferroelectrics behavior in electron-irradiated poly (vinylidene fluoride-tri-fluoroethylene) copolymer, Science 280, 1998, 2101-2104.

[52] Buschow, K. H. J. and De Boer, F. R., Physics of Magnetism and Magnetic Materials, Ch. 16, Vols 171-175, New York, Kluwer Academic/Plenum Publishers, 2003, 182.

[53] Joule, J. P., On the effects of magnetism upon the dimensions of iron and steel bars, The London, Edinburgh and Dublin philosophical magazine and journal of science (Taylor & Francis) 30, 1847, 225-241, Third Series: 76-87.

[54] Barrett, R. A. and Parsons, S. A., The influence of magnetic fields on calcium carbonate precipitation, Water Research 32, 1998, 609-612.

[55] Bar-Cohen, Y., Electroactive polymers as artificial muscles-Capabilities, Potentials and Challenges, HANDBOOK ON BIOMIMETICS, Yoshihito Osada (Chief ed.), Section 11, in Chapter 8, "Motion" paper #134, publisher: NTS Inc., August 2000, 1-13. Han, Y., Mechanics of magnetoactive polymers, Graduate Theses and Dissertations Paper 12929, Iowa State University, 2012, 112 pp.

[56] Aljanaideh, O., Rakheja, S. and Su, C.-Y., Experimental characterization and modeling of rate-dependent asymmetric hysteresis of magnetostrictive actuators, Smart Materials and Structures 23, 2014, 035002, 12. doi: 10.1088/0964-1726/23/3/035002.

[57] Zah, D. and Miehe, C., Variational-based computational homogenization of electro-magnetoactive polymer composite at large strains, 11 th World Congress on Computational Mechanics (WCCM XI), 5th European Conference on Computational Mechanics (ECCM V), 6th European Conference on Computational Fluid dynamics (ECFD VI), July 20-25, 204, Barcelona, Spain.

[58] Stoeckel, D., The shape memory effect-phenomenon, alloys and applications, Proceedings: Shape Memory Alloys for Power Systems, EPRI, 1995.

[59] Wei, Z. G., Sandström, R. and Miyazaki, S., Review shape-memory materials and hybrid composites for smart systems, Journal of Material Science 33, 1998, 3743-3762, 10 September 1997.

[60] Seelecke, S. and Müller, I., Shape memory alloy actuators in smart structures: Modeling and simulation,

Applied Mechanics Reviews 57 (1), January 2004, 23–46.

[61] Barbarino, S., Saavedra Flores, E. I., Ajaj, R. M., Dayyani, I. and Friswell, M. I., A review on shape memory alloys with applications to morphing aircraft, IOP Publishing, Smart Materials and Structures 23, 2014, 063001 (19 pp.), 1 August 2013, doi: 10.1088/09641726/23/6/063001.

[62] Winslow, W. M., Method and means for translating electrical impulses into mechanical force, U. S. Patent 2,417,850, 25 March 1947.

[63] Winslow, W. M., Induced fibration of suspensions, Journal of Applied Physics 20 (12), January 1950, 1137–1140.

[64] Ahn, Y. K., Yang, B. -S. and Morishita, S., Directionally controllable squeeze film damper using electrorheological fluid, Journal of Vibration and Acoustics 124, January 2002, 105–109.

[65] Choi, H. J. and Jhon, M. S., Electrorheology of polymers and nanocomposites, Soft Matter 5, 2009, 1562–1567.

[66] Bingham, E. C., An investigation of the laws of plastic flow, Bulletin of the U. S. Bureau of Standards Bulletin 13, 1916, 309–353, scientific paper 278 (S278).

[67] Zhu, X., Jing, X. and Cheng, L., Magnetorheological fluid dampers: A review on structure design and analysis, Journal of Intelligent Material Systems and Structures 23 (8), 2012, 839–873.

[68] Baranwal, D. and Deshmukh, T. S., MR-fluid technology and its application-A Review, International Journal of Emerging Technology and Advanced Engineering Website: www.ije tae.com (ISSN 2250-2459, ISO 9001: 2008Certified Journal 2 (12), December 2012).

[69] Sulakhe, V. N., Thakare, C. Y. and Aute, P. V., Review-MR fluid and its application, International Journal of Research in Aeronautical and Mechanical Engineering (IJRAME) 1 (7), ISSN (Online), 2321-3051, November 2013, 125–133.

[70] Kciuk, S., Turczyn, R. and Kciuk, M., Experimental and numerical studies of MR damper with prototype magnetorheological fluid, Journal of Achievements in Materials and manufacturing Engineering 39 (1), March 2010, 52–59.

[71] Furukawa, T., Uematsu, Y., Asakawa, K. and Wada, Y., Piezoelectricity, pyroelectricity, and thermoelectricity of polymer films, Journal of Applied Polymer Science 12 (12), December 1968, 2675–2689.

[72] Murayama, N., Oikawa, T., Katto, T. and Nakamura, K., Persistent polarization in poly (vinylidene fluoride). 2. Piezoelectricity of Poly (vinylidene fluoride) thermoelectrets, Journal of Polymer Science. Part B, Polymer physics 13 (5), May 1975, 1033–1047.

[73] Zimmerman, R. L. and Suchicital, C., Electric field-induced piezoelectricity in polymer film, Journal of Applied Polymer Science 19 (5), May 1975, 1373–1379.

[74] Shuford, R. J., Wilde, A. F., Ricca, J. J. and Thomas, G. R., Characterization of piezoelectric activity of stretched and poled poly (vinylidene fluoride) part I: effect of draw ratio and poling conditions, Polymer Engineering and Science 16 (1), January 1976, 25–35.

[75] Das-Gupta, D. K., Piezoelectricity and Pyroelectricity, Key Engineering Materials, Ferroelectric Polymers and Ceramic-Polymer Composites 92–93, February 1994.

[76] Simpson, J. O., Welch, S. S. and St. Clair, T. L., Novel piezoelectric polyimides, Materials Research Society MRS Proceedings 413, 1995.

[77] Hodges, R. V. and McCoy, L. E., Comparison of Polyvinylidene Fluoride (PVDF) gauge shock pressure

measurements with numerical shock code calculations, Propellants Explosives Pyrotechnics 24 (6), December 1999, 353-359.

[78] Hodges, R. V., McCoy, L. E. and Toolson, J. R., Polyvinylidene Fluoride (PVDF) gauges for measurement of output pressure of small ordnance devices, Propellants Explosives Pyrotechnics 25 (1), January 2000, 13-18.

[79] Stöbener, U. and Gaul, L., Modal vibration control for PVDF coated plates, Journal of Intelligent Material Systems and Structures 11 (4), April 2000, 283-293.

[80] Audrain, P., Masson, P., Berry, A., Pascal, J. -C. and Gazengel, B., The use of PVDF strain sensing in active control of structural intensity in beams, Journal of Intelligent Material Systems and Structures 15 (5), May 2004, 319-327.

[81] Williams, R. B., Austin, E. M. and Inman, D. J., Limitations of using membrane theory for modeling PVDF patches on inflatable structures, Journal of Intelligent Material Systems and Structures 12 (1), January 2001, 11-20.

[82] Williams, R. B., Austin, E. M. and Inman, D. J., Local effects of PVDF Patches on inflatable space-based structures, Proceedings of the 42nd AIAA/ASME/ASCE/AHS/ASC Structures Structural Dynamics and Materials Conference, Seattle, April 2001.

[83] Liao, W. H., Wang, D. H. and Huang, S. L., Wireless monitoring of cable tension of cablestayed bridges using PVDF piezoelectric films, Journal of Intelligent Material Systems and Structures 12, 2001, 331-339.

[84] Lee, Y. -C., Tein, Y. F. and Chao, Y. Y., A Point-Focus PVDF transducer for lamb wave measurements, Journal of Mechanics 18 (1), March 2002, 29-33.

[85] Wang, F., Tanaka, M. and Chonan, S., Development of a PVDF piezopolymer sensor for unconstrained in-sleep cardiorespiratory monitoring, Journal of Intelligent Material Systems and Structures 14 (3), March 2003, 185-190.

[86] Lee, Y. C., Yu, J. M. and Huang, S. W., Fabrication and characterization of a PVDF hydrophone array transducer, Key Engineering Materials 270-273, August 2004, 1406-1413.

[87] Chang, W. -Y., Cheng, Y. -H., Liu, S. -Y., Hu, Y. -C. and Lin, Y. -C., Dynamic behavior investigation of the piezoelectric PVDF for flexible fingerprint, Materials Research Society MRS Proceedings, 949, 2006.

[88] Lin, B. and Giurgiutiu, V., PVDF and PZT piezoelectric wafer active sensors for structural health monitoring, Proceedings of the ASME 2005 International Mechanical Engineering Congress and Exposition, 2005, 69-76.

[89] Lin, B. and Giurgiutiu, V., Modeling and testing of PZT and PVDF piezoelectric wafer active sensors, Smart Materials and Structures 15 (4), 2006, 1085-1093.

[90] Luo, D., Guo, Y., Hao, H., Liu, H. and Ouyang, S., Fabrication of energy storage media BST/PVDF-PAN, Materials Research Society MRS Proceedings, Vol. 949, 2006.

[91] Nunes, J. S., Sencadas, V., Wu, A., Kholkin, A. L., Vilarinho, P. M. and Lanceros-Méndez, S., Electrical and microstructural changes of β-PVDF under different processing conditions by scanning force microscopy, Materials Research Society MRS Proceedings, Vol. 949, 2006.

[92] Ren, X. and Dzenis, Y., Novel continuous Poly (vinylidene fluoride) nanofibers, Materials Research

Society MRS Proceedings, 920, 2006.

[93] Nasir, M., Matsumoto, H., Minagawa, M., Tanioka, A., Danno, T. and Horibe, H., Preparation of porous PVDF nanofiber from PVDF/PVP Blend by electrospray deposition, Polymer Journal 39 (10), May 2007, 1060-1064.

[94] Mago, G., Kalyon, D. M. and Fisher, F. T., Membranes of polyvinylidene fluoride and PVDF nanocomposites with carbon nanotubes via immersion precipitation, Journal of Nanomaterials 2008, September 2007, article ID 759825.

[95] Vickraman, P., Aravindan, V., Srinivasan, T. and Jayachandran, M., Polyvinylidenefluoride (PVDF) based novel polymer electrolytes complexed With $Mg(ClO_4)_2$, The European Physical Journal Applied Physics 45 (1), January 2009, article ID 11101.

[96] Costa, L. M. M., Bretas, R. E. S. and Gregorio, R. Jr., Effect of solution concentration on the electrospray/electrospinning transition and on the crystalline phase of PVDF, Materials Sciences and Applications 1 (4), October 2010, 247-252.

[97] Kim, Y. -J., Ahn, C. H., Lee, M. B. and Choi, M. -S., Characteristics of electrospun $PVDF/SiO_2$ composite nanofiber membranes as polymer electrolyte, Materials Chemistry and Physics 127 (1-2), May 2011, 137-142.

[98] Liu, F., Hashim, N. A., Liu, Y., Abed, M. R. M. and Li, K., Progress in the production and modification of PVDF membranes, Journal of Membrane Science 375 (1-2), June 2011.

[99] Hou, M., Tang, X. G., Zou, J. and Truss, R., Increase the mechanical performance of polyvinylidene fluoride (PVDF), Advanced Materials Research 393-395, November 2011, 144-148.

[100] Gupta, A. K., Tiwari, A., Bajpai, R. and Keller, J. M., Short circuit thermally stimulated discharge current measurement on PMMA: PEMA: PVDF ternary blends, Scientific Research: Materials Sciences and Application 2 (8), August 2011, 1041-1048.

[101] Li, Q., Zhou, B., Bi, Q. -Y. and Wang, X. -L., Surface modification of PVDF membranes with sulfobetaine polymers for a stably anti-protein-fouling performance, Journal of Applied Polymer Science 125 (5), March 2012, 4015-4027.

[102] Klimiec, E., Zaraska, W., Piekarski, J. and Jasiewicz, B., PVDF sensors-research on foot pressure distribution in dynamic conditions, Advances in Science and Technology 79, September 2012, 94-99.

[103] Chiu, -Y. -Y., Lin, W. -Y., Wang, H. -Y., Huang, S. -B. and Wu, M. -H., Development of a piezoelectric polyvinylidene fluoride (PVDF) polymer-based sensor patch for simultaneous heartbeat and respiration monitoring, Sensors and Actuators A, Physical 189, January 2013, 328-334.

[104] Hartono, A., Satira, S., Djamal, M., Ramli, R., Bahar, H. and Sanjaya, E., Effect of mechanical treatment temperature on electrical properties and crystallite size of PVDF film, Advances in Materials Physics and Chemistry AMPC 3 (1), January2013, 71-76.

[105] Jain, A., Kumar, J. S., Srikanth, S., Rathod, V. T. and Mahapatra, D. R., Sensitivity of polyvinylidene fluoride films to mechanical vibration modes and impact after optimizing stretching conditions, Polymer Engineering and Science 53 (4), April 2013, 707-715.

[106] Kang, G. -D. and Cao, Y. -M., Application and modification of poly (vinylidene fluoride) (PVDF) membranes-a review, Journal of Membrane Science 463, January 2014, 145-165.

[107] Saïdi, S., Mannaî, A., Bouzitoun, M. and Mohamed, A. B., Alternating current conductivity and die-

lectric relaxation of PANI: PVDF composites, The European Physical Journal Applied Physics 66 (1), April 2014, 10201. http: //dx. doi. org/10. 1051/epjap/2014130245.

[108] Sun, C. , Shi, J. , Bayerl, D. J. and Wang, X. , PVDF microbelts for harvesting energy from respiration, Energy & Environmental Science, 4, July 2011, 4508-4512.

[109] Chen, D. , Sharma, T. and Zhang, J. X. J. , Mesoporous surface control of PVDF thin films for enhanced piezoelectric energy generation, Sensors and Actuators A 216, September 2014, 196-201.

[110] Cady, W. G. , Piezoelectricity, McGraw-Hill Book Company, Inc. , New York, 1946, 842.

[111] Yaffe, B. , Cook, W. R. Jr. and Jaffe, H. , Piezoelectric Ceramics, Academic Press, London, 1971, 317.

[112] Gruver, R. M. , Buessem, W. R. , Dickey, C. W. and Anderson, J. W. , State-of-the-art review on ferroelectric ceramic materials, Technical Report AFML-TR-66-164, Air Force Materials Laboratory Research and Technology Division, Airforce Systems Command, Wright-Patterson Air Force Base, Ohio, USA, May 1966, 223.

[113] Haertling, G. H. , Ferroelectric ceramics: History and technology, Journal of the American Ceramic Society 82 (4), January 1999, 797-818.

[114] Muralt, P. , Ferroelectric thin films for micro-sensors and actuators: A review, Journal Micromechanics and Microengineering 10, 2000, 136-146.

[115] Niezrecki, C. , Brei, D. , Balakrishnan, S. and Moskalik, A. , Piezoelectric actuation: state of the art, The Shock and Vibration Digest 44 (4), July 2001, 269-280.

[116] Setter, N. , Damjanovic, D. , Eng, L. , Fox, G. , Gevorgian, S. , Hong, S. , Kingon, A. , Kohlstedt, H. , Park, N. Y. , Stephenson, G. B. , Stolitchnov, I. , Taganstev, A. K. , Taylor, D. V. , Yamada, T. and Streiffer, S. , Ferroelectric thin films: review of materials, properties, and applications, Journal of Applied Physics 100, 2006, article ID 051606. doi: 10. 1063/1. 2336999.

[117] Gopinathan, S. V. , Varadan, V. V. and Varadan, V. K. , A review and critique of theories for piezoelectric laminates, Smart Materials and Structures 9 (1), 2000, 24-48.

[118] Chopra, I. , Review of state-of-art of smart structures and integrated systems, AIAA Journal 40 (11), November 2002, 2145-2187.

[119] Gupta, V. , Sharma, M. and Thakur, N. , Mathematical modeling of actively controlled piezo smart structures: a review, Smart Structures and Systems An International Journal 8 (3), September 2011, 275-302.

[120] Li, T. , Ma, J. , Es-Souni, M. and Woias, P. , Advanced piezoelectrics: materials, devices, and their applications, Smart Materials Research 2012, March 2012, article ID 259275.

[121] Liu, X. , Zhang, K. and Li, M. , A survey on experimental characterization of hysteresis in piezoceramic actuators, Advanced Materials Research 694-697, May 2013, 1558-1564.

[122] Harne, R. L. and Wang, K. W. , A Review of the recent research on vibration energy harvesting via bistable systems, Smart Materials and Structures 22 (2), 2013, article ID 023001.

[123] Ramadan, K. S. , Sameoto, D. and Evoy, S. , A review of piezoelectric polymers as functional materials for electromechanical transducers, Smart Materials and Structures 23 (3), January 2014, article ID 033001.

[124] Aridogan, U. and Basdogan, I. , A review of active vibration and noise suppression of platelike structures

with piezoelectric transducers, Journal of Intelligent Material Systems and Structures 26 (12), August 2015, 1455-1476.

[125] Baz, A., Static deflection control of flexible beams by piezo-electric actuators, NASA Technical Report No: N87-13788, September 1986.

[126] Baz, A., Poh, S. and Studer, P., Optimum vibration control of flexible beams by piezo-electric actuators, Proceedings of 6th Conference on the Dynamics & Control of Large Structures, Blacksburg, June 1987, 217-234.

[127] Crawley, E. F. and Anderson, E. H., Detailed models of piezoceramic actuation of beams, Journal of Intelligent Material Systems and Structures 1 (1), January 1990, 4-25.

[128] Chandra, R. and Chopra, I., Structural modeling of composite beams with induced-strain actuation, AIAA Journal 31 (9), September 1993, 1692-1701.

[129] Lagoudas, D. C. and Bo, Z., The cylindrical bending of composite plates with piezoelectric and SMA layers, Journal of Smart Materials and Structures 3, 1994, 309-317.

[130] Kim, S. J. and Jones, J. D., Influence of piezo-actuator thickness on the active vibration control of a cantilever beam, Journal of Intelligent Material Systems and Structures 6 (5), September 1995, 610-623.

[131] Inman, D. J., Huang, S. -C. and Austin, E. M., Piezoceramic versus viscoelastic damping treatments, Proceedings of the 6th International Conference on Adaptive Structure Technology, November 1995, 241-252.

[132] Saravanos, D. A. and Heyliger, P. R., Coupled layerwise analysis of composite beams with embedded piezoelectric sensors and actuators, Journal of Intelligent Material Systems and Structures 6 (3), 1995, 350-363.

[133] Chen, P. C. and Chopra, I., Induced strain actuation of composite beams and rotor blades with embedded piezoceramic elements, Smart Materials and Structures 5 (1), February 1996, 35-48.

[134] Lee, H. J. and Saravanos, D. A., Coupled layerwise analysis of thermopiezoelectric smart composite beams, AIAA Journal 34 (6), June 1996, 1231-1237.

[135] Zapfe, J. A. and Lesieutre, G. A., Iterative calculation of the transverse shear distribution in laminated composite beams, AIAA Journal 34 (6), June 1996, 1299-1300.

[136] Park, C. and Chopra, I., Modeling piezoceramic actuation of beams in torsion, AIAA Journal 34 (12), December 1996, 2582-2589.

[137] Lesieutre, G. A. and Lee, U., A finite element model for beams having segmented active constrained layers with frequency-dependent viscoelastic material properties, Smart Materials and Structures 5, 1996, 615-627.

[138] Smith, C. B. and Wereley, N. M., Transient analysis for damping identification in rotating composite beams with integral damping layers, Smart Materials and Structures 5 (5), 1996, 540-550.

[139] Baz, A., Dynamic boundary control of beams using active constrained layer damping, Mechanical Systems and Signal Processing 11 (6), November 1997, 811-825.

[140] Abramovich, H. and Pletner, B., Actuation and sensing of piezolaminated sandwich type structures, Composite Structures 38 (1-4), 1997, 17-27.

[141] Lesieutre, G. A., Vibration damping and control using shunted piezoelectric materials, Shock and Vibration Digest 30, 1998, 187-195.

[142] Liao, W. H., Actuator location for active/passive piezoelectric control systems, Proceedings of the Symposium on Image Speech Signal Processing and Robotics ISSPR '98, 1, 1998, 341-346.

[143] Abramovich, H. and Meyer-Piening, H.-R., Induced vibrations of piezolaminated elastic beams, Composite Structures 43 (1), September 1998, 47-55.

[144] Abramovich, H., Deflection control of laminated composite beams with piezoceramic layersclosed form solutions, Composite Structures 43 (3), November 1998, 217-231.

[145] Saravanos, D. A. and Heyliger, P. R., Mechanics and computational models for laminated piezoelectric beams plates and shells, Applied Mechanics Reviews 52 (10), 1999, 305-320.

[146] Barboni, R., Mannini, A. and Gaudenzi, P., Optimal placement of PZT actuators for the control of beam dynamics, Smart Materials and Structures 9 (1), February 2000, 110-120.

[147] Ang, K. K., Reddy, J. N. and Wang, C. M., Displacement control of timoshenko beams via induced strain actuators, Smart Materials and Structures 9, 2000, PII: S0964-1726 (00) 16310-4, 981-984.

[148] Shih, H.-R., Distributed vibration sensing and control of a piezoelectric laminated curved beam, Smart Materials and Structures 9 (6), 2000, 761-766.

[149] Wang, Q. and Quek, S. T., Flexural vibration analysis of sandwich beam coupled with piezoelectric actuator, Smart Materials and Structures 9 (1), 2000, 103-109.

[150] Smith, C., Analytical modeling and equivalent electromechanical loading techniques for adaptive laminated piezoelectric beams and plates, Master's Thesis at Virginia Polytechnic Institute and State University, USA, etd-02062001-101444, January 2001.

[151] Krommer, M., On the correction of the Bernoulli-Euler beam theory for smart piezoelectric beams, Smart Materials and Structures 10 (4), August 2001, 668-680.

[152] Achuttan, A., Keng, A. K. K. and Ming, W. C., Shape control of coupled nonlinear piezoelectric beams, Smart Materials and Structures 10 (5), October 2001, 914-924.

[153] Cai, C., Liu, G. R. and Lam, K. Y., A technique for modelling multiple piezoelectric layers, Smart Materials and Structures 10 (4), August 2001, 689-694.

[154] Tong, L., Sun, D. and Atluri, S. N., Sensing and actuating behaviors of piezoelectric layers with debonding in smart beams, Smart Materials and Structures 10 (4), August 2001, 724-729.

[155] Sun, D., Tong, L. and Atluri, S. N., Effects of piezoelectric sensor/actuator debonding on vibration control of smart beams, International Journal of Solids and Structures 38 (50-51), December 2001, 9033-9051.

[156] Baz, A. and Ro, J., Vibration control of rotating beams with active constrained layer damping, Journal of Smart Materials and Structures 10 (1), 2001, 112-120.

[157] Ray, M. and Baz, A., Control of nonlinear vibration of beams using active constrained layer damping, Journal of Vibration and Control 7, 2001, 539-549.

[158] Trindade, M. A., Benjeddou, A. and Ohayon, R., Finite element modelling of hybrid activepassive vibration damping of multilayer piezoelectric sandwich beams-part i: formulation, International Journal for Numerical Methods in Engineering 51 (7), 2001, 835-854.

[159] Trindade, M. A., Benjeddou, A. and Ohayon, R., Finite element modelling of hybrid activepassive vibration damping of multilayer piezoelectric sandwich beams-part ii: system analysis, International Journal for Numerical Methods in Engineering 51 (7), 2001, 855-864.

[160] Trindade, M. A., Benjeddou, A. and Ohayon, R., Piezoelectric active vibration control of damped sandwich beams, Journal of Sound and Vibration 246 (4), 2001, 653-677.

[161] Wang, G. and Wereley, N. M., Spectral finite element analysis of sandwich beams with passive constrained layer damping, ASME Journal of Vibration and Acoustics 124 (3), July 2002, 376-386.

[162] Abramovich, H. and Livshits, A., Flexural vibrations of piezolaminated slender beams: a balanced model, Journal of Vibration and Control 8 (8), August 2002, 1105-1121.

[163] Waisman, H. and Abramovich, H., Variation of natural frequencies of beams using the active stiffening effect, Composites Part B: Engineering 33 (6), September 2002, 415-424.

[164] Yocum, M. and Abramovich, H., Static behavior of piezoelectric actuated beams, Computers & Structures 80 (23), September 2002, 1797-1808.

[165] Waisman, H. and Abramovich, H., Active stiffening of laminated composite beams using piezoelectric actuators, Composite Structures 58 (1), October 2002, 109-120.

[166] Gao, J. X. and Liao, W. H., Damping characteristics of beams with enhanced self-sensing active constrained layer treatments under various boundary conditions, ASME Journal of Vibration and Acoustics 127 (2), 2005, 173-187.

[167] Gao, J. X. and Liao, W. H., Vibration analysis of simply supported beams with enhanced self-sensing active constrained layer damping treatments, Journal of Sound and Vibration 280 (1-2), 2005, 329-357.

[168] Park, C. H. and Baz, A., Vibration control of beams with negative capacitive shunting of interdigital electrode piezoceramics, Journal of Vibration and Control 11 (3), 2005, 331-346.

[169] Edery-Azulay, L. and Abramovich, H., Active damping of piezo - composite beams, Composite Structures 74 (4), August 2006, 458-466.

[170] Edery-Azulay, L. and Abramovich, H., The integrity of piezo-composite beams under high cyclic electromechanical loads-experimental results, Smart Materials and Structures 16 (4), July 2007, 1226-1238.

[171] Adhikari, S. and Friswell, M. I., Shaped modal sensors for linear stochastic beams, Journal of Intelligent Material Systems and Structures 20 (18), December 2009, 2269-2284.

[172] Wang, T. R., Structural responses of surface-mounted piezoelectric curved beams, Journal of Mechanics 26 (4), December 2010, 439-451. Kim, J. S. and Wang, K. W., An Asymptotic Approach for the Analysis of Piezoelectric Fiber Composite Beams, Smart Materials and Structures 20, 2011. doi: 10.1088/0964-1726/20/2/025023.

[173] Bachmann, F., Bergamini, A. and Ermanni, P., Optimum piezoelectric patch positioning - a strain energy-based finite element approach, Journal of Intelligent Material Systems and Structures 23 (14), 2012, 1575-1591.

[174] Koroishi, E. H., Molina, F. A. L., Faria, A. W. and Steffen, V. Jr., Robust optimal control applied to a composite laminated beam, Journal of Aerospace Technology and Management 7 (1), January-March 2015, 70-80.

[175] Giurgiutiu, V., Chaudhry, Z. A. and Rogers, C. A., Active control of helicopter rotor blades with induced strain actuators, Proceedings of the 35th Structures Structural Dynamics and Materials Conference, Hilton Head, 1994, 288-297.

[176] Giurgiutiu, V., Chaudhry, Z. A. and Rogers, C. A., Engineering feasibility of induced strain actuators

## 第1章 智能材料与结构概述

for rotor blade active vibration control, Journal of Intelligent Material Systems and Structures 6 (5), 1995, 583-597.

[177] Chen, P. and Chopra, I. , Hover testing of smart rotor with induced-strain actuation of blade twist, AIAA Journal 35 (1), January 1997, 6-16.

[178] Chen, P. and Chopra, I. , Wind tunnel test of a smart rotor model with individual blade twist control, Journal of Intelligent Material Systems and Structures 8 (5), May 1997, 414-423.

[179] Koratkar, N. A. and Chopra, I. , Testing and validation of a Froude scaled helicopter rotor model with piezo-bimorph actuated trailing-edge flaps, Journal of Intelligent Material Systems and Structures 8 (7), July 1997, 555-570.

[180] Pletner, B. and Abramovich, H. , Adaptive suspensions of vehicles using piezoelectric sensors, Journal of Intelligent Material Systems and Structures 6 (6), November 1995, 744-756.

[181] Sirohi, J. and Chopra, I. , Fundamental understanding of piezoelectric strain sensors, Journal of Intelligent Material Systems and Structures 11 (April), 2000, 246-257.

[182] Giurgiutiu, V. and Zagrai, A. N. , Characterization of piezoelectric wafer active sensors, Journal of Intelligent Material Systems and Structures 11 (12), 2000, 959-976.

[183] Tabellout, M. , Raquois, A. , Emery, J. R. and Jayet, Y. , The inserted piezoelectric sensor method for monitoring thermosets cure, The European Physical Journal Applied Physics 13 (2), February 2001, 107-113.

[184] Giurgiutiu, V. , Bao, J. and Zhao, W. , Piezoelectric wafer active sensor embedded ultrasonics in beams and plates, Experimental Mechanics 43 (4), 2003, 428-449.

[185] Baz, A. , Poh, S. , Lin, S. and Chang, P. , Distributed sensing of rotating beams, Proceedings of the 1st International Workshop on Smart Materials and Structures Technology, Honolulu, January 2004.

[186] Chrysohoidis, N. A. and Saravanos, D. A. , Assessing the effects of delamination on the damped dynamic response of composite beams with piezoelectric actuators and sensors, Smart Materials and Structures 13 (4), May 2004, 733-742. doi: 10. 1088/0964-1726/13/4/01.

[187] Plagianakos, T. S. and Saravanos, D. A. , Coupled high-order shear layerwise analysis of adaptive sandwich composite beams with piezoelectric actuators and sensors, AIAA Journal 43 (4), April 2005, 885-894.

[188] Chrysochoidis, N. A. and Saravanos, D. A. , Generalized layerwise mechanics for the static and modal response of delaminated composite beams with active piezoelectric sensors, International Journal of Solids and Structures 44 (25-26), 2007, 8751-8768. doi: 10. 1016/j. ijsolstr. 2007. 07. 004.

[189] Chrysochoidis, N. A. and Saravanos, D. A. , High frequency dispersion characteristics of smart delaminated composite beams, Journal of Intelligent Material Systems and Structures 20 (9), June 2009. doi: 10. 1177/1045389X09102983.

[190] Abramovich, H. , Burgard, M. , Edery-Azulay, L. , Evans, K. E. , Hoffmeister, M. , Miller, W. , Scarpa, F. , Smith, C. W. and Tee, K. F. , Smart tetrachiral and hexachiral honeycomb: sensing and impact detection, Composites Science and Technology 70 (7), July 2010, 1072-1079.

[191] Lin, B. , Giurgiutiu, V. , Pollock, P. , Xu, B. and Doane, J. , Durability and survivability of piezoelectric wafer active sensors on metallic structure, AIAA journal 48 (3), 2010, 635-643.

[192] Martinez, M. and Artemev, A. , A novel approach to a piezoelectric sensing element, Journal of Sensors

2010, 2010, article ID 816068.

[193] Gresil, M. and Giurgiutiu, V., Guided wave propagation in composite laminates using piezoelectric wafer active sensors, Aeronautical Journal 117 (1196), 2013.

[194] Dosch, J. J., Inman, D. J. and Garcia, E., A self sensing piezoelectric actuator for collocated control, Journal of Intelligent Material Systems and Structures 3 (1), 1992, 166-185.

[195] Giurgiutiu, V. and Rogers, C. A., Large-amplitude rotary induced-strain (laris) actuator, Journal of Intelligent Material Systems and Structures 8 (1), 1997, 41-50.

[196] Giurgiutiu, V. and Rogers, C. A., Power and energy characteristics of solid-state induced strain actuators for static and dynamic applications, Journal of Intelligent Material Systems and Structures 8 (9), 1997, 738-750.

[197] Kowbel, W., Xia, X., Withers, J. C., Crocker, M. J. and Wada, B. K., PZT/PVDF composite for actuator/sensor application, Materials Research Society MRS Proceedings 493, 1997.

[198] Lesieutre, G. A. and Davis, C. L., Can a coupling coefficient of apiezoelectric actuator be higher than those of its active material, Journal of Intelligent Materials Systems and Structures 8, 1997, 859-867.

[199] Pietrzakowski, M., Dynamic model of beam-piezoceramic actuator coupling for active vibration control, mechanika teoretetyczna i stosowana, Journal of Theoretical and Applied Mechanics 35 (1), 1997, 3-20.

[200] Chandra, R. and Chopra, I., Actuation of trailing edge flap in a wing model Using Piezostack Device, Journal of Intelligent Material Systems and Structures 9 (10), October 1998, 847-853.

[201] Moskalik, A. J. and Brei, D., Force-deflection behavior of piezoelectric C-block actuator arrays, Smart Materials and Structures 8 (5), 1999, 531-543.

[202] Sirohi, J. and Chopra, I., Fundamental behavior of piezoceramic sheet actuators, Journal of Intelligent Material Systems and Structures 11 (1), January 2000, 47-61.

[203] Tylikowski, A., Influence of bonding layer on piezoelectric actuators of an axisymmetrical annular plate, Journal of Theoretical and Applied Mechanics 3 (38), May 2000, 607-621.

[204] Benjeddou, A., Trindade, M. A. and Ohayon, R., Piezoelectric Actuation mechanisms for intelligent sandwich structures, Smart Materials and Structures 9 (3), 2000, 328-335.

[205] Galante, T., Frank, J., Bernard, J., Chen, W., Lesieutre, G. A. and Koopmann, G. H., A high-force high-displacement piezoelectric inchworm actuator, Journal of Intelligent Materials Systems and Structures 10 (12), 2000, 962-972.

[206] Lee, T. and Chopra, I., Design of piezostack-driven trailing-edge flaps for helicopter rotors, Journal Smart Materialand Structures 10 (1), February 2001, 15-24.

[207] Monturet, V. and Nogarede, B., Optimal dimensioning of a piezoelectric bimorph actuator, The European Physical Journal Applied Physics 17 (2), February 2002, 107-118.

[208] Yocum, M., Abramovich, H., Grunwald, A. and Mall, S., Fully reversed electromechanical fatigue behavior of composite laminate with embedded piezoelectric actuator/sensor, Smart Materials and Structures 12 (4), June 2003, 556-564.

[209] Abramovich, H., Piezoelectric actuation for smart sandwich structures-closed form solutions, Journal of Sandwich Structures & Materials 5 (4), October 2003, 377-396.

[210] Law, W. W., Liao, W. H. and Huang, J., Vibration control of structures with self-sensing piezoelectric actuators incorporating adaptive mechanisms, Smart Materials and Structures 12, 2003, 720-730.

[211] Lesieutre, G. A., Rusovici, R., Koopmann, G. H. and Dosch, J. J., Modeling and characterization of a piezoceramic inertial actuator, Journal Sound and Vibration 261 (1), 2003, 93-107.

[212] Barrett, R., McMurtry, R., Vos, R., Tiso, P. and De Breuker, R., Post-Buckled Precompressed (PBP) elements: a new class of flight control actuators enhancing high-speed autonomous VTOL MAVs, smart structures and materials 2005: industrial and commercial applications of smart structures technologies, edited by Edward V. White, Proceedings of SPIE Vol. 5762, SPIE, Bellingham, WA, 2005.

[213] Abramovich, H., Weller, T. and Yeen-Ping, S., Dynamics response of a high aspect ratio wing equipped With PZT patches-a theoretical and experimental study, Journal of Intelligent Materials Systems and Structures 16 (11-12), December 2005, 919-923.

[214] Parsons, Z. and Staszewski, W. J., Nonlinear acoustics with low-profile piezoceramic excitation for crack detection in metallic structures, Smart Materials and Structures15, 2006, 1110-1118.

[215] Nir, A. and Abramovich, H., Design, analysis and testing of a smart fin, Composite Structures 92 (4), March 2010, 863-872.

[216] Haller, D., Paetzold, A., Losse, N., Neiss, S., Peltzer, I., Nitsche, W., King, R. and Woias, P., Piezo-Polymer-composite unimorph actuators for active cancellation of flow instabilities across airfoil, Journal of Intelligent Material Systems and Structures 22 (5), March 2011, 461-474.

[217] Wereley, N. M., Wang, G. and Chaudhuri, A., Demonstration of uniform cantilevered beam bending vibration using a pair of piezoelectric actuators, Journal of Intelligent Material Systems and Structures 22 (4), 2011, 307-316. doi: 10.1177/1045389X10379661.

[218] Pan, C. L. and Liao, W. H., A new two-axis optical scanner actuated by piezoelectric bimorphs, International Journal of Optomechatronics 6 (4), 2012, 336-349. doi: 10.1080/15599612.2012.721867.

[219] Arrieta, A. F., Bilgen, O., Friswell, M. I. and Ermanni, P., Modelling and configuration control of wing-shaped Bi-stable piezoelectric composites under aerodynamic loads, Aerospace Science and Technology 29 (1), August 2013, 453-461.

[220] Davis, J., Kim, N. H. and Lind, R., Control of the flexural axis of a wing with piezoelectric actuation, Journal of Aircraft 52 (2), March-April 2015, 584-594.

[221] Aldraihem, O. J. and Khdeir, A. A., Smart beams with extension and thickness-shear piezoelectric actuators, Smart Materials and Structures 9, 2000, PII: S0964-1726 (00) 07958-1.

[222] Przbylowicz, P. M., An application of piezoelectric shear effect to active damping of transverse vibration in beams, Journal of Theoretical and Applied Mechanics 3 (38), 2000, 573-589.

[223] Vel, S. S. and Batra, R. C., Exact solution for rectangular sandwich plates with embedded piezoelectric shear actuators, AIAA Journal 39 (7), July 2001, 1363-1373.

[224] Benjeddou, A., Gorge, V. and Ohayon, R., Use of piezoelectric shear response in adaptive sandwich shells of revolution-part 1: theoretical formulation, Journal of Intelligent Material Systems and Structures 12 (4), 2001, 235-245.

[225] Benjeddou, A., Gorge, V. and Ohayon, R., Use of piezoelectric shear response in adaptive sandwich shells of revolution-part 2: finite element implementation, Journal of Intelligent Material Systems and Structures 12 (4), 2001, 247-257.

[226] Vel, S. S. and Batra, R. C., Analysis of piezoelectric bimorphs and plates with segmented actuators, Thin Wall Structure 39, 2001, 23-44.

[227] Edery-Azulay, L. and Abramovich, H., Piezoelectric actuation and sensing mechanisms-closed form solutions, Composite Structures 64 (3-4), June 2004, 443-453.

[228] Senthil, S. V. and Baillargeon, B. P., Active vibration suppression of smart structures using piezoelectric shear actuators, Proceedings of the 15th International Conference on Adaptive Structures and Technologies, ICAST, Bar Harbor, October 2004.

[229] Edery-Azulay, L. and Abramovich, H., Augmented damping of a piezo-composite beam using extension and shear piezoceramic transducers, Composite B: Engineering 37 (4-5), 2006, 320-327.

[230] Sawano, M., Tahara, K., Orita, Y., Nakayama, M. and Tajitsu, Y., New design of actuator using shear piezoelectricity of a chiral polymer, and prototype device, Polymer International 59 (3), August 2009, 365-370.

[231] Aladwani, A., Aldraihem, O. and Baz, A., Single degree of freedom shear-mode piezoelectric energy harvester, ASME Journal of Vibration and Acoustics 135 (5), 2013, 051011, Paper No. VIB-11-1250, doi: 10.1115/1.4023950.

[232] Meressi, T. and Paden, B., Buckling control of a flexible beam using piezoelectric actuators, Journal of Guidance Control, and Dynamics 16 (5), 1993, 977-980.

[233] Alghamdi, A. A. A., Adaptive imperfect column with piezoelectric actuators, Journal of Intelligent Materials Systems and structures 12, March 2001, 183-189. doi: 10.1106/uaokqwxq-p8kl-g3k2.

[234] Rao, G. V. and Singh, G., A smart structures concept for the buckling load enhancement of columns, Smart Materials and Structures 10 (4), August 2001, 843-845.

[235] Batra, R. C. and Geng, T. S., Enhancement of the dynamic buckling load for a plate by using piezoceramic actuators, Smart Materials and Structures 10 (5), October 2001, 925-933.

[236] Varelis, D. and Saravanos, D. A., Nonlinear coupled mechanics and buckling analysis of composite plates with piezoelectric actuators and sensors, Smart Materials and Structures 11 (3), June 2002, 330-336.

[237] Mukherjee, A. and Chaudhuri, A. S., Active control of dynamic instability of piezolaminated imperfect columns, Smart Materials and Structures 11 (6), 2002, 874-879.

[238] Wang, Q. and Varadan, V. K., Transition of the buckling load of beams by the use of piezoelectric layers, Smart Materials and Structures 12 (5), 2003, 696-702.

[239] Fridman, Y. and Abramovich, H., Enhanced structural behavior of flexible laminatedcomposite beams, Composite Structures 82 (1), January 2008, 140-154.

[240] Zehetner, C. and Irschik, I., On the static and dynamic stability of beams with an axial piezoelectric actuation, Smart Structures and Systems 4 (1), January 2008, 67-84.

[241] Sridharan, S. and Kim, S., Piezoelectric control of columns prone to instabilities and nonlinear modal interaction, Smart Materials and Structures 17 (3), 2008, article ID 035001.

[242] Sridharan, S. and Kim, S., Piezoelectric control of stiffened panels subject to interactive buckling, International Journal of Solids and Structures 46 (6), 2009, 1527-1538.

[243] De Faria, A. R. and Donadon, M. V., The use of piezoelectric stress stiffening to enhance buckling of laminated plates, Latin American Journal of Solids and Structures 7 (March), 2010, 167-183.

[244] Wang, Q. S., Active buckling control of beams using piezoelectric actuators and strain gauge sensors, Smart Materials and Structures 19 (62010), May 2010, article ID 065022.

[245] Enss, G. C., Platz, R. and Hanselka, H., An approach to control the stability in an active loadcarrying

beam-column by one single piezoelectric stack actuator, Proceedings of International Conference on Noise and Vibration (ISMA) 2010 including International Conference on Uncertainty in Structural Dynamics (USD) 2010, Leuven Belgium, 535-546.

[246] Abramovich, H., A new insight on vibrations and buckling of a cantilevered beam under a constant piezoelectric actuation, Composite Structures 93 (2), January 2011, 1054-1057.

[247] Qishan, W., Active vibration and buckling control of piezoelectric smart structures, Ph. D. Thesis submitted to Civil Engineering and Applied Mechanics Dep., McGill University, Montreal, Quebec, Canada, 2012.

[248] Wluka, P. and Kubiak, T., Stability of cross-ply composite plate with piezoelectric actuators, Stability of Structures XIII-th Symposium, Zakopane, Poland, 2012, 667-686.

[249] Zenz, G. and Humer, A., Experimental investigations to enhance the buckling load of slender beams, 7th ECCOMAS Thematic Conference on Smart Structures and Materials, SMART 2015, Araúo, A. L., Mota Soares, C. A., et al. (eds.), 2015.

[250] Abramovich, H., Axial stiffness variation of thin walled laminated composite beams using piezoelectric patches-a new experimental insight, 26th International Conference on Adaptive Structures and Technologies at Kobe, Japan, on October 14-16, 2015 (ICAST2015).

[251] Ruan, X., Danforth, S. C., Safari, A. and Chou, T. -W., Saint-venant end effects in piezoceramic materials, International Journal of Solids and Structures 37 (19), May 2000, 2625-2637.

[252] Rovenski, V., Harash, E. and Abramovich, H., Saint-Venant's problem for homogeneous piezoelastic beams, Journal of Applied Mechanics 74 (6), December 2006, 1095-1103.

[253] Rovenski, V. and Abramovich, H., Behavior of piezoelectric beams under axially non-uniform distributed loading, Journal of Elasticity 88 (3), September 2007, 223-253.

[254] Rovenski, V. and Abramovich, H., Saint Venant's problem for compound piezoelastic beams, Journal of Elasticity 96, April 2009, 105-127.

[255] Krommer, M., Berik, P., Vetyukov, Y. and Benjeddou, A., Piezoelectric d15 Shear-responsebased torsion actuation mechanism: an exact 3D Saint-Venant type solution, International Journal of Smart and Nano Materials 3 (2), June 2012, 82-102.

[256] Barrett, R., Active Plate and Wing research Using EDAP Elements, Smart Materials and Structures 1 (3), September 1992, 214-226.

[257] Shah, D. K., Joshi, S. P. and Chan, W. S., Static structural response of plates with piezoceramic layers, Smart Materials and Structures 2, 1993, 172-180.

[258] Heyliger, P., Ramirez, G. and Saravanos, D. A., Coupled discrete-layer finite elements for laminated piezoelectric plates, Communications in Numerical Methods in Engineering 10, 1994, 971-981.

[259] Heyliger, P. R. and Saravanos, D. A., Exact free-vibration analysis of laminated plates with embedded piezoelectric layers, Journal of Acoustical Society of America 98 (3), 1995, 1547-1557.

[260] Miller, S. E. and Abramovich, H., A Self-Sensing Piezolaminated Actuator model for shells using a first order Shear deformation theory, Journal of Intelligent Material Systems and Structures 6 (5), 1995, 624-638.

[261] Mitchell, J. A. and Reddy, J. N., A refined hybrid plate theory for composite laminates with piezoelectric laminae, International Journal of Solids and Structure 32 (16), 1995, 2345-2367.

[262] Heyliger, P. R., Pei, K. C. and Saravanos, D. A., Layerwise mechanics and finite element model for laminated piezoelectric shells, AIAA Journal 34 (11), 1996, 2353-2360.

[263] Miller, S. E., Oshman, Y. and Abramovich, H., Modal control of piezolaminated anisotropic rectangular plates: part 1-modal transducer theory, AIAA Journal 34 (9), September 1996, 1868-1875.

[264] Miller, S. E., Oshman, Y. and Abramovich, H., Modal control of piezolaminated anisotropic rectangular plates: part 2-control theory, AIAA Journal 34 (9), September 1996, 1876-1884.

[265] Baz, A., Ro, J., Vibration control of plates with active constrained layer damping, Journal of Smart Materials & Structures 5, 1996, 272-280.

[266] Lin, C. -C., Hsu, C. -Y. and Huang, H. -N., Finite element analysis on deflection control of plate with piezoelectric actuators, Composite Structures 35 (4), 1996, 423-433.

[267] Kaljevic, I. and Saravanos, D. A., Steady-state response of acoustic cavities bounded by piezoelectric composite shell structures, Journal of Sound and Vibration 205 (3), July 1997, 459-476.

[268] Pletner, B. and Abramovich, H., A Consistent Methodology for the modeling of piezolaminated shells, AIAA Journal 35 (8), August 1997, 1316-1326.

[269] Saravanos, D. A., Mixed laminate theory and finite element for smart piezoelectric composite shell structures, AIAA Journal 35 (8), August 1997, 1327-1333.

[270] Miller, S. E., Abramovich, H. and Oshman, Y., Selective modal transducers for anisotropic rectangular plates: experimental validation, AIAA Journal 35 (10), October 1997, 1621-1629.

[271] Lee, H. J. and Saravanos, D. A., Generalized finite element formulation for smart multilayered thermal piezoelectric plates, International Journal of Solids and Structures 34 (26), 1997, 3355-3371.

[272] Shields, W., Ro, J. and Baz, A., Control of sound radiation from a plate into an acoustic cavity using active piezoelectric damping composites, Journal of Smart Materials & Structures 7, 1998, 1-11.

[273] Hong, C. H. and Chopra, I., Modeling and validation of induced strain actuation of composite coupled plates, AIAA Journal 37 (3), March 1999, 372-377.

[274] Saravanos, D. A., Damped vibration of composite plates with passive piezoelectric-resistor elements, Journal of Sound and Vibration 221 (5), April 1999, 867-885.

[275] Miller, S. E., Oshman, Y. and Abramovich, H., Selective Modal transducers for piezolaminated anisotropic shells, Journal of Guidance Controland Dynamics 22 (3), May-June 1999, 455-466.

[276] Zhang, X. D. and Sun, C. T., Analysis of a Sandwich plate containing a piezoelectric core, Smart Materials and Structures 8, 1999, 31-40.

[277] Chee, C. Y. K., Static shape control of laminated composite plate smart structure using piezoelectric actuators, PhD Thesis at The University of Sydney Aeronautical Engineering, Australia, thesis ID 2123/709, 2000.

[278] Saravanos, D. A., Passively damped laminated piezoelectric shell structures with integrated electric networks, AIAA Journal 38 (7), July 2000, 1260-1268.

[279] Chee, C. Y. K., Tong, L. and Steven, G. P., A mixed model for adaptive composite plates with piezoelectric for anisotropic actuation, Computers & Structures 77 (3), 2000. doi: 10.1016/S0045-7949(99)00225-4, 253-268.

[280] Abramovich, H. and Meyer-Piening, H. -R., Actuation and sensing of soft core sandwich plates with a built-in adaptive layer, Journal of Sandwich Structures & Materials 3 (1), January 2001. doi: 10.1106/

kx19-falt-x1b8-56gq, 75-86.

[281] Miller, S. E., Oshman, Y. and Abramovich, H., A Selective Modal Control Theory for piezolaminated anisotropic shells, Journal of Guidance Control and Dynamics 24 (4), July-August 2001, 844-852.

[282] Chee, C. K., Tong, L. and Steven, G. P., Static shape control of composite plates using a curvature-displacement based algorithm, International Journal of Solids and Structures38, 2001, 6381-6403.

[283] Saravanos, D. A. and Christoforou, A. P., Low-energy impact of adaptive cylindrical laminated piezoelectric-composite shells, International Journal of Solids and Structures 39 (8), May 2002, 2257-2279.

[284] Tzou, H. S. and Wang, D. W., Micro-sensing Characteristics and modal voltages of piezoelectric laminated linear and nonlinear toroidal shells, Journal of Sound & Vibration 254 (2), 2002, 203-218.

[285] Wu, C. Y., Chang, J. S. and Wu, K. C., Analysis of wave propagation in infinite piezoelectric plates, Journal of Mechanics 21 (2), June 2005, 103-108.

[286] Edery-Azulay, L. and Abramovich, H., A Reliable Plain Solution for rectangular plates with piezoceramic patches, Journal of Intelligent Material Systems and Structures 18 (5), May 2007, 419-433.

[287] Larbi, W., Deü, J.-F. and Ohayon, R., Vibration of axisymmetric composite piezoelectric shells coupled with internal fluid, International Journal for Numerical Methods in Engineering 71 (12), 2007, 1412-1435.

[288] Ghergu, M., Griso, G., Mechkour, H. and Miara, B., Homogenization of thin piezoelectric perforated shells, ESAIM: Mathematical Modeling and Numerical Analysis 41 (5), September-October 2007, 875-895.

[289] Edery-Azulay, L. and Abramovich, H., Piezolaminated plates-highly accurate solutions based on the extended kantorovich method, Composite Structures 84 (3), July 2008, 241-247.

[290] Chen, Z. G., Hu, Y. T. and Yang, J. S., Shear horizontal piezoelectric waves in a piezoceramic plate imperfectly bonded to two piezoceramic half-spaces, Journal of Mechanics 24 (3), September 2008, 229-239.

[291] Li, H., Chen, Z. B. and Tzou, H. S., Distributed actuation characteristics of clamped-free conical shells using diagonal piezoelectric actuators, Smart Materials and Structures 19 (11), 2010, article ID 115015.

[292] Messina, A. and Carrera, E., Three-dimensional free vibration of multi-layered piezoelectric plates through approximate and exact analyses, Journal of Intelligent Material Systems and Structures 26 (5), March 2015, 489-504.

[293] Sodano, H. A., Magliula, E., Park, G. and Inman, D. J., Electricpower generation using piezoelectric devices, Proceedings of the 13th International Conference on Adaptive Structures and Technologies, October 2002, 153-161.

[294] Ottman, G., Bhatt, A., Hofmann, H. and Lesieutre, G. A., Adaptive piezoelectric energy harvesting circuit for wireless remote power supply, IEEE Transactions on Power Electronics 17 (5), 2002, 669-676.

[295] Sunghwan, K., Low Power Energy Harvesting with Piezoelectric Generators, Ph. D. Thesis submitted to the School of Engineering at the University of Pittsburgh, PA, USA, December 2002, 136.

[296] Eggborn, T., Analytical Model to Predict Power Harvesting in Piezoelectric Material, Master's Thesis at

the Virginia Polytechnic Institute and State University, Blacksburg, Virginia, May 2003, 94.

[297] Lesieutre, G. A., Hofmann, H. and Ottman, G., Damping as a result of piezoelectric energy harvesting, Journal of Sound and Vibration 269, 2004, 991–1001.

[298] Sodano, H. A., Inman, D. J. and Park, G., Comparison of piezoelectric energy harvesting devices for recharging batteries, Journal of Intelligent Material Systems and Structures 16 (10), October 2005, 799–807.

[299] Ng, T. H. and Liao, W. H., Sensitivity Analysis and Energy Harvesting for a Self-Powered Piezoelectric Sensor, Journal of Intelligent Material Systems and Structures 16 (10), 2005, 785–797.

[300] Sodano, H. A., Lloyd, J. and Inman, D. J., An experimental comparison between several active composite actuators for power generation, Smart Materials and Structures 15, 2006, 1211–1216.

[301] Guan, M. J. and Liao, W. H., On the efficiencies of piezoelectric energy harvesting circuits towards storage device voltages, Smart Materials and Structures 16, 2007, 498–505.

[302] Erturk, A. and Inman, D. J., On mechanical modeling of cantilevered piezoelectric vibration energy harvesters, Journal of Intelligent Material Systems and Structures 19 (11), 2008, 1311–1325.

[303] Kauffman, J. L. and Lesieutre, G. A., A low-order model for the design of piezoelectric energy harvesting devices, Journal of Intelligent Material Systems and Structures 20, March 2009, 495–504.

[304] Liang, J. R. and Liao, W. H., Piezoelectric energy harvesting and dissipation on structural damping, Journal of Intelligent Material Systems and Structures 20 (5), 2009. doi: 10.1177/1045389X08098194, 515–527.

[305] Erturk, A., Tarazaga, P. A., Farmer, J. R. and Inman, D. J., Effect of strain nodes and electrode configuration on piezoelectric energy harvesting from cantilevered beams, ASME Journal of Vibration and Acoustics 131 (1), January 2009, 011010. doi: 10.1115/1.2981094.

[306] Friswell, M. I. and Adhikari, S., Sensor shape design for piezoelectric cantilever beams to harvest vibration energy, Journal of Applied Physics 108 (1), July 2010, 014901. http://dx.doi.org/10.1063/1.3457330.

[307] Rödig, T., Schönecker, A. and Gerlach, G., A survey on piezoelectric ceramics for generator applications, Journal of the American Ceramic Society 93 (4), April 2010, 901–912.

[308] Dietl, J. M., Wickenheiser, A. M. and Garcia, E., A Timoshenko Beam Model for cantilevered piezoelectric energy harvesters, Smart Materials and Structures 19 (5), 2010, article ID 055018.

[309] Erturk, A. and Inman, D. J., Broadband vibration energy harvesting using bistable beams and plates, Proceedings of the American Ceramic Society Symposium: ACerS Electronic Materials and Applications, Orlando, January 2011.

[310] Abramovich, H., Tsikhotsky, E. and Klein, G., An experimental determination of the maximal allowable stresses for high power piezoelectric generators, Journal of Ceramic Science and Technology 4 (3), 2013, 131–136.

[311] Abramovich, H., Tsikhotsky, E. and Klein, G., An experimental investigation on PZT behavior under mechanical and cycling loading, Journal of the Mechanical Behavior of Materials 22 (3-4), 2013, 129–136.

[312] Leinonen, M., Palosaari, J., Juuti, J. and Jantunen, H., Combined Electrical and Electromechanical simulations of a piezoelectric cymbal harvester for energy harvesting from walking, Journal of Intelligent Material Systems and Structures 25 (4), March 2014, 391–400.

[313] Xiaomin, X., Luqi, C., Xiaohong, W. and Qing, S., Study on electric-mechanical hysteretic model of macro-fiber composite actuator, Journal of Intelligent Material Systems and Structures 25 (12), August 2014, 1469-1483.

[314] Zhang, Y., Cai, S. C. S. and Deng, L., Piezoelectric-based energy harvesting in bridge systems, Journal of Intelligent Material Systems and Structures 25 (12), August 2014, 1414-1428.

[315] Bilgen, O., Friswell, M. I., Ali, S. F. and Litak, G., Broadband vibration energy harvesting from a vertical cantilever piezocomposite beam with tip mass, International Journal of Structural Stability and Dynamics 15 (2), March 2015, Paper 1450038.

[316] Vijayan, K., Friswell, M. I., Khodaparast, H. H. and Adhikari, S., Non-linear energy harvesting from coupled impacting beams, International Journal of Mechanical Sciences 96-97, June 2015, 101-109.

[317] Wang, J., Shi, Z., Xiang, H. and Song, G., Modeling on energy harvesting from a railway system using piezoelectric transducers, Smart Materials and Structures 24 (10), October 2015, article ID 105017.

[318] Ansari, M. H. and Karami, M. A., Energy harvesting from controlled buckling of piezoelectric beams, Smart Materials and Structures 24 (11), November 2015, article ID 115005.

[319] Rusovici, R., Dosch, J. J. and Lesieutre, G. A., Design of a single-crystal piezoceramic vibration absorber, Journal of Intelligent Materials Systems and Structures 13 (11), November 2002, 705-712.

[320] Tressler, J. F., A comparison of single crystal versus ceramic piezoelectric materials for acoustic applications, The Journal of the Acoustical Society of America 113 (4), April 2003. doi: org/10.1121/1.4780737, 2311.

[321] Lloyd, J. M., Williams, R. B., Inman, D. J. and Wilkie, W. K., An analytical model of the mechanical properties of single-crystal macro fiber composite actuators, Proceedings of the SPIE's 11th Annual International Symposium on Smart Structures and Materials, San Diego, Paper no. 5387-08, March 2004.

[322] Rusovici, R. and Lesieutre, G. A., Design of a single-crystal piezoceramic-driven, synthetic-jet actuator, Proceeding of SPIE's Symposium on Smart Structures and Materials, SPIE 5390, March 2004. doi: 10.1117/12.539576.

[323] Wilkie, W. K., Inman, D. J., Lloyd, J. M. and High, J. W., Anisotropic piezocomposite actuator incorporating machined PMN-PT single crystal, Journal of Intelligent Material Systems and Structures 17 (1), January 2006, 15-28.

[324] Erturk, A., Lee, H. Y. and Inman, D. J., Investigation of soft and hard ceramics and single crystals for resonant and off-resonant piezoelectric energy harvesting, Proceedings of the 3rd ASME Conference on Smart Materials Adaptive Structures and Intelligent Systems SMASIS, Philadelphia, 2010.

[325] Bilgen, O., Karami, M. A., Inman, D. J. and Friswell, M. I., Actuation characterization of cantilevered unimorph beams with single crystal piezoelectric materials, Smart Materials and Structures 20 (5), May 2011, article ID 055024.

[326] Karami, M. A., Bilgen, O., Inman, D. J. and Friswell, M. I., Experimental and analytical parametric study of single crystal unimorph beams for vibration energy harvesting, Transactions on Ultrasonics Ferroelectrics and Frequency Control 58 (7), July 2011, 1508-1520.

[327] Bilgen, O., Wang, Y. and Inman, D. J., Electromechanical comparison of cantilevered beams with multifunctional piezoceramic devices, Mechanical Systems and Signal Processing 27, February 2012, 763-777.

[328] Anton, S. R., Erturk, A. and Inman, D. J., Bending strength of piezoelectric ceramics and single crystals for multifunctional load-bearing applications, IEEE Transactions on Ultrasonics Ferroelectrics and Frequency Control 59 (6), June 2012, 1085-1092.

[329] Zhou, Q., Lamb, K. H., Zheng, H., Qiu, W. and Shung, K. K., Piezoelectric single crystal ultrasonic transducers for biomedical applications, Progress in Materials Science 66, October 2014, 87-111.

[330] Jiang, X., Kim, J. and Kim, K., Relaxor-PT-single crystal piezoelectric sensors, Crystals 4, 2014. doi: 10.3390/cryst4030351, 351-376.

[331] Patel, S. and Vaish, R., Design of PZT-PT functionally graded piezoelectric material for lowfrequency actuation applications, Journal of Intelligent Material Systems and Structures 26 (3), February 2015, 321-327.

[332] Park, G., Kabeya, K., Cudney, H. H. and Inman, D. J., Removing effects of temperature changes from piezoelectric impedance-based qualitative health monitoring, Proceedings of SPIE Conference on Sensory Phenomena and Measurement Instrumentation for Smart Structures and Materials, SPIE Vol. 3330, March 1998, 103-114.

[333] Schulz, M., Pai, P. F. and Inman, D. J., Health Monitoring and Active control of composite structures using piezoceramic patches, Composites Part B: Engineering 30 (7), 1999, 713-725.

[334] Nothwang, W. D., Hirsch, S. G., Demaree, J. D., Hubbard, C. W., Cole, M. W. and Lin, B., Direct integration of thin film piezoelectric sensors with structural materials for structural health monitoring, Integrated Ferroelectrics An International Journal 83 (1), 2006, 139-148.

[335] Grisso, B., Advancing autonomous structural health monitoring, Ph. D. Thesis at the Virginia Polytechnic Institute and State University, etd-12062007-105329, November 2007.

[336] Giurgiutiu, V., Piezoelectric wafer active sensors for structural health monitoring of composite structures using tuned guided waves, Journal of Engineering Materials and Technology 133 (4), 2011, article ID 041012.

[337] Wang, R. L., Gu, H. and Song, G., Active sensing based bolted structure health monitoring using piezoceramic transducers, International Journal of Distributed Sensor Networks 2013, July 2013, article ID 583205.

[338] Perkins, J. (ed.), Shape Memory Effects in Alloys, Springer US, 1975, ISBN: 978-1-4684-2211-5.

[339] Achenbach, M. and Muller, I., Simulation of material behavior of alloys with shape memory, Archives of Mechanics 37 (6), 1985, 573-585.

[340] Funakubo, H. (ed.), Shape Memory Alloys, Gordon and Breach Science Publication, New York, NY, USA, 1987.

[341] Tadaki, T., Otsuka, K. and Shimizu, K., Shape memory alloys, Annual Review of Material Science 18, 1988, 25-45.

[342] Achenbach, M., A model for an alloy with shape memory, International Journal of Plasticity 5, 1989, 371-395.

[343] Duerig, T., Melton, K., Stöckel, D. and Wayman, C. M., Engineering Aspectsof Shape Memory Alloys, Elsevier, 1990, ISBN: 978-0-7506-1009-4, 499.

[344] Wayman, C. M., Shape memory and related phenomena, Progress in Materials Science 36, 1992, 203-224.

# 第1章 智能材料与结构概述

[345] Wayman, C. M., Shape memory alloys, MRS Bulletin 18 (4), April 1993, 49-56.

[346] Birman, V., Review of mechanics of shape memory alloy structures, Applied Mechanics Reviews 50 (11), November 1997, 629-645.

[347] Humbeeck, J. V., Non-medical applications of shape memory alloys, Materials Science and Engineering 273-275, December 1999, 134-148.

[348] Otsuka, K. and Ren, X., Recent developments in the research of shape memory alloys, Intermetallics 7 (5), 1999, 511-528.

[349] Wu, M. H. and Schetky, L. M., Industrial applications for shape memory alloys, Proceedings of the International Conference on Shape Memory and Superelastic Technologies, Pacific Grove, California, USA, 2000, 171-182.

[350] Humbeeck, J. V., Shape memory alloys: a material and a technology, Advanced Engineering Materials 3 (11), November 2001, 837-850.

[351] Huang, W., On the selection of shape memory alloys for actuators, Materials & Design 23 (1), February 2002, 11-19.

[352] Otsuka, K. and Kakeshita, T., Science and technology of shape-memory alloys: new developments, MRS Bulletin, February 2002, 91-100.

[353] Frecker, M. I., Recent advances in optimization of smart structures and actuators, Journal of Intelligent material systems and structures 14, April-May 2003, 207-216.

[354] Yoneyama, T. and Miyazaki, S. (eds.), Shape Memory Alloys for Biomedical Applications, 1st edn., Woodhead Publishing, 28 November 2008, 352.

[355] Shabalovskaya, S., Anderegg, J. and Humbeeck, J. V., Critical overview of nitinol surfaces and their modifications for medical applications, Acta Biomaterialia 4, 2008, 447-467.

[356] Ozbulut, O. E., Hurlebaus, S. and Desroches, R., Seismic response control using shape memory alloys: a review, Journal of Intelligent Material Systems and Structures 22 (14), August 2011. doi: 10.1177/1045389X11411220, 1531-1549.

[357] Jani, J. M., Leary, M., Subic, A. and Gibson, M. A., A review of shape memory alloy research, Applications and Opportunities, Materials & Design 56, March 2014, 1078-1113.

[358] Lecce, L. and Concilio, A. (eds.), Shape Memory Alloy Engineering for Aerospace Structural and Biomedical Applications, Elsevier, 2015, 421.

[359] Baz, A. and Tampe, L., Active Control of Buckling of Flexible Beams, Proceeding of the ASME Conference on Failure Prevention and Reliability, Montreal, September 1989, 211-218.

[360] Baz, A., Poh, S., Ro, J., Mutua, M. and Gilheany, J., Active Control of Nitinol-Reinforced Composite Beam, in Intelligent Structural Systems, Tzou, H. S. and Anderson, G. L. (eds.), Dordrecht, Springer Science + Business Media, 1992, 169-212.

[361] Baz, A., Imam, K. and McCoy, J., Active Vibration Control of Flexible Beams Using Shape Memory Actuators, Journal of Sound and Vibration 140 (3), August 1990, 437-456.

[362] Lagoudas, D. C. and Tadjbakhsh, J. G., Active Flexible Rods with Embedded SMA Fibers, Smart Materials and Structures 1, 1992, 162-167.

[363] Brinson, L. C., Huang, M. S., Boller, C. and Brand, W., Analysis of Controlled Beam Deflections Using SMA Wires, Journal of Intelligent Material Systems and Structures 8, January 1997, 12-25.

[364] Birman, V., Theory and Comparison of the Effect of Composite and Shape Memory Alloy Stiffeners on Stability of Composite Shells and Plates, International Journal of Mechanical Science 39 (10), 1997, 1139-1149.

[365] Baz, A., Chen, T. and Ro, J., Shape Control of NITINOL-Reinforced Composite Beams, Composites Part B: Engineering 31 (8), 2000, 631-642.

[366] Tsai, X.-Y. and Chen, L.-W., Dynamic Stability of a Shape Memory Alloy Wire Reinforced Composite Beam, Composite Structures 56 (3), May-June, 2002, 235-241.

[367] Ikuta, K., Tsukumoto, M. and Hirose, S., Shape Memory Alloy Servo Actuator System with Electric Resistance Feedback and Application for Active Endoscope, Proceedings of the IEEE International Conference on Robotics and Automation, April 1988. doi: 10.1109/ROBOT.1988.12085, 427-430.

[368] Pelton, A. R., Stöckel, D. and Duerig, T. W., Medical Uses of Nitinol, Proceedings of the International Symposium on Shape Memory Materials, Kanazawa, Japan, May 1999, Material Science Forum, Vols. 327-328, 2000, 63-70.

[369] Filip, P, Titanium-Nickel Shape Memory Alloys in Medical Applications. In: Titanium in Medicine, Material Science, Surface Science, Engineering, Biological Responses and Medical Applications, Brunette, D. M., Tengvall, P. Textor, M. and Thomsen, P, (eds), Part II, 2001, 53-86.

[370] Duerig, T., Stöckel, D. and Johnson, D., SMA-Smart Materials for Medical Applications, Proceedings of SPIE 4763, 2002, 7-15.

[371] Machado, L. G. and Savi, M. A., Medical Applications of Shape Memory Alloys, Brazilian, Journal of Medical and Biological Research 36 (2003), 683-691.

[372] Morgan, N. B., Medical Shape Memory Alloy Applications-the Market and its Product, Material Science and Engineering: A 378 (1-2), July 2004, 16-23.

[373] Kleinstreuer, C., Li, Z., Basciano, C. A., Seelecke, S. and Farber, M. A., Computational Mechanics of Nitinol Stents Grafts, Journal of Biomechanics 41 (11), 2008, 2370-2378.

[374] De Miranda, R. L., Zamponi, C. and Quandt, E., Fabrication of TiNi Thin Film Stents, Smart Materials and Structures 18 (10), October 2009, article ID 104010.

[375] Schaffer, J., Mechanical Conditioning of Superelastic Nitinol Wire for Improved Fatigue Resistance, Journal of ASTM International 7 (5), 2010, 1-7.

[376] Petrini, L. and Migliavacca, F., Biomedical Applications of Shape Memory Alloys, Journal of Metallurgy, 2011 article ID 501483, 15.

[377] Zainal, M. A., Sahlan, S. and Ali, M. S. M., Micromachined Shape Memory Alloy Microactuators and Their Application in Biomedical Devices, Micromachines 6, 2015, 879-901.

[378] Baz, A. and Ro, J., Thermo-Dynamic Characteristics of Nitinol-Reinforced Composite Beams, Composites Engineering 2 (5-7), 1992, 527-542.

[379] Baz, A. and Chen, T., Torsional Stiffness of NITINOL-Reinforced Composite Drive Shafts, Composite Engineering 3 (12), 1993, 1119-1130.

[380] Boyd, J. G. and Lagoudas, D. C, Thermomechanical Response of Shape Memory Composites, Journal of Intelligent Material Systemsand Structures 5 (3), May 1994, 333-346.

[381] Zhang, C. and Zee, R. H., Development of Ni-Ti Based Shape Memory Alloys for Actuation and Control, Proceedings of the 31st Intersociety Energy Conversion Engineering Conference IECEC 1,

1996. doi: 10.1109/IECEC.1996.552877, 239-244.

[382] Chen, Q. and Levy, C., Active Vibration Control of Elastic Beam by Means of Shape Memory Alloy Layers, Smart Materials and Structures 5 (4), August 1996, 400-406.

[383] Dolce, M., Cardone, D. and Marnetto, R., Implementation and Testing of Passive Control Devices Based on Shape Memory Alloys, Earthquake Engineering & Structural Dynamics 29 (7), July 2000, 945-968.

[384] Prahlad, H. and Chopra, I., Experimental Characterization of Ni-Ti Shape Memory Alloy Wires Under Uniaxial Loading Conditions, Journal of Intelligent Material Systems and Structures 11 (4), April 2000, 263-271.

[385] Jonnalagadda, K. D., Sottos, N. R., Qidwai, M. A. and Lagoudas, D. C, Insitu-Displacement Measurements and Theoretical Prediction of Embedded SMA Actuation, Journal of Smart Materials and Structures 9, 2000, 701-710.

[386] Pae, S., Lee, H., Park, H. and Hwang, W., Realization of Higher-Mode Deformation of Beams Using Shape Memory Alloy Wires and Piezoceramics, Smart Materials and Structures 9 (6), December 2000, 848-854.

[387] Epps, J. J. and Chopra, I., In-Flight Tracking of Helicopter Rotor Blades Using Shape Memory Alloy Actuators, Structures Smart Material and Structures 10 (1), February 2001, 104-111.

[388] Prahlad, H. and Chopra, I., Comparative Evaluation of Shape Memory Alloy Constitutive Models with Experimental Data, Journal of Intelligent Material Systems and Structures 12 (6), June 2001, 383-397.

[389] Lammering, R. and Schmidt, I., Experimental Investigation on the Damping Capacity of NiTi Components, Smart Materials and Structures 10 (5), October 2001, 853-859.

[390] Matsuzaki, Y., Naito, H., Ikeda, T. and Funami, K., Thermo-Mechanical Behavior Associated with Pseudoelastic Transformation of Shape Memory Alloys, Smart Materials and Structures 10 (5), October 2001, 884-892.

[391] Chandra, R., Active Shape Control of Composite Blades Using Shape Memory Alloys, Smart Materials and Structures 10 (5), October 2001, 1018-1024.

[392] Mehrabi, R., Kadkhodaei, M., Andani, M. T. and Elahinia, M., Microplane Modeling of Shape Memory Alloy Tubes Under Tension, Torsion and Proportional Tension-Torsion Loading, Journal of Intelligent Material Systems and Structures 26 (2), January 2015, 144-155.

[393] Malukhin, K. and Ehmann, K., Model of a NiTi Shape Memory Alloy Actuator, Journal of Intelligent Material Systems and Structures 26 (4), March 2015, 386-399.

[394] Lacasse, S., Terriault, P., Simoneau, C. and Brailovski, V., Design, Manufacturing and Testing of an Adaptive Panel With Embedded Shape Memory Actuators, Journal of Intelligent Material Systems and Structures 26 (15), October 2015, 2055-2072.

[395] Ro, J. and Baz, A., NITINOL - Reinforced plates: Part I. Thermal Characteristics, Composite Engineering 5 (1), 1995, 61-75.

[396] Ro, J. and Baz, A., NITINOL-Reinforced plates: Part II. Static and buckling Characteristics, Composite Engineering 5 (1), 1995, 77-90.

[397] Ro, J. and Baz, A., NITINOL-Reinforced plates: Part III. Dynamic characteristics, Composite Engineering 5 (1), 1995, 91-106.

[398] Park, J. -S., Kim, J. -H. and Moon, S. -H., Vibration of Thermally Post-Buckled Composite Plates Embedded with Shape Memory Alloy Fibers, Composite Structures 63, 2004, 179-188.

[399] DeHaven, J. G. and Tzou, H. S., Forced Response of Cylindrical Shells Coupled with Nonlinear SMA Actuators Regulated by Sinusoidal and Saw-Tooth Temperature Profiles, Journal of Engineering Mathematics 61, 2008. doi: 10, 1007/S10665-008-9231-5.

[400] Boyd, J. G. and Lagoudas, D. C., A Thermodynamical Constitutive Model for Shape Memory Materials, Part I: The Monolithic Shape Memory Alloy, International Journal of Plasticity 12, 1996, 805-842.

[401] Boyd, J. G. and Lagoudas, D. C., A Thermodynamical Constitutive Model for Shape Memory Materials, Part II, The SMA Composite Material, International Journal of Plasticity 12, 1996, 843-873.

[402] Bo, Z. and Lagoudas, D. C., Thermomechanical Modeling of Polycrystalline SMAs Under Cyclic Loading, Part I: Theoretical Derivations, International Journal of Engineering Science 37 (9), July 1999, 1089-1140.

[403] Bo, Z. and Lagoudas, D. C., Thermomechanical Modeling of Polycrystalline SMAs Under Cyclic Loading, Part II: Material Characterization and Experimental Results for a Stable transformation Cycle, International Journal of Engineering Science 37 (9), July 1999, 1141-1173.

[404] Bo, Z. and Lagoudas, D. C., Thermomechanical Modeling of Polycrystalline SMAs Under Cyclic Loading, Part III: Evolution of Plastic Strains and Twoway Shape Memory Effect, International Journal of Engineering Science 37 (9), July 1999, 1175-1203.

[405] Bo, Z. and Lagoudas, D. C., Thermomechanical Modeling of Polycrystalline SMAs Under Cyclic Loading, Part IV: Modeling of Minor Hysteresis Loops, International Journal of Engineering Science 37 (9), July 1999, 1205-1249.

[406] Auricchio, F. and Sacco, E., A one-dimensional model for superelastic shape-memory alloys with different elastic properties between austenite and martensite, International Journal NonLinear Mechanics 32 (6), November 1997, 1101-1114.

[407] Lagoudas, D. C., Moorthy, D., Qidwai, M. A. and Reddy, J. N., Modeling of the Thermomechanical Response of Active Composite Laminates with SMA Layers, Journal of Intelligent Material Systems and Structures 8, 1997, 476-488.

[408] Xu, G. -M., Lagoudas, D. C., Hughes, D. and Wen, J. T., Modeling of a Flexible Beam Actuated by Shape Memory Alloys Wires, Journal of Smart Materials and Structures 6, 1997, 265-277.

[409] Webb, G. V., Lagoudas, D. C. and Kurdila, A. J., Hysteresis Modeling of SMA Actuators for Control Applications, Journal of Intelligent Materials Systems and Structures 9, 1998, 432-448.

[410] Matsuzaki, Y., Funami, K. and Naito, H., Inner Loops of Pseudoelastic Hysteresis of Shape Memory Alloys: Preisach Approach, Smart Structures and Materials 2002, Proceedings of the SPIE, 4699, 2002, 355-364.

[411] Seelecke, S. Modeling the Dynamic Behavior of Shape Memory Alloys, International Journal of Non-Linear Mechanics, 37, 2002, 1363-1374.

[412] Prahlad, H. and Chopra, I., Development of Strain-Rate Dependent Model for Uniaxial Loading of SMA Wires, Journal of Intelligent Material Systems and Structures 14 (5), July 2003, 429-442.

[413] Auricchio, F., Marfia, S. and Sacco, E., Modelling of SMA materials: Training and Two Way Memory Effects, Computers & Structures 81 (24-25), September 2003, 2301-2317.

[414] Singh, K., Sirohi, J. and Chopra, I., An Improved Shape-Memory Alloy Actuator for Rotor Blade Tracking, Journal of Intelligent Material Systems and Structures 14 (12), December 2003, 767-786.

[415] Lagoudas, D. C., Khan, M. M., Mayes, J. J. and Henderson, B. K., Pseudoelastic SMA Spring Elements for Passive Vibration Isolation, Part I: Modeling, Journal of Intelligent Material Systems and Structures 15, 2004, 415-441.

[416] Lagoudas, D. C., Khan, M. M., Mayes, J. J. and Henderson, B. K., Pseudoelastic SMA Spring Elements for Passive Vibration Isolation, Part II: Simulations and Experimental Correlations, Journal of Intelligent Material Systems and Structures 15, 2004, 443-470.

[417] Manzo, J., Garcia, E., Wickenheiser, A. M. and Horner, G., Design of a Shape-Memory Alloy Actuated Macro-Scale Morphing Aircraft Mechanism, Proceedings of SPIE Conference, San Diego, SPIE 5764, March 2005, 232-240.

[418] Prahlad, H. and Chopra, I., Modeling and Experimental Characterization of SMA Torsional Actuators, Journal of Intelligent Material Systems and Structures 18 (1), January 2007, 29-38.

[419] Hartl, D. and Lagoudas, D. C., Aerospace Applications of Shape Memory Alloys, Journal of Aerospace Engineering 221, 2007, 535-552.

[420] Popov, P. and Lagoudas, D. C., A 3-D Constitutive Model for Shape Memory Alloys Incorporating Pseudoelasticity and Detwinning of Self-Accommodated Martensite, International Journal of Plasticity 23, 2007, 1679-1720.

[421] Liang, C. and Rogers, C. A., Design of Shape Memory Alloy Actuators, Journal of Mechanical Design 114 (2), June 2008. doi: 10.1115/1.2916935.

[422] Churchill, C. B., Shaw, J. A. and Iadicola, M. A., Tips and Tricks for Characterizing Shape Memory Alloy Wire: Part 2-Fundamental Isothermal Responses, Experimental Techniques 33 (1), January-February 2009. doi: 10.1111/j.1747-1567.2008.00460.X, 51-62.

[423] Churchill, C. B., Shaw, J. A. and Iadicola, M. A., Tips and Tricks for Characterizing Shape Memory Alloy Wir: Part 3-Localization and Propagation Phenomena, Experimental Techniques 33 (5), September-October 2009. doi: 10.1111/j.1747-1567.2009.00558.X. 70-78.

[424] Bertacchini, O. W., Characterization and Modeling of Transformation Induced Fatigue of Shape Memory Alloy Actuators, Ph.D. Thesis at Texas A&M University, December 2009.

[425] Hartl, D., Modeling of Shape Memory Alloys Considering Rate-Independent and RateDependent Irrecoverable Strains, Ph.D. Thesis at Texas A&M University, December 2009.

[426] Hartl, D. J. and Lagoudas, D. C., Constitutive Modeling and Structural Analysis Considering Simultaneous Phase Transformation and Plastic Yield in Shape Memory Alloys, Smart Materials and Structures 18 (10), 2009, article ID 104017.

[427] Churchill, C. B., Shaw, J. A. and Iadicola, M. A., Tips and Tricks for Characterizing Shape Memory Alloy Wire: Part 4-Thermo-Mechanical Coupling, Society for Experimental Mechanics 34 (2), March-April 2010, 63-80.

[428] Furst, S. J. and Seelecke, S., Modeling and Experimental Characterization of the Stress, Strain, and Resistance of Shape Memory Alloy Actuator Wires with Controlled Power Input, Journal of Intelligent Material Systems and Structures 23 (11), July 2012, 1233-1247.

[429] Strelec, J. K. and Lagoudas, D. C., Fabrication and testing of a shape memory alloy actuated reconfigu-

rable wing, Proceeding of the SPIE Conference on Smart Structures and Materials, SPIE 4701, July 2002. doi: 10.1117/12.474664, 267-280.

[430] Strelec, J., Design and Implementation of a Shape Memory Alloy Actuated Reconfigurable Wing, Master's Thesis at the Texas A&M University, 2002.

[431] Strelec, J. K., Lagoudas, D. C., Khan, M. A. and Yen, J., Design and implementation of a shape memory alloy actuated reconfigurable wing, Journal of Intelligent Material Systems and Structures 14, 2003, 257-273.

[432] Peng, F., Jiang, X.-X., Hu, Y.-R. and Ng, A., Application of shape memory alloy actuators in active shape control of inflatable space structures, Proceedings of the IEEE Aerospace Conference, March 2005. doi: 10.1109/AERO.2005.1559577.

[433] Oehler, S., Developing methods for designing shape memory alloy actuated morphing aerostructures, Master's Thesis at the Texas A&M University, 2012.

[434] Naghashian, S., Fox, B. L. and Barnett, M. R., Actuation curvature limits for a composite beam with embedded shape memory alloy wires, Smart Materials and Structures 23 (6), June 2014, article ID 065002.

[435] Wuttig, M., Li, J. and Craciunescu, C., A new ferromagnetic shape memory alloy system, Scripta Materialia 44 (10), May 2001, 2393-2397.

[436] Ishida, A. and Martynov, V., Sputter-deposited shape-memory alloy thin films: properties and applications, MRS Bulletin, February 2002, 111-114.

[437] Ma, N. and Song, G., Control of shape memory alloy actuator using pulse width modulation, Smart Materials and Structures 12 (5), 2003, 712-719.

[438] Des Roches, R., McCormick, J. and Delemont, M., Cyclic properties of superelastic shape memory alloy wires and bars, Journal of Structural Engineering 130 (1), January 2004, 38-46.

[439] Nagai, H. and Oishi, R., Shape memory alloys as strain sensors in composites, Smart Materials and Structures 15 (2), 2006, 493-498.

[440] Schick, J., Transformation Induced fatigue of Ni-Rich NiTi shape memory alloy actuators, Master's Thesis at the Texas A&M University, December 2009.

[441] Lagoudas, D. C., Miller, D. A., Rong, L. and Kumar, P. K., Thermomechanical fatigue of shape memory alloys, Smart Materials and Structures 18, 2001, article ID 085021.

[442] Lan, C.-C. and Fan, C.-H., An accurate self-sensing method for the control of shape memory alloy actuated flexures, Sensors and Actuators 163 (1). doi: 10.1016/j.sna.2010.07.018, April 2010, 323-332.

[443] Moussa, M. O., Moumni, Z., Doare, O., Touze, C. and Zaki, W., Non-linear dynamic thermomechanical behavior of shape memory alloys, Journal of Intelligent Material Systems and Structures 23 (14), 2012, 1593-1611.

[444] Jordan, T. C. and Shaw, M. T., Electrorheology, Material Research Society MRS Bulletin 16 (8), August 1991, 38-43.

[445] Stanway, R., Sprostonz, J. L. and El-Wahed, A. K., Applications of electrorheological fluids in vibration control: a survey, Smart Materials and Structures 5 (4), August 1996, 464-482.

[446] Kamath, G. M. and Wereley, N. M., Modeling the damping mechanism in electro-rheological fluid based

dampers, ASTM Metals Test Methods and Analytical Procedures, STP1304-EB, January 1997, 331-348.

[447] Tao, R. (ed.), Electrorheological fluids and magnetorheological suspensions, Proceedings of the 7th International Conference on Electrorheological Fluids and Magnetorheological Suspensions, Honolulu, Hawaii, USA, July 12-23, 1999, World Scientific Publishing, 850 pp.

[448] Jolly, M. R., Properties and applications of magnetorheological fluids, symposium LL-materials for smart systems III. In: Fogle, M. W., Uchino, K., Ito, Y. and Gotthardt, R., (eds), Materials Research Society (MRS) Proceedings, Vol. 604, 1999.

[449] Phulé, P. P., Magnetorheological (MR) fluids: principles and applications, Smart Material Bulletin 2001 (2), February 2001, 7-10.

[450] Wen, W., Huang, X. and Sheng, P., Electrorheological fluids: structures and mechanism, Soft Matter 4, 2008, 200-210.

[451] Wereley, N. M. (ed.), Magnetorheology: Advances and Applications, RSC Smart Materials series, RSC Publishing, 26 November 2013, 396 pp.

[452] Kulkarni, A. N. and Patil, S. R., Magneto-Rheological (MR) and Electro-Rheological (ER) fluid damper: a review parametric study of fluid behavior, International Journal of Engineering Research and Applications 3 (6), November-December 2013, 1879-1882.

[453] Uejima, H., Dielectric mechanism and rheological properties of electro-fluids, Japanese Journal of Applied Physics 11 (3), March 1972, 319-326.

[454] Davis, L. C., Finite-element analysis of particle-particle forces in electrorheological fluids, Applied Physics Letters 60 (3), September 1991, 319-321.

[455] Halsey, T. C., Electrorheological fluids: structure formation relaxation and destruction, Materials Research Society MRS Proceedings, Vol. 248, 1991.

[456] Katsikopoulos, P. and Zukoski, C., Relaxation processes in the electrorheological response, Materials Research Society MRS Proceedings, Vol. 248, 1991.

[457] Davis, L. C., Polarization forces and conductivity effects in electrorheological fluids, Journal of Applied Physics 72, February 1992, 1334-1340.

[458] Choi, S.-B., Park, Y.-K. and Kim, J.-D., Vibration characteristics of hollow cantilevered beams containing an electrorheological fluid, International Journal of Mechanical Sciences 35 (9), August 1992, 757-768.

[459] Coulter, J. P., Weiss, K. D. and Carlson, J. D., Engineering applications of electrorheological materials, Journal of Intelligent Material Systems and Structures 4 (2), April 1993, 248-259.

[460] Powell, J. A., Modelling the Oscillatory Response of an Electrorheological Fluid, Smart Materials and Structures 3 (4), March 1994, 416-438.

[461] Rajagopal, K. R., Yalamanchili, R. C. and Wineman, A. S., Modeling electrorheological materials through mixture theory, International Journal of Engineering Science 32 (3), 1994, 481-500.

[462] Rajagopal, K. R. and Ruziicka, M., On the modeling of electrorheological materials Mechanics Research Communications 23 (4), April 1996, 401-407.

[463] Wu, S., Lu, S. and Shen, J., Electrorheological suspensions, Polymer International 41 (4), December 1996, 363-367.

[464] Gamota, D. R., Schubring, A. W., Mueller, B. L. and Filisko, F. E., Amorphous ceramics as the particulate phase in electrorheological materials systems, Journal of Materials Research 11 (1), 1996, 144-155.

[465] Kamath, G. M., Hurt, M. K. and Wereley, N. M., Analysis and testing of bingham plastic behavior in semi-active electrorheological fluid dampers, Smart Materials and Structures 5 (5), 1996, 576-590.

[466] Gordaninejad, F. and Bindu, R., A Scale Study of electrorheological fluid dampers, Journal of Structural Control 4 (2), December 1997, 5-17.

[467] Davis, L. C., Time-dependent and nonlinear effects in electrorheological fluids, Journal of Applied Physics 81 (4), 1997, 1985-1991.

[468] Kohl, J. G. and Tichy, J. A., Expressions for coefficients of electrorheological fluid dampers, Lubrication Science 10 (2), February 1998, 135-1143.

[469] Conrad, H., Properties and design of electrorheological suspensions, Materials Research Society MRS Bulletin 23 (8), August 1998, 35-42.

[470] Inoue, A., Ide, Y., Maniwa, S., Yamada, H. and Oda, H., ER Fluids Based on liquid-crystalline polymers, Materials Research Society MRS Bulletin 23 (8), August 1998, 43-49.

[471] Wen, W., Tam, W. Y. and Sheng, P., Electrorheological fluids using bi-dispersed particles, Journal of Material Research 13 (10), 1998, 2783-2786.

[472] Wu, C. W. and Conrad, H., Influence of mixed particle size on electrorheological response, Journal of Applied Physics 83 (7), 1998, 3880-3884.

[473] Kohl, J. G., Tichy, J. A., Craig, K. C. and Malcolm, S. M., Determination of the bingham parameters of an electrorheological fluid in an axial flow concentric-cylinder rheometer, Tribotest Journal 5 (3), March 1999, 221-224.

[474] Johnson, A. R., Bullough, W. A. and Makin, J., Dynamic Simulation and Performance of an electrorheological clutch based reciprocating mechanism, Smart Materials and Structures 8 (5), October 1999, 591-600.

[475] Lee, C.-Y. and Cheng, C.-C., Complex moduli of electrorheological material under oscillatory shear, International Journal of Mechanical Sciences 42 (3), March 2000, 561-573.

[476] Lee, C. Y. and Liao, W. C., Characteristics of an electrorheological fluid valve used in an inkjet print head, Smart Materials and Structures 9, March 2000, 839-847.

[477] Fukuda, T., Takawa, T. and Nakashima, K., Optimum vibration control of CFRP sandwich beam using electrorheological fluids and piezoceramic actuators, Smart Materials and Structures 9 (1), February 2000, 121-125.

[478] Wen, W., Ma, H., Tam, W. Y. and Sheng, P., Frequency-induced structure variation in electrorheological fluids, Applied Physics Letters 77 (23), December 2000, 3821-3823.

[479] Sakamoto, D., Oshima, N. and Fukuda, T., Tuned sloshing damper using electrorheological fluid, Smart Materials and Structures 10, January 2001, 963-969.

[480] Wang, B., Liu, Y. and Xiao, Z., Dynamical modelling of the chain structure formation in electrorheological fluids, International Journal of Engineering Science 39 (4), March 2001, 453-475.

[481] Hao, T., Electrorheological fluids, Advanced Materials 13 (24), December 2001, 1847-1857.

[482] Chen, S. H., Yang, G. and Liu, X. H., Response analysis of vibration systems with ER dampers, Smart

Materials and Structures 10 (5), October 2001, 1025-1030.

[483] Kang, Y. K., Kim, J. and Choi, S. -B., Passive and active damping characteristics of smart electrorheological composite beams, Smart Materials and Structures 10 (4), 2001, 724-729.

[484] Yoshida, K., Kikuchi, M., Park, J. -H. and Yokota, S., Fabrication of micro electrorheological valves (ER Valves) by Micromachining and experiments, Sensors and Actuators 95 (2-3), January 2002, 227-233.

[485] Hong, S. -R., Choi, S. -B. and Han, M. -S., Vibration control of a frame structure using electrorheological fluid mounts, International Journal of Mechanical Sciences 44, April 2002, 2027-2045.

[486] Noresson, V., Ohlson, N. G. and Nilsson, M., Design of electrorheological dampers by means of finite element analysis: theory and applications, Materials and Design 23 (4), June 2002, 361-369.

[487] Sproston, J. L., El Wahed, A. K. and Stanway, R., The rheological characteristics of electrorheological fluids in dynamic squeeze, Journal of Intelligent Material Systems and Structures 13 (10), October 2002, 655-660.

[488] Chen, S. H. and Yang, G., A method for determining locations of electrorheological dampers in structures, Smart Materials and Structures 12 (2), April 2003, 164-170.

[489] Lindler, J. and Wereley, N. M., Quasi-steady bingham-plastic analysis of an electrorheological flow mode bypass damper with piston bleed, Smart Materials and Structures 12 (3), 2003, 305-317.

[490] Barber, G. C., Jiang, Q. Y., Zou, Q. and Carlson, W., Development of a laboratory test device for electrorheological fluids in hydrostatic lubrication, Tribotest Journal 11 (3), March 2005, 185-191.

[491] Phani, A. S. and Venkatraman, K., Damping characteristics of electrorheological fluid sandwich beams, Acta Mechanica 180 (1), June 2005, 195-201.

[492] Choi, S. B., Choi, H. J., Choi, Y. T. and Wereley, N. M., Preparation and mechanical characteristics of poly-methylaniline based electrorheological fluid, Journal of Applied Polymer Science 96 (5), 2005, 1924-1929.

[493] Lim, S. C., Park, J. S., Choi, S. B., Choi, Y. T. and Wereley, N. M., Design and analysis program of electrorheological devices for vehicle systems, International Journal of Vehicle Autonomous Systems 3 (1), 2005, 15-33.

[494] Shulman, Z. P., Korobko, E. V., Levin, M. L., Binshtok, A. E., Bilyk, V. A. and Yanovsky, Yu. G., Energy dissipation in electrorheological damping devices, Journal of Intelligent Material Systems and Structures 17 (4), April 2006, 315-320.

[495] Liu, L., Chen, X., Niu, X., Wen, W. and Sheng, P., Electrorheological fluid-actuated microfluidic pump, Applied Physics Letters 89 (8), May 2006, 083.505-083.506.

[496] Yan, Q. S., Bi, F. F. and Wu, N. Q., Performance evaluation of electro-rheological fluid and its applications to micro-parts fine-machining, Journal Key Engineering Materials 315-316, July 2006, 352-356.

[497] Nikitczuk, J., Weinberg, B. and Mavroidis, C., Control of electrorheological fluid based resistive torque elements for use in active rehabilitation devices, Smart Materials and Structures 16 (2), February 2007, 418-428.

[498] Ramkumar, K. and Ganesan, N., Vibration and damping of composite sandwich box column with viscoelastic/electrorheological fluid core and performance comparison, Materials and Design 30 (8), September

2009, 2981-2994.

[499] Chen, Y. G. and Yan, H., The performance analysis of electro-rheological damper, Advanced Materials Research 179-180, January 2011, 443-448.

[500] Kaushal, M. and Joshi, Y. M., Self-similarity in electrorheological behavior, Soft Matter 7, May 2011, 9051-9060.

[501] El Wahed, A. K., The Influence of solid-phase concentration on the performance of electrorheological fluids in dynamic squeeze flow, Materials and Design 32 (3), 2011, 1420-1426.

[502] Hoppe, R. H. W. and Litvinov, W. G., Modeling simulation and optimization of electrorheological fluids, Handbook of Numerical Analysis, Vol. 16, 2011, 719-793.

[503] Liu, Y. D. and Choi, H. J., Electrorheological fluids: smart soft matter and characteristics, Soft Matter 8, May 2012, 11.961-11.978.

[504] Krivenkov, K., Ulrich, S. and Bruns, R., Extending the operation range of electrorheological actuators for vibration control through novel designs, Journal of Intelligent Material Systems and Structures 23 (12), 2012, 1323-1330.

[505] Mohammadi, F. and Sedaghati, R., Dynamic mechanical properties of an electrorheological fluid under large amplitude oscillatory shear strain, Journal of Intelligent Material Systems and Structures 23 (10), 2012, 1093-1105.

[506] Jiang, B. and Shi, W. K., The model simulation analysis of electro-rheological fluid engine mounting system, Advanced Materials Research 694-697, May 2013, 338-343.

[507] Allahverdizadeha, A., Mahjooba, M. J., Malekib, M., Nasrollahzadeha, N. and Naeia, M. H., Structural Modeling, Vibration Analysis and Optimal viscoelastic layer characterization of adaptive sandwich beams with electrorheological fluid core, Mechanics Research Communications 51, July 2013, 15-22.

[508] Wang, Z., Gong, X., Yang, F., Jiang, W. and Xuan, S., Dielectric relaxation effect on flow behavior of electrorheological fluids, Journal of Intelligent Material Systems and Structures 26 (10), May 2014, 1141-1149.

[509] Hoseinzadeh, M. and Rezaeepazhand, J., Vibration suppression of composite plates using smart electrorheological dampers, International Journal of Mechanical Sciences 84, July 2014, 31-40.

[510] Weiss, K. D., Carlson, J. D. and Nixon, D. A., Viscoelastic properties of magneto-and electrorheological fluids, Journal of Intelligent Material Systems and Structures 5 (6), November 1994, 772-775.

[511] Phulé, P. P. and Ginder, J. M., The materials science of field-responsive fluids, Materials Research Society MRS Bulletin 23 (8), August 1998, 19-22.

[512] Wereley, N. M., Pang, L. and Kamath, G. M., Idealized hysteresis modeling of electrorheological and magnetorheological dampers, Journal of Intelligent Material Systems and Structures 9 (8), August 1998, 642-649.

[513] Wereley, N. M. and Pang, L., Nondimensional analysis of semi-active electrorheological and magnetorheological dampers using approximate parallel plate models, Smart Materials and Structures 7 (5), 1998, 732-743.

[514] Rankin, P. J., Ginder, J. M. and Klingenberg, D. J., Electro-and magnetorheology, Current Opinion in Colloid & Interface Science 3 (4), August 1998, 373-381.

[515] El Wahed, A. K., Sproston, J. L. and Schleyer, G. K., A comparison between electrorheological and

magnetorheological fluids subjected to impulsive loads, Journal of Intelligent Material Systems and Structures 10 (9), September 1999, 695-700.

[516] Lee, D. Y. and Wereley, N. M., Quasi-steady Herschel-Bulkley analysis of electro-and magneto-rheological flow mode dampers, Journal of Intelligent Material Systems and Structures 10 (10), September 1999, 761-769.

[517] Yalcintas, M. and Dai, H., Magnetorheological and electrorheological materials in adaptive structures and their performance comparison, Smart Materials and Structures 8 (5), October 1999, 560-573.

[518] Xu, Y. L., Qu, W. L. and Ko, J. M., Seismic response control of frame structures using magnetorheological/electrorheological dampers, Earthquake Engineering Structural Dynamics 29 (5), April 2000, 557-575.

[519] Wang, Z., Fang, H., Lin, Z. and Zhou, L., Dynamic simulation studies of structural formation and transition in electro-magnetorheological fluids, International Journal of Modern Physics B 15 (6-7), 2001, 842-850.

[520] Butz, T. and Von Stryk, O., Modelling and simulation of electro-and magnetorheological fluid dampers, ZAMM Journal of Applied Mathematics and Mechanics 82 (1), 2002, 3-20.

[521] El Wahed, A. K., Sproston, J. L. and Schleyer, G. K., Electrorheological and magnetorheological fluids in blast resistant design applications, Materials and Design 23 (4), June 2002, 391-404.

[522] Choi, Y.-T. and Wereley, N. M., Comparative analysis of the time response of electrorheological and magnetorheological dampers using nondimensional parameters, Journal of Intelligent Material Systems and Structures 13 (7-8), July 2002, 443-451.

[523] Lee, D.-Y., Choi, Y.-T. and Wereley, N. M., Performance analysis of ER/MR impact damper systems using Herschel-Bulkley model, Journal of Intelligent Material Systems and Structures 13 (7-8), July-August 2002, 525-531.

[524] Dimock, G, Yoo, J.-H. and Wereley, N. M., Quasi-steady Bingham biplastic analysis of electrorheological and magnetorheological dampers, Journal of Intelligent Material Systems and Structures 13 (9), September 2002, 549-559.

[525] Widjaja, J. and Samali, B., Li, J., Electrorheological and magnetorheological duct flow in shear-flow mode using Herschel-Bulkley constitutive model, Journal of Engineering Mechanics 129 (12), November 2003, 1475-1477.

[526] Sims, N. D., Holmes, N. J. and Stanway, R., A unified modelling and model updating procedure for electrorheological and magnetorheological vibration dampers, Smart Materials and Structures 13 (1), December 2003, 100-121.

[527] Lindler, J., Choi, Y.-T. and Wereley, N. M., Double adjustable electrorheological and magnetorheological shock absorbers, International Journal of Vehicle Design 33 (1-3), 2003, 189-206.

[528] Rosenfeld, N. C. and Wereley, N. M., Volume-constrained optimization of magnetorheological and electrorheological valves and dampers, Smart Materials and Structures 13 (6), July 2004, 1303-1313.

[529] Gandhi, F. and Bullough, W. A., On the phenomenological modeling of electrorheological and magnetorheological fluid pre-yield behavior, Journal of Intelligent Material Systems and Structures 16 (3), March 2005, 237-248.

[530] Choi, Y. T., Cho, J. U., Choi, S. B. and Wereley, N. M., Constitutive models of electrorheological and

magnetorheological fluids using viscometers, Smart Materials and Structures 14 (5), October 2005, 1025-1034.

[531] Choi, Y. -T., Bitman, L. and Wereley, N. M., Nondimensional eyring analysis of electrorheological and magnetorheological dampers, Journal of Intelligent Material Systems and Structures 16 (5), 2005, 383-394.

[532] Yoo, J. -H. and Wereley, N. M., Nondimensional analysis of annular duct flow in ER/MR dampers, International Journal of Modern Physics Part B 19 (7-9), 2005, 1577-1583.

[533] Han, Y. M., Nguyen, Q. H., Choi, S. B. and Kim, K. S., Hysteretic behaviors of yield stress in smart ER/MR materials: experimental results, Key Engineering Materials 326-328, December 2006, 1459-1462.

[534] Wereley, N. M., Nondimensional Herschel-Bulkley analysis of magnetorheological and electrorheological dampers, Journal of Intelligent Material Systems and Structures 19 (3), March 2008, 257-268.

[535] Wereley, N., Quasi-steady Herschel-Bulkley analysis of magnetorheological and electrorheological dampers, Journal of Intelligent Material Systems and Structures 19 (3), 2008, 257-268.

[536] Goldasz, J. and Sapinski, B., Nondimensional characterization of flow-mode magnetorheological/electrorheological fluid dampers, Journal of Intelligent Material Systems and Structures 23 (14), September 2012, 1545-1562.

[537] Esteki, K., Bagchi, A. and Sedaghati, R., Dynamic analysis of electro-and magnetorheological fluid dampers using duct flow models, Smart Materials and Structures 23 (3), 2014, article ID 035016.

[538] Phulé, P. P., Ginder, J. M. and Jatkar, A. D., Synthesis and properties of magnetorheological mr fluids for active vibration control, Materials Research Society MRS Proceedings 459, 1996.

[539] Carlson, J. D., Catanzarite, D. M. and St. Clair, K. A., Commercial magnetorheological fluid devices, International Journal of Modern Physics B 10 (23&24), October 1996, 2857-2965.

[540] Ginder, J. M., Davis, L. C. and Elie, L. D., Rheology of magnetorheological fluids: models and measurements, International Journal of Modern Physics B 10 (23&24), October 1996, 3293-3303.

[541] Ashour, O. N., Kinder, D., Giurgiutiu, V. and Rogers, C. A., Manufacturing and characterization of magnetorheological fluids, Proceedings of the SPIE Conference on Smart Structures and Materials: Smart Materials Technologies, SPIE 3040, 1997. doi: 10.1117/12.267112, 174-184.

[542] Ginder, J. M., Behavior of magnetorheological fluids, Materials Research Society MRS Bulletin 23 (8), August 1998, 26-29.

[543] Phulé, P. P., Synthesis of novel magnetorheological fluids, Materials Research Society MRS Bulletin 23 (8), August 1998, 23-25.

[544] Jolly, M. R., Bender, J. W. and Carlson, J. D., Properties and applications of commercial magnetorheological fluids, Journal of Intelligent Material Systems and Structures 10 (1), January 1999, 5-13.

[545] Kamath, G. M., Wereley, N. M., Jolly, M. R., Characterization of magnetorheological helicopter lag dampers, Journal of the American Helicopter Society 44 (3), July 1999, 234-248.

[546] Wereley, N. M., Kamath, G. M. and Madhavan, V., Hysteresis modeling of semi-active magnetorheological helicopter lag dampers, Journal of Intelligent Material Systems and Structures 10 (8), August 1999, 624-633.

[547] Wang, D. H. and Liao, W. H., Application of MR dampers for semi-active suspension of railway vehi-

cles, Proceedings of the International Conference on Advances in Structural Dynamics, 2000, 1389-1396.

[548] Snyder, R., Kamath, G. M. and Wereley, N. M., Characterization and analysis of magnetorheological damper behavior under sinusoidal loading, AIAA Journal 39 (7), July 2001, 1240-1253.

[549] Bica, I., Damper with magnetorheological suspension, Journal of Magnetism and Magnetic Materials 241 (2-3), March 2002, 196-200.

[550] Li, W. H., Chen, G., Yeo, S. H. and Du, H., Dynamic properties of magnetorheological materials, Key Engineering Materials 227, August 2002, 119-124.

[551] Bossis, G., Lacis, S., Meunier, A. and Volkova, O., Magnetorheological fluids, Journal of Magnetism and Magnetic Materials 252, November 2002, 224-228.

[552] Bossis, G., Volkova, O., Lacis, S. and Meunier, A., Magnetorheology: Fluids, structures and rheology. In: Ferrofluids, Vol. 594 of the series Lecture Notes in Physics, Springer-Verlag, 2002, 202-230. doi: 10.1007/3-540-45646-5-11.

[553] Lais, C. Y. and Liao, W. H., Vibration control of a suspension system via a magnetorheological fluid damper, Journal of Vibration and Control 8, 2002, 527-547.

[554] Liaos, W. H. and Lai, C. Y., Harmonic analysis of a magnetorheological damper for vibration control, Smart Materials and Structures 11 (2), April 2002, 288-296.

[555] Yoos, J.-H. and Wereley, N. M., Design of a high-efficiency magnetorheological valve, Journal of Intelligent Material Systems and Structures 13 (10), 2002, 679-687.

[556] Bossisas, G., Khuzira, P., Lacisb, S. and Volkova, O., Yield behavior of magnetorheological suspensions, Journal of Magnetism and Magnetic Materials 258-259, 2003, 456-458.

[557] Lam, A. H. F. and Liao, W. H., Semi-active control of automotive suspension systems with magnetorheological dampers, International Journal of Vehicle Design 33 (1-3), 2003, 50-75.

[558] Liao, W. H. and Wang, D. H., Semiactive vibration control of train suspension systems via magnetorheological dampers, Journal of Intelligent Material Systems and Structures 14 (3), 2003, 161-172.

[559] Yoo, J. H., Sirohi, J. and Wereley, N. M., A Magnetorheological piezo hydraulic actuator, Journal of Intelligent Material Systems and Structures 16 (11-12), 2005, 945-954.

[560] Jean, P., Ohayon, R. and Le Bihan, D., Payload/Launcher vibration isolation: MR dampers modeling with fluid compressibility and inertia effects through continuity and momentum equations, International Journal of Modern Physics B 19 (7-9), 2005, 1534-1541.

[561] Lau, Y. K. and Liao, W. H., Design and analysis of magnetorheological dampers for train suspension, Journal of Rail and Rapid Transit 219 (4), 2005, 261-276.

[562] Brigley, M., Choi, Y.-T., Wereley, N. M. and Choi, S.-B., Magnetorheological isolators using multiple flow modes, Journal of Intelligent Material Systems and Structures 18 (12), 2007, 1143-1148.

[563] Guerrero-Sanchez, C., Lara-Ceniceros, T., Jimenez-Regalado, E., Ras, M. and Schubert, U. S., Magnetorheological fluids based on ionic liquids, Advanced Materials 19 (13), 1740-1747.

[564] Han, Y.-M., Choi, S.-B. and Wereley, N. M., Hysteretic behavior of magnetorheological fluid and identification using Preisach model, Journal of Intelligent Material Systems and Structures 18 (8), 2007, 973-981.

[565] John, S., Chaudhuri, A. and Wereley, N. M., A magnetorheological actuation system: test and model,

Smart Materials and Structures 17, March-April 2008, article ID 025023.

[566] Hong, S. -R. , Wereley, N. M. , Choi, Y. -T. and Choi, S. -B. , Analytical and experimental validation of a nondimensional bingham model for mixed mode magnetorheological dampers, Journal of Sound and Vibration 312 (3), May 2008, 399-417.

[567] Han, Y. M. , Kim, C. J. and Choi, S. B. , Design and control of magnetorheological fluid-based multifunctional haptic device for vehicle applications, Advanced Materials Research 47-50, June 2008, 141-144.

[568] Muc, A. and Barski, M. , Homogenization numerical analysis & optimal design of MR fluids, Advanced Materials Research 47-50, June 2008, 1254-1257.

[569] Hu, W. and Wereley, N. M. , Hybrid magnetorheological fluid elastomeric dampers for helicopter stability augmentation, Smart Materials and Structures 17 (4), August 2008, article ID 045021.

[570] Sapiński, B. and Snamina, J. , Modeling of an adaptive beam with MR fluid, Applied Mechanics and Materials 147-149, January 2009, 831-838.

[571] Mazlan, S. A. , Issa, A. , Chowdhury, H. A. and Olabi, A. G. , Magnetic circuit design for the squeeze mode experiments on magnetorheological fluids, Materials and Design 30 (6), June 2009, 1985-1993.

[572] Choi, Y. -T. and Wereley, N. M. , Self-powered magnetorheological dampers, ASME Journal Vibration and Acoustics 31 (4), 2009, article ID 044501.

[573] Zheng, L. , Li, Y. N. and Baz, A. , Fuzzy-sliding mode control of a full car semi-active suspension systems with MR dampers, Journal of Smart Structures & Systems 5 (3), 2009, 261-278.

[574] Zheng, G. X. , Huang, Y. J. and Gan, B. , The application of MR fluid damper to the vibrating screen as the enhance of screening efficiency, Advanced Materials Research 97-101 (March), 2010, 2628-2633.

[575] Park, B. J. , Fang, F. F. and Choi, H. J. , Magnetorheology: materials and application, Soft Matter 6, June 2010, 5246-5253.

[576] Han, K. , Feng, Y. T. and Owen, D. R. J. , Three-dimensional modelling and simulation of magnetorheological fluids, International Journal Numerical Methods in Engineering 84 (11), November 2010, 1273-1302.

[577] Wang, D. H. , Bai, X. X. and Liao, W. H. , An integrated relative displacement self-sensing magnetorheological damper: prototyping and testing, Smart Materials and Structures 19 (10), October 2010, article ID 105008.

[578] Snamina, J. , Energy dissipation in three-layered plate with magnetorheological fluid, Solid State Phenomena 177, July 2011, 143-150.

[579] Mazlan, S. A. , Ismail, I. , Fathi, M. S. , Rambat, S. and Anis, S. F. , An experimental investigation of magnetorheological MR fluids under quasi-static loadings, Key Engineering Materials 495, November 2011, 285-288.

[580] Zhao, D. M. and Liu, X. P. , Magnetorheological fluid test and application, Advanced Materials Research 395-398, November 2011, 2158-2161.

[581] DeVicente, J. , Klingenberg, D. J. and Hidalgo-Alvareza, R. , Magnetorheological fluids: a review, Soft Matter, 2011, 3701-3710.

[582] Kothera, C. S. , Ngatu, G. T. and Wereley, N. M. , Control evaluations of a magnetorheological fluid elas-

tomeric (MRFE) lag damper for helicopter rotor blades, AIAA Journal of Guidance Control and Dynamics 34 (4), 2011, 1143-1156.

[583] Wang, D. H. and Liao, W. H., Magnetorheological fluid dampers: a review of parametric modeling, Smart Materials and Structures 20 (2), February 2011, article ID 023001.

[584] Wereley, N. M., Choi, Y.-T. and Singh, H., Adaptive energy absorbers for drop-induced shock mitigation, Journal of Intelligent Material Systems and Structures 22 (6), 2011, 515-519.

[585] Choi, S. B. and Nguyen, Q. H., Selection of magnetorheological brakes via optimal design considering maximum torque and constrained volume, Smart Materials and Structures 21 (1), December 2011, article ID 015012.

[586] Chen, C. and Liao, W.-H., A self-sensing magnetorheological damper with power generation, Smart Materials and Structures 21 (2), January 2012, article ID 025014.

[587] Rodríguez-López, J., Segura, L. E. and De Espinosa Freijo, F. M., Ultrasonic velocity and amplitude characterization of magnetorheological fluids under magnetic fields, Journal of Magnetism and Magnetic Materials 324 (2), January 2012, 222-230.

[588] Zhu, S. X., Liu, X. and Ding, L., Modeling and analysis of magnetorheological fluid damper under impact load, Advanced Materials Research 452-453, January 2012, 1481-1485.

[589] Asthana, C. B. and Bhat, R. B., A novel design of landing gear oleo strut damper using MR fluid for aircraft and UAV's, Applied Mechanics and Materials 225, November 2012, 275-280.

[590] Wang, X. L., Hao, W. J. and Li, G. F., Study on magnetorheological fluid damper, Applied Mechanics and Materials 405-408, September 2013, 1153-1156.

[591] Rajamohan, V., Sundararaman, V. and Govindarajan, B., Finite element vibration analysis of a magnetorheological fluid sandwich beam, Procedia Engineering 64, 2013, 603-612.

[592] Yazid, I. I. M., Mazlan, S. A., Kikuchi, T., Zamzuri, H. and Imaduddin, F., Design of magnetorheological damper with a combination of shear and squeeze modes, Materials and Design 54, February 2014, 87-95.

[593] Singh, H. J., Hu, W., Wereley, N. M. and Glass, W., Experimental validation of a magnetorheological energy absorber optimized for shock and impact loads, Smart Materials and Structures 23 (12), April 2014, article ID 125033.

[594] Ismail, I. and Aqida, S. N., Fluid-particle separation of magnetorheological (MR) fluid in MR machining application, Key Engineering Materials 611-612, May 2014, 746-755.

[595] Ashtiani, M., Hashemabadi, S. H. and Ghaffari, A., A review on the magnetorheological fluid preparation and stabilization, Journal of Magnetism and Magnetic Materials 374, January 2015, 716-730.

[596] Becnel, A. C., Sherman, S. G., Hu, W. and Wereley, N. M., Nondimensional scaling of magnetorheological rotary shear mode devices using the mason number, Journal of Magnetism and Magnetic Materials 380, April 2015, 90-97.

[597] Sherman, S. G., Becnel, A. C. and Wereley, N. M., Relating mason number to bingham number in magnetorheological fluids, Journal of Magnetism and Magnetic Materials 380 (April), 2015, 98-104.

[598] Sapinski, B. and Goldasz, J., Development and performance evaluation of an MR squeezemode damper, Smart Materials and Structures 24 (11), November 2015, article ID 115007.

[599] Morrilas, J. R., González, E. C. and De Vicente, J., Effect of particle aspect ratio in magnetorheology,

Smart Materials and Structures 24 (12), December 2015, article ID 125005.

[600] Von Rensburg, R. J., Hunberstone, V., Close, J. A., Tavernert, A. W., Stevens, R., Greenough, R. D., Connor, K. P. and Gee, M. G., Review of Constitutive Description and measurement Methods for Piezoelectric, Electrostrictive and Magnetostrictive Materials, © Crown Copyright 1994, ISSN 0959 2423, National Physical Laboratory, Teddington, Middlesex, United Kingdom, TW11 0LW, July 1994, 44 pp.

[601] Calkins, F. T., Flatau, A. B. and Dapino, M. J., An Overview of Magnetostrictive Sensor Applications, Proceedings of the Structural Dynamics and Mechanics Conference, AIAA Paper 99-1551, April 1999.

[602] Dapino, M. J., Magnetostrictive Materials, Encyclopedia of Smart Materials, Schwartz, M., (ed.), New York, John Wiley and Sons, 2002, 600-620.

[603] Clark, A. E., Savage, H. T. and Spano, M. L., Effect of stress on magnetostriction and magnetization of single crystal, IEEE Transactions on Magnetics 20, 1984, 1443-1445.

[604] Butler, J. L., Butler, S. C. and Clark, A. E, Unidirectional magnetostrictive/piezoelectric hybrid transducer, Journal of the Acoustical Society of America 88 (1), 1990, 7-11.

[605] Flatau, A., Hall, D. L. and Schlesselman, J. M., magnetostrictive active vibration control systems, Proceedings of the AIAA 30th aerospace science meeting: recent advances in adaptive and sensory materials andtheir applications, Paper No. 92-0490, 1992, 419-429.

[606] Hall, D. L. and Flatau, A. B., Nonlinearities Harmonics and Trends in Dynamic Applications of Terfenol-D, Proceedings of the SPIE Conference on Smart Structures and Intelligent Systems, SPIE 1917, February 1993, 929-939.

[607] Pratt, J. and Flatau, A. B., Collocated Sensing and Actuation using magnetostrictive materials, Proceedings of the SPIE Conference on Smart Structures and Intelligent Systems, 1917, February 1993, 952-961.

[608] Hall, D. L. and Flatau, A. B., One-dimensional analytical constant parameter linear electromagnetic-magnetomechanical models of a cylindrical magnetostrictive terfenol-D transducer, Proceedings of the Second International Conference on Intelligent Materials, July 1994, 605-616.

[609] Hall, D. L. and Flatau, A. B, Broadband performance of a magnetostrictive shaker, active control of noise and vibration, Journal of Intelligent Material Systems and Structures 6 (1), January 1995, 109-116.

[610] Bothwell, C. M., Chandra, R. and Chopra, I., Torsional actuation with extension-torsional composite coupling and magnetostrictive actuators, AIAA Journal 33 (4), April 1995, 723-729.

[611] Calkins, F. T. and Flatau, A. B., Transducer Based Measurement of Terfenol-D Material Properties, Proceedings of the SPIE Conference on Smart Structures and Integrated Systems, SPIE 2717, February 1996.

[612] Dapino, M. J., Calkins, F. T., Hall, D. L., Flatau, A. B., Measured Terfenol-D material properties under varied operating conditions, Proceedings of the SPIE Conference on Smart Structures and Integrated Systems, SPIE 2717, February 1996, 697-708.

[613] Calkins, F. T., Flatau, A. B. and Hall, D. L., Characterization of the Dynamic Material Properties of Magnetostrictive Terfenol-D, Proceedings of the 4th Annual Workshop: Advances in Smart Materials for Aerospace Applications, NASA Technical Report Id 19960047691, March 1996.

[614] Calkins, F. T. and Flatau, A. B, Experimental evidence for maximum efficiency operation of a magnetos-

trictive transducer, Journal of the Acoustical Society of America 99 (4), April 1996, 2536-2536.

[615] Calkins, F. T., Dapino, M. J. and Flatau, A. B., Effect of Prestress on the Dynamic Performance of a Terfenol-D Transducer, Proceedings of the SPIE Conference on Smart Structures and Materials, SPIE 3041, March 1997, 293-304.

[616] Dapino, M. J., Calkins, F. T. and Flatau, A. B., Statistical Analysis of Terfenol-D Material Properties, Proceedings of the SPIE Conference on Smart Structures and Materials, SPIE 3041, March 1997, 256-267.

[617] Body, C., Reyne, G. and Meunier, G., Nonlinear finite element modelling of magnetomechanical phenomena in giant magnetostrictive thin films, IEEE Transactions on Magnetics 33 (2), 1997, 1620-1623.

[618] Snodgrass, J., Calkins, F. T., Dapino, M. J. and Flatau, A. B., Terfenol-D material property study, Journal of the Acoustical Society of America 101 (5), May 1997, 3094-3094.

[619] Giurgiutiu, V., Jichi, F., Quattrone, R. F. and Berman, J. B., Experimental study of magnetostrictive tagged composite strain sensing response for structure health monitoring, Proceedings of the International Workshop of Structural Health Monitoring, 1999, 8-10.

[620] Dapino, M. J., Smith, R. C., Faidley, L. E. and Flatau, A. B., Coupled structural-magnetic strain and stress model for magnetostrictive transducers, Journal of Intelligent Material Systems and Structures 11 (2000), 135-152.

[621] Butler, S. C. and Tito, F. A., A broadband hybrid magnetostrictive/piezoelectric transducer array, Oceans, 2000 MTS/IEEE, 3, 2000, 1469-1475.

[622] Reddy, J. N. and Barbosa, J. I, On vibration suppression of magnetostrictive beams, Smart Materials and Structures 9 (1), 2000, 49-58.

[623] Duenas, T. A. and Carman, G. P, Large magnetostrictive response of terfenol-d resin composites, Journal of Applied Physics 87 (9), 2000, 4696-4701.

[624] Clark, A. E., Wun-Fogle, M., Restorff, J. B., Lograsso, T. A., Ross, A. R. and Schlagel, D. L., Magnetostrictive galfenol/alfenol single crystal alloys under large compressive stresses, Proceedings of the 7th International Conference on New Actuators, Borgmann, H. (ed.), Bremen, Germany, 2000, 111-115.

[625] Ludwig, A. and Quandt, E., Giant magnetostrictive thin films for applications in microelectromechanical systems, Journal of Applied Physics 87 (9), 2000, 4691-4695.

[626] Flatau, A. and Kellogg, R. A., Magnetostrictive transducer performance characterization, The Journal of the Acoustical Society of America 109 (5), 2001, 2434-2434.

[627] Giurgiutiu, V., Jichi, F., Berman, J. B. and Kamphaus, J. M., Theoretical and experimental investigation of magnetostrictive composite beams, Smart Materials and Structures 10 (5), 2001, 934-945.

[628] Smith, R. C., Inverse compensation for hysteresis in magnetostrictive transducers, Mathematical and Computer Modelling 33, 2001, 285-298.

[629] Kakeshita, T. and Ullakko, K., Giant magnetostriction in ferromagnetic shape-memory alloys, MRS Bulletin, February 2002, 105-109.

[630] Smith, R. C., Dapino, M. J. and Seelecke, S., Free energy model for hysteresis in magnetostrictive transducers, Journal of Applied Physics 93, 2003, 458-466.

[631] Pomirleanu, R. and Giurgiutiu, V., High-field characterization of piezoelectric and magnetostrictive actuators, Journal of Intelligent Material Systems and Structures 15 (3), 2004, 161-180.

[632] Dapino, M. J., On magnetostrictive materials and their use in adaptive structures, International Journal of Structural Engineering and Mechanics 17 (3-4), 2004, 303-329.

[633] Tan, X. and Baras, J. S., Modeling and control of hysteresis in magnetostrictive actuators, Automatica 40 (9), 2004, 1469-1480.

[634] Tzou, H. S., Chai, W. K. and Hanson, M., Dynamic actuation and quadratic magnetoelastic coupling of thin magnetostrictive shells, ASME Journal of Vibration and Acoustics 128 (3), June 2006, 385-391.

[635] Chai, W. K., Tzou, H. S., Arnold, S. M. and Lee, H.-J., Magnetostrictive micro-actuations and modal sensitivities of thin cylindrical magnetoelastic shells, Journal of Pressure Vessel Technology 130 (1), February 2008, article ID 011206.

[636] Chaudhuri, A., Yoo, J. H. and Wereley, N. M., Design, rtest and model of a hybrid magnetostrictive hydraulic actuator, Smart Materials and Structures 18 (8), 2009, article ID 085019.

[637] Damjanovic, D. and Newnham, R. E., Electrostrictive and piezoelectric materials for actuator applications, Journal of Intelligent Material Systems and Structures 3, 1992, 190-208.

[638] Pilgrim, S. M., Massuda, M., Prodey, J. D. and Ritter, A. P., Electrostrictive Sonar Drivers for Flextensional Transducers, Transducers for Sonics and Ultrasonics, McCollum, M., Hamonic, B. F. and Wilson, O. B., (eds.), Lancaster, PA, Technomic, 1993.

[639] Hom, C. L. and Shankar, N., A Fully Coupled, Constitutive model for electrostrictive ceramic materials, Journal of Intelligent Material Systems and Structures 5, 1994, 795-801.

[640] Hom, C. L., Dean, P. D. and Winzer, S. R., Simulating electrostrictive deformable mirrors: i. nonlinear static analysis, Smart Materials and Structures 8, 1999, 691-699.

[641] Piquette, J. C. and Forsythe, S. E., Generalized material model for Lead Magnesium Niobate (PMN) and an associated electromechanical equivalent circuit, Journal of the Acoustical Society of America 104 (1998), 2763-2772.

[642] Hom, C. L., Simulating electrostrictive deformable mirrors: II, Nonlinear Dynamic Analysis, Smart Materials and Structures 8, 1999, 700-708.

[643] Chaudhuri, A. and Wereley, N. M., Experimental validation of a hybrid electrostrictive hydraulic actuator analysis, ASME Journal of Vibration and Acoustics 132 (2), 2010, article ID 021006.

[644] Pablo, F., Osmont, D. and Ohayon, R., A Plate Electrostrictive, Finite element-Part I: Modeling and variational formulations, Journal of Intelligent Material Systems and Structures 12 (11), 2001, 745-759.

[645] Pablo, F., Osmont, D. and Ohayon, R., Modeling of plate structures equipped with current driven electrostrictive actuators for active vibration control, Journal of Intelligent Material Systems and Structures 14 (3), 2003, 173-183.

[646] Tzou, H. S., Chai, W. K. and Arnold, S. M, Structronics and actuation of hybrid electrostrictive/ piezoelectric thin shells, ASME Journal of Vibration and Acoustics 128, February 2006, 79-87.

[647] Chai, W. K. and Tzou, H. S., Design and testing of a hybrid electrostrictive/piezoelectric polymeric beam with bang-bang control, Journal of Mechanical Systems and Signal Processing 21 (2007), 417-429.

[648] John, S., Sirohi, J., Wang, G. and Wereley, N. M., Comparison of piezoelectric, magnetostrictive and

## 第 1 章　智能材料与结构概述

electrostrictive hybrid hydraulic actuators, Journal Intelligent Material Systems and Structures 18 (10), 2007, 1035-1048.

[649] Baz, A, Magnetic constrained layer damping, Proceedings of the 11th Conference on Dynamics & Control of Large Structures, Blacksburg, May 1997, 333-344.

[650] Oh, J., Ruzzene, M. and Baz, A., Control of the dynamic characteristics of passive magnetic composites, Composites Part B: Engineering 30 (7), October 1999, 739-751.

[651] Oh, J., Poh, S., Ruzzene, M. and Baz, A., Vibration control of beams using electromagnetic compressional damping treatment, ASME Journal of Vibration & Acoustics 122 (3), 2000, 235-243.

[652] Ebrahim, A. and Baz, A., Vibration control of plates using magnetic constrained layer damping, Journal of Intelligent Material Systems and Structures 11 (10), October 2000, 791-797.

[653] Baz, A. and Poh, S., Performance characteristics of magnetic constrained layer damping, Journal of Shock & Vibration 7 (2), 2000, 18-90.

[654] Omer, A. and Baz, A., Vibration control of plates using electromagnetic damping treatment, Journal of Intelligent Material Systems & Structures 11 (10), 2000, 791-797.

[655] Ruzzene, M., Oh, J. and Baz, A., Finite element modeling of magnetic constrained layer damping, Journal of Sound & Vibration 236 (4), 2000, 657-682.

[656] Boyd, J., Lagoudas, D.C. and Seo, C., Arrays of micro-electrodes and electromagnets for processing of eectro-magneto-elastic multifunctional composite materials, Proceedings of SPIE's 10th Annual International Symposium on Smart Structures and Materials, San Diego, SPIE 5053, March 2003, 70-80.

[657] Sodano, H.A., Inman, D.J. and Belvin, W.K, Development of a new passive-active magnetic damper for vibration suppression, ASME Journal of Vibration and Acoustics 128 (3), 2006, 318-327.

[658] Joyce, B., Development of an Electromagnetic Energy Harvester for Monitoring Wind Turbine Blades, Master's Thesis at the Virginia Polytechnic Institute and State University, 10919/36354, December 2011.

[659] Han, Y., Mechanics of magneto-active polymers, Ph.D. Thesis at the Iowa State University, Thesis No. 12929, 2012.

[660] Cottone, F., Basset, P., Vocca, H., Gammaitoni, L. and Bourouina, T., Bistable electromagnetic generator based on buckled beams for vibration energy harvesting, Journal of Intelligent Material Systems and Structures, October 2013, 1484-1495. doi: 10.1177/1045389X13508330.

[661] Hobeck, J.D. and Inman, D.J., Magnetoelastic Metastructures for Passive Broadband Vibration Suppression, Proceedings of SPIE's Smart Structures & NDE Conference, San Diego, Paper No. 9431-43, March 2015.

[662] Gonzalez-Buelga, A., Clare, L.R., Cammarano, A., Neild, S.A., Burrow, S.G. and Inman, D.J., An electromagnetic vibration absorber with harvesting and tuning capabilities, Structural Control and Health Monitoring 22 (11), 2015. doi: 10.1002/stc.1748.

[663] Shen, W. and Zhu, S., Harvesting energy via electromagnet damper: application to bridge stay cables, Journal of Intelligent Material Systems and Structures 26 (1), January 2015, 156-171.

[664] Salas, E. and Bustamante, R., Numerical solution of some boundary value problems in nonlinear magnetoelasticity, Journal of Intelligent Material Systems and Structures 26 (2), January 2015, 3-19.

[665] Farjoud, A. and Bagherpour, E.A., Electromagnet design for magnetorheological devices, Journal of In-

telligent Material Systems and Structures 27 (1), January 2016, 51-70.

[666] Piezoelectric films-Technical information brochure, issued by PIÉZOTECH S. A. S., www.piezo tech.fr.

[667] Wadley, H. N. G., Characteristics and processing of smart materials, Paper presented at the AGARD SMP lecture Series on "Smart Structures and Materials: Implications for Military Aircraft of New Generation," held in Philadelphia, USA, 30-31 October 1996.

[668] Shahinpoor, M., Bar-Cohen, Y., Xue, T., Simpson, J. O. and Smith, J., Ionic polymer-metal composites (IPMC) as biomimetic sensors and actuators-artificial muscles, Proc. Of the SPIE's 5th Annual International Symposium on Smart Structures and materials, 1-5 March, 1998, San Diego, CA. Paper # 3324-3327.

[669] Chopra, I., Review of state of art of smart structures and integrated systems, AIAA Journal 40 (11), 2002, 2145-2187.

[670] Tzou, H. S., Lee, H. -J. and Arnold, S. M., Smart Materials, Precision Sensors/Actuators, Smart Structures, and Structronic Systems, NASA Publications, Paper 200, 2004. http: //digitalcom mons.unl.edu/nasapub/200.

[671] Claeyssen, F., Hermet, N. L., Barillot, F. and Le Letty, R., Giant dynamic strains in magnetostrictive actuators and transducers, paper presented at ISAGMM 2006 Conference, October 2006, China.

[672] Shaikh, A. M., Smart materials and its applications, International Journal of Emerging Technologies in Computational and Applied Sciences (IJETCAS) 12 (2), 2015, 193-197.

[673] Barbarino, S., Bilgen, O., Ajaj, R. M., Friswell, M. I. and Inman, D. J., A review of morphing aircraft, Journal of Intelligent Material Systems and Structures 22 (9), 2011, 823-877.

[674] Joo, J. J., Reich, G. W. and Westfall, J. T, Flexible skin development for morphing aircraft applications via topology optimization, Journal of Intelligent Material Systems and Structures 20 (16), 2009, 1969-1985.

[675] Bubert, E. A., Woods, B. K. S., Lee, K., Kothera, C. S. and Wereley, N. M., Design and fabrication of a passive 1D morphing aircraft skin, Journal of Intelligent Material Systems and Structures 21 (17), 2010, 1699-1717.

[676] De-Breuker, R., Abdalla, M. M. and Gurdal, Z, A generic morphing wing analysis and design framework, Journal of Intelligent Material Systems and Structures 22 (10), 2011, 1025-1039.

[677] Lesieutre, G. A., Browne, J. A. and Frecker, M. I., Scaling of performance, weight, and actuation of a 2-D compliant cellular frame structure for a morphing wing, Journal of Intelligent Material Systems and Structures 22 (10), 2011, 979-986.

[678] Min, Z., Kien, V. K. and Richard, L. J. Y., Aircraft morphing wing concepts with radical geometry change, The IES Journal Part A: Civil & Structural Engineering 3 (3), 2010, 188-195.

[679] Pecora, R., Barbarino, S., Concilio, A., Lecce, L. and Russo, S., Design and functional test of a morphing high-lift device for a regional aircraft, Journal of Intelligent Material Systems and Structures 22 (10), 2011, 1005-1023.

[680] Previtali, F., Arrieta, A. F. and Ermanni, P., Double-walled corrugated structure for bendingstiff anisotropic morphing skins, Journal of Intelligent Material Systems and Structures 26 (5), 2015, 599-613.

[681] Tadesse, Y., Zhang, S. and Priya, S., Multimodal energy harvesting system: piezoelectric an Electromagnetic, Journal of Intelligent Material Systems and Structures 20 (5), 2009, 625-632.

# 第1章 智能材料与结构概述

[682] Liao, Y. and Sodano, A., Structural effects and energy conversion efficiency of power harvesting, Journal of Intelligent Material Systems and Structures 25 (14), 2014, 505-514.

[683] Sun, C., Qin, L., Li, F. and Wang, Q-M., Piezoelectric energy harvesting using single crystal Pb (Mg1/3Nb2/3) O3-xPbTiO3 (PMN-PT) device, Journal of Intelligent Material Systems and Structures 20 (5), 2009, 559-568.

[684] Liu, Y., Tian, G., Wang, Y., Lin, J., Zhang, Q. and Hofmann, H. F., Active piezoelectric energy harvesting: general principle and experimental demonstration, Journal of Intelligent Material Systems and Structures 20 (5), 2009, 575-585.

[685] Erturk, A., Renno, J. M. and Inman, D. J., Modeling of piezoelectric energy harvesting from an L-shaped beam-mass structure with an application to UAVs, Journal of Intelligent Material Systems and Structures 20 (5), 2009, 529-544.

[686] Guyomar, D., Sebald, G., Pruvost, S., Lallart, M., Khodayari, A. and Richard, C., Energy harvesting from ambient vibrations and heat, Journal of Intelligent Material Systems and Structures 20 (5), 2009, 609-624.

[687] Farinholt, K. M., Pedrazas, N. A., Schluneker, D. M., Burt, D. W. and Farrar, C. R, An energy harvesting comparison of piezoelectric and ionically conductive polymers, Journal of Intelligent Material Systems and Structures 20 (5), 2009, 633-642.

[688] Olivier, J. M. and Priya, S., Design, fabrication, and modeling of a four-bar electromagnetic vibration power generator, Journal of Intelligent Material Systems and Structures 21 (16), 2010, 1303-1316.

[689] Aldraihem, O. and Baz, A., Energy harvester with a dynamic magnifier, Journal of Intelligent Material Systems and Structures 22 (6), 2011, 521-530.

[690] Zoric, N. D., Simonovic, A. M., Mitrovic, Z. S. and Stupar, S. N, Optimal vibration control of smart composite beams with optimal size and location of piezoelectric sensing and actuation, Journal of Intelligent Material Systems and Structures 24 (4), 2012, 499-526.

[691] Tang, L., Yang. Y. and Soh, C.-K., Improving functionality of vibration energy harvesters using magnets, Journal of Intelligent Material Systems and Structures 23 (13), 2012, 1433-1449.

[692] Elvin, N. G. and Elvin, A. A, Large deflection effects in flexible energy harvesters, Journal of Intelligent Material Systems and Structures 23 (13), 2012, 1475-1484.

[693] Mikoshiba, K., Manimala, J. M. and Sun, C. T., Energy harvesting using an array of multifunctional resonators, Journal of Intelligent Material Systems and Structures 24 (2), 2012, 168-179.

[694] Zhang, S., Lebrun, L., Randall, C. A. and Shrout, T. R., Growth and electrical properties of (Mn, F) Co-doped 0.92Pb ($Zn_{1/3}Nb_{2/3}$) $O_3$-0.08$PbTiO_3$ single crystal, Journal of Crystal Growth 267, 2004, 204-212.

[695] Green, P. L., Papatheou, E. and Sims, N. D., Energy harvesting from human motion and bridge vibrations: an evaluation of current nonlinear energy harvesting solutions, Journal of Intelligent Material Systems and Structures 24 (12), 2013, 1494-1505.

[696] Xiong, X. and Oyadiji, S. O., Modal electromechanical optimization of cantilevered piezoelectric vibration energy harvesters by geometric variation, Journal of Intelligent Material Systems and Structures 25 (10), 2014, 1177-1195.

[697] Anton, S. R., Farinholt, K. M. and Erturk, A., Piezoelectret foam-based vibration energy harvesting,

Journal of Intelligent Material Systems and Structures 25 (14), 2014, 1-12.

[698] Stoppa, M. and Chiolerio, A., Wearable electronics and smart textiles: a critical review, Sensors 14, 2014, 11.957-11.992. doi: 10.3390/S140711957.

[699] Yu, L. and Leckey, C. A. C., Lamb wave-based quantitative crack detection using a focusing array algorithm, Journal of Intelligent Material Systems and Structures 24 (9), 2012, 1138-1152.

[700] Annamdas, V. G. M. and Radhika, M. A, Electromechanical impedance of piezoelectric transducers for monitoring metallic and non-metallic structures: a review of wired, wireless and energy-harvesting methods, Journal of Intelligent Material Systems and Structures 24 (9), 2013, 1021-1042.

[701] Spaggiari, A., Dragoni, E. and Tuissi, A., Experimental characterization and modelling validation of shape memory alloy Negator springs, Journal of Intelligent Material Systems and Structures 26 (6), 2015, 619-630.

[702] Ghaffari, A., Hashemabadi, H. and Ashtiani, M., A review on the simulation and modeling of magnetorheological fluids, Journal of Intelligent Material Systems and Structures 26 (8), 2015, 881-904.

[703] Meisel, N., Elliott, A. M. and Williams, C. B., A procedure for creating actuated joints via embedding shape memory alloys in PolyJet 3D printing, Journal of Intelligent Material Systems and Structures 26 (12), 2015, 1498-1512.

# 第 2 章 层压复合材料

## 2.1 经典层压理论

### 2.1.1 简介

复合材料由两种成分组成：一种是纤维（增强材料），另一种是胶结成分（基体）。复合材料的定义之一指出这两种材料的结合将产生比单一组分单独使用时更好的性能。复合材料与其他现有材料（如金属或塑料）相比，主要优势是高强度、高刚度和低密度，可以减轻成品部件的重量。在本章中，复合材料指的是基体为胶结成分，其中嵌入连续增强纤维的材料。这种连续增强纤维的材料包括单向纤维、编织布和螺旋缠绕（图 2.1（a）），在文献中的名称如图 2.1（b）中的框图所示。一些最常用的连续纤维的特性见本章的附录 A。连续纤维复合材料通常是通过将不同方向的连续纤维堆叠成层压板，以获得所需的强度和刚度性能，纤维体积高达 60%~70%。由于纤维直径小，可用于生产高强度复合材料；与批量生产的材料相比，纤维包含的缺陷（通常是表面缺陷）要少得多。此外，由于纤维直径小，其具有柔韧性，适用于复杂的制造工艺，如小半径或编织产品。可用于生产纤维的材料包括玻璃、石墨、碳或芳纶等。基体的主要材料是聚合物，具有强度和刚度较低的特性。基体的主要功能是使纤维保持正确的方向和间距，并保护纤维免受磨损和环境的影响。在聚合物-基体复合材料中，基体和增强材料之间良好而牢固的结合可使基体通过界面处的剪切载荷将外部载荷从基体传递到纤维。

聚合物基体有热固性塑料和热塑性塑料两种类型。热固性塑料最初是一种低黏度树脂，在加工过程中会发生反应和固化，形成固体。热塑性塑料是一种高黏度树脂，通过将其加热到熔融温度以上进行加工。热固性树脂在加工过程中会凝固和固化，因此不能通过重新加热进行再加工。热塑性塑料可以在其熔化温度以上重新加热以进行额外加工。

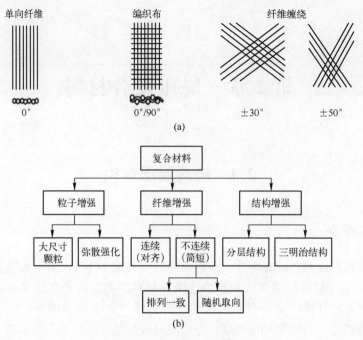

图 2.1 复合材料的类型

## 2.1.2 位移

在力的作用下,材料上的每一点都会发生变形。图 2.2 中描述坐标系的位移可写为

$$u = \{u_1, u_2, u_3\} \tag{2.1}$$

## 2.1.3 应变

线性应变张量 $\varepsilon_{ij}$ 的定义为

$$\varepsilon_{ij} = \frac{1}{2}(u_{i,j} + u_{j,i}), \quad i,j = 1,2,3$$

$$\varepsilon_{ij} = \varepsilon_{ji} \tag{2.2}$$

在扩展形式中,可以将应变写为

$$\varepsilon_{11} = \frac{\partial u_1}{\partial x_1} \equiv \varepsilon_1, \quad \varepsilon_{22} = \frac{\partial u_2}{\partial x_2} \equiv \varepsilon_2, \quad \varepsilon_{33} = \frac{\partial u_3}{\partial x_3} \equiv \varepsilon_3$$

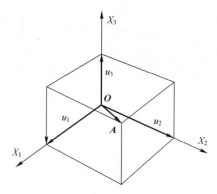

图 2.2　坐标系和位移的 3 个分量 $u_1$、$u_2$ 和 $u_3$

（矢量 $OA$ 是 3 个位移的合成，其大小为 $OA=\sqrt{x_1^2+x_2^2+x_3^2}$）

$$2\varepsilon_{23}=2\varepsilon_{32}=\gamma_4=\left(\frac{\partial u_2}{\partial x_3}+\frac{\partial u_3}{\partial x_2}\right)\equiv\varepsilon_4$$

$$2\varepsilon_{13}=2\varepsilon_{31}=\gamma_5=\left(\frac{\partial u_1}{\partial x_3}+\frac{\partial u_3}{\partial x_1}\right)\equiv\varepsilon_5$$

$$2\varepsilon_{12}=2\varepsilon_{21}=\gamma_6=\left(\frac{\partial u_1}{\partial x_2}+\frac{\partial u_2}{\partial x_1}\right)\equiv\varepsilon_6 \quad (2.3)$$

## 2.1.4　应力

应力张量可以使用与应变相同的符号来表示，即

$$[\sigma]=\begin{bmatrix}\sigma_{11}&\sigma_{12}&\sigma_{13}\\\sigma_{12}&\sigma_{22}&\sigma_{23}\\\sigma_{13}&\sigma_{23}&\sigma_{33}\end{bmatrix}=\begin{bmatrix}\sigma_1&\sigma_6&\sigma_5\\\sigma_6&\sigma_2&\sigma_4\\\sigma_5&\sigma_4&\sigma_3\end{bmatrix}$$

$$\boldsymbol{\sigma}_{ij}=\boldsymbol{\sigma}_{ji} \quad (2.4)$$

注意，在文献中有时为了区分法向应力（$\boldsymbol{\sigma}_{ii}$）和剪切应力（$\boldsymbol{\sigma}_{ij}$，$i\neq j$），剪切应力用 $\tau_{ij}$ 表示。

## 2.1.5　正交各向异性材料

具有两个垂直对称平面的均匀线性弹性材料称为正交各向异性材料。其本构关系为

$$\begin{Bmatrix} \varepsilon_{11} \\ \varepsilon_{22} \\ \varepsilon_{33} \\ \gamma_{23} \\ \gamma_{13} \\ \gamma_{12} \end{Bmatrix} = \begin{bmatrix} \dfrac{1}{E_1} & -\dfrac{v_{21}}{E_2} & -\dfrac{v_{31}}{E_3} & 0 & 0 & 0 \\ -\dfrac{v_{12}}{E_1} & \dfrac{1}{E_2} & -\dfrac{v_{32}}{E_3} & 0 & 0 & 0 \\ -\dfrac{v_{13}}{E_1} & -\dfrac{v_{23}}{E_2} & \dfrac{1}{E_3} & 0 & 0 & 0 \\ 0 & 0 & 0 & \dfrac{1}{G_{23}} & 0 & 0 \\ 0 & 0 & 0 & 0 & \dfrac{1}{G_{13}} & 0 \\ 0 & 0 & 0 & 0 & 0 & \dfrac{1}{G_{12}} \end{bmatrix} \begin{Bmatrix} \sigma_{11} \\ \sigma_{22} \\ \sigma_{33} \\ \sigma_{23} \\ \sigma_{13} \\ \sigma_{12} \end{Bmatrix} \quad (2.5)$$

式中：$E_1$、$E_2$、$E_3$ 为纵向弹性模量；$G_{23}$、$G_{13}$、$G_{12}$ 为剪切弹性模量；$v_{12}$、$v_{13}$、$v_{23}$、$v_{21}$、$v_{31}$ 和 $v_{32}$ 为泊松比。同样，由于柔度矩阵的对称性，以下关系成立：

$$\frac{v_{21}}{E_2} = \frac{v_{12}}{E_1}, \quad \frac{v_{31}}{E_3} = \frac{v_{13}}{E_1}, \quad \frac{v_{32}}{E_3} = \frac{v_{23}}{E_2} \quad (2.6)$$

应变和应力之间的关系可以简写为

$$\{\varepsilon\} = [S]\{\sigma\} \quad (2.7)$$

式中：矩阵 $[s]$ 称为柔度矩阵。如果把应力写成应变的函数，有以下表达式：

$$\{\sigma\} = [c]\{\varepsilon\} \quad (2.8)$$

式中：$[c]$ 为刚度矩阵，$[c] = [s]^{-1}$。

用刚度矩阵表示的式（2.5）具有以下形式：

$$\begin{Bmatrix} \sigma_{11} \\ \sigma_{22} \\ \sigma_{33} \\ \sigma_{23} \\ \sigma_{13} \\ \sigma_{12} \end{Bmatrix} = \begin{bmatrix} c_{11} & c_{12} & c_{13} & 0 & 0 & 0 \\ c_{12} & c_{22} & c_{23} & 0 & 0 & 0 \\ c_{13} & c_{23} & c_{33} & 0 & 0 & 0 \\ 0 & 0 & 0 & c_{44} & 0 & 0 \\ 0 & 0 & 0 & 0 & c_{55} & 0 \\ 0 & 0 & 0 & 0 & 0 & c_{66} \end{bmatrix} \begin{Bmatrix} \varepsilon_{11} \\ \varepsilon_{22} \\ \varepsilon_{33} \\ \gamma_{23} \\ \gamma_{13} \\ \gamma_{12} \end{Bmatrix} \quad (2.9)$$

刚度矩阵中的各项定义为

$$\begin{cases} c_{11} = \dfrac{1-v_{23}v_{32}}{E_2 E_3 \Pi}, \quad c_{22} = \dfrac{1-v_{13}v_{31}}{E_1 E_3 \Pi}, \quad c_{33} = \dfrac{1-v_{12}v_{21}}{E_1 E_2 \Pi} \\ c_{12} = \dfrac{v_{21}+v_{31}v_{23}}{E_2 E_3 \Pi} = \dfrac{v_{12}+v_{32}v_{13}}{E_1 E_3 \Pi}, \quad c_{13} = \dfrac{v_{31}+v_{21}v_{32}}{E_2 E_3 \Pi} = \dfrac{v_{13}+v_{12}v_{23}}{E_1 E_2 \Pi} \\ c_{23} = \dfrac{v_{32}+v_{12}v_{31}}{E_1 E_3 \Pi} = \dfrac{v_{23}+v_{21}v_{13}}{E_1 E_2 \Pi}, \quad c_{44} = G_{23}, \quad c_{55} = G_{13}, \quad c_{66} = G_{12} \end{cases} \quad (2.10)$$

式中：

$$\Pi = \dfrac{1-v_{12}v_{21}-v_{23}v_{32}-v_{31}v_{13}-2v_{21}v_{32}v_{13}}{E_1 E_2 E_3}$$

若将材料定义为横向各向同性[①]（如 PZT），则以下等价成立：$G_{12}=G_{13}$；$E_3=E_2$；$v_{12}=v_{13}$。

## 2.1.6 单向复合材料

单向复合材料通常由两种成分组成，即纤维和基体（基体是将两种成分黏合在一起的胶黏剂）。根据混合定律，可以由纤维和基体的性质以及它们的体积分数来计算单向层的性质。应用混合规则时，可以假设将两种成分结合在一起，它们的性能就像一个单一的物体。该层的纵向模量（或主模量）$E_{11}$ 可以写为

$$E_{11} = E_f V_f + E_m V_m \quad (2.11)$$

式中：$E_f$ 和 $E_m$ 分别为纤维和基体的纵向模量；$V_f$ 和 $V_m$ 为它们的体积分数[②]。

主要泊松比 $v_{12}$ 由下式给出：

$$v_{12} = v_f V_f + v_m V_m \quad (2.12)$$

式中：$v_f$ 和 $v_m$ 分别为纤维和基体的纵向模量。

应当注意，根据式（2.6），次要泊松比 $v_{21}$ 为

$$\dfrac{v_{12}}{E_{11}} = \dfrac{v_{21}}{E_{22}} \Rightarrow v_{21} = v_{12}\dfrac{E_{22}}{E_{11}} \quad (2.13)$$

层的横向模量（或次模量）$E_{22}$ 表示为

$$\dfrac{1}{E_{22}} = \dfrac{V_f}{E_f} + \dfrac{V_m}{E_m} \Rightarrow E_{22} = \dfrac{E_m}{V_f \dfrac{E_m}{E_f} + V_m} = \dfrac{E_m}{V_f \dfrac{E_m}{E_f} + (1-V_f)} \quad (2.14)$$

---

[①] 横向各向同性材料是一种具有各向同性平面的均匀线弹性材料，而在与该平面垂直的轴上，该材料是正交各向异性的。

[②] 注意：$V_f + V_m = 1$。

层的剪切模量 $G_{12}$ 表示为

$$\frac{1}{G_{12}}=\frac{V_f}{G_f}+\frac{V_m}{G_m} \Rightarrow G_{12}=\frac{G_m}{V_f\frac{G_m}{G_f}+V_m}=\frac{G_m}{V_f\frac{G_m}{G_f}+(1-V_f)} \qquad (2.15)$$

式中：$G_f$ 和 $G_m$ 分别为纤维和基体的剪切模量。

为了评估纤维特性之间的差异并将它们与基体的特性进行比较，读者可以参考表 2.1。

表 2.1 T300 碳纤维和 914 环氧树脂基体的典型特性

| 性　　质 | T300 碳纤维 | 914 环氧树脂基体 |
| --- | --- | --- |
| 杨氏模量 $E$/GPa | 220 | 3.1 |
| 剪切模量 $G$/GPa | 20 | 1.2 |
| 泊松比 $U$ | 0.15 | 0.37 |

表 2.2 提供了上述简化的微观力学模型与实验值之间的对比[①]。可以看到，该模型很好地预测了主杨氏模量 $E_{11}$（实验值的 0.998），主泊松比和次杨氏模量 $E_{22}$ 偏低，分别是实验值的 0.735 和 0.813。通过简化的微观力学模型剪切模量 $G_{12}$ 严重偏低，仅是实验值的 0.52。

表 2.2 通过简化的微观力学模型预测单向复合材料的性能并与实验值进行比较

| 公式 | 关系 | 预测值 | 实验值 |
| --- | --- | --- | --- |
| 式 (2.11) | $E_{11}=E_f V_f+E_m(1-V_f)$ | 124.7 | 125.0 |
| 式 (2.12) | $U_{12}=U_f V_f+U_m(1-V_f)$ | 0.25 | 0.34 |
| 式 (2.14) | $E_{22}=\dfrac{E_m}{V_f(E_m/E_f)+(1-V_f)}$ | 7.4 | 9.1 |
| 式 (2.15) | $G_{12}=\dfrac{G_m}{V_f(G_m/G_f)+(1-V_f)}$ | 2.6 | 5.0 |

## 2.1.7 单层的特性

一层有两个主要尺寸及厚度，其中厚度与两个主要尺寸相比非常小。因此，通过在式（2.5）中假设 $\sigma_{33}=0$，正交各向异性材料的三维表示将被简化为二维表示（平面应力）。因而柔度矩阵可简化为

---

① 改编自：Harris, B., Engineering composite materials, The Institute of Materials, London, UK, 1999, 193p.

$$\begin{Bmatrix} \varepsilon_{11} \\ \varepsilon_{22} \\ \gamma_{12} \end{Bmatrix} = \begin{bmatrix} \dfrac{1}{E_1} & -\dfrac{v_{21}}{E_2} & 0 \\ -\dfrac{v_{12}}{E_1} & \dfrac{1}{E_2} & 0 \\ 0 & 0 & \dfrac{1}{G_{12}} \end{bmatrix} \begin{Bmatrix} \sigma_{11} \\ \sigma_{22} \\ \sigma_{12} \end{Bmatrix} \tag{2.16}$$

第3个方程，对于厚度方向的应变 $\varepsilon_{33}$，其形式为

$$\varepsilon_{33} = -\dfrac{v_{13}}{E_1}\sigma_{11} - \dfrac{v_{23}}{E_2}\sigma_{22} \tag{2.17}$$

剩下的两个剪切应变方程为

$$\begin{Bmatrix} v_{23} \\ v_{13} \end{Bmatrix} = \begin{bmatrix} \dfrac{1}{G_{23}} & 0 \\ 0 & \dfrac{1}{G_{13}} \end{bmatrix} \begin{Bmatrix} \sigma_{23} \\ \sigma_{13} \end{Bmatrix} \tag{2.18}$$

以式（2.16）和式（2.18）可计算得到应力与应变的函数关系：

$$\begin{Bmatrix} \sigma_{11} \\ \sigma_{22} \\ \sigma_{12} \end{Bmatrix} = \begin{bmatrix} Q_{11} & Q_{12} & 0 \\ Q_{21} & Q_{22} & 0 \\ 0 & 0 & Q_{66} \end{bmatrix} \begin{Bmatrix} \varepsilon_{11} \\ \varepsilon_{22} \\ \gamma_{12} \end{Bmatrix} = \begin{bmatrix} \dfrac{E_1}{(1-v_{12}v_{21})} & \dfrac{v_{21}E_1}{(1-v_{12}v_{21})} & 0 \\ \dfrac{v_{12}E_2}{(1-v_{12}v_{21})} & \dfrac{E_2}{(1-v_{12}v_{21})} & 0 \\ 0 & 0 & G_{12} \end{bmatrix} \begin{Bmatrix} \varepsilon_{11} \\ \varepsilon_{22} \\ \gamma_{12} \end{Bmatrix}$$
$$\tag{2.19}$$

式中：$Q_{12} = Q_{21}$，$v_{12} \neq v_{21}$，且

$$\begin{Bmatrix} \sigma_{23} \\ \sigma_{13} \end{Bmatrix} = \begin{bmatrix} Q_{23} & 0 \\ 0 & Q_{13} \end{bmatrix} \begin{Bmatrix} v_{23} \\ v_{13} \end{Bmatrix} = \begin{bmatrix} G_{23} & 0 \\ 0 & G_{13} \end{bmatrix} \begin{Bmatrix} v_{23} \\ v_{13} \end{Bmatrix} \tag{2.20}$$

## 2.1.8 应力和应变的转换

对于图2.3中描述的两个坐标系，1和2描述了层的正交各向异性坐标系，而 $x$ 和 $y$ 是另一个任意坐标系，相对于1、2坐标系旋转了给定的角度 $\theta$。应力和应变从1、2坐标系到 $x$、$y$ 坐标系的转换是通过将层间的应力和应变乘以转换矩阵 $T$ 来完成的[①]，即

---

① 例如，文献[1]。

$$\left\{\begin{matrix}\sigma_1\\\sigma_2\\\tau_{12}\end{matrix}\right\}^k = [T]\left\{\begin{matrix}\sigma_x\\\sigma_y\\\tau_{xy}\end{matrix}\right\}^k \tag{2.21}$$

$$\left\{\begin{matrix}\varepsilon_1\\\varepsilon_2\\\dfrac{\gamma_{12}}{2}\end{matrix}\right\}^k = [T]\left\{\begin{matrix}\varepsilon_x\\\varepsilon_y\\\dfrac{\gamma_{xy}}{2}\end{matrix}\right\}^k \tag{2.22}$$

式中：$k$ 为进行应变和应力转换的层号①。转换矩阵 $T$ 由下式给出：

$$[T] = \begin{bmatrix} c^2 & s^2 & 2cs \\ s^2 & c^2 & -2cs \\ -cs & cs & c^2-s^2 \end{bmatrix} \tag{2.23}$$

式中

$$c \equiv \cos\theta$$
$$s \equiv \sin\theta$$

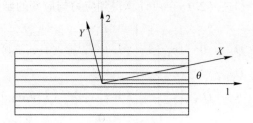

图 2.3 两个坐标系：1、2 为层的正交各向异性轴；$X$、$Y$ 为任意轴

为了获得矩阵 $T$ 的逆矩阵，只需在式（2.23）中插入 $-\theta$，即可得

$$[T]^{-1} = [T(-\theta)] = \begin{bmatrix} c^2 & s^2 & -2cs \\ s^2 & c^2 & 2cs \\ cs & -cs & c^2-s^2 \end{bmatrix} \tag{2.24}$$

转换到层板参考轴 $(x, y)$ 的层（或层板）应变-应力关系写为

$$\left\{\begin{matrix}\sigma_1\\\sigma_2\\\tau_{12}\end{matrix}\right\}^k = [T]^{-1}[Q]^k[T]\left\{\begin{matrix}\varepsilon_x\\\varepsilon_y\\\gamma_{xy}\end{matrix}\right\}^k \tag{2.25}$$

式中

---

① 注意：$\sigma_{11} \equiv \sigma_1$；$\sigma_{22} \equiv \sigma_2$；$\varepsilon_{11} \equiv \varepsilon_1$；$\varepsilon_{22} \equiv \varepsilon_2$。

$$[Q]^k = \begin{bmatrix} Q_{11} & Q_{12} & 0 \\ Q_{12} & Q_{22} & 0 \\ 0 & 0 & 2Q_{66} \end{bmatrix}^k$$

式中：$Q_{11}$、$Q_{12}$、$Q_{22}$ 和 $Q_{66}$ 的表达式在式（2.19）中给出。执行式（2.25）中的矩阵乘法可得

$$\begin{Bmatrix} \sigma_1 \\ \sigma_2 \\ \tau_{12} \end{Bmatrix}^k = [\overline{Q}]^k \begin{Bmatrix} \varepsilon_x \\ \varepsilon_y \\ \gamma_{xy} \end{Bmatrix}^k \text{ 且 } [\overline{Q}]^k = \begin{bmatrix} \overline{Q}_{11} & \overline{Q}_{12} & \overline{Q}_{16} \\ \overline{Q}_{12} & \overline{Q}_{22} & \overline{Q}_{26} \\ \overline{Q}_{16} & \overline{Q}_{26} & \overline{Q}_{66} \end{bmatrix}^k \quad (2.26)$$

式中

$$\begin{aligned}
\overline{Q}_{11} &= Q_{11}\cos^4\theta + 2(Q_{12}+2Q_{66})\sin^2\theta\cos^2\theta + Q_{22}\sin^4\theta \\
\overline{Q}_{12} &= (Q_{11}+Q_{22}-4Q_{66})\sin^2\theta\cos^2\theta + Q_{12}(\sin^4\theta+\cos^4\theta) \\
\overline{Q}_{22} &= Q_{11}\sin^4\theta + 2(Q_{12}+2Q_{66})\sin^2\theta\cos^2\theta + Q_{22}\cos^4\theta \\
\overline{Q}_{16} &= (Q_{11}-Q_{12}-2Q_{66})\sin\theta\cos^3\theta + (Q_{12}-Q_{22}+2Q_{66})\sin^3\theta\cos\theta \\
\overline{Q}_{26} &= (Q_{11}-Q_{12}-2Q_{66})\sin^3\theta\cos\theta + (Q_{12}-Q_{22}+2Q_{66})\sin\theta\cos^3\theta \\
\overline{Q}_{66} &= (Q_{11}+Q_{22}-2Q_{12}-2Q_{66})\sin^2\theta\cos^2\theta + Q_{66}(\sin^4\theta+\cos^4\theta)
\end{aligned} \quad (2.27)$$

另一种表示矩阵 $[\overline{Q}]^k$ 的各项的有用方法是 Tsai 和 Pagano[2] 提出的不变过程：

$$\begin{aligned}
\overline{Q}_{11} &= U_1 + U_2\cos(2\theta) + U_3\cos(4\theta) \\
\overline{Q}_{12} &= U_4 - U_3\cos(4\theta) \\
\overline{Q}_{22} &= U_1 - U_2\cos(2\theta) + U_3\cos(4\theta) \\
\overline{Q}_{16} &= -\frac{1}{2}U_2\sin(2\theta) - U_3\sin(4\theta) \\
\overline{Q}_{26} &= -\frac{1}{2}U_2\sin(2\theta) + U_3\sin(4\theta) \\
\overline{Q}_{66} &= U_5 - U_3\cos(4\theta)
\end{aligned} \quad (2.28)$$

式中

$$U_1 = \frac{1}{8}[3Q_{11}+3Q_{22}+2Q_{12}+4Q_{66}]$$

$$U_2 = \frac{1}{2}[Q_{11}-Q_{22}]$$

$$U_3 = \frac{1}{8}[Q_{11}+Q_{22}-2Q_{12}-4Q_{66}]$$

$$U_4 = \frac{1}{8}\left[Q_{11}+Q_{22}+6Q_{12}-4Q_{66}\right]$$

$$U_5 = \frac{1}{8}\left[Q_{11}+Q_{22}-2Q_{12}+4Q_{66}\right]$$

注意，$U_1$、$U_4$ 和 $U_5$ 对于相对于 3 个轴（垂直于 1-2 平面）的旋转是不变的。

下面将介绍每个层板的属性相加以形成层压板，图 2.4 所示为施加载荷时要研究的结构。

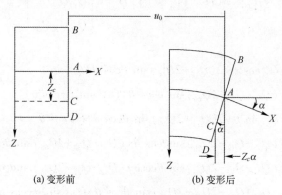

(a) 变形前　　　　　　　　(b) 变形后

图 2.4　变形前后的板截面

参照图 2.4，将距离中平面 $z$ 处的点在 $x$ 方向上的位移表示为（其中 $w$ 为 $z$ 方向的位移）：

$$u = u_0 - z\frac{\partial w}{\partial x} \tag{2.29}$$

类似地，$y$ 方向的位移为

$$v = v_0 - z\frac{\partial w}{\partial y} \tag{2.30}$$

则应变（$\varepsilon_x$、$\varepsilon_y$ 和 $\gamma_{xy}$）和曲率（$\kappa_x$、$\kappa_y$ 和 $\kappa_{xy}$）可以表示为

$$\begin{aligned} \varepsilon_x &\equiv \frac{\partial u}{\partial x} = \frac{\partial u_0}{\partial x} - z\frac{\partial^2 w}{\partial x^2} = \varepsilon_x^0 + z\kappa_x \\ \varepsilon_y &\equiv \frac{\partial v}{\partial y} = \frac{\partial v_0}{\partial y} - z\frac{\partial^2 w}{\partial y^2} = \varepsilon_y^0 + z\kappa_y \\ \gamma_{xy} &\equiv \frac{\partial u}{\partial y} + \frac{\partial v}{\partial x} = \frac{\partial u_0}{\partial y} + \frac{\partial v_0}{\partial x} - 2z\frac{\partial^2 w}{\partial x \partial y} = \gamma_{xy}^0 + z\kappa_{xy} \end{aligned} \tag{2.31}$$

式中：$\varepsilon_x^0$、$\varepsilon_y^0$、$\gamma_{xy}^0$ 是中性面的应变。在矩阵表示法中，式（2.31）可以表示为

$$\begin{Bmatrix} \varepsilon_x \\ \varepsilon_y \\ \gamma_{xy} \end{Bmatrix} = \begin{Bmatrix} \varepsilon_x^0 \\ \varepsilon_y^0 \\ \gamma_{xy}^0 \end{Bmatrix} + z \begin{Bmatrix} \kappa_x \\ \kappa_y \\ \kappa_{xy} \end{Bmatrix} \Rightarrow \{\varepsilon\} = \{\varepsilon^0\} + z\{\kappa\} \quad (2.32)$$

层间的应力表示为

$$\{\sigma\}^k = [\overline{Q}]^k \{\varepsilon^0\} + z[\overline{Q}]^k [\kappa] \quad (2.33)$$

下面将处理合力（$N_x$, $N_y$, $N_{xy}$）和合力矩（$M_x$, $M_y$, $M_{xy}$）间的关系，它们的定义如下（$h$ 是层压板的总厚度）：

$$\begin{aligned} N_x &\equiv \int_{-h/2}^{h/2} \sigma_x \mathrm{d}z; \quad N_y \equiv \int_{-h/2}^{h/2} \sigma_y \mathrm{d}z; \quad N_{xy} \equiv \int_{-h/2}^{h/2} \tau_{xy} \mathrm{d}z \\ M_x &\equiv \int_{-h/2}^{h/2} \sigma_x z \mathrm{d}z; \quad M_y \equiv \int_{-h/2}^{h/2} \sigma_y z \mathrm{d}z; \quad M_{xy} \equiv \int_{-h/2}^{h/2} \tau_{xy} z \mathrm{d}z \end{aligned} \quad (2.34)$$

代入应力表达式，可以得到力和力矩合力的表达式，为中平面上的应变 $\varepsilon^0$ 和曲率 $\kappa$ 的函数（另请参见文献[1]）。简短的表达式为

$$\begin{Bmatrix} \{N\} \\ \{M\} \end{Bmatrix} = \begin{bmatrix} [A] & [B] \\ [B] & [D] \end{bmatrix} \begin{Bmatrix} \{\varepsilon^0\} \\ \{\kappa\} \end{Bmatrix}$$

或

$$\begin{Bmatrix} \begin{Bmatrix} N_x \\ N_y \\ N_{xy} \end{Bmatrix} \\ \begin{Bmatrix} M_x \\ M_y \\ M_{xy} \end{Bmatrix} \end{Bmatrix} = \begin{bmatrix} \begin{bmatrix} A_{11} & A_{12} & A_{16} \\ A_{12} & A_{22} & A_{26} \\ A_{16} & A_{26} & A_{66} \end{bmatrix} & \begin{bmatrix} B_{11} & B_{12} & B_6 \\ B_{12} & B_{22} & B_6 \\ B_6 & B_6 & B_{66} \end{bmatrix} \\ \begin{bmatrix} B_{11} & B_{12} & B_{16} \\ B_{12} & B_{22} & B_{26} \\ B_{16} & B_{26} & B_{66} \end{bmatrix} & \begin{bmatrix} D_{11} & D_{12} & D_6 \\ D_{12} & D_{22} & D_{26} \\ D_{16} & D_6 & D_{66} \end{bmatrix} \end{bmatrix} \begin{Bmatrix} \begin{Bmatrix} \varepsilon_x^0 \\ \varepsilon_y^0 \\ \gamma_{xy}^0 \end{Bmatrix} \\ \begin{Bmatrix} \kappa_x \\ \kappa_y \\ \kappa_{xy} \end{Bmatrix} \end{Bmatrix} \quad (2.35)$$

式中：各种常数定义为

$$\begin{aligned} A_{ij} &\equiv \int_{-h/2}^{h/2} \overline{Q}_{ij}^k \mathrm{d}z = \sum_{k=1}^{n} \overline{Q}_{ij}^k (h_k - h_{k-1}) \\ B_{ij} &\equiv \int_{-h/2}^{h/2} \overline{Q}_{ij}^k z \mathrm{d}z = \frac{1}{2} \sum_{k=1}^{n} \overline{Q}_{ij}^k (h_k^2 - h_{k-1}^2) \\ D_{ij} &\equiv \int_{-h/2}^{h/2} \overline{Q}_{ij}^k z^2 \mathrm{d}z = \frac{1}{3} \sum_{k=1}^{n} \overline{Q}_{ij}^k (h_k^3 - h_{k-1}^3) \end{aligned} \quad (2.36)$$

式中：$i, j=1,1$；$1,2$；$2,2$；$1,6$；$2,6$；$6,6$。

式（2.36）中求和的方式与图 2.5 中的符号一致。从层压板厚度的积分得到总厚度的过程基于以下两个条件：单层厚度非常薄，并且假定每个层压板的

性能在厚度方向上是恒定的。

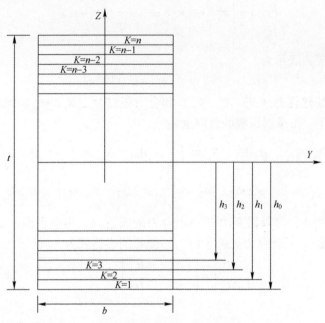

图 2.5　层压板内的层级符号

最后，静态情况下，对于由层压复合层板组成的薄板，其运动方程可利用经典层压理论（Classical Lamination Theory，CLT）表达如下[1,3]：

$$\frac{\partial N_x}{\partial x}+\frac{\partial N_{xy}}{\partial y}=I_1\frac{\partial^2 u_0}{\partial t^2}-I_2\frac{\partial^2}{\partial t^2}\left(\frac{\partial w_0}{\partial x}\right)$$

$$\frac{\partial N_{xy}}{\partial x}+\frac{\partial N_y}{\partial y}=I_1\frac{\partial^2 v_0}{\partial t^2}-I_2\frac{\partial^2}{\partial t^2}\left(\frac{\partial w_0}{\partial y}\right)$$

$$\frac{\partial^2 M_x}{\partial x^2}+2\frac{\partial^2 M_{xy}}{\partial x\partial y}+\frac{\partial^2 M_y}{\partial y^2}+\frac{\partial}{\partial x}\left[N_{xx}\frac{\partial w_0}{\partial x}+N_{xy}\frac{\partial w_0}{\partial y}\right]+\frac{\partial}{\partial y}\left[N_{yy}\frac{\partial w_0}{\partial y}+N_{xy}\frac{\partial w_0}{\partial x}\right]$$

$$=-p_z+I_1\frac{\partial^2 w_0}{\partial t^2}-I_3\frac{\partial^2}{\partial t^2}\left(\frac{\partial^2 w_0}{\partial x^2}+\frac{\partial^2 w_0}{\partial y^2}\right)+I_2\frac{\partial^2}{\partial t^2}\left(\frac{\partial w_0}{\partial x}+\frac{\partial w_0}{\partial y}\right)$$

(2.37)

式中：$p_z$ 为 $z$ 方向①上单位面积的载荷；下标 0 表示横截面中平面处的值；$N$ 代表平面内载荷。各种转动惯量 $I_1$、$I_2$ 和 $I_3$ 由下式给出（$\rho$ 代表质量/单位长度）：

---

① 注意：坐标 $z$ 通常用于厚度方向，而坐标 $x$ 和 $y$ 定义了板的面积。

$$I_j = \int_{-h/2}^{h/2} \rho z^{j-1} \mathrm{d}z, \quad j = 1, 2, 3 \tag{2.38}$$

可以通过式（2.26）来获得梁的方程，此时所有关于 $y$ 的导数都为零，可得以下形式的一维方程：

$$\frac{\partial^2 M_x}{\partial x^2} + \frac{\partial}{\partial x}\left(N_{xx}\frac{\partial w_0}{\partial x}\right) = -p_z + I_1\frac{\partial^2 w}{\partial t^2} - I_3\frac{\partial^4 w}{\partial t^2 \partial x^2} + I_2\frac{\partial^3 w}{\partial t^2 \partial x} \tag{2.39}$$

式中：$N_x$ 为轴向（平面内沿梁的长度方向）载荷。已知横向挠度 $w$ 和弯矩之间的关系，可以根据 $w$ 重写式（2.39），得

$$\begin{aligned} &-D_{11}\frac{\partial^2 w}{\partial x^2} = M_x \Rightarrow \\ &-\frac{\partial^2}{\partial x^2}\left(D_{11}\frac{\partial^2 w}{\partial x^2}\right) + \frac{\partial}{\partial x}\left(N_{xx}\frac{\partial w_0}{\partial x}\right) = -p_z + I_1\frac{\partial^2 w}{\partial t^2} - I_3\frac{\partial^4 w}{\partial t^2 \partial x^2} + I_2\frac{\partial^3 w}{\partial t^2 \partial x} \end{aligned} \tag{2.40}$$

及其相关的边界条件

$$\begin{cases} 几何边界：指定~w~和~\partial w/\partial x \\ 自然边界：指定~Q \equiv \partial M/\partial x~和~M \end{cases} \tag{2.41}$$

通常在文献中使用典型的边界条件可以表示为

$$\begin{cases} 简单支撑：w = 0~和~M = 0 \\ 夹紧：w = 0~和~\partial w/\partial x = 0 \\ 自由：Q \equiv \partial M/\partial x = 0~且~M = 0 \end{cases} \tag{2.42}$$

读者应了解与复合结构制造相关的热学问题，这是由于后固化阶段的热收缩差异，以及结构使用寿命期间任何温度变化的结果。这个问题是由于现有增强纤维的轴向热膨胀系数相对较小（对于碳纤维来说甚至略为负），而树脂基体具有较大的导热系数。当从典型的固化温度（如 140°C）冷却到室温时，层压复合材料的纤维将处于压缩状态，而基体将显示出拉应力[4]。由于层压板的两个组件之间的热失配导致的典型残余应力如表 2.3 所列。

表 2.3 一些常见单向复合材料的典型热应力[4]

| 基　体 | 纤维 | 纤维体积 $V_f$ /% | 温度范围 $\Delta T$/K | 纤维残余应力 /MPa | 基体残余应力 /MPa |
|---|---|---|---|---|---|
| 环氧树脂（高温固化） | T300 碳 | 65 | 120 | −19 | 36 |
| 环氧树脂（低温固化） | E 玻璃 | 65 | 100 | −15 | 28 |
| 环氧树脂（低温固化） | Kevlar-49 | 65 | 100 | −16 | 30 |

续表

| 基体 | 纤维 | 纤维体积 $V_f$ /% | 温度范围 $\Delta T$/K | 纤维残余应力 /MPa | 基体残余应力 /MPa |
|---|---|---|---|---|---|
| 硼硅玻璃 | T300 碳 | 50 | 520 | -93 | 93 |
| CAS① 玻璃陶瓷 | Nicalon 碳化硅 | 40 | 1000 | -186 | 124 |

① $CaO-Al_2O_3-SiO_2$。

另一个重要的设计数据是在实验室的各种测试中测得的实验拉伸强度和压缩强度,如表 2.4 所列[4]。

表 2.4 常见复合材料的典型实验拉伸强度和压缩强度[4]

| 材 料 | 层压板 | 纤维体积 $V_f$/% | 抗拉强 $\sigma_t$/GPa | 抗压强度 $\sigma_c$/GPa | $\sigma_c/\sigma_t$ 比值 |
|---|---|---|---|---|---|
| GRP | ud① | 60 | 1.3 | 1.1 | 0.85 |
| CFRP | ud | 60 | 2.0 | 1.1 | 0.55 |
| KFRP | ud | 60 | 1.0 | 0.4 | 0.40 |
| HTA/913 (CFRP) | $[(\pm45°,0°_2)_2]_S$ | 65 | 1.27 | 0.97 | 0.77 |
| T800/924 (CFRP) | $[(\pm45°,0°_2)_2]_S$ | 65 | 1.42 | 0.90 | 0.63 |
| T800/5245 (CFRP) | $[(\pm45°,0°_2)_2]_S$ | 65 | 1.67 | 0.88 | 0.53 |
| SiC/CAS (CMC②) | ud | 37 | 334 | 1360 | 4.07 |
| SiC/CAS (CMC) | $[0°,90°]_{3S}$ | 37 | 210 | 463 | 2.20 |

① ud,单向。
② CMC,陶瓷基复合材料。

## 2.2 一阶剪切变形理论模型

在经典层压理论中需要消除基于 Kirchhoff-Love 板理论[5-7]的某些限制性假设,如忽略剪切应变的影响以及变形前的平面在变形后仍为平面,从而推导出更高级的板弯曲理论,如包含平面内剪切应变的 Mindlin 板理论[8-10],是一阶剪切效应 Kirchhoff-Love 板理论的扩展。

Mindlin 理论假设在板厚度上的位移呈线性变化,但厚度在变形过程中没有变化。另一个假设是通过厚度的法向应力被忽略,该假设也称为平面应力条件。该理论通常称为板的一阶剪切变形理论(First-order Shear Deformation Theory,FSDT),接下来介绍其在复合材料中的应用。在 Mindlin 理论(与 Timoshenko 的梁理论[11-14]相似)的假设和限制下,位移场有 5 个未知数($u_0$、

$v_0$、$w_0$ 分别为中平面在 $x$、$y$ 和 $z$ 方向上的位移,另外还有分别围绕 $x$ 和 $y$ 方向的剪切引起的旋转),由下式给出(另请参见图 2.6,其与图 2.4 类似,但图 2.6 是基于 FSDT 方法):

$$\begin{cases} u(x,y,z,t) = u_0(x,y,t) + z\phi_x(x,y,t) \\ v(x,y,z,t) = v_0(x,y,t) + z\phi_y(x,y,t) \\ w(x,y,z,t) = w_0(x,y,t) \end{cases} \quad (2.43)$$

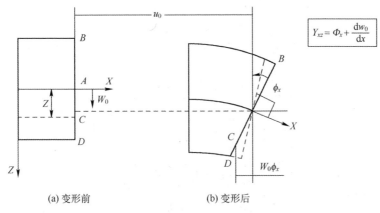

(a) 变形前　　　　(b) 变形后

图 2.6　变形前后的板截面(FSDT 法)

包括非线性项的相关应变[3]① 如下:

$$\begin{cases} \varepsilon_x = \dfrac{\partial u_0}{\partial x} + \dfrac{1}{2}\left(\dfrac{\partial w_0}{\partial x}\right)^2 + z\dfrac{\partial \phi_x}{\partial x} \\ \varepsilon_y = \dfrac{\partial u_0}{\partial x} + \dfrac{1}{2}\left(\dfrac{\partial w_0}{\partial x}\right)^2 + z\dfrac{\partial \phi_x}{\partial x} \\ \varepsilon_z = 0;\quad \gamma_{xz} = \dfrac{\partial w_0}{\partial x} + \phi_x;\quad \gamma_{yz} = \dfrac{\partial w_0}{\partial y} + \phi_x \\ \gamma_{xy} = \left(\dfrac{\partial u_0}{\partial y} + \dfrac{\partial v_0}{\partial x} + \dfrac{\partial^2 w_0}{\partial x \partial y}\right) + z\left(\dfrac{\partial \phi_x}{\partial y} + \dfrac{\partial \phi_y}{\partial x}\right) \end{cases} \quad (2.44)$$

将式(2.44)乘以刚度矩阵 $[\overline{Q}]$ 并通过层压板的厚度积分产生力和力矩合力(如式(2.34)中的定义),加上两个附加项,剪切合力定义为

---

① 注意,跨层压板高度的恒定剪切应变的假设是真实应变分布的粗略近似,其至少在厚度上是二次方的。然而,虽然是粗略的近似,但与实验结果相比,Mindlin 板理论的应用结果呈现出非常好的结果。

$$\begin{Bmatrix} Q_x \\ Q_y \end{Bmatrix} = \kappa \int_{-h/2}^{h/2} \begin{Bmatrix} \tau_{xz} \\ \tau_{yz} \end{Bmatrix} \mathrm{d}z \tag{2.45}$$

式中：$\kappa$ 为剪切修正系数，定义为由实际剪切应力分布计算的剪切应变能与 FSDT 理论中假设的常数分布之间的比例；矩形横截面 $\kappa$ 值取为 $5/6$[①]。

那么运动方程将具有如下形式：

$$\begin{cases} \dfrac{\partial N_x}{\partial x} + \dfrac{\partial N_{xy}}{\partial y} = I_1 \dfrac{\partial^2 u_0}{\partial t^2} + I_2 \dfrac{\partial^2 \phi_x}{\partial t^2} \\[6pt] \dfrac{\partial N_{xy}}{\partial x} + \dfrac{\partial N_y}{\partial y} = I_1 \dfrac{\partial^2 v_0}{\partial t^2} + I_2 \dfrac{\partial^2 \phi_y}{\partial t^2} \\[6pt] \dfrac{\partial Q_x}{\partial x} + \dfrac{\partial Q_y}{\partial y} + \dfrac{\partial}{\partial x}\left[ N_{xx}\dfrac{\partial w_0}{\partial x} + N_{xy}\dfrac{\partial w_0}{\partial y} \right] \\[6pt] + \dfrac{\partial}{\partial y}\left[ N_{yy}\dfrac{\partial w_0}{\partial y} + N_{xy}\dfrac{\partial w_0}{\partial x} \right] = -p_z + I_1 \dfrac{\partial^2 w_0}{\partial t^2} \\[6pt] \dfrac{\partial M_x}{\partial x} + \dfrac{\partial M_{xy}}{\partial y} - Q_x = I_3 \dfrac{\partial^2 \phi_x}{\partial t^2} + I_1 \dfrac{\partial^2 u_0}{\partial t^2} \\[6pt] \dfrac{\partial M_{xy}}{\partial x} + \dfrac{\partial M_y}{\partial y} - Q_y = I_3 \dfrac{\partial^2 \phi_y}{\partial t^2} + I_1 \dfrac{\partial^2 v_0}{\partial t^2} \end{cases} \tag{2.46}$$

式中：$p_z$ 为 $z$ 方向上单位面积的载荷[②]；$N$ 为面内载荷和各种转动惯量。$I_1$、$I_2$ 和 $I_3$ 由（$\rho$ 为质量/单位长度）式（2.38）给出。

式（2.46）将力和力矩的合力描述为刚度系数 $A_{ij}$、$B_{ij}$ 和 $D_{ij}$ 的函数，除此之外，剪切合力 $Q_x$ 和 $Q_y$ 定义如下：

$$\begin{Bmatrix} Q_y \\ Q_x \end{Bmatrix} = \kappa \begin{bmatrix} A_{44} & A_{45} \\ A_{45} & A_{55} \end{bmatrix} \begin{Bmatrix} \dfrac{\partial w_0}{\partial y} + \phi_y \\ \dfrac{\partial w_0}{\partial x} + \phi_x \end{Bmatrix} \tag{2.47}$$

式中

$$A_{44} \equiv \kappa \int_{-h/2}^{h/2} \overline{Q}_{44}^k \mathrm{d}z = \kappa \sum_{k=1}^{n} \overline{Q}_{44}^k (h_k - h_{k-1})$$

---

[①] 矩形截面的准确值为 $\kappa = \dfrac{10(1+\upsilon)}{12+11\upsilon}$，实心圆形截面的准确值为 $\kappa = \dfrac{6(1+\upsilon)}{7+6\upsilon}$。

[②] 注意：坐标 $z$ 通常用于厚度方向，而坐标 $x$ 和 $y$ 定义了板的面积。

$$A_{45} \equiv \kappa \int_{-h/2}^{h/2} \overline{Q}_{45}^k \mathrm{d}z = \kappa \sum_{k=1}^{n} \overline{Q}_{45}^k (h_k - h_{k-1}) \qquad (2.48)$$

$$A_{55} \equiv \kappa \int_{-h/2}^{h/2} \overline{Q}_{55}^k \mathrm{d}z = \kappa \sum_{k=1}^{n} \overline{Q}_{55}^k (h_k - h_{k-1})$$

并且

$$\begin{aligned}\overline{Q}_{44} &= Q_{44}\cos^2\theta + Q_{55}\sin^2\theta \\ \overline{Q}_{45} &= (Q_{55} - Q_{44})\cos\theta\sin\theta \\ \overline{Q}_{55} &= Q_{44}\sin^2\theta + Q_{55}\cos^2\theta\end{aligned} \qquad (2.49)$$

式中：$Q_{44} = G_{23}$，$Q_{55} = G_{13}$。

代入根据 5 个未知位移（$u_0$，$v_0$，$w_0$，$\phi_x$，$\phi_y$）定义的合力，得到 5 个未知位移的微分方程[3]：

$$\begin{aligned}&A_{11}\left[\frac{\partial^2 u_0}{\partial x^2} + \frac{\partial^3 w_0}{\partial x^3}\right] + A_{12}\left[\frac{\partial^2 v_0}{\partial x \partial y} + \frac{\partial^3 w_0}{\partial x \partial y^2}\right] + \\ &A_{16}\left[2\frac{\partial^2 u_0}{\partial x \partial y} + \frac{\partial^2 v_0}{\partial x^2} + 3\frac{\partial^3 w_0}{\partial x^2 \partial y}\right] + A_{26}\left[\frac{\partial^2 v_0}{\partial y^2} + \frac{\partial^3 w_0}{\partial y^3}\right] + \\ &A_{66}\left[\frac{\partial^2 u_0}{\partial y^2} + \frac{\partial^2 v_0}{\partial x \partial y} + 2\frac{\partial^3 w_0}{\partial x \partial y^2}\right] + B_{11}\frac{\partial^2 \phi_x}{\partial x^2} + B_{12}\frac{\partial^2 \phi_y}{\partial x \partial y} + \\ &B_{16}\left[2\frac{\partial^2 \phi_x}{\partial x \partial y} + \frac{\partial^2 \phi_y}{\partial x^2}\right] + B_{26}\frac{\partial^2 \phi_y}{\partial y^2} + B_{66}\left[\frac{\partial^2 \phi_x}{\partial y^2} + \frac{\partial^2 \phi_y}{\partial x \partial y}\right] = I_1 \frac{\partial^2 u_0}{\partial t^2} + I_2 \frac{\partial^2 \phi_x}{\partial t^2}\end{aligned} \qquad (2.50)$$

$$\begin{aligned}&A_{22}\left[\frac{\partial^2 v_0}{\partial y^2} + \frac{\partial^3 w_0}{\partial y^3}\right] + A_{12}\left[\frac{\partial^2 u_0}{\partial x \partial y} + \frac{\partial^3 w_0}{\partial x^2 \partial y}\right] + \\ &A_{16}\left[\frac{\partial^2 u_0}{\partial x^2} + \frac{\partial^3 w_0}{\partial x^3}\right] + A_{26}\left[\frac{\partial^2 u_0}{\partial y^2} + 2\frac{\partial^2 v_0}{\partial x \partial y} + 3\frac{\partial^3 w_0}{\partial x \partial y^2}\right] + \\ &A_{66}\left[\frac{\partial^2 u_0}{\partial x \partial y} + \frac{\partial^2 v_0}{\partial x^2} + 2\frac{\partial^3 w_0}{\partial x^2 \partial y}\right] + B_{22}\frac{\partial^2 \phi_y}{\partial y^2} + B_{12}\frac{\partial^2 \phi_x}{\partial x \partial y} + \\ &B_{16}\frac{\partial^2 \phi_x}{\partial x^2} + B_{26}\left[\frac{\partial^2 \phi_x}{\partial y^2} + 2\frac{\partial^2 \phi_y}{\partial x \partial y}\right] + B_{66}\left[\frac{\partial^2 \phi_y}{\partial x^2} + \frac{\partial^2 \phi_x}{\partial x \partial y}\right] = I_1 \frac{\partial^2 v_0}{\partial t^2} + I_2 \frac{\partial^2 \phi_y}{\partial t^2}\end{aligned} \qquad (2.51)$$

$$\begin{aligned}&\kappa A_{55}\left[\frac{\partial^2 w_0}{\partial x^2} + \frac{\partial \varphi_x}{\partial x}\right] + \kappa A_{45}\left[2\frac{\partial^2 w_0}{\partial x \partial y} + \frac{\partial \varphi_x}{\partial y} + \frac{\partial \varphi_y}{\partial x}\right] + \kappa A_{44}\left[\frac{\partial^2 w_0}{\partial y^2} + \frac{\partial \varphi_y}{\partial y}\right] + \\ &\frac{\partial}{\partial x}\left[N_{xx}\frac{\partial w_0}{\partial x} + N_{xy}\frac{\partial w_0}{\partial y}\right] + \frac{\partial}{\partial y}\left[N_{yy}\frac{\partial w_0}{\partial y} + N_{xy}\frac{\partial w_0}{\partial x}\right] = -p_z + I_1 \frac{\partial^2 w_0}{\partial t^2}\end{aligned} \qquad (2.52)$$

$$B_{11}\left[\frac{\partial^2 u_0}{\partial x^2} + \frac{\partial^3 w_0}{\partial x^3}\right] + B_{12}\left[\frac{\partial^2 v_0}{\partial x \partial y} + \frac{\partial^3 w_0}{\partial x \partial y^2}\right] +$$

$$B_{16}\left[2\frac{\partial^2 u_0}{\partial x\partial y}+\frac{\partial^2 v_0}{\partial x^2}+3\frac{\partial^3 w_0}{\partial x^2\partial y}\right]+B_{26}\left[\frac{\partial^2 v_0}{\partial y^2}+\frac{\partial^3 w_0}{\partial y^3}\right]+$$

$$B_{66}\left[\frac{\partial^2 u_0}{\partial y^2}+\frac{\partial^2 v_0}{\partial x\partial y}+2\frac{\partial^3 w_0}{\partial x\partial y^2}\right]+D_{11}\frac{\partial^2 \phi_x}{\partial x^2}+D_{12}\frac{\partial^2 \phi_y}{\partial x\partial y}+$$

$$D_{16}\left[2\frac{\partial^2 \phi_x}{\partial x\partial y}+\frac{\partial^2 \phi_y}{\partial x^2}\right]+D_{26}\frac{\partial^2 \phi_x}{\partial y^2}+D_{66}\left[\frac{\partial^2 \phi_x}{\partial y^2}+\frac{\partial^2 \phi_y}{\partial x\partial y}\right]-$$

$$\kappa A_{55}\left[\frac{\partial w_0}{\partial x}+\phi_x\right]-\kappa A_{45}\left[\frac{\partial w_0}{\partial y}+\phi_y\right]=I_3\frac{\partial^2 u_0}{\partial t^2}+I_3\frac{\partial^2 \phi_x}{\partial t^2} \quad (2.53)$$

$$B_{22}\left[\frac{\partial^2 v_0}{\partial y^2}+\frac{\partial^3 w_0}{\partial y^3}\right]+B_{12}\left[\frac{\partial^2 u_0}{\partial x\partial y}+\frac{\partial^3 w_0}{\partial x^2\partial y}\right]+$$

$$B_{16}\left[\frac{\partial^2 u_0}{\partial x^2}+\frac{\partial^3 w_0}{\partial x^3}\right]+B_{26}\left[\frac{\partial^2 u_0}{\partial y^2}+2\frac{\partial^2 v_0}{\partial x\partial y}+3\frac{\partial^3 w_0}{\partial x\partial y^2}\right]+$$

$$B_{66}\left[\frac{\partial^2 u_0}{\partial x\partial y}+\frac{\partial^2 v_0}{\partial x^2}+2\frac{\partial^3 w_0}{\partial x^2\partial y}\right]+D_{22}\frac{\partial^2 \phi_y}{\partial y^2}+D_{12}\frac{\partial^2 \phi_x}{\partial x\partial y}+$$

$$D_{16}\frac{\partial^2 \phi_x}{\partial x^2}+D_{26}\left[\frac{\partial^2 \phi_x}{\partial y^2}+2\frac{\partial^2 \phi_y}{\partial x\partial y}\right]+D_{66}\left[\frac{\partial^2 \phi_y}{\partial x^2}+\frac{\partial^2 \phi_x}{\partial x\partial y}\right]-$$

$$\kappa A_{44}\left[\frac{\partial w_0}{\partial y}+\phi_y\right]-\kappa A_{45}\left[\frac{\partial w_0}{\partial x}+\phi_x\right]=I_1\frac{\partial^2 v_0}{\partial t^2}+I_2\frac{\partial^2 \phi_y}{\partial t^2} \quad (2.54)$$

要求解这 5 个微分方程，需要以几何边界和自然边界条件的形式提供 10 个边界条件①。

为了获得梁的运动方程，使用前面介绍的板的一阶剪切变形理论，假设消除 $y$ 方向上的所有导数，并且 $v$ 和 $\phi_y$ 都为零。这样得到通用情况下的 3 个耦合运动方程（假设各种属性沿梁保持恒定）：

$$\begin{cases} A_{11}\dfrac{\partial^2 u_0}{\partial x^2}+B_{11}\dfrac{\partial^2 \phi_x}{\partial x^2}=I_1\dfrac{\partial^2 u_0}{\partial t^2}+I_2\dfrac{\partial^2 \phi_x}{\partial t^2} \\[2mm] B_{11}\dfrac{\partial^2 u_0}{\partial x^2}+D_{11}\dfrac{\partial^2 \phi_x}{\partial x^2}-\kappa A_{55}\left[\dfrac{\partial w_0}{\partial x}+\phi_x\right]=I_3\dfrac{\partial^2 u_0}{\partial t^2}+I_3\dfrac{\partial^2 \phi_x}{\partial t^2} \\[2mm] \kappa A_{55}\left[\dfrac{\partial^2 w_0}{\partial x^2}+\dfrac{\partial \varphi_x}{\partial x}\right]+N_{xx}\dfrac{\partial^2 w_0}{\partial x}=-p_z+I_1\dfrac{\partial^2 w_0}{\partial t^2} \end{cases} \quad (2.55)$$

并具有以下边界条件：

---

① 关于施加的边界条件类型的进一步讨论，请参阅文献 [3]。

$$\begin{cases} A_{11}\dfrac{\partial u_0}{\partial x}+B_{11}\dfrac{\partial \phi_x}{\partial x}=-N_{xx} \quad \text{或} \quad u_0=0 \\ A_{55}\left(\phi_x+\dfrac{\partial w_0}{\partial x}\right)-N_{xx}\dfrac{\partial w_0}{\partial x}=0 \quad \text{或} \quad w_0=0 \\ B_{11}\dfrac{\partial u_0}{\partial x}+D_{11}\dfrac{\partial \phi_x}{\partial x}=0 \quad \text{或} \quad \phi_x=0 \end{cases} \quad (2.56)$$

读者可以参阅文献[15-82]，其中包括利用经典层压理论和 FSDT 理论处理复合梁、板和壳结构（包括夹层结构）的弯曲、振动、分层和其他方面的文章。文献[83-98]展示了引入压电贴片、形状记忆合金、光纤等智能材料以及先进的控制理论减少振动、提高复合体结构的可靠性和弯曲载荷。

## 2.3 附 录 一

### 2.3.1 附录 A

最常用的增强连续纤维的典型特性如表 2.a 所列。

表 2.a 最常用的增强连续纤维的典型特性

| 材 料 | 商品名称 | 密度$\rho$/($kg/m^3$) | 典型纤维直径/$\mu m$ | 杨氏模量 $E$/GPa | 抗拉强度/GPa |
| --- | --- | --- | --- | --- | --- |
| $\alpha-Al_2O_3$（氧化铝） | FP（美国） | 3960 | 20 | 385 | 1.8 |
| $Al_2O_3+SiO_2+B_2O_3$（莫来石） | Nextel480（美国） | 3050 | 11 | 224 | 2.3 |
| $Al_2O_3+SiO_2$（氧化铝-二氧化硅） | Altex（日本） | 3300 | 10~15 | 210 | 2.0 |
| 硼（钨上的 CVD[①]） | VMC（日本） | 2600 | 140 | 410 | 4.0 |
| 碳（PAN[②] 前体） | T300（日本） | 1800 | 7 | 230 | 3.5 |
| 碳（PAN[②] 前体） | T800（日本） | 1800 | 5.5 | 295 | 5.6 |
| 碳（沥青[③] 前体） | Thorne IP755（美国） | 2060 | 10 | 517 | 2.1 |
| SiC（+O）（碳化硅） | Nicalon（日本） | 2600 | 15 | 190 | 2.5~3.3 |
| SiC（低 O）（碳化硅） | Hi-Nicalon（日本） | 2740 | 14 | 270 | 2.8 |

续表

| 材　　料 | 商品名称 | 密度 $\rho$/ (kg/m³) | 典型纤维直径 /μm | 杨氏模量 $E$ /GPa | 抗拉强度 /GPa |
|---|---|---|---|---|---|
| SiC（+O+Ti）（碳化硅） | Tyranno（日本） | 2400 | 9 | 200 | 2.8 |
| SiC（单丝）（碳化硅） | Sigma | 3100 | 100 | 400 | 3.5 |
| E-玻璃（二氧化硅） |  | 2500 | 10 | 70 | 1.5~2.0 |
| 石英（二氧化硅） |  | 2200 | 3~15 | 80 | 3.5 |
| 芳香族聚酰胺 | Kevlar 49（美国） | 1500 | 12 | 130 | 3.6 |
| 聚乙烯（UHMW）[④] | Spectra 100（美国） | 970 | 38 | 175 | 3.0 |
| 高碳钢 | 如钢琴线 | 7800 | 250 | 210 | 2.8 |
| 铝 | 电线 | 2680 | 1670 | 75 | 0.27 |
| 钛 | 线 | 4700 | 250 | 115 | 0.434 |

资料来源：Harris, B., Engineering composite materials, The Institute of Materials, London, UK, 1999, 193p. and Jones, R. M., Mechanics of composite materials, 2nd edition, Taylor & Francis, Philadelphia, PA 19106, USA, 1999, 519p.

① CVD，化学气相沉积。
② PAN，聚丙烯腈，大约90%的碳纤维是由PAN制成的。
③ 沥青表示由芳烃组成的黏弹性材料。沥青是通过对碳基材料（如植物、原油和煤炭）进行蒸馏而产生的。
④ UHMW，超高分子量聚乙烯（或聚乙烯，世界上最常见的塑料）是热塑性聚乙烯的子集。

# 参 考 文 献

[1] Ashton, J. E., Halpin, J. C. and Petit, P. H., Primer on Composite Materials: Analysis, Progress in Material Science Series, Vol. III, Technomic publication, Library of Congress Catalog Card No. 72-81344, 1969.

[2] Tsai, S. W. and Pagano, N. J., Invariant properties of composite materials. In: Composite Materials Workshop, Tsai, S. W., Halpin, J. C. and Pagano, N.J. (eds.), St. Louis, Missouri, 1967, Technomic Publishing Company, 1968, 233-253.

[3] Reddy, J. N., Mechanics of Laminated Composite Plates and Shells, 2nd edn., CRC Press LLC, 2004.

[4] Harris, B., Engineering Composite Materials, London, UK, The Institute of Materials, 1999, 193 p.

[5] Love, A. E. H., On the small free vibrations and deformations of elastic shells, Philosophical Transaction of the Royal Society (London) Series A 17, 1888, 491-549.

[6] Reddy, J. N., Theory and Analysis of Elastic Plates and Shells, CRC Press, Taylor and Francis,

2007, 568.

[7] Timoshenko, S. and Woinowsky-Krieger, S., Theory of Plates and Shells, New York, McGrawHill, 1959, 580.

[8] Mindlin, R. D., Influence of rotatory inertia and shear on flexural motions of isotropic, elastic plates, ASME Journal of Applied Mechanics 18, 1951, 31-38.

[9] Reddy, J. N., Theory and Analysis of Elastic Plates, Philadelphia, Taylor and Francis, 1999.

[10] Lim, G. T. and Reddy, J. N., On canonical bending relationships for plates, International, Journal of Solids and Structures 40, 2003, 3039-3067.

[11] Timoshenko, S. P., On the correction factor for shear of the differential equation for transverse vibrations of bars of uniform cross-section, Philosophical Magazine, 1921, 744.

[12] Timoshenko, S. P., On the transverse vibrations of bars of uniform cross-section, Philosophical Magazine, 1922, 125.

[13] Rosinger, H. E. and Ritchie, I. G., On Timoshenko's correction for shear in vibrating isotropic beams, Journal of Physics D: Applied Physics 10, 1977, 1461-1466.

[14] Timoshenko, S. P. and Gere, J. M., Mechanics of Materials, Van Nostrand Reinhold Co, 1972.

[15] Chailleux, A., Hans, Y. and Verchery, G., Experimental study of the buckling of laminated composite columns and plates, International Journal of Mechanical Sciences 17 (8), August 1975, 489-498.

[16] Wilson, D. W. and Vinson, J. R., Viscoelastic Buckling Analysis of Laminated Composite Columns, Recent Advances in Composites in the United States and Japan, ASTM STP 864, 1985.

[17] Tanigawa, Y., Murakami, H. and Ootao, Y., Transient thermal stress analysis of a laminated composite beam, Journal of Thermal Stresses 12 (1), 1989, 25-39.

[18] Ootao, Y., Tanigawa, Y. and Murakami, H., Transient thermal stress and deformation of a laminated composite beam due to partially distributed heat supply, Journal of Thermal Stresses 13 (2), 1990, 193-206.

[19] Chandrashekhara, K., Krishnamurthy, K. and Roy, S., Free vibration of composite beams including rotary inertia and shear deformation, Composite Structures 14 (4), 1990, 269-279.

[20] Singh, G. and Venkateswara, R., Analysis of the nonlinear vibrations of unsymmetrically laminated composite beams, AAIA Journal 29 (10), October 1991, 1727-1735.

[21] Barbero, E. and Tomblinj, J., Buckling testing of composite columns, AIAA Journal 30 (11), November 1991, 2798-2800.

[22] Qatu, M. S., In-plane vibration of slightly curved laminated composite beams, Journal of Sound and Vibration 159 (2), 8 December 1992, 327-338. doi: 10.1016/0022-460x(92) 90039-z.

[23] Barbero, E. J. and Raftoyiannis, I. G., Euler buckling of pultruded composite columns, Composite Structures 24 (2), 1993, 139-147.

[24] Abramovich, H. and Livshits, A., Free vibrations of non-symmetric cross-ply laminated composite beams, Journal of Sound and Vibration 176 (5), 6 October 1994, 597-612.

[25] Owen, B., Gurdal, Z. and Lee, J., Buckling analysis of laminated composite beams with multiple delaminations, AAIA-94-1573-CP paper, 35th Structures, Structural Dynamics, and Materials Conference Hilton Head, SC, U.S.A., 1994. doi: 10.2514/6.1994-1573.

[26] Maiti, D. K. and Sinha, P. K., Bending and free vibration analysis of shear deformable laminated com-

posite beams by finite element method, Composite Structures 29 (4), 1994, 421-431.

[27] Abramovich, H., Eisenberger, M. and Shulepov, O., Vibrations and buckling of non-symmetric laminated composite beams via the exact element method, AIAA/ASME/ASCE/AHS/ASC Structures, Structural Dynamics, and Materials Conference, 36th, and AIAA/ASME Adaptive Structures Forum, New Orleans, LA, USA, 1995. doi: http://avc.aiaa.org/doi/abs/10.2514/6.1995-1459.

[28] Kim, Y., Davalos, J. F. and Barbero, E. J., Progressive Failure Analysis of Laminated Composite Beams, Journal of Composite Materials 30 (5), March 1996, 536-560.

[29] Abramovich, H., Eisenberger, M. and Shulepov, O., Vibrations and Buckling of Cross-Ply Nonsymmetric Laminated Composite Beams, AIAA Journals 34 (5), May 1996, 1064-1069.

[30] Gadelrab, R. M., The effect of delamination on the natural frequencies of a laminated composite beam, Journal of Sound and Vibration 197 (3), 31 October 1996, 283-292.

[31] Song, S. J. and Waas, A. M., Effects of shear deformation on buckling and free vibration of laminated composite beams, Composite Structures 37 (1), January 1997, 33-43.

[32] Bhattacharya, P., Suhail, H. and Sinha, P. K., Finite Element Free Vibration Analysis of Smart Laminated Composite Beams and Plates, Journal of Intelligent Material Systems and Structures 9 (1), January 1998, 20-28.

[33] Kadivar, M. H. and Mohebpour, S. R., Forced vibration of unsymmetric laminated composite beams under the action of moving loads, Composites Science and Technology 58 (10), October 1998, 1675-1684.

[34] Yıldırım, V., Governing equations of initially twisted elastic space rods made of laminated composite materials, International Journal of Engineering Science 37 (8), June 1999, 1007-1035.

[35] Moradi, S. and Taheri, F., Delamination buckling analysis of general laminated composite beams by differential quadrature method, Composites Part B: Engineering 30 (5), July 1999, 503-511.

[36] Lee, J. J. and Choi, S., Thermal buckling and post-buckling analysis of a laminated composite beam with embedded SMA actuator, Composite Structures 47 (1-4), December 1999, 695-703.

[37] Rehfield, L. W. and Mueller, U., Design of thin-walled laminated composite to resist buckling, AIAA paper 99-1375, 40th Structures, Structural Dynamics, and Materials Conference and Exhibit St. Louis, MO, U.S.A, 1999. http://arc.aiaa.org/doi/pdf/10.2514/6.1999-1375.

[38] Khdeir, A. A. and Reddy, J. N., Jordan canonical form solution for thermally induced deformations of cross-ply laminated composite beams, Journal of Thermal Stresses 22 (3), 1999, 331-346.

[39] Khdeir, A. A., Thermal buckling of cross-ply laminated composite beams, Acta Mechanica 149 (1-4), 2001, 201-213.

[40] Lee, S., Park, T. and Voyiadjis, G. Z., Free vibration analysis of axially compressed laminated composite beam-columns with multiple delaminations, Composites Part B: Engineering 33 (8), December 2002, 605-617.

[41] Fares, M. E., Youssif, Y. G. and Hafiz, M. A., Optimization control of composite laminates for maximum thermal buckling and minimum vibrational response, Journal of Thermal Stresses 25 (11), 2002, 1047-1064.

[42] Goyal, V. K. and Kapania, R. K., Dynamic Stability of Laminated Composite Beams Subject to Subtangential Loads, 44th AIAA/ASME/ASCE/AHS Structures, Structural Dynamics, and Materials Conference, AAIA Journals, April 2003. doi: 10.2514/6.2003-1930.

## 第2章 层压复合材料

[43] Anilturk, D. and Chan, W. S., Structural stability of composite laminated column exposed to high temperature or fire, Journal of Composite Materials 37 (8), April 2003, 687-700.

[44] Zhang, Z. and Taheri, F., Dynamic pulse buckling and postbuckling of composite laminated beam using higher order shear deformation theory, Composites Part B: Engineering 34 (4), 1 June 2003, 391-398.

[45] Zhang, Z. and Taheri, F., Dynamic pulse-buckling behavior of 'quasi-ductile' carbon/epoxy and E-glass/epoxy laminated composite beams, Composite Structures 64 (3-4), June 2004, 269-274.

[46] Çallioğlu, H., Tarakcilar, A. R. and Bektaş, N. B., Elastic-plastic stress analysis of laminated composite beams under linear temperature distribution, Journal of Thermal Stresses 27 (11), 2004, 1075-1088.

[47] Ganesan, R. and Kowda, V. K., Free-vibration of composite beam-columns with stochastic material and geometric properties subjected to random axial loads, Journal of Reinforced Plastics and Composites 24 (1), January 2005, 69-91.

[48] Ganesan, R. and Kowda, V. K., Buckling of composite beam-columns with stochastic properties, Journal of Reinforced Plastics and Composites 24 (5), March 2005, 513-543.

[49] Vengallatore, S., Analysis of thermoelastic damping in laminated composite micromechanical beam resonators, Journal of Micromechanics and Microengineering 15 (12), 2005, 2398-2404.

[50] Lee, J., Lateral buckling analysis of thin-walled laminated composite beams with monosymmetric sections, Engineering Structures 28 (14), December 2006, 1997-2009.

[51] Aydogdu, M., Thermal buckling analysis of cross-ply laminated composite beams with general boundary conditions, Composites Science and Technology 67 (10), May 2007, 1096-1104.

[52] Kiral, Z. and Kiral, B. G., Dynamic analysis of a symmetric laminated composite beam subjected to a moving load with constant velocity, Journal of Reinforced Plastics and Composites 27 (1), January 2008, 19-32.

[53] Li, J., Hua, H. and Shen, R., Dynamic stiffness analysis for free vibrations of axially loaded laminated composite beams, Composite Structures 84 (1), June 2008, 87-89.

[54] Chai, G. B. and Yap, C. W., Coupling effects in bending, buckling and free vibration of generally laminated composite beams, Composites Science and Technology 68 (7-8), June 2008, 1664-1670.

[55] Zhen, W. and Wanji, C., An assessment of several displacement-based theories for the vibration and stability analysis of laminated composite and sandwich beams, Composite Structures 84 (4), August 2008, 337-349.

[56] Atlihan, G., Çallioğlu, H., Conkur, E. Ş., Topcu, M. and Yücel, U., Free vibration analysis of the laminated composite beams by using DQM, Journal of Reinforced Plastics and Composites 28 (7), April 2009, 881-892.

[57] Kiral, B. G. and Kiral, Z., Effect of elastic foundation on the dynamic response of laminated composite beams to moving loads, Journal of Reinforced Plastics and Composites 28 (8), April 2009, 913-935.

[58] Kiral, Z., Malgaca, L., Malgaca, M. and Kiral, B. G., Experimental investigation of the dynamic response of a symmetric laminated composite beam via laser vibrometry, Journal of Composite Materials 43 (24), November 2009, 2943-2962.

[59] Chai, G. B., Yap, C. W. and Lim, T. M., Bending and buckling of a generally laminated composite beam-column, Proceedings of the Institution of Mechanical Engineers, Part L: Journal of Materials:

Design and Applications 224 (1), 01 January 2010, 1-7.
[60] Campbell, F. C., Introduction to Composite Materials, ASM International, #05287G documents/10192/1849770/05287G, November 2010, 599 pp.
[61] Baghani, M., Jafari-Talookolaei, R. A. and Salarieh, H., Large amplitudes free vibrations and post-buckling analysis of unsymmetrically laminated composite beams on nonlinear elastic foundation, Applied Mathematical Modelling 35 (1), January 2011, 130-138.
[62] Lezgy-Nazargah, M., Shariyat, M. and Beheshti-Aval, S. B., A refined high-order global-local theory for finite element bending and vibration analyses of laminated composite, Acta Mechanica 217 (3-4), March 2011, 219-242.
[63] Jafari-Talookolaei, R. A., Salarieh, H. and Kargarnovin, M. H., Analysis of large amplitude free vibrations of unsymmetrically laminated composite beams on a nonlinear elastic foundation, Acta Mechanica 219 (1-2), June 2011, 65-75.
[64] Vidal, P. and Polit, O., A sine finite element using a zig-zag function for the analysis of laminated composite beams, Composites Part B: Engineering 42 (6), September 2011, 1671-1682.
[65] Jun, L. and Hongxing, H., Free vibration analyses of axially loaded laminated composite beams based on higher-order shear deformation theory, Meccanica 46 (6), December 2011, 1299-1317.
[66] Vosoughi, A. R., Malekzadeh, P., Banan, M. R. and Banan, M. R., Thermal buckling and postbuckling of laminated composite beams with temperature-dependent properties, International Journal of Non-Linear Mechanics 47 (3), April 2012, 96-102.
[67] Jafari-Talookolaei, R. A., Abedi, M., Kargarnovin, M. H. and Ahmadian, M. T., An analytical approach for the free vibration analysis of generally laminated composite beams with shear effect and rotary inertia, International Journal of Mechanical Sciences 65 (1), December 2012, 97-104.
[68] Kargarnovin, M. H., Ahmadian, M. T., Jafari-Talookolaei, R. -A. and Abedi, M., Semi-analytical solution for the free vibration analysis of generally laminated composite Timoshenko beams with single delamination, Composites Part B: Engineering 45 (1), February 2013, 587-600.
[69] Kim, N. -I. and Lee, J., Lateral buckling of shear deformable laminated composite I-beams using the finite element method, International Journal of Mechanical Sciences 68, March 2013, 246-257.
[70] Erkliğ, A., Yeter, E. and Bulut, M., The effects of cut-outs on lateral buckling behavior of laminated composite beams, Composite Structures 104, October 2013, 54-59.
[71] Fu, Y., Wang, J. and Hu, S., Analytical solutions of thermal buckling and postbuckling of symmetric laminated composite beams with various boundary conditions, Acta Mechanica 225 (1), January 2014, 13-29.
[72] Li, J., Wu, Z., Kong, X., Li, X. and Wu, W., Comparison of various shear deformation theories for free vibration of laminated composite beams with general lay-ups, Composite Structures 108, February 2014, 767-778.
[73] Abadi, M. M. and Daneshmehr, A. R., An investigation of modified couple stress theory in buckling analysis of micro composite laminated Euler-Bernoulli and Timoshenko beams, International Journal of Engineering Science 75, February 2014, 40-53.
[74] Huo, J. L., Li, X., Kong, X. and Wu, W., Vibration analyses of laminated composite beams using refined higher-order shear deformation theory, International Journal of Mechanics and Materials in Design 10

(1), March 2014, 43-52.

[75] Lanc, D., Turkalj, G. and Pesic, I., Global buckling analysis model for thin-walled composite laminated beam type structures, Composite Structures 111, May 2014, 371-380.

[76] Mohandes, M. and Ghasemi, A. R., Finite strain analysis of nonlinear vibrations of symmetric laminated composite Timoshenko beams using generalized differential quadrature method, Journal of Vibration and Control, 30 June 2014. doi: 10.1177/1077546314538301.

[77] Kuehn, T., Pasternak, H. and Mittelstedt, C., Local buckling of shear-deformable laminated composite beams with arbitrary cross-sections using discrete plate analysis, Composite Structures 113, July 2014, 236-248.

[78] Carrera, E., Filippi, M. and Zappino, E., Free vibration analysis of laminated beam by polynomial, trigonometric, exponential and zig-zag theories, Journal of Composite Materials 48 (19), August 2014, 2299-2316.

[79] Panahandeh-Shahraki, D., Mirdamadi, H. R. and Vaseghi, O., Thermoelastic buckling analysis of laminated piezoelectric composite plates, International Journal of Mechanics and Materials in Design, 11, October 2014, 371. doi: 10.1007/S10999-014-9284-8.

[80] Sahoo, R. and Singh, B. N., A new trigonometric zigzag theory for buckling and free vibration analysis of laminated composite and sandwich plates, Composite Structures 117, November 2014, 316-332.

[81] Qu, Y., Wu, S., Li, H. and Meng, G., Three-dimensional free and transient vibration analysis of composite laminated and sandwich rectangular parallelepipeds: Beams, plates and solids, Composites Part B: Engineering 73, May 2015, 96-110.

[82] Li, Z.-M. and Qiao, P., Buckling and postbuckling behavior of shear deformable anisotropic laminated beams with initial geometric imperfections subjected to axial compression, Engineering Structures 85 (15), February 2015, 277-292.

[83] Raja, S., Rohwer, K. and Rose, M., Piezothermoelastic modeling and active vibration control of laminated composite beams, Journal of Reinforced Plastics and Composites 10 (11), November 1999, 890-899.

[84] Choi, S., Lee, J. J., Seo, D. C. and Choi, S. W., The active buckling control of laminated composite beams with embedded shape memory alloy wires, Composite Structures 47 (1-4), December 1999, 679-686.

[85] Jeon, B. S., Lee, J. J., Kim, J. K. and Huh, J. S., Low velocity impact and delamination buckling behavior of composite laminates with embedded optical fibers, Smart Materials and Structures 8 (1), 1999, 41-48.

[86] Lee, H. J. and Lee, J. J., A numerical analysis of the buckling and post-buckling behavior of laminated composite shells with embedded shape memory alloy wire actuators, Smart Materials and Structures 9 (6), 2000, 780-788.

[87] Sun, B. and Huang, D., Vibration suppression of laminated composite beams with a piezoelectric damping layer, Composite Structures 53 (4), September 2001, 437-447.

[88] Subramanian, P., Vibration suppression of symmetric laminated composite beams, Smart Materials and Structures 11 (6), 2002, 880-886.

[89] Ray, M. C. and Mallik, N., Active control of laminated composite beams using a piezoelectric fiber reinforced composite layer, Smart Materials and Structures 13 (1), 2004, 146-153.

[90] Pradhan, S. C. and Reddy, J. N., Vibration control of composite shells using embedded actuating layers, Smart Materials and Structures 13 (5), 2004, 1245-1258.

[91] Zabihollah, A., Ganesan, R. and Sedaghati, R., Sensitivity analysis and design optimization of smart laminated beams using layerwise theory, Smart Materials and Structures 15 (6), 2006, 1775-1785.

[92] Zhou, H.-M. and Zhou, Y.-H., Vibration suppression of laminated composite beams using actuators of giant magnetostrictive materials, Smart Materials and Structures 16 (1), 2007, 198-207.

[93] Zabihollah, A., Sedagahti, R. and Ganesan, R., Active vibration suppression of smart laminated beams using layerwise theory and an optimal control strategy, Smart Materials and Structures 16 (6), 2007, 2190-2202.

[94] Foda, M. A., Almajed, A. A. and El Madany, M. M., Vibration suppression of composite laminated beams using distributed piezoelectric patches, Smart Materials and Structures 19 (11), 2010, article ID 115018.

[95] Sarangi, S. K. and Ray, M. C., Smart damping of geometrically nonlinear vibrations of laminated composite beams using vertically reinforced 1-3 piezoelectric composites, Smart Materials and Structures 19 (7), 2010, article ID 075020.

[96] Asadi, H., Bodaghi, M., Shakeri, M. and Aghdam, M. M., An analytical approach for nonlinear vibration and thermal stability of shape memory alloy hybrid laminated composite beams, European Journal of Mechanics-A/Solids 42, November-December 2013, 454-468.

[97] Gei, M., Springhetti, R. and Bortot, E., Performance of soft dielectric laminated composites, Smart Materials and Structures 22 (10), 2013, article ID 104014.

[98] Mareishi, S., Rafiee, M., He, X. Q. and Liew, K. M., Nonlinear free vibration, postbuckling and nonlinear static deflection of piezoelectric fiber-reinforced laminated composite beams, Composites Part B: Engineering 59, March 2014, 123-132.

# 第 3 章 压电材料

## 3.1 本构方程

Arthur R. von Hippel 在他的开创性著作《介电材料与应用》(Artech House 出版社，1995 年 12 月 19 日，485 页）中定义介电材料为"电介质……不是狭义的所谓绝缘体，而是从其与电场、磁场或电磁场相互作用的角度来考虑的广泛的非金属。因此，我们不仅要考虑气体，而且要考虑液体和固体，还要研究电能和磁能的储存与耗散"。

介电材料是一种能被外加电场极化的电绝缘体（图 3.1）。当电介质置于电场中时，电荷不会像在导体中那样流过材料，而是只会从它们的平均平衡位置轻微偏移，从而引起电介质极化。由于介电极化，正电荷沿电场方向移动，而负电荷向相反的方向移动。这就产生了一个内部电场，减小了电介质自身内部的整体电场。如果电介质是由弱键分子组成的，那么这些分子不仅会被极化，还会重新定向，从而使它们的对称轴与电场对齐[①]。

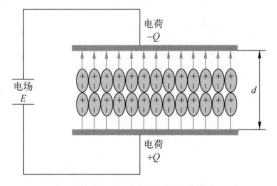

图 3.1 介电材料的示意图

---

① 摘自维基百科，en.wikipedia.org/wiki/Dielectric。

### 3.1.1 介电材料分类

所有介电材料在受到外部电场作用时都会发生尺寸变化。这是由于材料内部正负电荷的位移造成的。电介质晶格可以认为是由阳离子（失去电子而带正电的原子）和阴离子（获得电子而带负电的原子）通过弹簧（离子间化学键）连接起来的。当材料受到外部电场作用时，阳离子在电场方向发生位移，阴离子在相反方向发生位移，从而导致材料产生净变形。这样的尺寸变化可能非常小，也可能非常显著，这取决于电介质所属的晶体类型。

在所有的32种晶体类型中，11个为中心对称的（即具有对称中心或反转中心），21个为非中心对称的（即不具有对称中心）。

当具有对称中心的介电材料受到外部电场作用时，其对称性（反转中心）、阳离子和阴离子的位移使相邻弹簧（化学键）的伸展和收缩相互抵消，因此晶体的净变形在理想状态下为零。由于非完全谐振以及非谐振性，化学键出现二阶效应，导致晶格具有小的净变形。这种情况下的变形与电场的平方成正比。也就是说，形变与外加电场的方向无关。这种效应称为电致伸缩效应。非谐振效应存在于所有电介质中，因此可以说，所有的电介质都是电致伸缩材料。

当属于非中心对称类型（除八面体类型外）的介电材料受到外部电场作用时，相邻离子会发生不对称运动，导致晶体发生显著的变形，而这种变形与所施加的电场成正比。由于键的非谐振性，这些材料表现出电致伸缩效应，但其被更显著的非对称位移所掩盖。这种材料称为压电材料。根据对外部刺激的响应对介电材料进行分类，如图3.2所示。

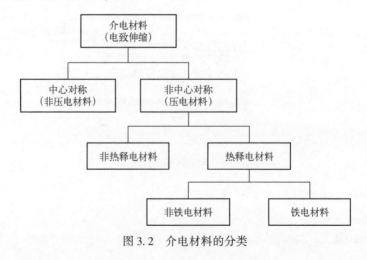

图3.2 介电材料的分类

每一种这类材料均表现出某些特殊的特性,使其成为重要的工程材料。因为这些材料表现出固有的传感器特性,所以它们属于智能材料。

## 3.1.2 重要的介电参数

1. 电偶极矩

在原子或者分子中,当正负电荷的中心相隔一定距离 $d$ 时,原子或者分子具有的电偶极矩为

$$p = qd \tag{3.1}$$

式中:$q$ 为电荷;$d$ 为正负电荷中心的距离;$p$ 为从负电荷到正电荷方向的矢量(C·m)。

2. 极性和非极性介电材料

介电材料可分为极性与非极性两种。在非极性介电材料中,由于正负电荷的中心重合,原子通常不具有电偶极矩。典型的非极性电介质包括氧气、氮气、苯、甲烷等。当这些材料受到外部电场作用时,正电荷和负电荷的中心就会分开,从而产生偶极矩。一旦电场移除,诱导偶极矩就会消失。

在极性介电材料中,由于正负电荷的中心不重合,每个原子或分子都具有偶极矩。典型的极性电介质包括水、HCl、乙醇和 $NH_3$。大部分的陶瓷和聚合物都属于这一类。当外加电场作用于这些材料时,电偶极子趋于朝向电场的方向。

3. 电极化强度

极性介电材料由大量具有电偶极矩的原子或分子组成。介电材料的总偶极矩或净偶极矩是所有单个偶极矩的矢量和,可表示为

$$P_{total} = \sum_i p_i \tag{3.2}$$

电极化强度 $P$ 定义为单位体积的总偶极矩,可表示为

$$P = \frac{\sum_i p_i}{Vol} \tag{3.3}$$

式中:Vol 为材料的体积;$P$ 的单位是库仑/平方米($C/m^2$),也称为表面电荷密度,$P$ 为垂直于材料表面的矢量。

通常,在极性电介质中,每个电偶极子都是随机取向的,所以净极化为零。当施加电场时,单个偶极子朝向趋于电场方向,材料会产生有限的极化(图3.1)。极化随着电场的增大而增加,当所有的偶极矩都朝向外加电场的方向时达到饱和。

4. 介电位移（通量密度）、介电常数和电极化率

当电场 $E$ 作用于介电材料时，材料产生有限极化 $P$（非极性材料中的诱导极化和极性材料中的取向极化）。由外场 $E$ 产生的材料内部的电通量密度 $D$ 为

$$D = \varepsilon_0 E + P \tag{3.4}$$

式中：$\varepsilon_0$ 为自由空间的介电常数。

$D$ 也可以表示为

$$D = \bar{\varepsilon} E = \varepsilon_0 \varepsilon_r E \tag{3.5}$$

式中：$\varepsilon_r$ 为介电材料的相对介电常数或介电常数；$D$ 的单位为 $C/m^2$（与 $P$ 相同）。

介电常数 $\varepsilon_r$ 也可以由下式的比值来定义：

$$\varepsilon_r = \frac{D}{\varepsilon_0 E} \tag{3.6}$$

注意：$\varepsilon_r$ 是一个无量纲量，并且总是大于 1。

极化矢量 $P$ 通过以下关系与外加电场直接相关：

$$P = \varepsilon_0 \chi E \tag{3.7}$$

式中：$\varepsilon_0$ 为自由空间的介电常数（式（3.4））；$\chi$ 为材料的电极化率。

由上述方程可以得到介电常数 $\varepsilon_r$ 与电极化率 $\chi$ 的关系为

$$\varepsilon_r = 1 + \chi \tag{3.8}$$

5. 压电效应

非中心对称晶体类的介电材料属于压电材料。当这些材料受到外部电场的作用时，阴离子和阳离子会发生不对称的位移，从而导致晶体产生相当大的净变形。与电致伸缩材料中应变与电场的平方（$E^2$）成正比不同，压电材料所得到的应变与所施加的电场成正比。压电材料中的应变可能是伸长的或压缩的，这取决于所施加外场的极性。这种效应称为压电效应，或者更准确地说，是逆压电效应。

压电材料还有另一个独特的性质，当它们受到压力/应力而产生外部应变时，晶体中的电偶极子会取向，使晶体在相反的表面上产生正电荷和负电荷，从而在整个晶体中产生电场。这正好是上面提到的逆压电效应的逆效应。1880年，皮埃尔·居里（Pierre Curie）和雅克·居里（Jacques Curie）兄弟第 1 次在石英晶体中观察到这种效应并称为压电效应。压电一词源自希腊语，意思是压力产生的电（piesi 在希腊语中是压力的意思），这种效应称为正压电效应。

图 3.3 和图 3.4 分别为正压电效应和逆压电效应的示意图。在正压电效应中，当极化的压电材料受到平行于极化方向的拉应力时，其表面会产生正电压（图 3.3（b））。当材料受到极化方向上的压应力时，会在其表面产生负电压

(图 3.3（c））。在逆压电效应中，当材料受到外部电场作用时，若电压的极性与极化时施加电场的极性相同，则材料会伸长（图 3.4（b））；而当施加相反方向的电压时，材料会被压缩（图 3.4（c））。

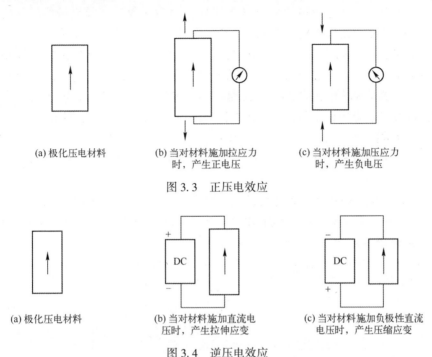

(a) 极化压电材料　　(b) 当对材料施加拉应力时，产生正电压　　(c) 当对材料施加压应力时，产生负电压

图 3.3　正压电效应

(a) 极化压电材料　　(b) 当对材料施加直流电压时，产生拉伸应变　　(c) 当对材料施加负极性直流电压时，产生压缩应变

图 3.4　逆压电效应

图 3.5 所示为交流电场对极化压电材料的影响。交流电场使材料以与所施加电场相同的频率交替伸缩。该振动在材料附近产生一个声场（声音或超声波场）。这种效应被用于声场的产生。

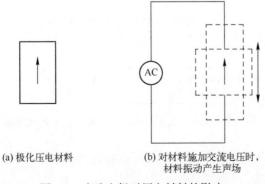

(a) 极化压电材料　　(b) 对材料施加交流电压时，材料振动产生声场

图 3.5　交流电场对压电材料的影响

图 3.4 中所描述的行为是针对施加直流电压的。在施加交流电压的情况下（图 3.5），材料会根据所施加交流电压的频率交替伸缩，并在其附近产生一个声场。

正压电效应和逆压电效应有许多应用，因为该效应涉及机械能转换为电能，反之亦然。其应用包括超声波的产生和检测、压力传感器和致动器，超声在工程和医疗领域都有广泛的应用。在工程中，它被用于材料的无损检测（Nondestructive Testing，NDT）、水下声学（声纳）、超声波钻孔、能量获取等；而在医疗领域中，它被用于诊断（超声检查）、治疗（药物输送）和手术。作为传感器和致动器，它们在工程和医疗领域都有广泛的应用。

6. 压电系数

压电材料将机械能转化为电能（正压电效应）和将电能转化为机械能（逆压电效应）。

在正压电效应中，输入是机械能，输出是电能。机械能输入的形式可以是外应力（$\sigma$）或应变（$\varepsilon$）。电能输出的形式为表面电荷密度（$D$ 或 $P$）、电场（$E$）或电压（$V$）（图 3.6）。

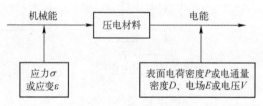

图 3.6　正压电效应：输入是机械能，输出是电能

在逆压电效应中，输入是电能，输出是机械能。电能输入的形式是表面电荷密度（$P$ 或 $D$）、电场（$E$）或电压（$V$），机械能输出的形式是材料上的应变（$\varepsilon$）或应力（$\sigma$）（图 3.7）。

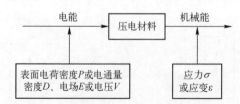

图 3.7　逆压电效应：输入是电能，输出是机械能

描述压电材料灵敏度的参数是与输入和输出参数相关的压电系数。接下来定义各种压电系数。

在正压电效应中，将机械输入应变 $\varepsilon$ 与电输出（形式为 $D$ 和 $E$）关联起来的公式为

$$D = e\varepsilon \quad (3.9)$$

$$E = h\varepsilon \quad (3.10)$$

而将机械输入应力 $\sigma$ 与电输出（形式为 $D$ 和 $E$）关联起来的公式为

$$D = d\sigma \quad (3.11)$$

$$E = g\sigma \quad (3.12)$$

式中：$e$、$h$、$d$ 和 $g$ 是描述正压电效应的压电系数。

描述逆压电效应的压电系数用 $e^*$、$d^*$、$h^*$ 和 $g^*$ 表示。它们由以下关系式定义：

$$\sigma = e^* E \quad (3.13)$$

$$\varepsilon = d^* E \quad (3.14)$$

$$\sigma = h^* D \quad (3.15)$$

$$\varepsilon = g^* D \quad (3.16)$$

式（3.13）和式（3.14）将输入电场（$E$）与机械输出（$\sigma$ 和 $\varepsilon$）关联起来，而式（3.15）和式（3.16）将输入电荷密度（$D$）与机械输出（$\sigma$ 和 $\varepsilon$）关联起来。

利用 $D$ 与 $E$ 的关系（式（1.5）），得到各压电系数之间的关系如下：

$$h = \frac{e}{\varepsilon}, \quad h^* = \frac{e^*}{\varepsilon}, \quad g = \frac{d}{\varepsilon}, \quad g^* = \frac{d^*}{\varepsilon} \quad (3.17)$$

压电效应是一种瞬态效应，即观测到的参数不是绝对值，而是参数的变化。在正压电效应中，应变的变化 $\partial\varepsilon$（或应力的变化 $\partial\sigma$）导致极化的变化 $\partial D$（或电场的变化 $\partial E$），而在逆压电效应中，外加电场的变化 $\partial E$（或极化的变化 $\partial D$）导致应变的变化 $\partial\varepsilon$（或应力的变化 $\partial\sigma$）。由于 $D$ 和 $P$ 通过式（3.4）相关联，因此，对于恒定的 $E$，$D$、$\partial D$ 的变化可以由 $P$、$\partial P$ 的变化来代替。

上面定义的各种压电系数用下面的偏导数来定义更恰当，对于正压电系数可表示为

$$d = \left(\frac{\partial D}{\partial X}\right)_E, \quad g = -\left(\frac{\partial E}{\partial X}\right)_D, \quad e = \left(\frac{\partial D}{\partial x}\right)_E, \quad h = -\left(\frac{\partial E}{\partial x}\right)_D \quad (3.18)$$

而对逆压电系数，则为

$$d^* = \left(\frac{\partial \varepsilon}{\partial E}\right)_\sigma, \quad g^* = \left(\frac{\partial \varepsilon}{\partial D}\right)_\sigma, \quad e^* = -\left(\frac{\partial \sigma}{\partial E}\right)_\varepsilon, \quad h^* = -\left(\frac{\partial \sigma}{\partial D}\right)_\varepsilon \quad (3.19)$$

其中，括号外的字母表示对字母的常数值进行了推导（$\sigma$ 为应力、$\varepsilon$ 为应变、$E$ 为电场、$D$ 为电位移）。

从热力学角度，可以证明下列关系是正确的：
$$d=d^*, \quad g=g^*, \quad e=e^*, \quad h=h^* \tag{3.20}$$

表 3.1 总结了各压电系数及其各自单位的定义。压电材料用下列参数进行表征：

(1) 压电系数：$d$、$g$、$e$、$h$。
(2) 电参数：介电常数 $\overline{\varepsilon}$。
(3) 弹性参数：由字母 $s$ 和/或刚度常数 $c$ 定义的刚度常数。

表 3.1 压电系数：定义和单位

| 压电系数 | 名称 | 定义 | 数学定义 | 单位 |
| --- | --- | --- | --- | --- |
| $d$ | 压电电荷或应变常数 | 极化/应力 | $d=-\left(\dfrac{\partial D}{\partial X}\right)_E$ | $\left[\dfrac{C}{N}\right]$ |
| $g$ | 压电电压常数 | 电场/应力 | $g=-\left(\dfrac{\partial E}{\partial X}\right)_D$ | $\left[\dfrac{V\cdot m}{N}\right]$ |
| $e$ | 与应变和极化有关的压电常数 | 极化/应变 | $e=-\left(\dfrac{\partial D}{\partial x}\right)_E$ | $\left[\dfrac{C}{m^2}\right]$ |
| $h$ | 与应变与电场有关的压电常数 | 电场/应变 | $h=-\left(\dfrac{\partial E}{\partial X}\right)_D$ | $\left[\dfrac{V}{m}\right]$ |
| $d$ | 压电电荷或应变常数 | 应变/电场 | $d^*=\left(\dfrac{\partial \varepsilon}{\partial E}\right)_\sigma$ | $\left[\dfrac{m}{V}\right]$ |
| $g$ | 与应变和极化有关的压电常数 | 应变/极化 | $g^*=\left(\dfrac{\partial \varepsilon}{\partial E}\right)_\sigma$ | $\left[\dfrac{m^2}{C}\right]$ |
| $e$ | 与应力和电场有关的压电常数 | 应力/电场 | $e^*=-\left(\dfrac{\partial \sigma}{\partial E}\right)_\varepsilon$ | $\left[\dfrac{N}{V\cdot m}\right]$ |
| $h$ | 与应力和极化有关的压电常数 | 应力/极化 | $h^*=-\left(\dfrac{\partial \sigma}{\partial D}\right)_E$ | $\left[\dfrac{N}{C}\right]$ |

注：很容易验证这 4 对系数 $d$、$d^*$，$g$、$g^*$，$e$、$e^*$，$h$、$h^*$ 的单位相同（可以使用以下标识：$V=N\cdot m/C$）。

正压电效应的控制方程取决于机械和电的边界条件。

如果材料处于能够随着所施加电场改变尺寸的"自由"状态下，这将是一个正常条件（恒应力条件），其控制公式如下：
$$D=d\sigma+\overline{\varepsilon}^\sigma E \tag{3.21}$$
式中：$\overline{\varepsilon}^\sigma$ 为恒定应力 $\sigma$ 下的介电常数。

对于处于物理上被夹紧或者被足够高频率驱动的"夹紧"条件下的材料，器件不能对变化的电场做出响应。这称为恒应变条件，其控制方程如下：
$$D=e\sigma+\overline{\varepsilon}^\varepsilon E \tag{3.22}$$

式中：$\bar{\varepsilon}^\varepsilon$ 为恒定应变 $\varepsilon$ 下的介电常数。

逆压电效应的控制方程有以下形式（在短路材料的情况下，即恒定的电场）：

$$\varepsilon = s^E\sigma + dE$$
$$\text{或 } \sigma = c^E\varepsilon - eE \tag{3.23}$$

式中：$s^E$ 和 $c^E$ 分别为恒定电场下的弹性柔度常数和弹性刚度常数。

当材料处于开路条件（恒电荷密度）时，逆压电效应的控制方程为

$$\varepsilon = s^D\sigma + g$$
$$\text{或 } \sigma = c^D\varepsilon - hD \tag{3.24}$$

式中：$s^D$ 和 $c^D$ 分别为恒定电荷密度（恒定电位移）下的弹性柔度常数和弹性刚度常数。

有时，式（3.23）和式（3.24）表示热膨胀的贡献时可表示为

$$\varepsilon = s^E\sigma + dE + \alpha\Delta T$$
$$\text{或 } \sigma = c^E\varepsilon - eE + \alpha\Delta T \tag{3.25}$$

$$\varepsilon = s^D\sigma + gD + \alpha\Delta T$$
$$\text{或 } \sigma = c^D\varepsilon - hD + \alpha\Delta T \tag{3.26}$$

式中：$\Delta T$ 为温度升高（或降低）；$\alpha$ 为包含坐标系（$\alpha_1$、$\alpha_2$、$\alpha_3$、0、0、0）三个方向 1、2、3 的热膨胀系数的矢量。

7. 压电方程的张量形式

压电系数、弹性常数和介电常数都是张量，因为它们与矢量和张量相关。

将两个矢量 $D$ 和 $E$ 联系起来的介电常数矩阵是一个二阶张量。将两个二阶张量 $\sigma$ 和 $\varepsilon$ 联系起来的弹性柔度常数是一个四阶张量。压电系数与应力/应变（二阶张量）和电场/极化（矢量）有关，所以它们是三阶张量。坐标系为右手笛卡儿坐标系，$X$、$Y$ 和 $Z$ 轴分别用 1、2 和 3 表示，$X$、$Y$ 和 $Z$ 轴的旋转用 4、5 和 6 表示，如图 3.8 所示。

8. 压电系数 $d$ 的矩阵

正压电效应通过 $D = d\sigma$ 关系将极化 $D$ 和应力 $\sigma$ 联系起来（式（3.11））。

应力 $[\sigma]$ 是一个具有以下分量的二阶张量：

$$[\sigma] = \begin{bmatrix} \sigma_{11} & \tau_{12} & \tau_{13} \\ \tau_{21} & \sigma_{22} & \tau_{23} \\ \tau_{31} & \tau_{32} & \sigma_{33} \end{bmatrix} \tag{3.27}$$

因为 $\tau_{12} = \tau_{21}$，$\tau_{13} = \tau_{31}$，$\tau_{23} = \tau_{32}$，所以 $[\sigma]$ 是对称的，只有 6 个独立分量。

$D$ 是一个具有以下分量的矢量（其中 1、2 和 3 表示图 3.8 所示的坐

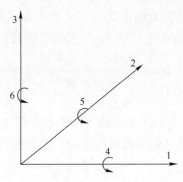

图3.8 右手笛卡儿坐标系，方向1、2和3分别代表$X$、$Y$和$Z$轴，方向4、5和6分别代表围绕$X$、$Y$和$Z$三个轴的旋转（逆时针）

标系）：

$$\begin{Bmatrix} D_1 \\ D_2 \\ D_3 \end{Bmatrix} \quad (3.28)$$

为了简单起见，对二阶张量 $\sigma_{ij}$ 采用了简化矩阵符号。取值从1到3的两个指数 $i$ 和 $j$ 被一个取值从1到6的单一指数替代，即

$$11 \equiv 1, \quad 22 \equiv 2, \quad 33 \equiv 3, \quad 23 = 32 \equiv 4, \quad 31 = 13 \equiv 5, \quad 12 = 21 \equiv 6 \quad (3.29)$$

下标1、2和3表示垂直的拉伸或压缩应力（或应变），4、5和6表示剪切应力（或剪切应变）分别围绕轴1、2和3旋转（图3.8）。因此，下列关系成立：

$$\begin{cases} \sigma_{11} = \sigma_1, \quad \sigma_{22} = \sigma_2, \quad \sigma_{33} = \sigma_3 \\ \tau_{23} = \tau_{32} = \sigma_4 \\ \tau_{31} = \tau_{13} = \sigma_5 \\ \tau_{12} = \tau_{21} = \sigma_6 \end{cases} \quad (3.30)$$

所以，式（3.11）可以写成矩阵形式，即

$$\begin{Bmatrix} D_1 \\ D_2 \\ D_3 \end{Bmatrix} = \begin{bmatrix} d_{11} & d_{12} & d_{13} & d_{14} & d_{15} & d_{16} \\ d_{21} & d_{22} & d_{23} & d_{24} & d_{25} & d_{26} \\ d_{31} & d_{32} & d_{33} & d_{34} & d_{35} & d_{36} \end{bmatrix} \begin{Bmatrix} \sigma_1 \\ \sigma_2 \\ \sigma_3 \\ \sigma_4 \\ \sigma_5 \\ \sigma_6 \end{Bmatrix} \quad (3.31)$$

因而压电系数 $d$ 用 $3\times6$ 矩阵表示。

在逆压电效应中,电场 $E$ 与应力 $\sigma$ 的关系式用矩阵形式表示为

$$\begin{Bmatrix}\sigma_1\\\sigma_2\\\sigma_3\\\sigma_4\\\sigma_5\\\sigma_6\end{Bmatrix}=\begin{bmatrix}e_{11}&e_{12}&e_{13}\\e_{21}&e_{22}&e_{23}\\e_{31}&e_{32}&e_{33}\\e_{41}&e_{42}&e_{43}\\e_{51}&e_{52}&e_{53}\\e_{61}&e_{62}&e_{63}\end{bmatrix}\begin{Bmatrix}E_1\\E_2\\E_3\end{Bmatrix} \tag{3.32}$$

因此,在逆压电效应中,压电系数 $e$ 用 $6\times3$ 矩阵表示。更多信息可见文献 [1-12]。

9. 力学参数:弹性柔度常数和弹性刚度常数

压电材料中关注的主要力学参数是弹性柔度和弹性刚度常数。这两个常数联系着两个二阶张量应力和应变,所以它们是四阶张量。

弹性柔度常数 $s$ 由下式定义:

$$\varepsilon_i = s_{ij}\sigma_j \tag{3.33}$$

弹性刚度常数 $c$ 由下式定义:

$$\sigma_i = c_{ij}\varepsilon_j \tag{3.34}$$

式中:下标 $i$ 和 $j$ 的取值都为 $1\sim6$。

10. 电介质参数:介电常数

压电材料中被关注的电介质参数是介电常数 $\bar{\varepsilon}$,它将矢量 $\boldsymbol{D}$ 和 $\boldsymbol{E}$ 联系起来,因此是一个二阶张量:

$$D_i = \bar{\varepsilon}_{ij}E_j \tag{3.35}$$

下标 $i$ 和 $j$ 取值都为 1、2 和 3。这个方程的张量形式为

$$\begin{Bmatrix}D_1\\D_2\\D_3\end{Bmatrix}=\begin{bmatrix}\bar{\varepsilon}_{11}&\bar{\varepsilon}_{12}&\bar{\varepsilon}_{13}\\\bar{\varepsilon}_{21}&\bar{\varepsilon}_{22}&\bar{\varepsilon}_{23}\\\bar{\varepsilon}_{31}&\bar{\varepsilon}_{32}&\bar{\varepsilon}_{33}\end{bmatrix}\begin{Bmatrix}E_1\\E_2\\E_3\end{Bmatrix} \tag{3.36}$$

因此,介电常数 $\bar{\varepsilon}$ 是一个 $3\times3$ 张量。

值得注意的是,由于最常见的压电材料之一锆钛酸铅(PZT)的特殊晶体结构,以及压电陶瓷材料在垂直于极化方向的平面上是各向同性的假设,上面提到的各种矩阵中的一些项会等于零或彼此相等。恒定电场 $E$(式(3.25))下 PZT 的柔度矩阵为

$$s^E = \begin{bmatrix} \dfrac{1}{E_1} & -\dfrac{v_{12}}{E_1} & -\dfrac{v_{13}}{E_1} & 0 & 0 & 0 \\ -\dfrac{v_{12}}{E_1} & \dfrac{1}{E_1} & -\dfrac{v_{13}}{E_1} & 0 & 0 & 0 \\ -\dfrac{v_{13}}{E_1} & -\dfrac{v_{13}}{E_1} & \dfrac{1}{E_3} & 0 & 0 & 0 \\ 0 & 0 & 0 & \dfrac{2(1+v_{13})}{E_3} & 0 & 0 \\ 0 & 0 & 0 & 0 & \dfrac{2(1+v_{13})}{E_3} & 0 \\ 0 & 0 & 0 & 0 & 0 & \dfrac{2(1+v_{12})}{E_1} \end{bmatrix} \quad (3.37)$$

而恒定电场 $E$（式（3.25））下，其刚度矩阵为

$$c^E = \begin{bmatrix} \dfrac{1-v_{13}^2}{E_1 E_3 \theta} & \dfrac{v_{12}+v_{13}^2}{E_1 E_3 \theta} & \dfrac{v_{13}+v_{12}v_{13}}{E_1 E_3 \theta} & 0 & 0 & 0 \\ \dfrac{v_{12}+v_{13}^2}{E_1 E_3 \theta} & \dfrac{1-v_{13}^2}{E_1 E_3 \theta} & \dfrac{v_{13}+v_{12}v_{13}}{E_1 E_3 \theta} & 0 & 0 & 0 \\ \dfrac{v_{13}+v_{12}v_{13}}{E_1 E_3 \theta} & \dfrac{v_{13}+v_{12}v_{13}}{E_1 E_3 \theta} & \dfrac{1-v_{13}^2}{E_3^2 \theta} & 0 & 0 & 0 \\ 0 & 0 & 0 & \dfrac{E_3}{2(1+v_{13})} & 0 & 0 \\ 0 & 0 & 0 & 0 & \dfrac{E_3}{2(1+v_{13})} & 0 \\ 0 & 0 & 0 & 0 & 0 & \dfrac{E_1}{2(1+v_{12})} \end{bmatrix}$$

(3.38)

式中

$$\theta = \dfrac{(1+v_{12})(1-v_{12}-2v_{13}^2)}{E_1^2 E_3} \quad (3.39)$$

则 PZT 的压电系数矩阵为

$$d = \begin{bmatrix} 0 & 0 & d_{31} \\ 0 & 0 & d_{32} \\ 0 & 0 & d_{33} \\ 0 & d_{24} & 0 \\ d_{15} & 0 & 0 \\ 0 & 0 & 0 \end{bmatrix} \quad (3.40)$$

11. 压电材料中的应变与电场的关系

图 3.9 所示为典型的铁电压电材料（PZT）的应变-电场曲线。纵向和横向模式的应变都表现出迟滞效应。开始时，应变随外加电场的增加呈线性增加，随后趋于饱和。PZT 是最著名的压电陶瓷，当电场在 1~1.5kV 范围内时，其应变约为 $10^{-3}$ 数量级，广泛应用于传感器领域。对于其他压电材料，应变则要小得多。当外加电场逐渐减小时，应变曲线并不相同。应变的变化滞后于电场的变化。当电场为零时，在材料中可观察到残余应变。若电场向相反的方向增大，则应变在一个特定的负电场处为零。若电场在同一方向上增加，则应变增加并达到饱和。因此，应变与电场之间的关系是一条对称的曲线。当反向电场再次变为零并向正方向增大时，就会形成闭环。

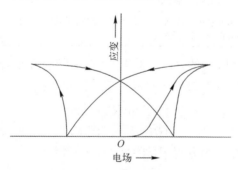

图 3.9 典型压电材料的应变与电场的关系

12. 压电耦合系数 $k$

压电耦合系数是衡量压电材料作为传感器效率的一个指标。它量化了压电材料将一种形式的能量（机械能或电能）转换为另一种形式的能量（电能或机械能）的能力。其定义为

$$k^2 = \frac{(压电能量密度)^2}{电能密度 \times 机械能密度} \quad (3.41)$$

若电能密度为 $W_e$，机械能密度为 $W_m$，则 $k^2$ 为

$$k^2 = \frac{(W_{em})^2}{W_e W_m} \tag{3.42}$$

式中：$W_{em}$ 为压电能量密度。

$k^2$ 的表达式可用压电系数来获得（详细的推导过程参见本章附录A）：

$$k^2 = \frac{d^2}{\varepsilon^\sigma S^E} \tag{3.43}$$

耦合系数是由压电元件提供的可用能量与该元件所占用的总能量之比。压电元件制造商通常指定的理论 $k$ 值在 30%~75% 的范围内。在实践中，$k$ 值取决于设备的设计、所施加刺激的方向和测量的响应。

耦合系数就像压电常数一样用下标表示外部刺激的方向和测量的方向。压电耦合系数的定义如表3.2所列。

表3.2 压电耦合系数的定义

| 符 号 | 定 义 |
| --- | --- |
| $k_{33}$ | 电场方向与机械振动方向相同且均为3时的耦合系数 |
| $k_{31}$ | 电场方向为3、机械振动方向为1时的耦合系数 |
| $k_t$ | 薄圆盘电场方向（穿过圆盘极化所在的圆盘厚度）与机械振动方向相同且均为3时的耦合系数 |
| $k_p$ | 薄圆盘电场方向（穿过圆盘极化所在的圆盘厚度）为3，机械振动方向沿径向方向时的耦合系数 |

为了理解耦合系数 $k_{ij}$ 的含义，对图3.10中所示的元素进行简单的一维分析。式（3.21）和式（3.23）可写成一维分析如下：

$$\begin{cases} \varepsilon_1 = s_{11}^E \sigma_1 + d_{31} E_3 \\ D_3 = d_{31}\sigma_1 + \overline{\varepsilon}_{33}^\sigma E_3 \end{cases} \tag{3.44}$$

从式（3.44）中的第二个方程中排除掉 $E_3$ 项，得

$$\varepsilon_1 = s_{11}^E \left[1 - \frac{d_{31}^2}{s_{11}^E \overline{\varepsilon}_{33}^\sigma}\right]\sigma_1 + \frac{d_{31}}{\overline{\varepsilon}_{33}^\sigma} D_3 \tag{3.45}$$

而将式（3.26）重写，进行一维分析，得

$$\varepsilon_1 = s_{11}^D \sigma_1 + g_{31} D_3 \tag{3.46}$$

使这两个方程相等，可得

$$\begin{cases} s_{11}^D = s_{11}^E \left[1 - \frac{d_{31}^2}{s_{11}^E \overline{\varepsilon}_{33}^\sigma}\right] \\ g_{31} = \frac{d_{31}}{\overline{\varepsilon}_{33}^\sigma} \end{cases} \tag{3.47}$$

式(3.47)的第1个方程说明了压电材料边界条件的重要性。它可以改写为开路($s_{11}^D$)与短路($s_{11}^E$)之间的关系,即

$$s_{11}^D = s_{11}^E [1-k_{31}^2]$$

或 $\quad k_{31}^2 = \dfrac{d_{31}^2}{s_{11}^E \bar{\varepsilon}_{33}^\sigma}$ \hfill (3.48)

相应地,在三个方向上也施加电场时,沿三个方向的延伸耦合系数为

$$k_{33}^2 = \dfrac{d_{33}^2}{s_{33}^E \bar{\varepsilon}_{33}^\sigma} \tag{3.49}$$

对于1-3平面的剪切模式,耦合系数为

$$k_{15}^2 = \dfrac{d_{15}^2}{s_{55}^E \bar{\varepsilon}_{11}^\sigma} \tag{3.50}$$

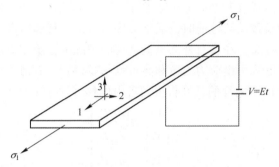

图 3.10 单轴应力下的压电元件($t$ 为厚度,$E$ 为电场)

对于正压电效应,建议使用传感器方程,因为它们是用来感知以电荷 $q$ 形式的压电输出。因此,对于 PZT 材料,式(3.21)改写为

$$D_k = d_{ki}\sigma_i + \bar{\varepsilon}_{kl}^\sigma E_l, \quad i=1,2,\cdots,6; \quad k,l=1,2,3 \tag{3.51}$$

在矩阵形式中,式(3.51)将有以下形式:

$$\begin{Bmatrix} D_1 \\ D_2 \\ D_3 \end{Bmatrix} = \begin{bmatrix} 0 & 0 & 0 & 0 & d_{15} & 0 \\ 0 & 0 & 0 & d_{15} & 0 & 0 \\ d_{31} & d_{31} & d_{33} & 0 & 0 & 0 \end{bmatrix} \begin{Bmatrix} \sigma_1 \\ \sigma_2 \\ \sigma_3 \\ \sigma_4 \\ \sigma_5 \\ \sigma_6 \end{Bmatrix} + \begin{bmatrix} \bar{\varepsilon}_{11}^\sigma & 0 & 0 \\ 0 & \bar{\varepsilon}_{22}^\sigma & 0 \\ 0 & 0 & \bar{\varepsilon}_{33}^\sigma \end{bmatrix} \begin{Bmatrix} E_1 \\ E_2 \\ E_3 \end{Bmatrix} \tag{3.52}$$

压电传感器的输出是电荷,其与电位移之间的关系如下:

$$q = \iint_A [D_1 \ D_2 \ D_3] \begin{Bmatrix} dA_1 \\ dA_2 \\ dA_3 \end{Bmatrix} \tag{3.53}$$

式中：$dA_1$、$dA_2$ 和 $dA_3$ 是电极在 2-3、1-3 和 1-2 平面的分量（注意，对于图 3.10，只有 $D_3$ 出现在等式中）。则电压可通过下式计算：

$$V_{\text{sensor}} = \frac{q}{C_{\text{sensor}}} \tag{3.54}$$

式中：$C_{\text{sensor}}$ 为压电传感器的电容（F）。若压电传感器是由长度为 $l_s$、宽度为 $w_s$、厚度为 $t_s$ 的薄板形成的，则电容为

$$C_{\text{sensor}} = \frac{\overline{\varepsilon}_{33}^\sigma l_s w_s}{t_s} \tag{3.55}$$

逆压电效应由式（3.25）（参见式（3.37）和式（3.40））表示，参考致动器方程。在谈及致动器性能时，通常会提出自由位移和阻滞力两个重要的概念。自由位移被定义为在没有外部机械载荷的情况下，给定电场中压电效应的最大位移。它的形式可由应力矢量和热贡献归零得到。其形式将由 $\boldsymbol{\Lambda}$（自由应变）给出，其乘以压电片的长度将得到自由位移（对于 PZT 材料而言）：

$$\boldsymbol{\Lambda} = \begin{Bmatrix} \Lambda_1 \\ \Lambda_2 \\ \Lambda_3 \\ \Lambda_4 \\ \Lambda_5 \\ \Lambda_6 \end{Bmatrix} = \begin{Bmatrix} d_{31} E_3 \\ d_{31} E_3 \\ d_{33} E_3 \\ d_{15} E_2 \\ d_{15} E_1 \\ 0 \end{Bmatrix} \tag{3.56}$$

对于 PVDF 压电薄膜，方程的形式为（假设表面不像 PZT 材料那样是各向同性的）

$$\boldsymbol{\Lambda} = \begin{Bmatrix} \Lambda_1 \\ \Lambda_2 \\ \Lambda_3 \\ \Lambda_4 \\ \Lambda_5 \\ \Lambda_6 \end{Bmatrix} = \begin{Bmatrix} d_{31} E_3 \\ d_{32} E_3 \\ d_{33} E_3 \\ d_{24} E_2 \\ d_{15} E_1 \\ 0 \end{Bmatrix} \tag{3.57}$$

阻滞力被定义为完全约束压电致动器（在电场作用下无变形）所需要的力。将应变矢量归零（而热贡献也为零）即可得到其形式。

对于图 3.10 中的压电片，阻滞力（$F_{bl}$）和自由位移（$\delta_{free}$）为

$$\begin{cases} \varepsilon_1 = s_{11}^E \sigma_1 + d_{31} E_3 \\ 若\ \sigma_1 = 0, \quad \varepsilon_1 = \Lambda_1 = d_{31} E_3 \Rightarrow \delta_{free} = d_{31}\left(\dfrac{l}{t}\right)_p V \\ 若\ \varepsilon_1 = 0, \quad \sigma_1 = -\dfrac{d_{31} E_3}{s_{11}^E} \Rightarrow F_{bl} = -\dfrac{d_{31} w_p}{s_{11}^E} V \end{cases} \quad (3.58)$$

式中

$$\varepsilon_1 \equiv \Lambda_1 = \frac{\delta_{free}}{l_p}, \quad \sigma_1 \equiv \frac{F_{bl}}{w_p t_p}, \quad E_3 = \frac{V}{t_p}$$

如果定义致动器的刚度为

$$k_{actuator} = \frac{w_p t_p}{s_{11}^E l_p} \equiv \frac{A_p}{s_{11}^E l_p} \quad (3.59)$$

可以将阻滞力（取其绝对值）与自由位移的关系写成

$$F = F_{bl}\left(1 - \frac{\delta}{\delta_{free}}\right) \quad 或 \quad \delta = \delta_{free}\left(1 - \frac{F}{F_{bl}}\right) = \delta_{free} - \frac{F}{k_{actuator}} \quad (3.60)$$

$$\Rightarrow F = F_{bl} - \delta \cdot k_{actuator}$$

式（3.60）的示意图如图 3.11 所示。

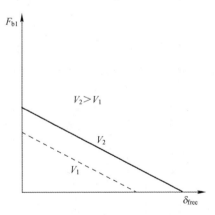

图 3.11　压电致动器负载线（示意图）

假设压电致动器可用刚度（$k_{actuator}$）表示，当压电致动器作用于一个由刚度（$k_{ext}$）表示的外部结构时，致动器负载线上的点（图 3.11）可用如下形式表示。

利用式（3.60）可得到以下等式：

$$F_{\text{ext}} = F_{\text{bl}} - \delta_{\text{ext}} k_{\text{actuator}} \qquad (3.61)$$

然而

$$F_{\text{ext}} = \delta_{\text{ext}} k_{\text{ext}} \Rightarrow \delta_{\text{ext}} = \frac{F_{\text{bl}}}{k_{\text{ext}} + k_{\text{actuator}}}$$

式（3.61）给出了两个并联弹簧的位移。可以看出，当 $k_{\text{ext}} = k_{\text{actuator}}$ 时，压电致动器达到做功最大的加载状态，这意味着该致动器通过匹配外部结构刚度与其自身刚度，使其所能提取的能量最大。

13. 压电材料的动态行为

当压电材料受到交流电场作用时，材料的尺寸会发生周期性的变化。换而言之，材料以与外场频率相同的频率振动。在正压电效应中，当振动力作用于压电材料时，它会产生一个相同频率的振荡电场。

为了分析振动压电材料的动态行为，本书采用了等效电路将其类比为机械和电气元件，如图 3.12 所示。施加在材料上的振动力类似于交流电压。压电元件作为电容器，其电容为 $C_0 = \varepsilon A/d$，其中 $\varepsilon$ 为材料的介电常数，$A$ 和 $d$ 分别为元件的面积和厚度。压电元件的质量 $M$（惯性）等效于电感 $L$，而柔度常数等效于电容器 $C$。由于摩擦而造成的能量损失相当于电路中由于电阻 $r$ 而造成的能量损失。

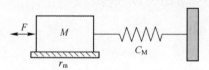

(a) 机械系统：$F$ 为施加在质量系统 $M$（惯性）上的振动力

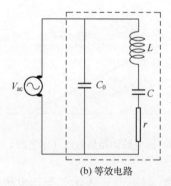

(b) 等效电路

图 3.12 振动压电材料可以由一个机械系统或等效电路表示

在图 3.12 中，$r_{\text{m}}$ 表示由于摩擦造成能量损失而产生的机械阻力。$C_{\text{M}}$ 表示机械系统的弹性常数。对于等效电路，所施加的交流电压 $V$ 等效于力 $F$，电感

$L$ 等效于质量 $M$，电阻 $r$ 等效于机械摩擦（能量损失），电容 $C$ 等效于材料的柔度（与弹性常数有关），$C_0$ 等效于压电材料的电容。

振动系统的阻抗是频率的函数。这种系统的阻抗随频率的变化如图 3.13 所示。阻抗具有一个最小值和一个最大值。最小阻抗频率称为共振频率，最大阻抗频率称为反共振频率。

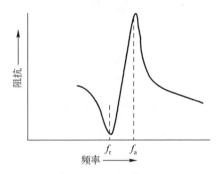

图 3.13 压电元件的频率响应。阻抗在 $f_r$（共振频率）处最小，在 $f_a$（反共振频率）处最大

在共振频率 $f_m$ 处，压电系统以最大振幅振动。假设机械损耗产生的阻力为零，共振频率 $f_m$ 等于等效电路阻抗为零时的串联共振频率 $f_s$：

$$f_m = f_s = \sqrt{1/L \cdot C}/(2\pi)$$

假设机械损耗产生的阻力为零，反共振频率 $f_a$ 等于等效电路的并联共振频率 $f_p$：

$$f_a = f_p = \sqrt{(C+C_0)/(L \cdot C \cdot C_0)}/(2\pi)$$

压电元件的共振和反共振频率可以通过实验测量。$f_r$ 和 $f_a$ 的值可用于估算机电耦合系数 $k$。耦合系数与频率 $f_r$ 和 $f_a$ 的关系取决于压电元件的形状。

## 3.2 销力梁模型

本节旨在推导带有压电层梁的运动方程，如图 3.14 所示。

对于一个具有以下尺寸的压电贴片（图 3.15）：长度为 $l_p$、宽度为 $w_p$、厚度为 $t_p$，固定到一个承载梁上，假设其为各向同性（以方便接下来的计算），见图 3.14。

在压电贴片沿厚度方向的电极上施加电压 $V$。大多数压电材料的压电常数 $d_{31}$ 为负，因此电压方向与压电片的极化方向相反，在 1 方向上产生正应变，

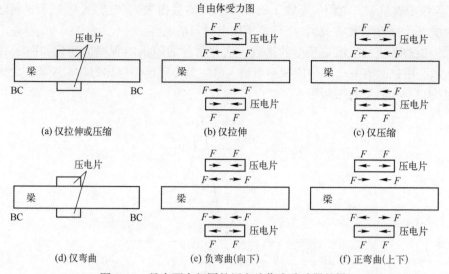

图 3.14 具有两个相同的压电片作为致动器的梁

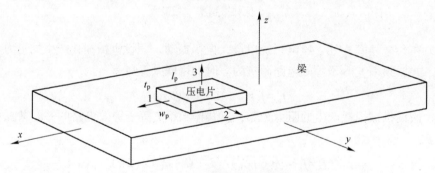

图 3.15 具有表面黏结压电片的承载结构

关系如下：

$$\varepsilon_{\max} = \frac{V}{t_p}d_{31} \equiv \Lambda \tag{3.62}$$

需要注意的是，式（3.62）中所示的最大应变是不附着在任何结构上的自由压电片的应变。

将式（3.62）乘以压电片的杨氏模量 $E_p$ 和横截面 ($w_p \times t_p = A_p$)，得到作用在 $x$ 方向上的力，即

$$F_{\max} = V w_p E_p d_{31} = w_p t_p E_p \Lambda = F_{bl} \tag{3.63}$$

注意：式（3.63）中给出的最大力是对于压电片两端被限制在 $x$ 方向上运动的情况，而这就是它通常在文献中称为"阻滞力" $F_{bl}$ 的原因。

当压电片与承载结构（在本例中为梁（图3.15））连接时，压电片所受到的应变和力受结构轴向刚度的影响。当电压施加在压电片电极上时，由于压电片连接在承载结构上，结构将受到一个表面力 $F$，这将导致压电片上的反作用力 $-F$，从而产生应变 $\varepsilon_\mathrm{p}$。为了估算力 $F$，压电片中的应变 $\varepsilon_\mathrm{p}$ 可写成自由应变 $\Lambda$（式（3.62））与贴片中反作用力 $-F$ 引入的机械应变的差值，得到以下公式：

$$\varepsilon_\mathrm{p} = \Lambda - \frac{F}{w_\mathrm{p} t_\mathrm{p} E_\mathrm{p}} = \Lambda\left(1 - \frac{F}{F_\mathrm{bl}}\right)$$

$$\text{或 } F = F_\mathrm{bl}\left(1 - \frac{\varepsilon_\mathrm{p}}{\Lambda}\right) \tag{3.64}$$

作用在梁上的力 $F$ 可以通过将应变 $\varepsilon_\mathrm{p}$ 乘以梁的截面（因为梁本身也会遇到相同的应变）及其杨氏模量来计算得到，即

$$F = w_\mathrm{b} h_\mathrm{b} E_\mathrm{b} \varepsilon_\mathrm{p} = (AE)_\mathrm{b} \varepsilon_\mathrm{p} \tag{3.65}$$

式中：下标 b 为梁的性质（尺寸和杨氏模量），$A_\mathrm{b} = w_\mathrm{b} \times h_\mathrm{b}$ 为梁的截面（$w_\mathrm{b}$ 为梁的宽度，$h_\mathrm{b}$ 为梁的高度）；$(AE)_\mathrm{b}$ 为梁的轴向刚度。

式（3.64）给出了众所周知的致动器关系，即致动器的无量纲力 $F/F_\mathrm{bl}$ 与其应变比 $\varepsilon_\mathrm{p}/\Lambda$ 是反线性的；从而得到零应变时的最大力 $F = F_\mathrm{bl}$ 和零应力时的最大应变 $\varepsilon_\mathrm{p} = \Lambda$。

在给出力与应变的关系后，对于单个压电片的梁，将推导出被两个压电片夹住的梁的轴向位移的表达式（图3.14）。

第1种情况是图3.14中所示的模型，解决的是由于两个压电片（或致动器）的符号和幅值相等的电压（同相电压）而导致梁的延伸或压缩。这里假设两个压电片（或致动器）在尺寸和性质（机械和电）上相同。

运用梁的轴向位移与黏接片的轴向位移之间的位移协调原则，根据图3.14（b）和（c），梁的轴向位移可以写成（应该注意梁受到 $2F$ 的牵引）

$$\Delta_\mathrm{b} = \frac{2F}{w_\mathrm{b} h_\mathrm{b} E_\mathrm{b}} l_\mathrm{p} \tag{3.66}$$

需要注意的是，在计算式（3.66）中梁的轴向位移时，应将梁的长度作为压电片的长度 $l_\mathrm{p}$。

压电片的轴向位移可由式（3.64）的第1部分得到，即

$$\varepsilon_\mathrm{p} = \frac{\Delta l_\mathrm{p}}{l_\mathrm{p}} = \Lambda - \frac{F}{w_\mathrm{p} t_\mathrm{p} E_\mathrm{p}} \rightarrow \Delta l_\mathrm{p} = \left(\frac{V}{t_\mathrm{p}} d_{31} - \frac{F}{w_\mathrm{p} t_\mathrm{p} E_\mathrm{p}}\right) l_\mathrm{p} \tag{3.67}$$

将两个位移等值，如式（3.66）和式（3.67）所示，得

$$\Delta L_{\mathrm{b}} = \Delta l_{\mathrm{p}} \to \frac{2F}{w_{\mathrm{b}} h_{\mathrm{b}} E_{\mathrm{b}}} l_{\mathrm{p}} = \left( \frac{V}{t_{\mathrm{p}}} d_{31} - \frac{F}{w_{\mathrm{p}} t_{\mathrm{p}} E_{\mathrm{p}}} \right) l_{\mathrm{p}}$$

或 $F = \dfrac{\dfrac{V}{t_{\mathrm{p}}} d_{31}}{\dfrac{2}{(EA)_{\mathrm{b}}} + \dfrac{2}{(2EA)_{\mathrm{p}}}} = \Lambda \dfrac{(EA)_{\mathrm{b}} (EA)_{\mathrm{p}}}{(EA)_{\mathrm{b}} + (2EA)_{\mathrm{p}}} = F_{\mathrm{bl}} \dfrac{(EA)_{\mathrm{b}}}{(EA)_{\mathrm{b}} + (2EA)_{\mathrm{p}}}$ (3.68)

将式（3.68）中给出的 $F$ 的表达式代入式（3.66），即可得到梁所承受的应变为

$$\varepsilon_{\mathrm{b}} = \frac{\Delta_{\mathrm{b}}}{l_{\mathrm{p}}} = \frac{2F}{(EA)_{\mathrm{b}}} = \Lambda \frac{(2EA)_{\mathrm{p}}}{(EA)_{\mathrm{b}} + (2EA)_{\mathrm{p}}} \tag{3.69}$$

由式（3.68）和式（3.69）可知，当梁的轴向刚度相对于压电片（$(2EA)_{\mathrm{p}} \gg (EA)_{\mathrm{b}}$）的刚度可以忽略时，力 $F$ 将趋于零，而梁所经受的应变值为 $\varepsilon_{\mathrm{b}} \approx \Lambda$。反之，即与梁的轴向刚度相比，压电片的轴向刚度可以忽略时（通常是更常见的情况）（$(2EA)_{\mathrm{p}} \ll (EA)_{\mathrm{b}}$），梁的应变将消失，而轴向力将趋于 $F_{\mathrm{bl}}$。

接下来要处理的情况是图 3.14（d）~（f）所示的梁的诱导弯曲。对贴片的两个电极施加异相电压（幅值相同，电压符号相反），会导致其中一个贴片膨胀，而另一个收缩，从而导致梁的净弯曲（图 3.14（e）和（f））。与上述方法一样，计算力 $F$、梁上的诱导弯矩、应变和位移。图 3.14（f）为施加负电压在顶部的压电片（由于 $d_{31}$ 为负值，导致其膨胀）、正电压在底部压电片（导致其收缩）而产生的正弯矩（梁的顶面受压缩，底面受拉伸）。参照图 3.14（f），梁上部纤维的应变可以写成诱导力矩 $M$ 的函数：

$$\varepsilon_{\mathrm{b}}^{\mathrm{top}} = -\frac{M \dfrac{h_{\mathrm{b}}}{2}}{I_{\mathrm{b}} E_{\mathrm{b}}} = -\frac{(F h_{\mathrm{b}}) \dfrac{h_{\mathrm{b}}}{2}}{I_{\mathrm{b}} E_{\mathrm{b}}} = -\frac{F h_{\mathrm{b}}^2}{2 I_{\mathrm{b}} E_{\mathrm{b}}} \tag{3.70}$$

式中：$I_{\mathrm{b}}$ 为梁的转动惯量；$I_{\mathrm{b}} E_{\mathrm{b}}$ 为其抗弯刚度。

梁的收缩可以通过将顶面应变乘以压电片的长度 $l_{\mathrm{p}}$ 来计算得到：

$$\Delta_{\mathrm{b}}^{\mathrm{top}} = -\frac{F h_{\mathrm{b}}^2}{2 I_{\mathrm{b}} E_{\mathrm{b}}} l_{\mathrm{p}} \tag{3.71}$$

顶面压电片长度的变化见式（3.67）。令两个绝对位移相等，得到诱导力 $F$ 的必要方程：

$$\frac{F h_{\mathrm{b}}^2}{2 I_{\mathrm{b}} E_{\mathrm{b}}} l_{\mathrm{p}} = \left( \frac{V}{t_{\mathrm{p}}} d_{31} - \frac{F}{(AE)_{\mathrm{p}}} \right) l_{\mathrm{p}} \to F = \frac{\dfrac{V}{t_{\mathrm{p}}} d_{31}}{\dfrac{h_{\mathrm{b}}^2}{2(EI)_{\mathrm{b}}} + \dfrac{1}{(AE)_{\mathrm{p}}}} \tag{3.72}$$

式 (3.72) 中 $F$ 的表达式可以化简为

$$F = \frac{\dfrac{V}{t_p} d_{31}}{\dfrac{h_b^2}{2(EI)_b} + \dfrac{1}{(AE)_p}} = F_{bl} \frac{(EI)_b}{(EI)_b + (EI)_p} \tag{3.73}$$

式中：两个压电片的弯曲刚度为 $(EI)_p = \dfrac{h_b^2}{2}(EA)_p$。考虑到"阻滞力矩"为 $M_{bl} = F_{bl} h_b$，我们可以将弯矩的表达式写成

$$M = M_{bl} \frac{(EI)_b}{(EI)_b + (EI)_p} \tag{3.74}$$

在纯弯曲下，应变沿梁的截面呈线性分布，可表示为 $z$ 坐标的函数，即

$$\varepsilon_b = -\frac{M}{(EI)_b} z = -\frac{M_{bl}}{(EI)_b + (EI)_p} z \tag{3.75}$$

对于正力矩，$z = +h_b/2$（梁的上部纤维）和 $z = -h_b/2$（梁的底部纤维）；我们可将顶部和底部梁的应变写成自由压电应变 $\Lambda$ 的函数为

$$\begin{cases} \varepsilon_b^{top} = -\dfrac{(EI)_p}{(EI)_b + (EI)_p} \Lambda \\ \varepsilon_b^{bottom} = \dfrac{(EI)_p}{(EI)_b + (EI)_p} \Lambda \end{cases} \tag{3.76}$$

正如前述纯拉伸或压缩情况一样，评估两个压电片夹在中间的梁在诱导弯曲下的性能。

由式 (3.74) 和式 (3.76) 可知，当梁的弯曲刚度相对于贴片的刚度可以忽略时（$(EI)_p \gg (EI)_b$），该力矩 $M$ 将趋于零，而梁所承受的应变值将为 $\varepsilon_b^{top} \approx -\Lambda$ 和 $\varepsilon_b^{bottom} \approx +\Lambda$。相反，即当压电片的弯曲刚度与梁的弯曲刚度相比可以忽略时（$(EI)_p \ll (EI)_b$）（通常是更常见的情况），梁的应变将消失，而弯矩趋于 $M_{block}$。

现在，以一对贴片位置和梁边界条件的函数来表示梁的诱导弯曲挠度。首先从最简单的悬臂梁情况开始，其夹紧端的根部有一对压电片（图 3.16 (a)）。

根据伯努利-欧拉梁理论，得出弯曲挠度与施加力矩的关系为

$$\frac{d^2 w}{dx^2} = \frac{M}{(EI)_b} \tag{3.77}$$

由于力矩沿梁分布是不均匀的，梁被分成若干区域：第 1 个区域包括靠近梁夹紧端的一对压电片，第 2 个区域仅包括梁，见图 3.16 (a)。对于得到的

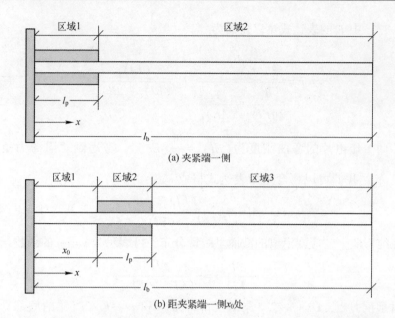

图 3.16 具有一对压电片的悬臂梁位于夹紧端一侧和距夹紧端一侧 $x_0$ 处

第 1 个区域，两次积分式（3.76）可得

$$w_1(x) = \frac{Mx^2}{2(EI)_b} + A_1 x + B_1 \tag{3.78}$$

式中：$A_1$ 和 $B_1$ 为常数，由边界条件和两个区域之间的连续性要求来确定。

考虑到沿着第 2 个区域的力矩为零，其弯曲挠度有以下表达式：

$$w_2(x) = A_2 x + B_2 \tag{3.79}$$

式中：$A_2$ 和 $B_2$ 为常数，由边界条件和两个区域之间的连续性要求确定。

由应用梁的夹紧边界条件（仅对第 1 个区域）得

$$\begin{aligned}
& w_1(0) = 0 \rightarrow B_1 = 0 \\
& \frac{w_1(0)}{\mathrm{d}x} = 0 \rightarrow A_1 = 0 \\
& \rightarrow w_1(x) = \frac{Mx^2}{2(EI)_b} = \frac{M_{b1} x^2}{2((EI)_b + (EI)_p)}
\end{aligned} \tag{3.80}$$

为了得到第 2 个区域弯曲挠度的形状，在两个区域之间的边界挠度和斜率应用连续性要求，得

$$\begin{cases} w_1(l_p) = w_2(l_p) \to \dfrac{Ml_p^2}{2(EI)_b} = A_2 l_p + B_2 \\ \dfrac{dw_1(l_p)}{dx} = \dfrac{dw_2(l_p)}{dx} \to \dfrac{Ml_p}{(EI)_b} = A_2 \\ \to B_2 = -\dfrac{Ml_p^2}{2(EI)_b} \\ \to w_2(x) = \dfrac{Ml_p}{(EI)_b}\left(x - \dfrac{l_p}{2}\right) = \dfrac{M_{bl}l_p}{((EI)_b + (EI)_p)}\left(x - \dfrac{l_p}{2}\right) \end{cases} \quad (3.81)$$

式（3.80）和式（3.81）描述了梁的挠度形状。由式（3.81）可知，第2个区域内梁的斜率（$B_2$）是常数，即在压电片之后的区域，梁只旋转而不弯曲（其保持挺直）。

现在分析图 3.16（b）中描述的情况。需要注意的是，力矩只在区域 2 产生，而其他两个区域的弯矩为零。这意味着关于挠度有下列表达式：

$$\begin{cases} 0 < x < x_0 \to w_1(x) = A_1 x + B_1 \\ x_0 < x < x_0 + l_p \to w_2(x) = \dfrac{Mx^2}{2(EI)_b} + A_2 x + B_2 \\ x_0 + l_p < x < l_b \to w_3(x) = A_3 x + B_3 \end{cases} \quad (3.82)$$

式中：各个常数 $A_1$、$B_1$、$A_2$、$B_2$、$A_3$ 和 $B_3$ 由两个相邻区域边界的边界条件和连续性要求确定。

对于第 1 个区域，应用梁的几何边界条件，得

$$\begin{cases} w_1(0) = 0 \to B_1 = 0 \\ \dfrac{dw_1(0)}{dx} = 0 \to A_1 = 0 \end{cases} \quad (3.83)$$

因此，在区域 1 中，挠度和斜率均为零，即

$$\begin{cases} w_1(x) = 0 \\ \dfrac{dw_1(x)}{dx} = 0 \end{cases} \quad (3.84)$$

对于区域 2，在 $x = x_0$ 处有下面的连续性要求：

$$\begin{cases} w_1(x_0) = w_2(x_0) \to 0 = \dfrac{Mx_0^2}{2(EI)_b} + A_2 x_0 + B_2 \\ \dfrac{dw_1(x_0)}{dx} = \dfrac{dw_2(x_0)}{dx} \to 0 = \dfrac{Mx_0}{(EI)_b} + A_2 \end{cases} \quad (3.85)$$

由式（3.85），可以求出两个未知数 $A_2$ 和 $B_2$，得到第 2 个区域的挠度和斜率为

$$\begin{cases} w_2(x) = \dfrac{M(x-x_0)^2}{2(EI)_b} \\ \dfrac{dw_2(x)}{dx} = \dfrac{M(x-x_0)}{(EI)_b} \end{cases} \tag{3.86}$$

对于最后一个部分区域 3，在 $x = x_0 + l_p$ 处有下面的连续性要求：

$$\begin{cases} w_2(x_0+l_p) = w_3(x_0+l_p) \rightarrow \dfrac{Ml_p^2}{2(EI)_b} = A_3(x_0+l_p) + B_3 \\ \dfrac{dw_2(x_0+l_p)}{dx} = \dfrac{dw_3(x_0+l_p)}{dx} \rightarrow \dfrac{Ml_p}{(EI)_b} = A_3 \end{cases} \tag{3.87}$$

由式（3.87），可以求出两个未知数 $A_3$ 和 $B_3$，即可得到区域 3 的挠度和斜率为

$$\begin{cases} w_3(x) = \dfrac{Ml_p(x-x_0-0.5l_p)}{(EI)_b} \\ \dfrac{dw_3(x)}{dx} = \dfrac{Ml_p}{(EI)_b} = \text{const} \end{cases} \tag{3.88}$$

将力矩 $M$ 代入表达式，可得沿梁的挠度及其伴随斜率的分布为

$$\begin{cases} 0 < x < x_0 \rightarrow w_1(x) = 0 \quad \dfrac{dw_1(x)}{dx} = 0 \\ x_0 < x < x_0+l_p \rightarrow w_2(x) = \dfrac{M_{bl}(x-x_0)^2}{2((EI)_b+(EI)_p)} \quad \dfrac{dw_2(x)}{dx} = \dfrac{M_{bl}(x-x_0)}{(EI)_b+(EI)_p} \\ x_0+l_p < x < l_b \rightarrow w_3(x) = \dfrac{M_{bl}l_p(x-x_0-0.5l_p)}{(EI)_b+(EI)_p} \quad \dfrac{dw_3(x)}{dx} = \dfrac{M_{bl}l_p}{(EI)_b+(EI)_p} \end{cases} \tag{3.89}$$

接下来分析在简支边界条件下具有图 3.17 所示的一对压电片的梁沿梁的诱导挠度和斜率。

可以看到，图 3.17 所示情况的边界条件只能对区域 1 和 3 实施，而对梁中部区域 2 可以实施连续性要求。考虑到诱导力矩 $M$ 仅作用于区域 1，则梁各部分的横向位移方程可写为（类似于式（3.82））

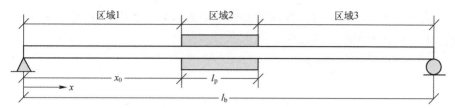

图 3.17 具有一对压电片的简支梁

$$\begin{cases} 0<x<x_0 \to w_1(x) = \overline{A}_1 x + \overline{B}_1 \\ x_0<x<x_0+l_p \to w_2(x) = \dfrac{Mx^2}{2(EI)_b} + \overline{A}_2 x + \overline{B}_2 \\ x_0+l_p<x<l_b \to w_3(x) = \overline{A}_3 x + \overline{B}_3 \end{cases} \quad (3.90)$$

对于第 1 个区域,得

$$\begin{cases} 0<x<x_0 \quad w_1(x) = \overline{A}_1 x + \overline{B}_1 \\ w_1(0) = 0 \to \overline{B}_1 = 0 \\ \to w_1(x) = \overline{A}_1 x \quad \dfrac{dw_1(x)}{dx} = \overline{A}_1 \end{cases} \quad (3.91)$$

对于第 2 个区域,可写成

$$\begin{cases} x_0<x<x_0+l_p \quad w_2(x) = \dfrac{Mx^2}{2(EI)_b} + \overline{A}_2 x + \overline{B}_2 \\ w_1(x_0) = w_2(x_0) \to \overline{A}_1 x_0 = \dfrac{Mx_0^2}{2(EI)_b} + \overline{A}_2 x_0 + \overline{B}_2 \to (\overline{A}_1 - \overline{A}_2) x_0 = \dfrac{Mx_0^2}{2(EI)_b} + \overline{B}_2 \\ \dfrac{dw_1(x_0)}{dx} = \dfrac{dw_2(x_0)}{dx} \to \overline{A}_1 = \dfrac{MX_0}{(EI)_b} + \overline{A}_2 \to (\overline{A}_1 - \overline{A}_2) = \dfrac{MX_0}{(EI)_b} \end{cases} \quad (3.92)$$

应用第 2 和第 3 部分边界处的边界条件与位移和斜率的连续性,对最后一个区域 3,有

$$\begin{cases} x_0+l_p<x<l_b \quad w_3(x) = \overline{A}_3 x + \overline{B}_3 \\ w_3(l_b) = 0 \to \overline{A}_3 l_b + \overline{B}_3 = 0 \\ \to w_3(x) = \overline{A}_3(x-l_b) \quad \dfrac{dw_3(x)}{dx} = \overline{A}_3 \\ w_2(x_0+l_p) = w_3(x_0+l_p) \to \dfrac{M(x_0+l_p)^2}{2(EI)_b} + \overline{A}_2(x_0+l_p) + \overline{B}_2 = \overline{A}_3(x_0+l_p-l_b) \\ \dfrac{dw_2(x_0+l_p)}{dx} = \dfrac{dw_3(x_0+l_p)}{dx} \to \dfrac{M(x_0+l_p)}{(EI)_b} + \overline{A}_2 = \overline{A}_3 \to \dfrac{M(x_0+l_p)}{(EI)_b} = (\overline{A}_3 - \overline{A}_2) \end{cases} \quad (3.93)$$

式（3.91）~式（3.93）提供了找到 6 个未知数所必需的 6 个方程，即 $\bar{A}_1$、$\bar{B}_1$、$\bar{A}_2$、$\bar{B}_2$、$\bar{A}_3$ 和 $\bar{B}_3$，其表达式如下：

$$\begin{cases} \bar{A}_1 = \dfrac{M}{2(EI)_b} \dfrac{l_p}{l_b}(2x_0 + 2l_b - l_p) \quad \bar{B}_1 = 0 \\ \bar{A}_2 = \dfrac{M(l_p - l_b)}{(EI)_b} \dfrac{x_0}{l_b} - \dfrac{M(l_p - 2l_b)}{2(EI)_b} \dfrac{l_p}{l_b} \bar{B}_2 = \dfrac{Mx_0^2}{2(EI)_b} \\ \bar{A}_3 = \dfrac{M(2x_0 + l_p)}{2(EI)_b} \dfrac{l_p}{l_b} \bar{B}_3 = -\dfrac{M(2x_0 + l_p)}{2(EI)_b} l_p \end{cases} \quad (3.94)$$

式中：$\dfrac{M}{(EI)_b} = \dfrac{M_{bl}}{(EI)_b + (EI)_p}$。

利用式（3.94）中给出的常数，可以计算梁的 3 个区域中任何一个的挠度和相应的斜率。

之前例子中压电片的位置是对称的，这使得当压电层作为致动器时，会导致梁的纯弯曲或纯拉伸/收缩。

图 3.18 所示的情形是将压电片附加到梁上的一系列非对称位置的可能性。这些情况的共同之处在于在承载梁上同时引入了弯曲和拉伸/收缩。

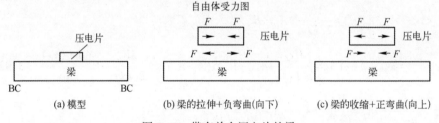

图 3.18 带有单个压电片的梁

为了简化计算，假设单个压电片的厚度可以忽略不计，从而假定梁的中轴仍然位于梁截面中间。如果这个假设不成立，就应该使用一种常见的方法计算中轴的位置，如在每本处理梁弯曲应力的书中均可找到模量加权中轴①。

由图 3.18 可知，梁的顶部纤维受压缩，其应变可表示为两个表达式的和：第 1 个是由于纤维的收缩，第 2 个是由于正弯曲力矩导致的压缩。可得

$$\varepsilon_b^{top} = -\dfrac{F}{(wh)_b E_b} - \dfrac{Mh}{2(EI)_b} \quad (3.95)$$

---

① 参见 Popov, E. P., Mechanics of Materials (2nd edn.) Hardcover-Prentice Hall；April 7, 1976。

力矩 $M$ 和转动惯量 $I$ 可以写成

$$M = F\frac{h_b}{2}, \quad I = \frac{w_b h_b^3}{12} \tag{3.96}$$

因此,式(3.95)可看作

$$\varepsilon_b^{top} = -\frac{4F}{(wh)_b E_b} \tag{3.97}$$

而顶端纤维沿着压电片长度 $l_p$ 的绝对收缩量为(负号省略)

$$\Delta l_b^{top} = \varepsilon_b^{top} l_p \tag{3.98}$$

将压电片长度的变化(式(3.64))与顶部纤维的长度变化等值,则式(3.97)导致轴向力 $F$ 的表达式为

$$\begin{cases} \Delta l_b^{top} = \Delta l_p \to \dfrac{4F}{(wh)_b E_b} = \Lambda - \dfrac{F}{(wh)_p E_p} \\ \to F = \Lambda \dfrac{(wh)_p E_p (wh)_b E_b}{4(wh)_p E_p + (wh)_b E_b} \\ F_{blocking} = \Lambda E_p (wt)_p \end{cases} \tag{3.99}$$

考虑到 $F_{blocking} = \Lambda E_p(wt)_p$,可以将式(3.99)重写为

$$F = F_{bl}\frac{(wh)_b E_b}{4(wh)_p E_p + (wh)_b E_b} \tag{3.100}$$

当 $M_{bl} = F_{bl}\dfrac{h_b}{2}$ 时,弯曲力矩的表达式为

$$M = M_{bl}\frac{(wh)_b E_b}{4(wh)_p E_p + (wh)_b E_b} \tag{3.101}$$

其他非对称问题,如涉及两个不同的压电片,或连接在承载梁的两侧(图 3.19),这里差别可能是电压不相等、厚度不同、压电常数 $d_{31}$ 不同,或所有的差异一起导致的。解决这些问题应该使用以下方法。

假设力分布如图 3.19 所示,可以定义以下两种关系:

$$\begin{cases} F_1 = \dfrac{F_{top} + F_{bottom}}{2} \\ F_2 = \dfrac{F_{top} - F_{bottom}}{2} \end{cases} \tag{3.102}$$

图 3.20 所示为在两个确定力 $F_1$ 和 $F_2$ 作用下的梁。只有当 $F_{top} > F_{bottom}$ 且 $F_{bottom}$ 为正时,$F_{top}$ 和 $F_{bottom}$ 之间的平均值,即力 $F_1$,会导致轴向拉伸(图 3.20)。相反,当 $F_{bottom}$ 为负且大于 $F_{top}$ 时,梁会收缩。$F_{top}$ 和 $F_{bottom}$ 差值除

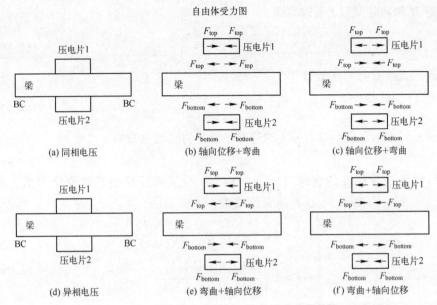

图 3.19 具有两个不同压电片作为致动器的梁

以 2,即力 $F_2$,导致梁的净弯曲。弯曲的方向,即向上(正弯曲)或向下(负弯曲)取决于这两个力差值的符号。

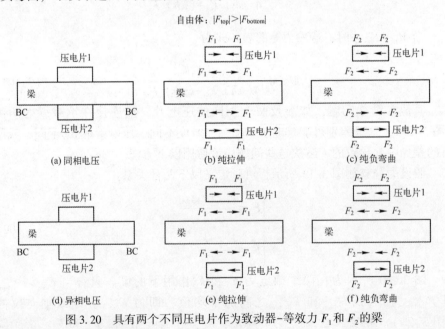

图 3.20 具有两个不同压电片作为致动器-等效力 $F_1$ 和 $F_2$ 的梁

图 3.20 所示的每种情况的解,都可以通过前面提出的对称情况的方程求解①。

## 3.3　均匀应变梁模型

均匀应变梁模型得名于图 3.21 (a) 和 (b) 中的假设,即由于压电层的厚度小,所以假设压电层横截面上的应变分布是恒定的。

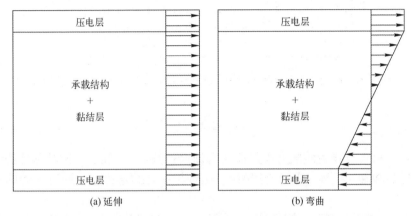

图 3.21　跨承载结构、黏结层和表面黏结压电层的应变分布

本节将介绍不能忽略结合层厚度的情况下具有表面黏结压电层承载结构 (在本例中为梁) 的解决方案 (图 3.22)。为了得到这个问题的运动方程,将使用微分元素 d$x$ (图 3.22)。

推导过程是根据文献 [2] 中给出的结果进行的 (更多关于该主题的文献见本章参考文献列表)。

假设结合层中为纯一维剪切,压电层和承载子结构中只有拉伸应变,则应变-位移关系可表示为

$$\varepsilon_\mathrm{p} = \frac{\mathrm{d} u_\mathrm{p}}{\mathrm{d} x} \qquad (3.103)$$

$$\varepsilon_\mathrm{b}^\mathrm{BL} = \frac{\mathrm{d} u_\mathrm{b}^\mathrm{BL}}{\mathrm{d} x} \qquad (3.104)$$

---

① 参见 Chopra, I. and Sirohi, J. *Smart Structures Theory*, Chapter 4.3, Cambridge University Press, 2014。

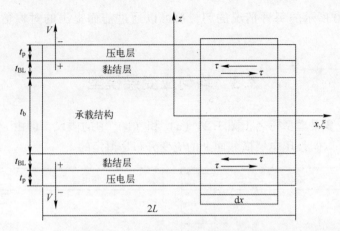

图 3.22 具有表面黏结压电层和有限厚度黏结层的承载结构的几何形状

$$\gamma_{BL} = \frac{u_p - u_b}{t_{BL}} \tag{3.105}$$

式中：$\varepsilon_p$ 和 $u_p$ 分别为压电层中的应变和轴向位移；$\varepsilon_b^{BL}$ 和 $u_b^{BL}$ 分别为承载结构中的应变和轴向位移，在结合层与 $\gamma_{BL}$ 结合界面处的梁为结合层所承受的剪切应变。

假设在弯曲情况下，在压电致动器的激励下，应变分布为线性（图 3.21（b）），而对于拉伸（或收缩）的情况（图 3.21（a）），应变分布是均匀的，基元 $dx$ 的平衡方程为

$$\frac{d\sigma_p}{dx} - \frac{\tau}{t_p} = 0 \tag{3.106}$$

$$\frac{dM}{dx} + \tau t_b w_p = 0 \tag{3.107}$$

式中：$w_p$ 为压电致动器（层）的宽度；$t_b$ 为梁（承载结构）的高度。

梁内的应力分布可用弯矩 $M$ 来表示为[13]

$$\sigma_b(z) = -\frac{M}{I_b} z \tag{3.108}$$

式中：$I_b = w_b t_b^3 / 12$ 是梁的横截面惯量与中轴的面积乘积（与梁的高度 $t_b$ 相比，压电材料和结合层的厚度小，贡献可以忽略，$w_b$ 是梁的宽度）。梁顶面（$z = t_b/2$）的应力可写为

$$\sigma_b^{top} = -\frac{M \cdot (t_b/2)}{\dfrac{w_b t_b^3}{12}} = -\frac{6M}{w_b t_b^2} \tag{3.109}$$

定义力矩 $M$ 为

$$M = -\frac{w_b t_b^2}{6}\sigma_b^{\text{top}} \tag{3.110}$$

借助式（3.109）并代入式（3.107），可得

$$\frac{\mathrm{d}\sigma_b^{\text{top}}}{\mathrm{d}x} - 6\frac{w_p}{w_b t_b}\tau = 0$$

对梁底面（$z = -t_b/2$）进行同样的计算，得到类似于式（3.110）的方程，即

$$\frac{\mathrm{d}\sigma_b^{\text{bottom}}}{\mathrm{d}x} + 6\frac{w_p}{w_b t_b}\tau = 0 \tag{3.111}$$

类似地，对于纯拉伸（或压缩）的情况，得到梁的上、下表面表达式如下：

$$\frac{\mathrm{d}\sigma_b^{\text{top}}}{\mathrm{d}x} + 2\frac{w_p}{w_b t_b}\tau = 0 \tag{3.112}$$

$$\frac{\mathrm{d}\sigma_b^{\text{bottom}}}{\mathrm{d}x} + 2\frac{w_p}{w_b t_b}\tau = 0 \tag{3.113}$$

式（3.110）~式（3.113）可以用一个通用表达式来代替，其形式如下：

$$\begin{cases} \dfrac{\mathrm{d}\sigma_b}{\mathrm{d}x} + \alpha\dfrac{w_p}{w_b t_b}\tau = 0 \\ \alpha = 2, \text{延伸} \\ \alpha = 6, \text{底部弯曲} \\ \alpha = -6, \text{顶部弯曲} \end{cases} \tag{3.114}$$

压电层的应力-应变表达式由下式给出：

$$\sigma_p = E_p\left(\varepsilon_p - \frac{d_{31}V}{t_p}\right) = E_p(\varepsilon_p - \Lambda) \tag{3.115}$$

式中：$V$ 为施加在压电致动器电极上的电压；$E_p$ 为压电材料的杨氏模量；$d_{31}$ 为压电常数。

梁和结合层的另外两个应力-应变方程为

$$\sigma_b^{\text{BL}} = E_b u_b^{\text{BL}} \tag{3.116}$$

$$\tau_{\text{BL}} = G_{\text{BL}}\gamma_{\text{BL}} \tag{3.117}$$

式中：$G_{\text{BL}}$ 为结合层的剪切模量。

式（3.103）~式（3.106）和式（3.114）~式（3.117）构成了当前问题的 8 个未知数的 8 个控制方程：3 个应力（$\sigma_p, \sigma_b^{\text{BL}}, \tau_{\text{BL}}$）、两个位移（$u_b, u_b^{\text{BL}}$）和 3 个

应变 ($\varepsilon_p, \varepsilon_b^{BL}, \gamma_{BL}$)。将式（3.103）~式（3.105）代入式（3.115）~式（3.117），再代入式（3.106）和式（3.114），并假设压电层的宽度与梁的宽度相匹配，即 $w_p = w_b$，得到未知数 $\varepsilon_b^{BL}$ 和 $\varepsilon_p$ 的两个耦合方程为

$$\frac{d\varepsilon_p}{dx^2} - \frac{G_{BL}}{E_p} \cdot \frac{\varepsilon_p - \varepsilon_b^{BL}}{t_{BL} t_p} = 0 \tag{3.118}$$

$$\frac{d\varepsilon_b^{BL}}{dx^2} + \frac{G_{BL}}{E_b} \cdot \frac{\varepsilon_p - \varepsilon_b^{BL}}{t_{BL} t_p} \alpha = 0 \tag{3.119}$$

为解耦上述两个方程，定义若干无量纲量，分别为

$$\begin{cases} \xi = \frac{\chi}{L}; \bar{G} = \frac{G}{E_p}; \bar{t}_{BL} = \frac{t_{BL}}{L}; \theta_b = \frac{t_b}{t_p} \\ \bar{G} = \frac{G}{E_p} \frac{E_b}{E_p} = \bar{E}_b; \psi = \frac{E_b t_b}{E_p t_p} = \bar{E}_b \theta_b; \theta_{BL} = \frac{t_{BL}}{t_p} \end{cases} \tag{3.120}$$

解耦式（3.118）和式（3.119）得到一对四阶微分方程，其形式如下：

$$\varepsilon_p^{iv} - \Gamma^2 \varepsilon_p'' = 0 \tag{3.121}$$

$$(\varepsilon_b^{BL})^{iv} - \Gamma^2 (\varepsilon_b^{BL})'' = 0 \tag{3.122}$$

式中

$$\Gamma^2 = \frac{\bar{G} \theta_{BL}}{\bar{t}_{BL}^2} \left( \frac{\psi + \alpha}{\psi} \right) \tag{3.123}$$

并在式（3.121）和式（3.122）对无量纲长度 $\xi$ 进行微分。式（3.121）和式（3.122）的解形式如下：

$$\begin{Bmatrix} \varepsilon_p \\ \varepsilon_b^{BL} \end{Bmatrix} = \begin{Bmatrix} 1 \\ 1 \end{Bmatrix} B_1 + \begin{Bmatrix} 1 \\ 1 \end{Bmatrix} B_2 \xi + \begin{Bmatrix} -\psi/\alpha \\ 1 \end{Bmatrix} B_3 \sinh \Gamma \xi + \begin{Bmatrix} -\psi/\alpha \\ 1 \end{Bmatrix} B_4 \cosh \Gamma \xi \tag{3.124}$$

应用这 4 种应变边界条件可以确定 $B_1 \sim B_4$ 这 4 个未知数。压电应变 $\Lambda$ 没有清晰地在式（3.124）中表现出来，但是它会通过问题的边界条件来反映出来。图 3.22 所示的分段式致动器具有无应力的末端，$\varepsilon_p$ 项必须等于两端的压电应变 $\Lambda$。在承载结构的两端，应变可以等于任意的非零应变，这是由非压电作用引起的，如机械载荷和变形。因此，4 个边界条件为

$$\begin{cases} \xi = +1: \varepsilon_p = \Lambda, \varepsilon_b^{BL} = \varepsilon_b^{BL+} \\ \xi = -1: \varepsilon_p = \Lambda, \varepsilon_b^{BL} = \varepsilon_b^{BL-} \end{cases} \tag{3.125}$$

式中：$\varepsilon_b^{BL+}$ 和 $\varepsilon_b^{BL-}$ 分别为承载结构在其右端（+）和左端（-）的应变。将上述 4 个边界条件代入式（3.124），得

$$B_1 = \frac{\psi}{\alpha + \psi} \left( \frac{\varepsilon_b^{BL+} + \varepsilon_b^{BL-}}{2} + \Lambda \frac{\alpha}{\psi} \right) = \frac{\psi}{\alpha + \psi} \left( \varepsilon^+ + \Lambda \frac{\alpha}{\psi} \right) \tag{3.126}$$

$$B_2 = \frac{\psi}{\alpha+\psi}\left(\frac{\varepsilon_b^{BL+}-\varepsilon_b^{BL-}}{2}\right) = \frac{\psi}{\alpha+\psi}\varepsilon^- \qquad (3.127)$$

$$B_3 = \frac{\alpha}{(\alpha+\psi)\sinh\varGamma}\left(\frac{\varepsilon_b^{BL+}-\varepsilon_b^{BL-}}{2}\right) = \frac{\alpha}{(\alpha+\psi)\sinh\varGamma}\varepsilon^- \qquad (3.128)$$

$$B_4 = \frac{\alpha}{(\alpha+\psi)\cosh\varGamma}\left(\frac{\varepsilon_b^{BL+}+\varepsilon_b^{BL-}}{2}-\varLambda\right) = \frac{\alpha}{(\alpha+\psi)\cosh\varGamma}(\varepsilon^+-\varLambda) \qquad (3.129)$$

为了获得剪切应力 $\tau_{BL}$,首先通过将方程(3.103)和(3.104)对 $d\xi$ 进行积分,求得位移 $u_p$ 和 $u_b^{BL}$,然后代入方程(3.105)得到剪切应力 $\gamma$,其形式如下:

$$\gamma = \frac{\tau_{BL}}{G_{BL}} = \frac{u_p-u_b^{BL}}{t_{BL}} = \frac{-B_3\dfrac{\cosh\varGamma\xi}{\varGamma}\left(\dfrac{\psi}{\alpha}+1\right)-B_4\dfrac{\sinh\varGamma\xi}{\varGamma}\left(\dfrac{\psi}{\alpha}+1\right)}{\bar{t}_{BL}} \qquad (3.130)$$

将式(3.128)和式(3.129)中 $B_3$、$B_4$ 的值代入式(3.130),得

$$\frac{\tau_{BL}}{E_b} = \frac{\overline{G}}{\overline{E}_b \bar{t}_{BL}\varGamma}\left[\varepsilon^-\frac{\cosh\varGamma\xi}{\sinh\varGamma}+(\varepsilon^+-\varLambda)\frac{\sinh\varGamma\xi}{\cosh\varGamma}\right] \qquad (3.131)$$

注意:式(3.131)中去掉了负号(-),因为负的剪切应力在力学中没有意义。

对于有限厚度结合的情况,压电层和承载结构中应变的显性表达式为

$$\begin{Bmatrix}\varepsilon_p\\\varepsilon_b^{BL}\end{Bmatrix} = \frac{\psi}{\alpha+\psi}\begin{Bmatrix}\varepsilon^++\varepsilon^-\xi\\\varepsilon^++\varepsilon^-\xi\end{Bmatrix} - \frac{\psi}{\alpha+\psi}\begin{Bmatrix}\varepsilon^+\dfrac{\cosh\varGamma\xi}{\cosh\varGamma}+\varepsilon^-\dfrac{\sinh\varGamma\xi}{\sinh\varGamma}\\-\varepsilon^+\dfrac{\cosh\varGamma\xi}{\cosh\varGamma}-\varepsilon^-\dfrac{\sinh\varGamma\xi}{\sinh\varGamma}\end{Bmatrix} + \frac{\alpha}{\alpha+\psi}\begin{Bmatrix}1+\dfrac{\psi}{\alpha}\dfrac{\cosh\varGamma\xi}{\cosh\varGamma}\\1-\dfrac{\cosh\varGamma\xi}{\cosh\varGamma}\end{Bmatrix}\varLambda \qquad (3.132)$$

对于完美的结合($\varGamma\to\infty$),得

$$\begin{Bmatrix}\varepsilon_p\\\varepsilon_b^{BL}\end{Bmatrix} = \frac{\psi}{\alpha+\psi}\begin{Bmatrix}\varepsilon^++\varepsilon^-\xi\\\varepsilon^++\varepsilon^-\xi\end{Bmatrix} + \frac{\alpha}{\alpha+\psi}\begin{Bmatrix}1\\1\end{Bmatrix}\varLambda \qquad (3.133)$$

得到 $\varepsilon_p = \varepsilon_b^{BL}$。

对于完美结合的情况,可以计算从压电层传递到承载结构的力。该力在压电致动器的两端传递,在 $\xi=\pm 1$ 时具有以下形式:

$$F = \frac{E_b t_b^{BL} w_b}{\alpha+\psi}(\varepsilon^++\varepsilon^-\xi) + \frac{\alpha E_b t_p w_p}{\alpha+\psi}\varLambda \qquad (3.134)$$

若弯曲是由两个压电层引起的,则弯矩可简单表达为
$$M = Ft_b^{BL} \tag{3.135}$$

可以看到,在有限黏结厚度的情况下,应变(式(3.132))和应力(式(3.131))取决于压电电极上施加的电压($\Lambda = d_{31}V/t_p$)和承载结构边界的应变($\varepsilon_b^{BL+}$ 和 $\varepsilon_b^{BL-}$),与 $\varepsilon_b^{BL+}$ 和 $\varepsilon_b^{BL-}$ 有关的项仅仅是额外的被动刚度。因此,为了显示压电层存在的情况下有限结合厚度的影响,可以将它们取为零,且不失一般性,即

$$\begin{Bmatrix} \dfrac{\varepsilon_p}{\Lambda} \\ \dfrac{\varepsilon_b^{BL}}{\Lambda} \end{Bmatrix} = +\dfrac{\alpha}{\alpha+\psi} \begin{Bmatrix} 1 + \dfrac{\psi}{\alpha}\dfrac{\cosh\Gamma\xi}{\cosh\Gamma} \\ 1 - \dfrac{\cosh\Gamma\xi}{\cosh\Gamma} \end{Bmatrix} \tag{3.136}$$

对于不同的 $\Gamma$ 值,图 3.23 和图 3.24 中分别给出了在 $\psi=15$ 和 $\alpha=6$ 及 $\alpha=2$ 时(通常用于将压电陶瓷黏接在其厚度 10 倍的承载铝梁上)式(3.134)所示的两个应变值。

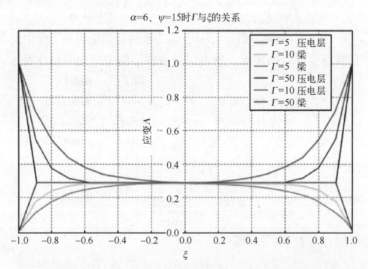

图 3.23 当 $\alpha=6$、$\psi=15$ 时,对于 $\Gamma=5$、10、50 情况下,压电层和沿梁长度承载结构中的无量纲应变(见彩插)

由式(3.123)定义的 $\Gamma$ 称为无量纲剪切滞后参数,表示相邻两个黏结层间剪切应力传递的有效性。由式(3.123)可以看出,它是结合层的剪切模量 $G_{BL}$、厚度以及 $\psi$ 和 $\alpha$ 的函数。增大 $G_{BL}$ 的值或减小厚度会使 $\Gamma$ 值增大,这意味着剪切滞后更不明显,而剪切应力在压电层两端得到有效传递(图 3.23)。

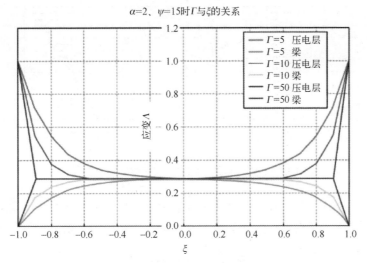

图 3.24　当 $\alpha=2$、$\psi=15$ 时，对于 $\Gamma=5$、10、50 情况下，
压电层和沿梁长度承载结构中的无量纲应变（见彩插）

$\psi$ 是影响无量纲剪切滞后参数的另一个参数，它是承载梁与压电层的厚度和模量之比的乘积。当 $\psi$ 趋于 0，即承载结构的厚度非常小时，就会形成一个完美的结合（$\Gamma\to\infty$），而压电层的应变将与承载结构的应变相等（式（3.136））。对于其他情况，大的 $\psi$ 意味着与压电层的性能相比，承载结构相对较厚且具有高的杨氏模量，意味着承载结构所承受的应变相对较小。根据前文所述，表面黏结的压电致动器应该有一个大的 $\Gamma$（即与承载结构几乎完美的结合）和低的 $\psi$（即与承载结构刚度相比，致动器的刚度更高）。

还应该注意到图 3.22 中的两个压电层具有相同方向的电压，导致无论拉伸还是收缩（取决于 $d_{31}$ 的符号），都有 $\alpha=2$。

如果其中一个电压反转，压电层将在承载结构中引起弯曲，$\alpha=6$。值得注意的是，较高的 $\alpha$ 值使得黏接压电层的致动器在弯曲时比在伸展或收缩时效率更高。

在介绍了表面黏接压电层的情况后，将介绍嵌入式压电致动器的情况，如图 3.25（a）和（b）所示。

压电层可以嵌入在由层状复合材料制成的承载结构中。在这种情况下，因为结合层的厚度通常是碳纤维直径的 5 倍（$\approx 1\mu m$），假设其可以忽略不计，如文献［2］所述，剪切滞后参数 $\Gamma$ 的值较大（约为 1000），这充分证明了完美结合的假设。

与表面黏接的压电致动器为应激弯曲不同，嵌入式压电致动器的应变会随

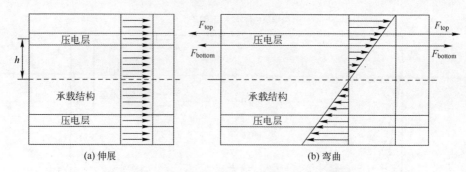

图 3.25 跨承载结构和嵌入式压电层的应变分布

其厚度线性变化,并且在致动器的上下表面都要求承载结构与压电层之间的应变兼容。考虑上述假设后,承载结构和压电层中的应力平衡方程可写为

$$\sigma_{\mathrm{p}} = \frac{6F_{\mathrm{DIF}}}{w_{\mathrm{p}} t_{\mathrm{p}}^2}(z-h) + \frac{F_{\mathrm{SUM}}}{w_{\mathrm{p}} t_{\mathrm{p}}} \quad (3.137)$$

$$\sigma_{\mathrm{b}} = \frac{(M_{\mathrm{p}} + M_M)}{I_{\mathrm{eq.}}} z \quad (3.138)$$

式中:$M_M$ 为由于压电层以外其他载荷产生的沿承载结构的力矩;$F_{\mathrm{DIF}} = F_2 - F_1$,$F_{\mathrm{SUM}} = F_1 + F_2$,$w_{\mathrm{p}}$ 和 $t_{\mathrm{p}}$ 分别为压电层的宽度和厚度;$I_{\mathrm{eq.}}$ 为梁截面(包括压电材料和承载结构材料,假设为层压复合材料)的等效惯性矩;$M_{\mathrm{p}}$(压电诱导弯矩)定义为

$$M_{\mathrm{p}} = 2\left[\frac{F_{\mathrm{DIF}}}{2} t_{\mathrm{p}} + F_{\mathrm{SUM}} h\right] \quad (3.139)$$

利用压电层与承载结构的应变-应力关系(式(3.115)和式(3.116)),并使压电层与承载结构边界处(即 $z=(h-t_{\mathrm{p}}/2)$ 和 $z=(h+t_{\mathrm{p}}/w)$)的应变相等,得到对应未知数 $F_{\mathrm{DIF}}$ 和 $F_{\mathrm{SUM}}$ 的两个方程。它们的表达式为[2]

$$F_{\mathrm{SUM}} = -\frac{12\theta_z}{24\theta_z^2 + \psi \theta_{\mathrm{b}}^2 \hat{I} + 2} \frac{M_M}{t_{\mathrm{p}}} - \frac{2E_{\mathrm{p}} t_{\mathrm{p}} w_{\mathrm{p}} + E_{\mathrm{b}} t_{\mathrm{b}} w_{\mathrm{p}} \theta_{\mathrm{b}}^2 \hat{I}}{24\theta_z^2 + \psi \theta_{\mathrm{b}}^2 \hat{I} + 2} \Lambda \quad (3.140)$$

$$F_{\mathrm{DIF}} = -\frac{2}{24\theta_z^2 + \psi \theta_{\mathrm{b}}^2 \hat{I} + 2} \frac{M_M}{t_{\mathrm{p}}} + \frac{4E_{\mathrm{p}} t_{\mathrm{p}} w_{\mathrm{p}} \theta_{\mathrm{b}}}{24\theta_z^2 + \psi \theta_{\mathrm{b}}^2 \hat{I} + 2} \Lambda \quad (3.141)$$

式中

$$w_{\mathrm{p}} = w_{\mathrm{b}}, \quad \theta_{\mathrm{b}} = \frac{t_{\mathrm{b}}}{t_{\mathrm{p}}}, \quad \theta_z = \frac{h}{t_{\mathrm{p}}}, \quad \psi = \frac{E_{\mathrm{b}} t_{\mathrm{b}}}{E_{\mathrm{p}} t_{\mathrm{p}}} = \bar{E}_{\mathrm{b}} \theta_{\mathrm{b}}, \quad \hat{I} = \frac{I_{EQ}}{w_{\mathrm{p}} t_{\mathrm{p}}^3 / 12} \quad (3.142)$$

$M_M$ 可以表示为在压电层两端沿承载结构的已知应变 $\varepsilon_{\mathrm{b}}^{\mathrm{top}}$ 和 $\varepsilon_{\mathrm{b}}^{\mathrm{bottom}}$ 的线性

函数：

$$M_{\mathrm{M}}=-\frac{I_{EQ}E_{\mathrm{b}}}{t_{\mathrm{b}}}(\varepsilon_{\mathrm{b}}^{\mathrm{top}}+\varepsilon_{\mathrm{b}}^{\mathrm{bottom}})-\frac{I_{EQ}E_{\mathrm{b}}}{t_{\mathrm{b}}}(\varepsilon_{\mathrm{b}}^{\mathrm{top}}-\varepsilon_{\mathrm{b}}^{\mathrm{bottom}})\xi \qquad (3.143)$$

利用式（3.141）~式（3.143）和梁的应变-应力关系，可得致动器和梁中的应变表达式为

$$\begin{aligned}\varepsilon_{\mathrm{b}}=\varepsilon_{\mathrm{p}}=&-\frac{2\psi\theta_{\mathrm{b}}\hat{I}}{(24\theta_z^2+\psi\theta_{\mathrm{b}}^2\hat{I}+2)}\frac{z}{t_{\mathrm{p}}}\left(\frac{\varepsilon_{\mathrm{b}}^{\mathrm{top}}+\varepsilon_{\mathrm{b}}^{\mathrm{bottom}}}{2}+\frac{\varepsilon_{\mathrm{b}}^{\mathrm{top}}-\varepsilon_{\mathrm{b}}^{\mathrm{bottom}}}{2}\xi\right)\\ &+\frac{24\theta_z}{(24\theta_z^2+\psi\theta_{\mathrm{b}}^2\hat{I}+2)}\frac{z}{t_{\mathrm{p}}}\Lambda\end{aligned} \qquad (3.144)$$

则施加在压电层两端的归一化力矩为

$$\begin{aligned}\frac{M_{\mathrm{p}}}{E_{\mathrm{b}}t_{\mathrm{b}}^2w_{\mathrm{b}}}=&\frac{(12\theta_z^2+1)\hat{I}}{3(24\theta_z^2+\psi\theta_{\mathrm{b}}^2\hat{I}+2)}\left(\frac{\varepsilon_{\mathrm{b}}^{\mathrm{top}}+\varepsilon_{\mathrm{b}}^{\mathrm{bottom}}}{2}+\frac{\varepsilon_{\mathrm{b}}^{\mathrm{top}}-\varepsilon_{\mathrm{b}}^{\mathrm{bottom}}}{2}\xi\right)-\\ &\frac{2\theta_z\theta_{\mathrm{b}}\hat{I}}{(24\theta_z^2+\psi\theta_{\mathrm{b}}^2\hat{I}+2)}\Lambda\end{aligned} \qquad (3.145)$$

对内嵌致动器拉伸或压缩的简单情况进行分析，得到致动器和梁中的应变表达式如下：

$$\varepsilon_{\mathrm{b}}=\varepsilon_{\mathrm{p}}=\frac{\overline{E}(\theta_{\mathrm{b}}-2)}{2+\overline{E}(\theta_{\mathrm{b}}-2)}\left(\frac{\varepsilon_{\mathrm{b}}^{\mathrm{top}}+\varepsilon_{\mathrm{b}}^{\mathrm{bottom}}}{2}+\frac{\varepsilon_{\mathrm{b}}^{\mathrm{top}}-\varepsilon_{\mathrm{b}}^{\mathrm{bottom}}}{2}\xi\right)+\frac{2}{2+\overline{E}(\theta_{\mathrm{b}}-2)}\Lambda \qquad (3.146)$$

式中：$\overline{E}=\dfrac{E_{\mathrm{b}}}{E_{\mathrm{P}}}$。施加在压电层两端的力为

$$\begin{aligned}\frac{F}{E_{\mathrm{b}}t_{\mathrm{b}}w_{\mathrm{b}}}=&-\frac{(\theta_{\mathrm{b}}-2)}{2\theta_{\mathrm{b}}+\overline{E}(\theta_{\mathrm{b}}-2)\theta_{\mathrm{b}}}\left(\frac{\varepsilon_{\mathrm{b}}^{\mathrm{top}}+\varepsilon_{\mathrm{b}}^{\mathrm{bottom}}}{2}+\frac{\varepsilon_{\mathrm{b}}^{\mathrm{top}}-\varepsilon_{\mathrm{b}}^{\mathrm{bottom}}}{2}\xi\right)-\\ &\frac{2}{2\theta_{\mathrm{b}}+\overline{E}(\theta_{\mathrm{b}}-2)\theta_{\mathrm{b}}}\Lambda\end{aligned} \qquad (3.147)$$

## 3.4 伯努利-欧拉梁模型

伯努利-欧拉梁模型得名于同名的梁理论，通常用于细长梁。文献[14-16]表明，当结合层较薄时，该模型比均匀应变模型的预测更准确。根据伯努利-欧拉梁模型，假设应变在整个承载结构、梁、压电层和结合层中都是线性的。

忽略横向剪切应变的影响，在变形前垂直于梁的中轴平面，在变形之后仍然保持平面（这些是伯努利-欧拉梁模型的常用假设）。使用经典层压理论（见第 2 章）对层压复合梁进行推导，在层各向同性的情况下得到伯努利-欧拉梁模型。图 3.26 所示为弯曲梁及其相关坐标系的示意图。

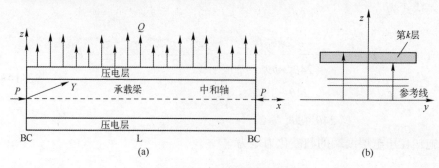

图 3.26 带有压电层的梁

定义一个坐标系 $X$、$Y$、$Z$，其中坐标轴 $X$ 沿梁的中间平面，坐标轴 $Z$ 垂直于梁的纵轴。因此，坐标轴 $Y$ 将在梁的宽度方向，从而形成一个右手坐标系（图 3.26）。

伯努利-欧拉梁模型仅有的两个变量分别是梁中间平面上一个点的轴向位移（在 $x$ 方向），指定为 $u(x)$；以及梁中间平面上一点的横向位移（在 $z$ 方向），指定为 $w(x)$。

考虑到这一点，可以将通过整个结构截面（梁、压电层和结合层）的轴向位移表示为

$$u(x,z) = u_0(x) - z \frac{\partial w(x)}{\partial x} \tag{3.148}$$

$$w(x,z) = w_0(x) \tag{3.149}$$

式中：$u_0(x)$ 和 $w_0(x)$ 为承载梁中轴上一个点的位移。

相应地，应变矢量包括在同一点上的机械应变 $\varepsilon^m$ 和压电电极上施加电压引起的诱导应变 $\varepsilon^a$。注意，对于非压电层，诱导应变为零。

由式（3.148）和式（3.149）所示的位移推导得到机械应变为

$$\varepsilon_x^m(x,z) = \frac{\partial u(x,z)}{\partial x} = \frac{\partial u_0(x)}{\partial x} - z \frac{\partial^2 w(x)}{\partial x^2} \equiv \varepsilon_0(x) - z\kappa(x) \tag{3.150}$$

根据伯努利-欧拉理论的假设，所有其他的机械应变为 $\varepsilon_y^m(x) = \varepsilon_z^m(x) = \gamma_{xy}^m(x) = \gamma_{yz}^m(x) = \gamma_{zx}^m(x) = 0$，$\kappa(x)$ 是梁的曲率。

诱导应变 $\varepsilon_x^a(x)$，$\varepsilon_z^a(x)$，$\gamma_{xy}^a(x)$ 为[17-18]

$$[\varepsilon_x^a(x)]_k = \frac{V_k(x)(d_{31})_k}{z_{k+}^a - z_{k-}^a} \tag{3.151}$$

$$\varepsilon_z^a(x) = \gamma_{xy}^a(x) = 0 \tag{3.152}$$

在式（3.151）中，$V_k(x)$ 为施加在第 $k$ 个压电层上的电压，厚度为 $(z_{k+}^a - z_{k-}^a)$（图 3.26（b）），而 $(d_{31})_k$ 为第 $k$ 层的压电系数。如上所述，式（3.151）和式（3.152）仅适用于压电层。

将应变乘以刚度系数 $(\overline{Q}_{11})_k$，对于每一个压电层[19]可得

$$(\sigma_x)_k = (\overline{Q}_{11})_k (\varepsilon_x^m - \varepsilon_x^a)_k \tag{3.153}$$

式中：$(\overline{Q}_{11})$（对每个单一层）被定义为

$$\overline{Q}_{11} = Q_{11}\cos^4\theta + Q_{22}\sin^4\theta + 2(Q_{12} + 2Q_{66})\sin^2\theta\cos^2\theta \tag{3.154}$$

式中：$\theta$ 为特定层上纤维的方向与纵轴 $x$ 之间的夹角。系数 $Q_{11}$、$Q_{22}$、$Q_{12}$ 和 $Q_{66}$ 被定义为每层弹性特性的函数（第 2 章），即

$$Q_{11} = \frac{\overline{E}_{11}}{1 - v_{12}v_{21}}, \quad Q_{12} = \frac{v_{21}\overline{E}_{11}}{1 - v_{12}v_{21}} = \frac{v_{12}\overline{E}_{22}}{1 - v_{12}v_{21}}, \quad Q_{22} = \frac{\overline{E}_{22}}{1 - v_{12}v_{21}}, \quad Q_{66} = G_{12} \tag{3.155}$$

式中：$\overline{E}_{11}$、$\overline{E}_{22}$、$G_{12}$ 分别为单层的拉伸、压缩和剪切模量；$v_{12}$ 为主要泊松比，其关系为 $\dfrac{\overline{E}_{11}}{\overline{E}_{22}} = \dfrac{v_{12}}{v_{21}}$（第 2 章）。

沿梁的高度 $h$ 积分，并乘以宽度 $b$，得到力和力矩为

$$N_x = b\int_{-h/2}^{+h/2} \sigma_x \mathrm{d}z, \quad M_y = b\int_{-h/2}^{+h/2} \sigma_x z \mathrm{d}z \tag{3.156}$$

执行式（3.156）中的积分得

$$\begin{Bmatrix} N_x \\ M_y \end{Bmatrix} = \begin{bmatrix} A_{11} & B_{11} \\ B_{11} & D_{11} \end{bmatrix} \begin{Bmatrix} \dfrac{\partial u}{\partial x} \\ -\dfrac{\partial^2 w}{\partial x^2} \end{Bmatrix} - \begin{Bmatrix} E_{11} \\ F_{11} \end{Bmatrix} \tag{3.157}$$

式中：系数 $A_{11}$、$B_{11}$、$D_{11}$ 分别为通常用于层压复合梁的轴向、弯曲-拉伸耦合、弯曲和横向剪切刚度的表达式。根据经典层压理论[19]，将积分替换为求和（图 3.26（b）），得

$$A_{11} = b\sum_{k=1}^{N} (\overline{Q}_{11})_k (z_k - z_{k-1}) \tag{3.158}$$

$$B_{11} = \frac{b}{2}\sum_{k=1}^{N} (\overline{Q}_{11})_k (z_k^2 - z_{k-1}^2) \tag{3.159}$$

$$D_{11} = \frac{b}{3}\sum_{k=1}^{N}(\overline{Q}_{11})_k(z_k^3 - z_{k-1}^3) \tag{3.160}$$

式中：$N$ 为包括压电层的层堆叠数。

式（3.157）中的 $E_{11}$ 和 $F_{11}$ 分别表示由压电层引起的诱导轴向力和诱导力矩，定义为

$$E_{11} = b\sum_{k=1}^{N_a}(\overline{Q}_{11})_k^a V_k(x,t)(d_{31})_k \tag{3.161}$$

$$F_{11} = \frac{b}{2}\sum_{k=1}^{N_a}(\overline{Q}_{11})_k^a V_k(x,t)(d_{31})_k(z_{k+}^a + z_{k-}^a) \tag{3.162}$$

式中：$N_a$ 为压电层层数。

为了得到运动方程和相关的边界条件，使用虚功原理，得到势的变化为[17]

$$\delta\pi = \int_{t_1}^{t_2}\int_{\text{volume}}(\delta E_k - \delta E_p + \delta\overline{W})\cdot d(\text{volume})dt = 0 \tag{3.163}$$

式中：$E_k$ 为梁的动能；$E_p$ 为梁的应变能；$\overline{W}$ 为作用在梁上外力所做的功。它们的表达式为

$$E_k = \frac{1}{2}\int_{\text{volume}}\rho\cdot\left[\left(\dot{u} - z\frac{\partial\dot{w}}{\partial x}\right)^2 + \dot{w}^2\right]\cdot d(\text{volume}) \tag{3.164}$$

$$E_p = \frac{1}{2}\int_{\text{volume}}(\sigma_x\varepsilon_x)\cdot d(\text{volume}) \tag{3.165}$$

$$\overline{W} = \int_0^L Q\cdot w\cdot x\cdot dx - P\cdot u(L) + P\cdot u(0) \tag{3.166}$$

在图 3.26（a）中，$L$ 为梁的长度，$P$ 为作用于梁两端的压缩载荷，$Q$ 为沿梁分布的侧向载荷。（·）表示对时间的微分。

将式（3.164）~式（3.166）代入式（3.163）进行分部积分，而要求电势 $\delta\pi$ 的变化应消失（只有当积分和梁两端的各种表达式都等于零时（假设位移是任意的）才成立），得到以下压电层梁的运动方程及其相关的边界条件：

$$\frac{\partial N_x}{\partial x} = \frac{\partial}{\partial t}\left[I_1\dot{u} + I_2\frac{\partial\dot{w}}{\partial x}\right] \tag{3.167}$$

$$\frac{\partial}{\partial x}\left[\frac{\partial M_y}{\partial x} + N_x\frac{\partial w}{\partial x}\right] = \frac{\partial}{\partial t}[I_1\dot{w}] + Q \tag{3.168}$$

而 $(I_1, I_2) = \int\rho\cdot(1,z)\cdot dz$，$\rho$ 是每层梁的密度。相关的可能边界条件为

$$N_x + P = 0 \quad \text{或} \quad u = 0 \tag{3.169}$$

$$\frac{\partial M_y}{\partial x}+N_x\frac{\partial w}{\partial x}=0 \quad 或 \quad w=0 \tag{3.170}$$

$$M_y=0 \quad 或 \quad \frac{\partial w}{\partial x}=0 \tag{3.171}$$

注意，正如文献中通常提到的，$N_x$ 只在上述非线性表达式中被 $-P$ 取代：

$$\left[N_x\frac{\partial w}{\partial x}\right]$$

将式（3.157）中力和力矩表达式代入式（3.167）和式（3.168）中，可得到含有压电层的梁的运动方程，以 $u$ 和 $w$ 这两个假定的挠度表示：

$$\frac{\partial}{\partial x}\left(A_{11}\frac{\partial u}{\partial x}-B_{11}\frac{\partial^2 w}{\partial x^2}-E_{11}\right)=\frac{\partial}{\partial t}\left[I_1\dot{u}+I_2\frac{\partial \dot{w}}{\partial x}\right] \tag{3.172}$$

$$\frac{\partial}{\partial x}\left[\frac{\partial}{\partial x}\left(B_{11}\frac{\partial u}{\partial x}-D_{11}\frac{\partial^2 w}{\partial x^2}-F_{11}\right)-P\frac{\partial w}{\partial x}\right]=\frac{\partial}{\partial t}[I_1\dot{w}]+Q \tag{3.173}$$

与它们相关的可能边界条件为

$$A_{11}\frac{\partial u}{\partial x}-B_{11}\frac{\partial^2 w}{\partial x^2}=-P+E_{11} \text{ 或 } u=0 \tag{3.174}$$

$$\frac{\partial}{\partial x}\left(B_{11}\frac{\partial u}{\partial x}-D_{11}\frac{\partial^2 w}{\partial x^2}-F_{11}\right)-P\frac{\partial w}{\partial x}=0 \text{ 或 } w=0 \tag{3.175}$$

$$B_{11}\frac{\partial u}{\partial x}-D_{11}\frac{\partial^2 w}{\partial x^2}=F_{11} \text{ 或} \frac{\partial w}{\partial x}=0 \tag{3.176}$$

假设沿梁的性质恒定，式（3.172）~式（3.176）有如下形式：

$$\left(A_{11}\frac{\partial^2 u}{\partial x^2}-B_{11}\frac{\partial^3 w}{\partial x^3}\right)=I_1\ddot{u}+I_2\frac{\partial \ddot{w}}{\partial x} \tag{3.177}$$

$$\left(B_{11}\frac{\partial^3 u}{\partial x^3}-D_{11}\frac{\partial^4 w}{\partial x^4}-P\frac{\partial^2 w}{\partial x^2}\right)=I_1\ddot{w}+Q \tag{3.178}$$

而字母上面的两个点表示对时间的二次微分。相关的可能边界条件形式如下：

$$A_{11}\frac{\partial u}{\partial x}-B_{11}\frac{\partial^2 w}{\partial x^2}=-P+E_{11} \quad 或 \quad u=0 \tag{3.179}$$

$$B_{11}\frac{\partial^2 u}{\partial x^2}-D_{11}\frac{\partial^3 w}{\partial x^3}-P\frac{\partial w}{\partial x}=0 \quad 或 \quad w=0 \tag{3.180}$$

$$B_{11}\frac{\partial u}{\partial x}-D_{11}\frac{\partial^2 w}{\partial x^2}=F_{11} \quad 或 \quad \frac{\partial w}{\partial x}=0 \tag{3.181}$$

应该注意的是，压电层的影响只通过边界条件出现。对于对称层压板，

$B_{11}$ 和 $I_2$ 为 0，得到以下解耦运动方程：

$$A_{11}\frac{\partial^2 u}{\partial x^2}=I_1\ddot{u} \tag{3.182}$$

$$\left(-D_{11}\frac{\partial^4 w}{\partial x^4}-P\frac{\partial^2 w}{\partial x^2}\right)=I_1\ddot{w}+Q \tag{3.183}$$

边界条件将有以下更简单的表达式：

$$A_{11}\frac{\partial u}{\partial x}=-P+E_{11} \quad \text{或} \quad u=0 \tag{3.184}$$

$$D_{11}\frac{\partial^3 w}{\partial x^3}+P\frac{\partial w}{\partial x}=0 \quad \text{或} \quad w=0 \tag{3.185}$$

$$D_{11}\frac{\partial^2 w}{\partial x^2}=-F_{11} \quad \text{或} \quad \frac{\partial w}{\partial x}=0 \tag{3.186}$$

根据供应给压电层的电压，首先分析两种类型诱导牵引力的静态情况。对所有压电层施加相同的电压（包括振幅和符号）将导致整个结构的净拉伸或压缩。对于这种加载情况，只使用式（3.182）（当 $I_1=0$ 时）及其由式（3.184）所示的关联的平面边界条件。式（3.182）的解具有一般形式：

$$u(x)=ax+b \tag{3.187}$$

式中：$a$ 和 $b$ 为根据该情况的边界条件确定的常数。

设 $u(0)=u(L)=0$（梁两端无轴向位移，见图 3.27），得到一个 $u(x)=0$ 形式的解。将结果代入式（3.157）的第 1 行，得

$$N_x=-E_{11}^0 \tag{3.188}$$

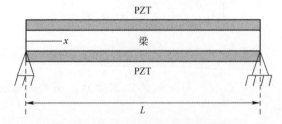

图 3.27 具有两个压电层的轴向约束梁

代入式（3.169）后得

$$P=E_{11}^0 \tag{3.189}$$

式中：$E_{11}^0$ 为静态诱导轴向力。将供给压电层的电压提高至与整个梁的屈曲载荷相等的诱导力，将导致其屈曲。导致这种屈曲的电压 $V_{cr}$ 可以定义为（在使用式（3.161）时）沿梁恒定的电压情况（因此它不依赖于 $x$ 和时间 $t$）：

$$V_{cr} = \frac{P_{cr}}{b\sum_{k=1}^{Na}(\overline{Q}_{11})_k^a(d_{31})_k}, \quad P_{cr} = \frac{\pi(EI)_{eq.}}{L^2} \quad (3.190)$$

式中：$(EI)_{eq.}$ 为含压电层梁的抗弯刚度。关于这个问题的更多信息可以在下一节中找到。

当压电层的上、下两层施加异相电压（振幅相同但符号不同）时，梁上会产生弯矩。首先来解决诱导弯曲的静态情况，即消除包含加速度的项：

$$D_{11}\frac{d^4w}{dx^4} + P\frac{d^2w}{dx^2} = -Q \quad (3.191)$$

将式（3.191）除以 $D_{11}$，假设 $Q$ 为常数，并使 $k^2 = \frac{P}{D_{11}}$，得到其通解为

$$w(x) = A_1\sin(kx) + A_2\cos(kx) - \frac{Q}{P}\chi^2 + A_3\chi + A_4 \quad (3.192)$$

通过在复合层梁的两端应用 4 个边界条件，可以求出 4 个常数 $A_1 \sim A_4$。当没有轴向力时（$P=0$），解的形式如下（当 $Q$ 为常数）：

$$w(x) = -\frac{Q}{24D_{11}}x^4 + A_1x^3 + A_2x^2 + A_3x + A_4 \quad (3.193)$$

如前所述，通过应用梁的适当边界条件得到了 4 个常数 $A_1 \sim A_4$。如果横向载荷 $Q(x)$ 是轴向坐标的函数，那么积分 4 次将得到梁的面外挠度，有 4 个常数将由满足梁的边界条件来确定。求解无横向载荷的梁的式（3.193），如图 3.28 所示。提供给压电层的电压是异相的，即振幅相同但符号不同，从而诱导梁上的弯曲。梁的边界条件为

$$\begin{cases} w(0) = 0 \text{ 时}, D_{11}\dfrac{d^2w(0)}{dx^2} = -F_{11} \\ w(L) = 0 \text{ 时}, D_{11}\dfrac{d^2w(L)}{dx^2} = -F_{11} \end{cases} \quad (3.194)$$

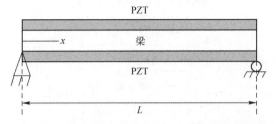

图 3.28　具有两个压电层的简支边界条件的梁

应用边界条件，可得诱导弯曲梁的横向挠度表达式如下：

$$w(x) = -\frac{Q}{24D_{11}}x^4 + \frac{QL}{12D_{11}}x^3 - \frac{F_{11}}{2D_{11}}x^2 + \left(\frac{F_{11}L}{2D_{11}} - \frac{QL^3}{24D_{11}}\right)x \quad (3.195)$$

需要注意的是，对于 $Q=0$ 的情况，式（3.195）可简化为如下表达式：

$$w(x) = -\frac{F_{11}}{2D_{11}}x(x-L) \quad (3.196)$$

这正是两端有两个弯矩 $M_0$ 作用的简支边界条件下的梁的挠度方程（这里 $M_0 = -F_{11}$、$D_{11}=EI$）。

同样值得一提的是，由于压电致动器的弯曲仅受梁的边界条件的影响，两端只有几何边界条件的梁（如两端夹紧的梁）将不会受到压电层从一端延伸到另一端的承载结构的影响。在这种情况下，只有贴片才会引起梁的弯曲。

静态情况解决后，式（3.183）中提出的梁没有机械载荷的动态情况，将解决 $Q=0$。它的形式为

$$D_{11}\frac{\partial^4 w}{\partial x^4} + P\frac{\partial^2 w}{\partial x^2} + I_1\ddot{w}\partial = 0 \quad (3.197)$$

除以 $D_{11}$，并假设谐振振动为 $w(x,t) = W(x)\sin(\omega t)$，得到只有一个变量 $x$ 的微分方程：

$$\frac{d^4 W}{dx^4} + k^2\frac{d^2 W}{dx^2} - \omega^2 \bar{I_1} W = 0 \quad (3.198)$$

式中

$$k^2 = \frac{P}{D_{11}}, \quad \bar{I_1} = \frac{I_1}{D_{11}} \quad (3.199)$$

$\omega$ 为梁的固有频率。

设 $W = Re^{sx}$，并代入式（3.198），得到其解为

$$s^4 + k^2 s^2 - \omega^2 \bar{I_1} = 0 \quad (3.200)$$

方程（3.200）有两个实解和两个复解，即

$$s_1 = \pm\sqrt{\left[\frac{-k^2+\sqrt{k^4+4\omega^2 \bar{I_1}}}{2}\right]} = \pm\lambda_1$$

$$s_2 = \pm\sqrt{\left[\frac{-k^2-\sqrt{k^4+4\omega^2 \bar{I_1}}}{2}\right]} = \pm i\lambda_2 \quad (3.201)$$

通解具有以下形式：

$$W(x) = A_1\sinh(\lambda_1 x) + A_2\cosh(\lambda_1 x) + A_3\sin(\lambda_2 x) + A_4\cos(\lambda_2 x) \quad (3.202)$$

通过应用适当的边界条件，可以得到常数 $A_1 \sim A_4$。

利用简支梁的边界条件（式（3.194）），需要修改恒定力矩-$F_{11}$的边界条件，以便能够计算梁两端的压缩和由压电层引起的两端诱导弯矩诱导下的固有频率。力矩的表达式可以写成

$$\begin{cases} -F_{11} = k_\theta \dfrac{\mathrm{d}w(x)}{\mathrm{d}x} \\ k_\theta = -\dfrac{F_{11}}{\dfrac{\mathrm{d}w(x)}{\mathrm{d}x}} \quad x = 0, L \end{cases} \quad (3.203)$$

则边界条件方程为

$$\begin{cases} w(0) = 0, D_{11} \dfrac{\mathrm{d}^2 w(0)}{\mathrm{d}x^2} = -k_\theta \dfrac{\mathrm{d}w(0)}{\mathrm{d}x} \\ w(L) = 0, D_{11} \dfrac{\mathrm{d}^2 w(L)}{\mathrm{d}x^2} = -k_\theta \dfrac{\mathrm{d}w(L)}{\mathrm{d}x} \end{cases} \quad (3.204)$$

将 4 个边界条件代入式（3.202），得到以下表达式：

$$\begin{bmatrix} a_{11} & a_{12} & a_{13} & a_{14} \\ a_{21} & a_{22} & a_{23} & a_{24} \\ a_{31} & a_{32} & a_{33} & a_{34} \\ a_{41} & a_{42} & a_{43} & a_{44} \end{bmatrix} \begin{Bmatrix} A_1 \\ A_2 \\ A_3 \\ A_4 \end{Bmatrix} = \begin{Bmatrix} 0 \\ 0 \\ 0 \\ 0 \end{Bmatrix} \quad (3.205)$$

式中：矩阵的系数由本章附录 B 给出。

如果系数的矩阵为 0，则可以计算由压电层引起弯曲和压缩载荷的梁的固有频率。

为了比较均匀梁模型和伯努利-欧拉模型，表 3.3 总结了压电层或承载梁中的应变表达式（见文献［14］和 3.3 节）。梁的配置见图 3.28。

表 3.3 均匀梁模型（无剪力滞后）与伯努利-欧拉梁模型的比较[①]

| 均匀梁模型 | 伯努利-欧拉梁模型 |
| --- | --- |
| 拉伸或压缩引起的应变 | |
| $\varepsilon_{\text{piezo}} = \dfrac{2\Lambda}{2 + \psi_e}$ | $\varepsilon_{\text{piezo}} = \dfrac{2\Lambda}{2 + \overline{\psi}_e}$ |
| $\varepsilon_{\text{beam}} = \dfrac{2\Lambda}{2 + \psi_e}$ | $\varepsilon_{\text{beam}} = \dfrac{2\Lambda}{2 + \overline{\psi}_e}$ |

续表

| 均匀梁模型 | 伯努利-欧拉梁模型 |
|---|---|
| 弯曲引起的应变 ||
| $\varepsilon_{\text{pizeo}}^{*} = -\dfrac{t_{\text{b}}}{2}k = \dfrac{6\Lambda}{6+\psi_{\text{b}}}$ | $\varepsilon_{\text{pizeo}}^{*} = -\dfrac{t_{\text{b}}}{2}k = \dfrac{6\left(1+\dfrac{1}{\theta_{\text{b}}}\right)}{(6+\psi_{\text{e}})+\dfrac{12}{\theta_{\text{b}}}+\dfrac{8}{\theta_{\text{b}}^{2}}}\Lambda$ |
| $\varepsilon_{\text{beam}}^{*} = -\dfrac{t_{\text{b}}}{2}k = \dfrac{6\Lambda}{6+\psi_{\text{b}}}$ | $\varepsilon_{\text{beam}}^{*} = -\dfrac{t_{\text{b}}}{2}k = \dfrac{6\left(1+\dfrac{1}{\theta_{\text{b}}}\right)}{(6+\psi_{\text{e}})+\dfrac{12}{\theta_{\text{b}}}+\dfrac{8}{\theta_{\text{b}}^{2}}}\Lambda$ |

① $\psi_{\text{e}} = \dfrac{(EA)_{\text{beam}}}{(EA)_{\text{piezo}}}$，$\psi_{\text{b}} = \dfrac{12(EI)_{\text{beam}}}{t_{\text{beam}}^{2}(EA)_{\text{piezo}}}$。

对于一个矩形的横截面，得

$\psi_{\text{e}} = \psi_{\text{b}}$；$\overline{\psi}_{\text{e}} = \dfrac{(EI)_{\text{beam}}}{(EA)_{\text{piezo}}} = \dfrac{E_{\text{beam}}\left[(w\cdot t)_{\text{beam}} - 2(w\cdot t)_{\text{piezo}}\right]}{(E\cdot w\cdot t)_{\text{piezo}}}$；

$\varepsilon_{\text{beam}}^{*}$是梁与压电层界面处的应变，$\theta_{\text{b}} = \dfrac{t_{\text{beam}}}{t_{\text{piezo}}}$，$t_{\text{beam}}$为梁的高度，$w$为梁的宽度和压电层的宽度。

图 3.29 所示为 $E_{\text{beam}}/E_{\text{piezo}} = 1.2$ 和 3 时的应变（$\Lambda$）与厚度比（$\theta_{\text{b}}$）的关系。从图 3.29 中可以清楚地看出，对于较大的厚度比，使用均匀梁模型或伯努利-欧拉模型预测应变的结果几乎是相同的。在 $\theta_{\text{b}}$ 值较低时，均匀梁模型不能正确地预测应变，而伯努利-欧拉模型则能较好地预测。

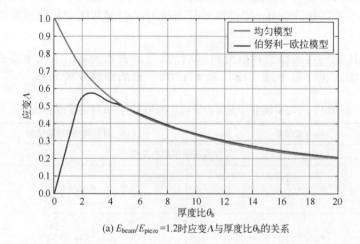

(a) $E_{\text{beam}}/E_{\text{piezo}} = 1.2$ 时应变 $\Lambda$ 与厚度比 $\theta_{\text{b}}$ 的关系

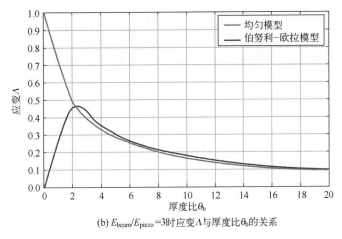

(b) $E_{beam}/E_{piezo}=3$ 时应变 $\Lambda$ 与厚度比 $\theta_b$ 的关系

图 3.29 弯曲应变对比（见彩插）

## 3.5 一阶剪切变形（Timoshenko 型）梁模型

本节旨在推导带有压电层的层压梁运动方程，如图 3.30 所示。

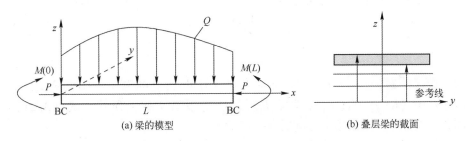

图 3.30 梁的模型和叠层梁的截面

首先，使用一阶剪切变形理论（FSDT）表示梁的可能位移，这与各向同性梁的 Timoshenko 模型相同。

定义一个坐标系 $X$、$Y$、$Z$，其中 $X$ 轴沿着梁的中间平面，$Z$ 轴垂直于梁的纵轴。因此，$Y$ 坐标将在梁的宽度方向，形成一个右手坐标系（图 3.30（a））。

模型的变量如下：

(1) 梁中间平面上一点的轴向位移（沿 $x$ 方向），设为 $U(x,t)$。

(2) 梁中间平面上一点的横向位移（沿 $z$ 方向），设为 $W(x,t)$。

(3) 梁截面的旋转（绕 $y$ 轴）$\Phi(x,t)$，顺时针为正。

字母 $t$ 代表时间。同时，假设各层之间的结合（包括压电层）是完美的，且可以忽略胶黏剂的厚度，而且在 $z$ 方向上没有压缩。

根据上述假设，梁截面上任意一点的位移场为

$$\widetilde{U}(x,z,t) = U(x,t) + z \cdot \Phi(x,t) \tag{3.206}$$

$$\widetilde{W}(x,z,t) = W(x,t) \tag{3.207}$$

相应地，应变矢量将包括在同一点上的机械应变 $\varepsilon^m$ 和压电电极上施加电压引起的诱导应变 $\varepsilon^a$。需记住，对于非压电层，诱导应变为零。

机械应变 $\varepsilon_x^m$，$\varepsilon_z^m$，$\gamma_{xz}^m$（法向和剪切）由上述式（3.206）和式（3.207）中位移的推导得

$$\varepsilon_x^m = \frac{\partial \widetilde{U}}{\partial x} + \frac{1}{2}\left(\frac{\partial W}{\partial x}\right)^2 = \frac{\partial U}{\partial x} + z\frac{\partial \Phi}{\partial x} + \frac{1}{2}\left(\frac{\partial W}{\partial x}\right)^2 \tag{3.208}$$

注意，式（3.208）还包括一个非线性项，由于其相互作用，对受轴向压缩（如果存在）的梁是非常重要的。

$$\varepsilon_z^m = \frac{\partial \widetilde{W}}{\partial z} = 0 \tag{3.209}$$

$$\gamma_{xz}^m = \frac{\partial \widetilde{U}}{\partial z} + \frac{\partial \widetilde{W}}{\partial x} = \Phi + \frac{\partial W}{\partial x} \tag{3.210}$$

诱导应变 $\varepsilon_x^a$，$\varepsilon_z^a$，$\gamma_{xz}^a$ 表示为[17-18]

$$(\varepsilon_x^a)_k = \frac{V_k(x,t)(d_{31})_k}{z_{k+}^a - z_{k-}^a} \tag{3.211}$$

$$\varepsilon_z^a = \gamma_{xz}^a = 0 \tag{3.212}$$

式（3.211）中 $V_k(x,t)$ 为施加在第 $k$ 压电层上的电压，厚度为 $z_{k+}^a - z_{k-}^a$（图 3.30（b）），而 $(d_{31})_k$ 为第 $k$ 层的压电系数。如上所述，式（3.211）和式（3.212）仅适用于压电层。

将每层的应变乘以刚度系数 $(\overline{Q}_{ij})_k$ 得[19]

$$(\sigma_x)_k = (\overline{Q}_{11})_k (\varepsilon_x^m - \varepsilon_x^a)_k \quad (\tau_{xz})_k = (\overline{Q}_{55})_k (\varepsilon_{xz})_k \tag{3.213}$$

其中 $\overline{Q}_{11}$ 和 $\overline{Q}_{55}$ 项（对于各单一层）被定义为

$$\overline{Q}_{11} = Q_{11}\cos^4\theta + Q_{22}\sin^4\theta + 2(Q_{12} + 2Q_{66})\sin^2\theta\cos^2\theta \tag{3.214}$$

$$\overline{Q}_{55} = G_{13}\cos^2\theta + G_{23}\sin^2\theta \tag{3.215}$$

式中：$\theta$ 为特定层上纤维方向与纵轴 $x$ 之间的夹角。系数 $Q_{11}$、$Q_{22}$、$Q_{12}$ 和 $Q_{66}$ 定义为薄层弹性特性的函数（第 2 章），即

$$Q_{11} = \frac{\overline{E}_{11}}{1-v_{12}v_{21}}, \quad Q_{12} = \frac{v_{21}\overline{E}_{11}}{1-v_{12}v_{21}} = \frac{v_{12}\overline{E}_{22}}{1-v_{12}v_{21}}$$
$$Q_{22} = \frac{\overline{E}_{22}}{1-v_{12}v_{21}}, \quad Q_{66} = G_{12} \tag{3.216}$$

式中：$\overline{E}_{11}$，$\overline{E}_{22}$ 和 $G_{12}$ 分别为单层的拉伸、压缩和剪切模量；$v_{12}$ 为主泊松比，其关系为 $\overline{E}_{11}/\overline{E}_{22} = v_{12}/v_{21}$（第 2 章）。

沿梁的高度 $h$ 积分，并乘以宽度 $b$，得到力和力矩为

$$N_x = b\int_{-h/2}^{+h/2}\sigma_x \mathrm{d}z, \quad M_y = b\int_{-h/2}^{+h/2}\sigma_x z \mathrm{d}z, \quad Q_{xz} = b\kappa\int_{-h/2}^{+h/2}\tau_{xz}\mathrm{d}z \tag{3.217}$$

对式（3.217）进行积分，得到如下关系：

$$\begin{Bmatrix} N_x \\ M_y \\ Q_{xz} \end{Bmatrix} = \begin{bmatrix} A_{11} & B_{11} & 0 \\ B_{11} & D_{11} & 0 \\ 0 & 0 & A_{55} \end{bmatrix} \begin{Bmatrix} \frac{\partial U}{\partial x} \\ \frac{\partial \Phi}{\partial x} \\ \Phi + \frac{\partial W}{\partial x} \end{Bmatrix} - \begin{Bmatrix} E_{11} \\ F_{11} \\ 0 \end{Bmatrix} \tag{3.218}$$

系数 $A_{11}$、$B_{11}$、$D_{11}$ 和 $A_{55}$ 通常分别用于层压复合梁的轴向、弯曲-拉伸耦合、弯曲和横向剪切刚度的表达式。根据经典层压理论[19]，将积分替换为求和（图 3.30（b）），得到如下表达式：

$$A_{11} = b\sum_{k=1}^{N}(\overline{Q}_{11})_k(z_k - z_{k-1}) \tag{3.219}$$

$$B_{11} = \frac{b}{2}\sum_{k=1}^{N}(\overline{Q}_{11})_k(z_k^2 - z_{k-1}^2) \tag{3.220}$$

$$D_{11} = \frac{b}{3}\sum_{k=1}^{N}(\overline{Q}_{11})_k(z_k^3 - z_{k-1}^3) \tag{3.221}$$

$$D_{55} = kb\sum_{k=1}^{N}(\overline{Q}_{55})_k(z_k - z_{k-1}) \tag{3.222}$$

式中：$\kappa$ 为剪切修正系数（取值 5/6）；$N$ 为包括叠层的压电层的层数。

式（3.218）中的 $E_{11}$ 和 $F_{11}$ 分别表示由压电层引起的诱导轴向力和诱导力矩，定义为

$$E_{11} = b\sum_{k=1}^{N}(\overline{Q}_{11})_k^a V_k(x,t)(d_{31})_k \tag{3.223}$$

$$F_{11} = \frac{b}{2}\sum_{k=1}^{N}(\overline{Q}_{11})_k^a V_k(x,t)(d_{31})_k(z_{k+}^a + z_{k-}^a) \tag{3.224}$$

为了得到运动方程和相关的边界条件，使用虚功原理，得到电势的变化为[20]

$$\delta \pi = \int_{t_1}^{t_2} \int_{volume} (\delta E_k - \delta E_p + \delta \overline{W}) \cdot d(volume) dt = 0 \quad (3.225)$$

式中：$E_k$ 为梁的动能；$E_p$ 为梁的应变能；$\overline{W}$ 为作用在梁上的外力的功。它们的表达式为

$$E_k = \frac{1}{2} \int_{volume} \rho \cdot [(\dot{U} + z\dot{\Phi})^2 + \dot{W}^2] \cdot d(volume) \quad (3.226)$$

$$E_p = \frac{1}{2} \int_{volume} (\sigma_x \varepsilon_x + \tau_{xz} \gamma_{xz}) \cdot d(volume) \quad (3.227)$$

$$\overline{W} = -\int_0^L Q \cdot W \cdot dx - P \cdot U(L) + P \cdot U(0) - M(L) \cdot \Phi(L) + M(0) \cdot \Phi(0) \quad (3.228)$$

见图 3.30（a），$L$ 为梁的长度，$P$ 和 $M$ 分别为作用于梁两端的压缩载荷和弯矩，$Q$ 为沿着梁分布的横向载荷。（·）代表对时间的微分。

将式（3.226）~式（3.228）代入式（3.225），在 $z$ 和 $y$ 方向上积分，而 $(I_1, I_2, I_3) = \int \rho \cdot (1, z, z^2) \cdot dz$ 和 $\rho$ 为梁各层的密度，得到如下表达式：

$$\delta \pi = \int_{t_1}^{t_2} \int_0^L (\delta E_k - \delta E_p + \delta \overline{W}) dx dt = 0 \quad (3.229)$$

式中

$$\delta E_k = (I_1 \dot{U} + I_2 \dot{\Phi}) \delta \dot{U} + I_1 \dot{W} \delta \dot{W} + (I_2 \dot{U} + I_3 \dot{\Phi}) \delta \dot{\Phi} \quad (3.230)$$

$$\delta \overline{W} = -Q \cdot \delta W + P \cdot \delta U(x=0) - P \cdot \delta U(x=L) + M(0) \cdot \delta \Phi(x=0) - M(L) \cdot \delta \Phi(x=L) \quad (3.231)$$

$$\delta E_p = N_x \frac{\partial(\delta U)}{\partial x} + N_x \frac{\partial W}{\partial x} \frac{\partial(\delta W)}{\partial x} + M_y \frac{\partial(\delta \Phi)}{\partial x} + Q_{xz} \left[ \delta \Phi + \frac{\partial(\delta W)}{\partial x} \right] \quad (3.232)$$

将式（3.230）~式（3.232）代入式（3.229）进行分部积分，忽略电势 $\delta \pi$ 的变化（只有当积分和梁两端的各种表达式都等于零时（假设位移是任意的）才成立），得到以下压电层梁的运动方程及其相关的边界条件：

$$\frac{\partial N_x}{\partial x} = \frac{\partial(I_1 \dot{U} + I_2 \dot{\Phi})}{\partial t} \quad (3.233)$$

$$\frac{\partial}{\partial x}\left(Q_{xz} + N_x \frac{\partial W}{\partial x}\right) = \frac{\partial(I_1 \dot{W})}{\partial t} + Q \quad (3.234)$$

$$\frac{\partial M_y}{\partial x} - Q_{xz} = \frac{\partial (I_3 \dot{\Phi} + I_2 \dot{U})}{\partial t} \tag{3.235}$$

相关的可能边界条件为

$$N_x + P = 0 \quad 或 \quad U = 0 \tag{3.236}$$

$$Q_{xz} + N_x \frac{\partial W}{\partial x} = 0 \quad 或 \quad W = 0 \tag{3.237}$$

$$M_y = M \quad 或 \quad \Phi = 0 \tag{3.238}$$

值得注意的是，非线性表达式 $N_x \frac{\partial W}{\partial x}$ 出现在式（3.234）和式（3.237）中。假设梁具有小的挠度，根据式（3.236），$N_x$ 近似等于 $-P$：

$$N_x \approx -P \tag{3.239}$$

因此，正如文献中通常提到的，$N_x$ 只在上述非线性表达式中被 $-P$ 取代。

将式（3.218）的力和力矩结果表达式代入式（3.233）~式（3.235），可得含压电层的梁的运动方程，以三种假设挠度 $U$、$W$ 和 $\Phi$ 表示：

$$\frac{\partial}{\partial x}\left( A_{11}\frac{\partial U}{\partial x} + B_{11}\frac{\partial \Phi}{\partial x} - E_{11} \right) = \frac{\partial (I_1 \dot{U} + I_2 \dot{\Phi})}{\partial t} \tag{3.240}$$

$$\frac{\partial}{\partial x}\left[ A_{55}\left(\Phi + \frac{\partial W}{\partial x}\right) - P\frac{\partial W}{\partial x} \right] = \frac{\partial (I_1 \dot{W})}{\partial t} + Q \tag{3.241}$$

$$\frac{\partial}{\partial x}\left( B_{11}\frac{\partial U}{\partial x} + D_{11}\frac{\partial \Phi}{\partial x} - F_{11} \right) - A_{55}\left(\Phi + \frac{\partial W}{\partial x}\right) = \frac{\partial (I_3 \dot{\Phi} + I_2 \dot{U})}{\partial t} \tag{3.242}$$

与它们相关的可能边界条件为

$$A_{11}\frac{\partial U}{\partial x} + B_{11}\frac{\partial \Phi}{\partial x} = -P + E_{11} \quad 或 \quad U = 0 \tag{3.243}$$

$$A_{55}\left(\Phi + \frac{\partial W}{\partial x}\right) - P\frac{\partial W}{\partial x} = 0 \quad 或 \quad W = 0 \tag{3.244}$$

$$B_{11}\frac{\partial U}{\partial x} + D_{11}\frac{\partial \Phi}{\partial x} = F_{11} + M \quad 或 \quad \Phi = 0 \tag{3.245}$$

值得注意的是，对于性质均匀的梁，诱导应变的影响并不出现在运动方程中，而只出现在梁的边界上。

为了符合文献中通常给出的销力梁模型的表示方法[13]，将边界条件改写为

$$A_{11}\frac{\partial U}{\partial x} + B_{11}\frac{\partial \Phi}{\partial x} = -P_m \quad 或 \quad U = 0 \tag{3.246}$$

$$A_{55}\left(\varPhi+\frac{\partial W}{\partial x}\right)-P\frac{\partial W}{\partial x}=0 \quad \text{或} \quad W=0 \tag{3.247}$$

$$B_{11}\frac{\partial U}{\partial x}+D_{11}\frac{\partial \varPhi}{\partial x}=m_m \quad \text{或} \quad \varPhi=0 \tag{3.248}$$

式中：$p_m$ 和 $m_m$ 分别为考虑压电层影响的广义轴向载荷和力矩，即

$$p_m = P - E_{11} \tag{3.249}$$

$$m_m = M + F_{11} \tag{3.250}$$

注意在式（3.241）和式（3.244）中，应使用非线性表达式 $P$ 而不是 $p_m$：

$$P = p_m + E_{11} \tag{3.251}$$

对于均质梁，运动方程为

$$A_{11}\frac{\partial^2 U}{\partial x^2}+B_{11}\frac{\partial^2 \varPhi}{\partial x^2}=I_1\ddot{U}+I_2\ddot{\varPhi} \tag{3.252}$$

$$A_{55}\left(\frac{\partial \varPhi}{\partial x}+\frac{\partial^2 W}{\partial x^2}\right)-P\frac{\partial^2 W}{\partial x^2}=I_1\ddot{W}+Q \tag{3.253}$$

$$B_{11}\frac{\partial^2 U}{\partial x^2}+D_{11}\frac{\partial^2 \varPhi}{\partial x^2}-A_{55}\left(\varPhi+\frac{\partial W}{\partial x}\right)=I_3\ddot{\varPhi}+I_2\ddot{U} \tag{3.254}$$

对于对称层叠（$B=I_2=0$），式（3.252）~式（3.254）在 $x$ 方向退化为一个非耦合方程和两个耦合方程，即

$$A_{11}\frac{\partial^2 U}{\partial x^2}=I_1\ddot{U} \tag{3.255}$$

$$A_{55}\left(\frac{\partial \varPhi}{\partial x}+\frac{\partial^2 W}{\partial x^2}\right)-P\frac{\partial^2 W}{\partial x^2}=I_1\ddot{W}+Q \tag{3.256}$$

$$D_{11}\frac{\partial^2 \varPhi}{\partial x^2}-A_{55}\left(\varPhi+\frac{\partial W}{\partial x}\right)=I_3\ddot{\varPhi} \tag{3.257}$$

值得注意的是，文献［19］中也提供了类似的表达式[①]。

对于各向同性材料，式（3.255）~式（3.257）可写为

$$EA\frac{\partial^2 U}{\partial x^2}=I_1\ddot{U} \tag{3.258}$$

$$\kappa GA\left(\frac{\partial \varPhi}{\partial x}+\frac{\partial^2 W}{\partial x^2}\right)-P\frac{\partial^2 W}{\partial x^2}=I_1\ddot{W}+Q \tag{3.259}$$

$$EI\frac{\partial^2 \varPhi}{\partial x^2}-\kappa GA\left(\varPhi+\frac{\partial W}{\partial x}\right)=I_3\ddot{\varPhi} \tag{3.260}$$

---

① 文献［22］的第 214 页，在方程（4.5.2（a）~（d））中插入 $A_{16}=B_{16}=0$ 和 $V_0=0$。

式（3.259）和式（3.260）与文献［21］第175页的式（5.10）和式（5.11）相同。

### 3.5.1 静态和动态情况的解

压电层或压电片对梁的小振动影响可分为两类：

（1）压电片上施加交变谐波电压。这将导致 $E_{11}$ 或 $F_{11}$ 项的谐波变化。出现在边界条件（式（3.243）~式（3.245））中的这些项将引起梁的小谐波振动。实际上，$F_{11}$ 的作用更有趣，因为它会引起横向弯曲振动[22-31]。

（2）对压电层施加恒定电压。对于所有压电层电压相等的情况，在约束梁两端轴向位移的同时，由于 $E_{11}$ 在梁内会诱导生成恒定的轴向力 $P$，从而影响横向小幅振动[32-33]。

相应地，诱导轴向力 $E_{11}$ 与诱导弯矩 $F_{11}$ 的关系可写为

$$E_{11} = E_{11}^0 + E_{11}'(t) \tag{3.261}$$

$$F_{11} = F_{11}^0 + F_{11}'(t) \tag{3.262}$$

式中：$(\ )^0$ 代表恒定的贡献；$(\ )'$ 表示谐波的、与时间相关的贡献。如前所述，有趣的情况是

$$F_{11}^0 = E_{11}'(t) = 0, \quad E_{11}^0 \neq 0, \quad F_{11}'(t) \neq 0 \tag{3.263}$$

### 3.5.2 由 $E_{11}^0$ 产生的轴向诱导力计算

第1种情况是沿梁的连续压电层，如图3.31所示。

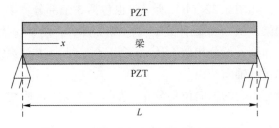

图 3.31 夹在两个连续压电层之间的梁

在图3.31中，对梁两端的轴向位移进行约束。施加相同的电压（幅值和符号都相同）将对梁产生诱导压缩或拉伸（取决于电压的符号）。

对于对称的与时间无关的情况，式（3.255）~式（3.257）有如下形式：

$$A_{11} \frac{\partial^2 U}{\partial x^2} = 0, \quad W = \Phi = 0 \tag{3.264}$$

式（3.255）解的形式如下：
$$U(x)=ax+b \tag{3.265}$$
式中：$a$ 和 $b$ 是由该情况的边界条件确定的常数。

令 $U(0)=U(L)=0$（梁两端无轴向位移），得到 $U(x)=0$ 形式的解。将这个结果代入式（3.218）的第1行，得
$$N_x=-E_{11}^0 \tag{3.266}$$
将其代入式（3.239）后得
$$P=E_{11}^0 \tag{3.267}$$

第2种要处理的情况是带有两个相同压电片的梁，如图3.32（a）和（b）所示。

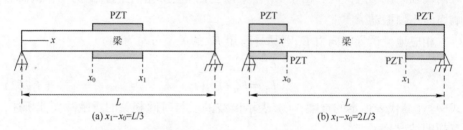

图 3.32　装有几对压电片的梁

图3.32（a）中所示的例子由3个部分组成：第1部分，仅有梁，从 $x=0$ 延伸到 $x=x_0$；第2部分，夹在两个相同压电片之间的梁，从 $x=x_0$ 开始，到 $x=x_1$ 结束；第3部分，仅有梁，与第1部分相似，从 $x_1$ 延伸到 $x=L$。

第2种情况，如图3.32（b）所示，也包括3个部分：第1部分，一个带有一对压电片的梁，从 $x=0$ 延伸到 $x=x_0$。第2部分，从 $x=x_0$ 到 $x=x_1$，仅有一个梁。第3部分和第1部分相似，从 $x=x_1$ 开始，到 $x=L$ 结束。对于这两种情况，梁的两端都受到轴向约束。

因此，可以写出梁的轴向位移分布，如式（3.265）所述，对于梁的3个部分：
$$\begin{cases} U_1(x)=a_1x+b_1, & 0 \leqslant x<x_0 \\ U_2(x)=a_2x+b_2, & x_0<x<x_1 \\ U_3(x)=a_3x+b_3, & x_1<x \leqslant L \end{cases} \tag{3.268}$$

式中：常数 $a_1$、$b_1$、$a_2$、$b_2$、$a_3$ 和 $b_3$ 将由梁边界条件，$U(0)=U(L)=0$，以及点 $x_0$ 和 $x_1$ 的连续性要求来确定。

对于第1种情况（图3.32（a）），得到如下等式：

$$\begin{cases} U_1(0)=a_1\cdot 0+b_1=0\\ U_3(L)=a_3\cdot L+b_3=0\\ U_1(x_0)=U_2(x_0)\Rightarrow a_1\cdot x_0+b_1=a_2\cdot x_0+b_2\\ U_2(x_1)=U_3(x_1)\Rightarrow a_2\cdot x_1+b_2=a_3\cdot x_1+b_3\\ N_1(x_0)=N_2(x_0)\Rightarrow (A_{11})_{\text{I}}\cdot a_1=(A_{11})_{\text{II}}\cdot a_2-E_{11}\\ N_2(x_1)=N_3(x_1)\Rightarrow (A_{11})_{\text{II}}\cdot a_2-E_{11}=(A_{11})_{\text{I}}\cdot a_3 \end{cases} \quad (3.269)$$

对于第2种情况（图3.32（b）），得到如下等式：

$$\begin{cases} U_1(0)=a_1\cdot 0+b_1=0\\ U_3(L)=a_3\cdot L+b_3=0\\ U_1(x_0)=U_2(x_0)\Rightarrow a_1\cdot x_0+b_1=a_2\cdot x_0+b_2\\ U_2(x_1)=U_3(x_1)\Rightarrow a_2\cdot x_1+b_2=a_3\cdot x_1+b_3\\ N_1(x_0)=N_2(x_0)\Rightarrow (A_{11})_{\text{II}}\cdot a_1-E_{11}=(A_{11})_{\text{I}}\cdot a_2\\ N_2(x_1)=N_3(x_1)\Rightarrow (A_{11})_{\text{I}}\cdot a_2=(A_{11})_{\text{II}}\cdot a_3-E_{11} \end{cases} \quad (3.270)$$

式中：$(A_{11})_{\text{I}}$ 为只有梁时的轴向刚度；$(A_{11})_{\text{II}}$ 为梁和两个压电片的轴向刚度。对式（3.269）或式（3.270）求解，并代入式（3.268），得到第1种情况（$x_1-x_0=L/3$，见图3.32（a））沿梁的轴向位移分布：

$$\begin{cases} U_1(x)=-\dfrac{E_{11}}{(A_{11})_{\text{I}}+2(A_{11})_{\text{II}}}x,\ 0\leqslant x<x_0\\ U_2(x)=-\dfrac{E_{11}}{(A_{11})_{\text{I}}+2(A_{11})_{\text{II}}}(3x_0-2x),\ x_0<x<x_1\\ U_3(x)=-\dfrac{E_{11}}{(A_{11})_{\text{I}}+2(A_{11})_{\text{II}}}(x-L),\ x_1<x\leqslant L \end{cases} \quad (3.271)$$

对于第2种情况（对于 $x_1-x_0=2L/3$，见图3.32（b）），得到如下等式：

$$\begin{cases} U_1(x)=\dfrac{2E_{11}}{(A_{11})_{\text{I}}+2(A_{11})_{\text{II}}}x,\ 0\leqslant x<x_0\\ U_2(x)=\dfrac{E_{11}}{(A_{11})_{\text{I}}+2(A_{11})_{\text{II}}}(3x_0-x),\ x_0<x<x_1\\ U_3(x)=\dfrac{2E_{11}}{(A_{11})_{\text{I}}+2(A_{11})_{\text{II}}}(x-L),\ x_1<x\leqslant L \end{cases} \quad (3.272)$$

类似地，对于前面处理的连续情况，将 $U(x)$ 的表达式代入式（3.218）的第1行，将得到沿着梁的力 $P$ 的轴向分布：

$$\begin{cases} P = \dfrac{E_{11} \cdot (A_{11})_{\text{I}}}{(A_{11})_{\text{I}} + 2(A_{11})_{\text{II}}}, & 0 \leq x < x_0 \\ P = \dfrac{E_{11} \cdot (A_{11})_{\text{I}}}{(A_{11})_{\text{I}} + 2(A_{11})_{\text{II}}}, & x_0 < x < x_1 \\ P = \dfrac{E_{11} \cdot (A_{11})_{\text{I}}}{(A_{11})_{\text{I}} + 2(A_{11})_{\text{II}}}, & x_1 < x \leq L \end{cases} \quad (3.273)$$

正如所预期的，轴向力 $P$ 沿梁是恒定的。

### 3.5.3 小横向振动的运动方程的解

本小节将推导一个装有压电片的梁在恒定的诱导轴向力 $P$ 作用下的振动情况，该振动情况与提供给压电片的电压有关。

假设该问题的 3 个变量的表达式如下：

$$W(x,t) = w(x)\mathrm{e}^{\mathrm{i}\omega t}, \quad \varPhi(x,t) = \phi(x)\mathrm{e}^{\mathrm{i}\omega t}, \quad U(x,t) = u(x)\mathrm{e}^{\mathrm{i}\omega t} \quad (3.274)$$

式中：$u(x)$、$w(x)$ 和 $\phi(x)$ 是频率为 $w$ 振动的小振幅。

将式（3.274）代入式（3.252）~式（3.254），得

$$A_{11}u'' + B_{11}\phi'' = -\omega^2 I_1 u - \omega^2 I_2 \phi \quad (3.275)$$

$$A_{55}(\phi' + w'') - Pw'' = -\omega^2 I_1 w \quad (3.276)$$

$$B_{11}u'' + D_{11}\phi'' - A_{55}(\phi + w') = -\omega^2 I_3 \phi - \omega^2 I_2 u \quad (3.277)$$

式中：$(\ )'$ 为对 $x$ 的微分。

对于对称情况，式（3.275）~式（3.277）具有如下形式：

$$\begin{cases} A_{11}u'' = -\omega^2 I_1 u & (3.278) \\ A_{55}(\phi' + w'') - Pw'' = -\omega^2 I_1 w & (3.279) \\ D_{11}\phi'' - A_{55}(\phi + w') = -\omega^2 I_3 \phi & (3.280) \end{cases}$$

式（3.279）和式（3.280）耦合，而式（3.279）有以下解：

$$u(x) = A\sin(kx) + B\cos(kx), \quad k = \omega\sqrt{\dfrac{I_1}{A_{11}}} \quad (3.281)$$

接下来提出并解决 $w$ 和 $\varphi$ 之间耦合的横向振动。

为了解决这个问题，必须解耦式（3.279）和式（3.280）两个方程。通过以下步骤解决：

由式（3.279）可得

$$\phi' = \left(\dfrac{P}{A_{55}} - 1\right)w'' - \dfrac{I_1}{A_{55}}\omega^2 w \quad (3.282)$$

式（3.280）对 $x$ 求导，并代入结果中，再将式（3.282）对 $x$ 二次求导，得到如下表达式：

$$(D^4 a_2 + D^2 a_1 + a_0)w = 0, \quad D \equiv \frac{\mathrm{d}}{\mathrm{d}x} \tag{3.283}$$

式中

$$\begin{cases} a_2 = D_{11}\left(1 - \dfrac{P}{A_{55}}\right) \\ a_1 = D_{11}I_1\dfrac{\omega^2}{A_{55}} - I_3\omega^2\left(\dfrac{P}{A_{55}} - 1\right) + P \\ a_0 = I_3 I_1 \dfrac{\omega^4}{A_{55}} - I_1\omega^2 \end{cases} \tag{3.284}$$

假设 $w = \bar{w}\mathrm{e}^{sx}$，可得式（3.283）的解。

对式（3.283）进行推导，得到特征方程如下：

$$a_2 s^4 + a_1 s^2 + a_0 = 0 \tag{3.285}$$

式（3.285）有两个实解和两个复解，即

$$\pm\lambda_1, \quad \pm\mathrm{i}\lambda_2 \tag{3.286}$$

这就推导出了梁的横向位移的通解形式为

$$w(x) = C_1 \sinh(\lambda_1 x) + C_2 \cosh(\lambda_1 x) + C_3 \sin(\lambda_2 x) + C_4 \cos(\lambda_2 x) \tag{3.287}$$

为了获得第 2 个变量 $\varphi$（梁的横截面的旋转角度）的通解，对该变量再次进行上述过程求解。式（3.280）中，变量 $w'$ 表示为其中 $\phi''$ 和 $\phi$ 的函数。该表达式对 $x$ 求导一次，并在对 $x$ 求导一次后代入式（3.279）。同样的表达式对 $x$ 求导两次，并再次代入方程（3.279）对 $x$ 求导。所有这些求导结果是 $\phi$ 的微分方程，类似于式（3.283），其有以下解：

$$\phi(x) = E_1 \sinh(\lambda_1 x) + E_2 \cosh(\lambda_1 x) + E_3 \sin(\lambda_2 x) + E_4 \cos(\lambda_2 x) \tag{3.288}$$

将式（3.287）和式（3.288）代入耦合式（3.279）和式（3.280）将得到常数（$C_1 \sim C_4$）和（$E_1 \sim E_4$）之间的关系：

$$m_1 = -\frac{1}{\lambda_1}\left(\frac{I_1}{A_{55}}\omega^2 + \lambda_1^2\left(\frac{P}{A_{55}} - 1\right)\right), \quad m_2 = -\frac{1}{\lambda_2}\left(\frac{I_1}{A_{55}}\omega^2 - \lambda_2^2\left(\frac{P}{A_{55}} - 1\right)\right) \tag{3.289}$$

得

$$E_1 = m_1 C_2, \quad E_2 = m_1 C_1, \quad E_3 = m_2 C_4, \quad E_4 = -m_2 C_3 \tag{3.290}$$

运用简支边界条件 $W(0) = M(0) = W(L) = M(L) = 0$，即对于压电层连续的情况，位移 $W(0) = \varphi'(0) = W(L) = \varphi'(L) = 0$，如图 3.31 所示，得到如下关系式：

$$\begin{bmatrix} 0 & 1 & 0 & 1 \\ F_{11} & D_{11}m_1\lambda_1+F_{11} & F_{11} & D_{11}m_2\lambda_2+F_{11} \\ \sinh(\lambda_1 L) & \cosh(\lambda_1 L) & \sin(\lambda_2 L) & \cos(\lambda_2 L) \\ A & B & C & D \end{bmatrix} \begin{Bmatrix} C_1 \\ C_2 \\ C_3 \\ C_4 \end{Bmatrix} = \begin{Bmatrix} 0 \\ 0 \\ 0 \\ 0 \end{Bmatrix}$$

(3.291)

式中

$$\begin{cases} A=D_{11}m_1\lambda_1\sinh(\lambda_1 L)+F_{11}, & B=D_{11}m_1\lambda_1\cosh(\lambda_1 L)+F_{11} \\ C=D_{11}m_2\lambda_2\sin(\lambda_2 L)+F_{11}, & D=D_{11}m_2\lambda_2\cos(\lambda_2 L)+F_{11} \end{cases}$$

(3.292)

要得到方程（3.291）的唯一解，行列式应该变为 0，从而得到固有频率 $\omega_i$。

值得注意的是，在目前情况下，$F_{11}$ 消失（因为上下压电层有相等的电压），由于 $E_{11}^0$ 不随时间而变化以及梁的两端轴向约束，产生一个诱导轴向压缩力 $-P$，如上所述。

对于压电片的情况，见图 3.32（a）所示，并如前假设梁的两端为简支边界条件，由于沿梁方向的性质不同，将梁分为 3 部分：

第 1 部分，仅有梁，从 $x=0$ 到 $x=x_0$，位移表达式如下：

$$w_1(x)=C_{11}\sinh(\lambda_{11}x)+C_{21}\cosh(\lambda_{11}x)+C_{31}\sin(\lambda_{21}x)+C_{41}\cos(\lambda_{21}x)$$

(3.293)

$$\phi_1(x)=C_{11}m_{11}\cosh(\lambda_{11}x)+C_{21}m_{11}\sinh(\lambda_{11}x)-C_{31}m_{21}\cos(\lambda_{21}x)+C_{41}m_{21}\sin(\lambda_{21}x)$$

(3.294)

第 2 部分，梁和两个压电片，从 $x=x_0$ 到 $x=x_1$，位移表达式如下：

$$w_2(x)=C_{12}\sinh(\lambda_{12}x)+C_{22}\cosh(\lambda_{12}x)+C_{32}\sin(\lambda_{22}x)+C_{42}\cos(\lambda_{22}x)$$

(3.295)

$$\phi_2(x)=C_{12}m_{12}\cosh(\lambda_{12}x)+C_{22}m_{12}\sinh(\lambda_{12}x)-C_{32}m_{22}\cos(\lambda_{22}x)+C_{42}m_{22}\sin(\lambda_{22}x)$$

(3.296)

第 3 部分，仅有梁，从 $x=x_0$ 到 $x=L$，位移表达式如下：

$$w_3(x)=C_{13}\sinh(\lambda_{13}x)+C_{23}\cosh(\lambda_{13}x)+C_{33}\sin(\lambda_{23}x)+C_{43}\cos(\lambda_{23}x)$$

(3.297)

$$\phi_3(x)=C_{13}m_{13}\cosh(\lambda_{13}x)+C_{23}m_{13}\sinh(\lambda_{13}x)-C_{33}m_{23}\cos(\lambda_{23}x)+C_{43}m_{23}\sin(\lambda_{23}x)$$

(3.298)

要求 $x_0$ 和 $x_1$ 两点的横向位移、转角、剪力和力矩具有连续性，即

$$\begin{cases} x=x_1, & x=x_0 \\ w_2(x_1)=w_3(x_1), & w_1(x_0)=w_2(x_0) \\ \phi_2(x_1)=\phi_3(x_1), & \phi_1(x_0)=\phi_2(x_0) \\ Q_2(x_1)=Q_3(x_1), & Q_1(x_0)=Q_2(x_0) \\ M_2(x_1)=M_3(x_1), & M_1(x_0)=M_2(x_0) \end{cases}$$

并应用边界条件如下:

$$\begin{cases} x=0, w_1(0)=0, M_1(0)=0, \dfrac{\mathrm{d}\varphi_1(0)}{\mathrm{d}x}=0 \\ x=L, w_3(L)=0, M_3(L)=0, \dfrac{\mathrm{d}\varphi_3(L)}{\mathrm{d}x}=0 \end{cases} \quad (3.299)$$

这就得到了有 12 个未知数的 12 个方程,用矩阵符号表示为

$$[A]\{C\}=\{0\} \quad (3.300)$$

使式(3.300)中矩阵 $[A]$ 为 0 可得到带有一对压电片梁的特征方程,得到固有频率及其相关模态振型。$A$ 矩阵的各项见本章附录 E。

### 3.5.4 静态情况的解

在上一节中,对于向顶部和底部压电层电极提供等电压(同相),从而在平面上产生可能造成梁屈曲(对于不可移动的轴向边界条件)的力的情况,已经给出了解。现在,介绍关于梁的诱导弯曲的情况,这是由于压电层电极的电压不相等(异相,图 3.33(d))造成的。图 3.33(a)~(d)中的模型与之前图 3.30 中展示的模型相同,其中 $P$ 为机械轴向力(图 3.33(a)),而对于异相电压(图 3.33(d)),连续压电材料和压电片(图 3.33(b)和(c))的可能性均被解决。在文献[30]中详细讨论了这个主题,这里仅给出主要结果。

运动方程式(3.47)~式(3.49),早先是针对均质梁而提出的,现在被改写为无量纲形式,去掉了右边的惯性项,得到以下几种形式。

对于均质梁,运动方程有

$$\alpha_0 \overline{U}''+\beta_0 \overline{\Phi}''=0 \quad (3.301)$$

$$c_0(\overline{\Phi}'+\overline{W}'')-\lambda \overline{W}''=q_0 \quad (3.302)$$

$$\beta_0 \overline{U}''+\overline{\Phi}''-c_0(\overline{\Phi}+\overline{W}')=0 \quad (3.303)$$

有以下无量纲边界条件:

$$\alpha_0 \overline{U}'+\beta_0 \overline{\Phi}'=-\lambda+\hat{E}_{11} \quad \text{或} \quad \overline{U}=0 \quad (3.304)$$

$$c_0(\overline{\Phi}+\overline{W}')-\lambda \overline{W}'=0 \quad \text{或} \quad \overline{W}=0 \quad (3.305)$$

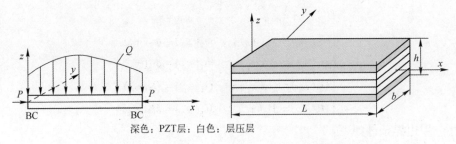

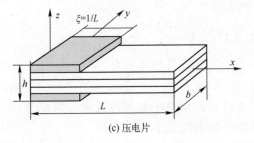
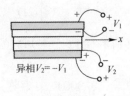

(c) 压电片　　　　　　　　　　　　(d) 压电层的电连接

图 3.33　压电层压梁

$$\beta_0 \overline{U}' + \overline{\Phi}' = \hat{F}_{11} \quad 或 \quad \overline{\Phi} = 0 \tag{3.306}$$

无量纲项为

$$\begin{cases} \alpha_0 = \dfrac{A_{11}L^2}{D_{11}}, \quad \beta_0 = \dfrac{B_{11}L}{D_{11}}, \quad c_0 = \dfrac{A_{55}L^2}{D_{11}}, \quad \hat{E}_{11} = \dfrac{E_{11}L^2}{D_{11}} \\ \hat{F}_{11} = \dfrac{F_{11}L}{D_{11}}, \quad q_x 0 = \dfrac{QL^3}{D_{11}}, \quad \lambda = \dfrac{PL^2}{D_{11}}, \quad \overline{U} = \dfrac{U}{L} \\ \overline{W} = \dfrac{W}{L}, \quad \overline{\Phi} = \Phi, \quad \xi = \dfrac{\chi}{L}, \quad [\ ]' = \dfrac{\mathrm{d}[\ ]}{\mathrm{d}\xi} \end{cases} \tag{3.307}$$

将式（3.301）~式(3.303) 解耦得到以下等式：

$$\beta \overline{W}'''' + \lambda \overline{W}'' = -q_0 + \mu q''_0 \tag{3.308}$$

$$\beta \overline{\Phi}''' + \lambda \overline{\Phi}' = q_0 \tag{3.309}$$

$$\beta \overline{U}'''' + \lambda \overline{U}'' = -\chi q'_0 \tag{3.310}$$

式中

$$\beta = \mu(c_0 - \lambda), \quad \mu = \dfrac{c_0}{(1 - \beta_0 \chi)}, \quad \chi = \dfrac{\beta_0}{\alpha_0} \tag{3.311}$$

对于给定分布的无量纲横向分布载荷 $q_0$，通过寻找齐次解和特解来求解式（3.308）~式（3.310）。对于 $q_0$ 为常数的情况，那些方程解的一般形式如下：

$$\overline{W}=A_1\sin(\gamma\xi)+A_2\cos(\gamma\xi)+A_3\xi^2+A_4\xi+A_5 \qquad (3.312)$$

$$\overline{\Phi}=B_1\sin(\gamma\xi)+B_2\cos(\gamma\xi)+B_3\xi+B_4 \qquad (3.313)$$

$$\overline{U}=C_1\sin(\gamma\xi)+C_2\cos(\gamma\xi)+C_3\xi^2+C_4\xi+C_5 \qquad (3.314)$$

式中

$$\gamma=\sqrt{\frac{\lambda}{\beta}} \qquad (3.315)$$

尽管有 14 个未知数需要确定（$A_1 \sim A_5$、$B_1 \sim B_4$ 和 $C_1 \sim C_5$），但只有 6 个可用的边界条件（梁的两端各 3 个）。将式（3.312）~式（3.314）代入非耦合运动方程式（3.301）~式（3.303），并要求方程两边相等而不考虑变量 $x$，从而得到另外 8 个关系。

注意，对于没有轴向压缩载荷 $P$ 的情况，式（3.312）~式（3.314）的形式如下：

$$\overline{W}=A_1\xi^4+A_2\xi^3+A_3\xi^2+A_4\xi+A_5 \qquad (3.316)$$

$$\overline{\Phi}=B_1\xi^3+B_2\xi^2+B_3\xi+B_4 \qquad (3.317)$$

$$\overline{U}=C_1\xi^4+C_2\xi^3+C_3\xi^2+C_4\xi+C_5 \qquad (3.318)$$

不同边界条件和载荷的不同解如表 3.4 所列[30]。表 3.5 给出了带有压电片的情况下的解（与表 3.4 所示的连续压电层不同）。

表 3.4 $\overline{W}$、$\overline{\Phi}$ 和 $\overline{\Phi}$-连续压电层（横向载荷为常数，$q_0 \neq 0$）的解析解①

| 名称 | 边界条件 | | 解 析 解 |
|---|---|---|---|
| | $\xi=0$ | $\xi=1$ | |
| | 无轴向力-$\lambda=0$ | | |
| $S-S^{②}(m)$ | $\overline{W}=0$ | $\overline{W}=0$ | $\overline{W}=-\dfrac{a}{4}\xi^4+a\xi^3+(b-g)\xi^2-(a+b-g)\xi$ |
| | $\beta_0\overline{U}'+\overline{\Phi}'=\hat{F}_{11}$ | $\beta_0\overline{U}'+\overline{\Phi}'=\hat{F}_{11}$ | $\overline{\Phi}=+a\xi^3-3a\xi^2+2g\xi+(a-g)$ |
| | $\overline{U}=0$ | $\alpha_0\overline{U}'+\beta_0\overline{\Phi}'=0$ | $\overline{U}=-\delta a\xi^3+3a\delta\xi^2-2\delta g\xi$ |
| $S-S^{②}(u)$ | $\overline{W}=0$ | $\overline{W}=0$ | $\overline{W}=-\dfrac{a}{4}\xi^4+a\xi^3+\left(b-a+\dfrac{\hat{F}_{11}}{2}\right)\xi^2-\varepsilon\xi$ |
| | $\beta_0\overline{U}'+\overline{\Phi}'=\hat{F}_{11}$ | $\beta_0\overline{U}'+\overline{\Phi}'=\hat{F}_{11}$ | $\overline{\Phi}=+a\xi^3-3a\xi^2-\left(\dfrac{a}{2}+\hat{F}_{11}\right)\xi+(\varepsilon-b)$ |
| | $\overline{U}=0$ | $\overline{U}=0$ | $\overline{U}=-\delta a\xi^3+3a\delta\xi^2-\delta a\dfrac{\xi}{2}$ |

| 名称 | 边界条件 $\xi=0$ | 边界条件 $\xi=1$ | 解析解 |
|---|---|---|---|
| $C\text{-}S^{②}(m)$ | $\overline{W}=0$ | $\overline{W}=0$ | $\overline{W}=-\dfrac{a}{4}\xi^4+\alpha_0\theta\dfrac{\xi^3}{6}+(b-\theta_1)\xi^2+\theta_2\xi$ |
|  | $\overline{\Phi}=0$ | $\beta_0\overline{U}'+\overline{\Phi}'=\hat{F}_{11}$ | $\overline{\Phi}=+a\xi^3-\alpha_0\theta\dfrac{\xi^2}{2}+2\theta_1\xi$ |
|  | $\overline{U}=0$ | $\alpha_0\overline{U}'+\beta_0\overline{\Phi}'=0$ | $\overline{U}=-\delta a\xi^3+\delta\alpha_0\theta\dfrac{\xi^2}{2}+\left(\dfrac{d}{\delta}-\delta\theta\right)\xi$ |
| $C\text{-}S^{②}(u)$ | $\overline{W}=0$ | $\overline{W}=0$ | $\overline{W}=-\dfrac{a}{4}\xi^4+\alpha_0\theta\dfrac{\xi^3}{6}+\left(b-\dfrac{\mu}{2}\right)\xi^2+\mu_1\xi$ |
|  | $\beta_0\overline{U}'+\overline{\Phi}'=\hat{F}_{11}$ | $\beta_0\overline{U}'+\overline{\Phi}'=\hat{F}_{11}$ | $\overline{\Phi}=+a\xi^3-\eta\xi^2+\mu\xi$ |
|  | $\overline{U}=0$ | $\overline{U}=0$ | $\overline{U}=-\delta a\xi^3+\delta\eta\xi^2+(\eta-\delta a)\xi$ |
| $C\text{-}F$ | $\overline{W}=0$ | $c_0(\overline{\Phi}+\overline{W}')=\lambda\overline{W}'$ | $\overline{W}=-\dfrac{a}{4}\xi^4+a\xi^3+(b-d)\xi^2-2b\xi$ |
|  | $\overline{\Phi}=0$ | $\beta_0\overline{U}'+\overline{\Phi}'=\hat{F}_{11}$ | $\overline{\Phi}=+a\xi^3-3a\xi^2+2d\xi$ |
|  | $\overline{U}=0$ | $\alpha_0\overline{U}'+\beta_0\overline{\Phi}'=0$ | $\overline{U}=-\delta a\xi^3+3a\delta\xi^2-2\delta d\xi$ |
|  |  | 有轴向力 $-\lambda\neq 0$ |  |
| $C\text{-}F$ | $\overline{W}=0$ | $c_0(\overline{\Phi}+\overline{W}')=\lambda\overline{W}'$ | $\overline{W}=B_1\sin(\gamma\xi)+B_2\cos(\gamma\xi)+B_0\xi^2+B_3+B_4$ |
|  | $\overline{\Phi}=0$ | $\beta_0\overline{U}'+\overline{\Phi}'=\hat{F}_{11}$ | $\overline{\Phi}=A_1\sin(\gamma\xi)+A_2\cos(\gamma\xi)+A_0\xi+A_3$ |
|  | $\overline{U}=0$ | $\alpha_0\overline{U}'+\beta_0\overline{\Phi}'=-\lambda$ | $\overline{W}=C_1\sin(\gamma\xi)+C_2\cos(\gamma\xi)+C_3\xi+C_4$ $\lambda_{cr}=\dfrac{\dfrac{\pi^4}{4}\left(1-\dfrac{\beta_0^2}{\alpha_0}\right)}{1+\dfrac{\pi^2}{4c_0}\left(1-\dfrac{\beta_0^2}{\alpha_0}\right)}$ |

① 右列中各种术语的表达式可以在本章附录 C 中找到。
② 施加电压 V 的符号与边界条件 C-F 相反,其中,S-S(m)= 简单支撑可移动;S-S(u)= 简单支撑不可移动;C-S(m)= 简单支撑可移动;C-S(u)= 加紧简单支撑不可移动;C-F= 未加紧。

表 3.5 $\overline{W}$、$\overline{\Phi}$ 和 $\overline{\Phi}$-压电片(无侧向载荷 $q_0=0$;无轴向力 $\lambda=0$)的解析解[①]

| 边界条件 | 解析解 第 I 部分 $0\leqslant\xi\leqslant\xi_0$ | 解析解 第 II 部分 $\xi_0\leqslant\xi\leqslant 1$ |
|---|---|---|
| $S\text{-}S^{②}(m)$ | $\overline{W}=-A[\xi^2+\xi(\xi_0^2-2\xi_0)]$ | $\overline{W}=-A\xi_0^2(\xi-1)$ |
|  | $\overline{\Phi}=2A\left[(\xi-\xi_0)+\dfrac{\xi_0^2}{2}\right]$ | $\overline{\Phi}=A\xi_0^2$ |
|  | $\overline{U}=A_1\xi$ | $\overline{U}=A_1\xi_0$ |

续表

| 边界条件 | 解析解 | |
|---|---|---|
| | 第Ⅰ部分 $0 \leq \xi \leq \xi_0$ | 第Ⅱ部分 $\xi_0 \leq \xi \leq 1$ |
| S-S(u) | $\overline{W}=(\beta_0 B+\hat{F}_{11})\dfrac{\xi^2}{2}-F\xi$ | $\overline{W}=\dfrac{\beta_0 B_1(\xi^2-1)}{2}-F_1(\xi-1)$ |
| | $\overline{\Phi}=-(\beta_0 B+\hat{F}_{11})\xi+F$ | $\overline{\Phi}=-\beta_0 B_1\xi+F_1$ |
| | $\overline{U}=B\xi$ | $\overline{U}=B_1(\xi-1)$ |
| C-S(m) | $\overline{W}=-D\dfrac{\xi^3}{3}-E\dfrac{\xi^2}{2}+I\xi$ | $\overline{W}=-D_1\dfrac{\xi^3}{3}+D_1\dfrac{\xi^2}{2}+I_1\xi+J$ |
| | $\overline{\Phi}=D\xi^2+E\xi$ | $\overline{\Phi}=D_1\xi^2-2D_1\xi+E_1$ |
| | $\overline{U}=-\delta D\xi^2+C\xi$ | $\overline{U}=-\delta D_1\xi^2+2\delta_1 D_1\xi+C_1$ |
| C-S③(u) | $\overline{W}=-d\dfrac{\xi^3}{3}-e\dfrac{\xi^2}{2}+i\xi$ | $\overline{W}=-d_1\dfrac{\xi^3}{3}+d_1\dfrac{\xi^2}{2}+i_1\xi+j$ |
| | $\overline{\Phi}=d\xi^2+e\xi$ | $\overline{\Phi}=d_1\xi^2+2e_1\xi+e_2$ |
| | $\overline{U}=-\delta d\xi^2+c\xi$ | $\overline{U}=-\delta d_1\xi^2+c_1\xi+c_2$ |
| C-F | $\overline{W}=-A\xi^2$ | $\overline{W}=-A\xi_0\left(\xi-\dfrac{\xi_0}{2}\right)$ |
| | $\overline{\Phi}=2A\xi$ | $\overline{\Phi}=2A\xi_0$ |
| | $\overline{U}=A_1\xi$ | $\overline{U}=A_1\xi_0$ |

| | 第Ⅰ部分 $0\leq\xi\leq 0.5(1-\xi_0)$ | 第Ⅱ部分 $0.5(1-\xi_0)\leq\xi\leq 0.5(1+\xi_0)$ | 第Ⅲ部分 $0.5(1+\xi_0)\leq\xi\leq 1$ |
|---|---|---|---|
| S-S(m) | $\overline{W}=A\xi_0\xi$ | $\overline{W}=-A\left[(\xi^2-\xi)+\dfrac{(1-\xi)^2}{4}\right]$ | $\overline{W}=-A\xi_0(\xi-1)$ |
| | $\overline{\Phi}=-A\xi_0$ | $\overline{\Phi}=2A\left(\xi-\dfrac{1}{2}\right)$ | $\overline{\Phi}=A\xi_0$ |
| | $\overline{U}=0$ | $\overline{U}=A_1\left[\xi+\dfrac{(\xi_0-1)}{2}\right]$ | $\overline{U}=A_1\xi_0$ |

① 右列中各种术语的表达式可以在本章附录 D 中找到。
② 施加电压 $V$ 的符号与边界条件 C-F 相反。其中：$S$-$S(m)$ = 简单支撑可移动；$S$-$S(u)$ = 简单支撑不可移动；$C$-$S(m)$ = 简单支撑可移动；$C$-$S(u)$ = 加紧简单支撑不可移动；$C$-$F$ = 未加紧。
③ 考虑到常量 $c$、$d$、$e$、$i$、$c_1$、$c_2$、$d_1$、$e_1$、$e_2$、$i_2$ 和 $j_2$ 的长度和复杂的性质，这里省略了它们的显式表达式，读者可以自行查询；$\xi_0$ 为压电片的无量纲长度。

## 3.5.5 诱导轴向力的实验验证

文献 [34] 提供了一个有趣的应用，它改变了装有压电片悬臂梁的固有频率。众所周知，对于自由端有一个自由轴向位移的悬臂梁情况，由于在梁每个截面上的合力等于零，诱导轴向力不能改变其横向振动（在梁上没有施加外部轴向力）。然而，在某些情况下，可以改变压电叠层梁的固有频率。这是由以下实验模型来完成的（更多细节可见文献［34］），该模型主要由玻璃钢制成，具有均匀的矩形空心截面。在截面的根部和顶端粘上两块玻璃钢板（图 3.34）。这些板的中心，即截面根部和尖端的中心，由一根 0.6mm 的钢丝连接。

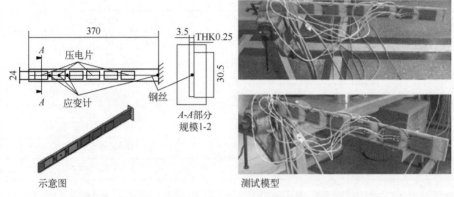

图 3.34　实验模型的示意图[34]（所有尺寸均以 mm 为单位）

6 对压电片被粘在梁的宽表面上。为了能够施加较高的电压（在 1000V 范围内），压电片采用了由 Smart Material Corporation[35] 制造的宏观纤维复合材料。压电片是电气并联的，因此所有压电片施加的电压相同。为了监测梁的行为，沿其表面黏接了 4 对应变测量片（它们的位置见图 3.34）。

改变悬臂梁固有频率的方法是连接梁两侧的导线。尽管梁内部会诱导产生内应力，但在没有导线的情况下，对连接在悬臂梁上的压电片施加恒定电压而对横向振动的影响可以忽略不计。另一个重要的特点是梁是空心的，所以导线可以在里面自由移动，防止导线在梁截面内的自由移动不会改变悬臂梁的固有频率。

图 3.35 所示为测试期间测量的频率与计算值之间的对比。总体来看，这两个结果一致性较好。刚开始实验结果低于计算结果。原因可能是夹紧不完全，以及与连接补丁到电压源和连接应变片到应变记录仪的电线有关的附加质量。在更高电压下，实验结果与计算结果吻合得较好。频率的平方随电压的增加而几乎呈线性减小。固有频率为 0 表示梁的屈曲，可以确定产生"屈曲"电压。实验测量值[34]为 1168V，其与计算值 1135V 非常吻合（误差仅为 2.9%）。

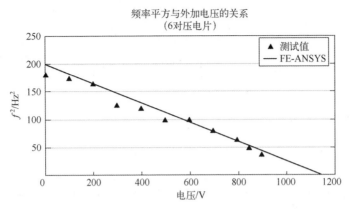

图 3.35 频率平方与施加电压的关系——实验和数值（FE-ANSYS）结果[34]

## 3.6 压电片复合板

前几节讨论了装有压电片梁的性能。在本节中，首先使用复合材料的经典层压理论①研究带有压电片薄板的性能。假设在笛卡儿坐标系中，一个板块的 $x$ 轴和 $y$ 轴平行于板块边界。为了简化模型，假设一个由层压结构材料和压电层组成的板（压电层压板）以中间平面对称，因此 $B_{ij}=0$，没有像 $(\ )_{16}=(\ )_{26}=0$ 这样的其他耦合项，也没有表面剪切应力。为模型[36]假设平面应力条件。位移场根据经典板理论确定，忽略横向剪切变形[31,36]（参见 2.1 节）。基于这些假设，并对包括压电效应在内的各层板的应力分量进行积分，得到单位长度上的力矩如下：

$$\begin{cases} M_x = -D_{11}\dfrac{\partial^2 w(x,y,t)}{\partial x^2} - D_{12}\dfrac{\partial^2 w(x,y,t)}{\partial y^2} + F_x(x,y,t) \\ M_y = -D_{12}\dfrac{\partial^2 w(x,y,t)}{\partial x^2} - D_{22}\dfrac{\partial^2 w(x,y,t)}{\partial y^2} + F_y(x,y,t) \\ M_{xy} = -2D_{66}\dfrac{\partial^2 w(x,y,t)}{\partial x \partial y} \end{cases} \quad (3.319)$$

式中：$D_{ij}$ 为复合板的弯曲刚度；压电诱导力矩 $F_x(x,y,t)$ 和 $F_y(x,y,t)$ 的定义为

---

① 本节的第一部分是基于文献：Edery-Azulay, L. and Abramovich, H., "Piezolaminated Plates-Highly Accurate Solutions Based on the Extended Kantorovich Method," Composite Structures 84 (3), 2008, 241-247.

$$\begin{cases} F_x(x,y,t) = \int_{-h/2}^{h/2} \left( \dfrac{Q_{13}e_{33}}{Q_{33}} - e_{33} \right) E_3 z \mathrm{d}z \\ F_y(x,y,t) = \int_{-h/2}^{h/2} \left( \dfrac{Q_{23}e_{33}}{Q_{33}} - e_{23} \right) E_3 z \mathrm{d}z \end{cases} \quad (3.320)$$

式中

$$\begin{cases} e_{31} = d_{31}Q_{11} + d_{32}Q_{12} \\ e_{32} = d_{31}Q_{12} + d_{32}Q_{22} \\ e_{33} = d_{33}Q_{33} \end{cases} \quad (3.321)$$

并且，$d_{31}$、$d_{32}$ 和 $d_{33}$ 为已知的压电场应变常数（单位为 m/V 或 C/N）；$e_{31}$、$e_{32}$ 和 $e_{33}$ 为已知的压电场应力常数（单位为 N/mV 或 C/m²）；$E_3$ 为电场；$h$ 为压电层的总厚度；$Q_{ij}$ 为弹性常数；$e_{ij}$ 为压电常数[36]。式（3.320）的诱导力矩定义是基于假定厚度方向的应力为零，即 $\sigma_z = 0$。更多详细信息，参见文献[31]。

众所周知，受横向压力影响的薄板运动方程 $q(x,y)$ 为

$$\frac{\partial^2 M_x}{\partial x^2} + 2\frac{\partial^2 M_{xy}}{\partial x \partial y} + \frac{\partial^2 M_y}{\partial y^2} = q(x,y) \quad (3.322)$$

将式（3.319）代入式（3.322），并假设仅为静态行为（去掉时间依赖性），得

$$D_1 \frac{\partial^4 w(x,y)}{\partial x^4} + D_3 \frac{\partial^4 w(x,y)}{\partial x^2 \partial y^2} + D_2 \frac{\partial^4 w(x,y)}{\partial y^4} = P(x,y) \quad (3.323)$$

式中

$$D_1 \equiv D_{11}, \quad D_2 \equiv D_{22}, \quad D_3 \equiv (2D_{12} + 4D_{66})$$

$$P(x,y) = \frac{\partial^2 F_x(x,y)}{\partial x^2} + \frac{\partial^2 F_y(x,y)}{\partial y^2} + q(x,y)$$

式（3.323）无法用解析法求解。因此，寻求使用 Galerkin 法，即要求

$$\int_0^a \int_0^b \left[ D_1 \frac{\partial^4 w(x,y)}{\partial x^4} + D_3 \frac{\partial^4 w(x,y)}{\partial x^2 \partial y^2} + D_2 \frac{\partial^4 w(x,y)}{\partial y^4} - P(x,y) \right] w(x,y) \mathrm{d}x \mathrm{d}y = 0$$

$$(3.324)$$

式中：$a$ 和 $b$ 分别为板的长度和宽度。正如 Edery-Azulay 和 Abramovich 所做的[36]，式（3.324）采用 Kantorovich 法求解[37-41]，该方法假设解 $w(x,y)$ 是可分离的，即

$$w(x,y) = W(x)W(y) \quad (3.325)$$

值得留意的是，根据经典的 Kantorovich 方法，假设两个方向中的一个解是已知的。因此，假设 $W(y)$ 是一个先验选择的已知函数，将其代入式 (3.325) 和对应的 Galerkin 方程 (式 (3.324))，得

$$\int_0^a \int_0^b \left[ D_1 \frac{d^4 W(x)}{dx^4} W(y) + D_3 \frac{d^2 W(x)}{dx^2} \frac{d^2 W(y)}{dy^2} + D_2 W(x) \frac{d^4 W(y)}{dy^4} - P(x,y) \right] W(y) dy W(x) dx = 0 \tag{3.326}$$

为了满足式 (3.326)，方括号中的表达式必须为零。这就得到了一个确定未知 $W(x)$ 函数的常微分方程①，其形式如下：

$$A_4 \frac{d^4 W(x)}{dx^4} + A_2 \frac{d^2 W(x)}{dx^2} + A_0 W(x) = F(x) \tag{3.327}$$

式中

$$A_4 \equiv D_1 \int_0^b W^2(y) dy, \quad A_2 \equiv D_3 \int_0^b \frac{d^2 W(y)}{dy^2} W(y) dy$$

$$A_0 \equiv D_2 \int_0^b \frac{d^4 W(y)}{dy^4} W(y) dy, \quad F(x) = \int_0^b P(x,y) W(y) dy$$

给出的积分可以使用众多可用的商业数学程序如 Maple 代码[42]进行数值方法求解。为给出式 (3.327) 的解，可以考虑压电引起的弯矩和机械载荷按双正弦形式分布的情况，如下式②：

$$\begin{Bmatrix} q(x,y) \\ F_x(x,y) \\ F_y(x,y) \end{Bmatrix} = \begin{Bmatrix} q_0 \sin\left(\frac{m\pi}{a}x\right) \sin\left(\frac{n\pi}{b}y\right) \\ f_0 \sin\left(\frac{m\pi}{a}x\right) \sin\left(\frac{n\pi}{b}y\right) \\ f_0 \sin\left(\frac{m\pi}{a}x\right) \sin\left(\frac{n\pi}{b}y\right) \end{Bmatrix} \tag{3.328}$$

PZT 陶瓷致动器通常是横向各向同性的，其中 $e_{31} = e_{32}$，因此在 $x$ 和 $y$ 方向上产生的诱导弯矩相等。将式 (3.328) 代入式 (3.327) 得到微分方程：

$$A_4 \frac{d^4 W(x)}{dx^4} + A_2 \frac{d^2 W(x)}{dx^2} + A_0 W(x) = [q_0 - f_0(k_x^2 + k_y^2)] \sin(k_x x) \tag{3.329}$$

式中

$$k_x = \frac{m\pi}{a}, \quad k_y = \frac{n\pi}{b}$$

---

① 同样地，当 $W(x)$ 是一个先验已知函数时，可以得到一个确定 $W(y)$ 未知函数的常微分方程。
② 对于加载直流电压的压电层，压电诱导力矩是恒定的。

其解如下：

$$W(x) = \sum_{j=1}^{4} B_j e^{\lambda_j x} + \frac{f_0(k_x^2 + k_y^2) - q_0}{-A_4 k_x^4 + A_2 k_x^2 - A_0} \sin(k_x x) \quad (3.330)$$

式中

$$\lambda_j = \pm \sqrt{\frac{-A_2 \pm \sqrt{A_2^2 - 4A_4 A_0}}{2A_4}}, \quad j = 1,2,3,4$$

4个常数 $B_1$、$B_2$、$B_3$ 和 $B_4$ 是通过在 $x$ 方向上施加边界条件来确定的。

需要注意的是，在得到第1个 Kantorovich 解后，应用扩展的 Kantorovich 解时需要将得到的解析解作为指定函数，并将另一个方向释放出来进行解析，从而改变板的方向才能求解。这个迭代过程可以重复，直到结果收敛到期望的精度。Edery-Azulay 和 Abramovich 报道了一个非常有趣的结论[36]，即初始试验函数既不需要满足几何边界条件也不需要满足自然边界条件，因为迭代过程使解满足所有的板边界条件。这意味着，对于板边界附近复杂的边界条件，运用扩展的 Kantorovich 方法是一种以较少的计算工作量获得精确结果的优异工具，如图 3.36 所示。

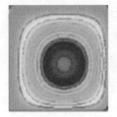 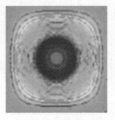 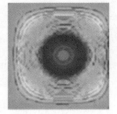 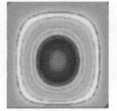

(a) 夹紧-简支-夹紧-简支　　(b) 全部夹紧　　(c) 夹紧-夹紧-夹紧-简支　　(d) 夹紧-简支-夹紧-夹紧

图 3.36　连续压电层致动板的面外位移模式-不同的边界条件[36]

接下来，使用一阶剪切变形理论处理另一个薄板模型。图 3.37 所示为笛卡儿坐标系中的矩形层压板。该结构包括具有与坐标系重合的对称平面弹性层。延伸型压电材料①黏结在主体结构表面，并沿厚度方向极化；剪切型压电材料②嵌入在板芯中，并沿平面结构方向极化。在厚度方向上施加电场会导致表面致动器的平面尺寸增加或减少，并使嵌入式致动器产生横向挠度；这些变形在主体结构上引起面外位移。

---

① 在本章中，延伸型压电材料是指压电层在厚度方向施加电场作用，导致材料的延伸或压缩。
② 顾名思义，剪切型压电材料是指压电层在厚度方向施加电场作用，引起剪切应变。

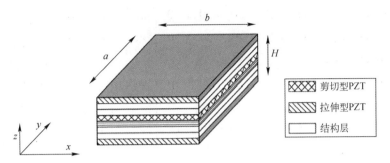

图 3.37　笛卡儿坐标系中的矩形层压板

在剪切型压电材料的厚度方向施加电场，会产生剪切应变，从而使主板发生横向挠度。

本章提出的模型适用于在厚度上具有任意方向的矩形压电层压板。然而，对于本章提出的推导，假定材料主坐标与所分析问题的坐标重合。逆压电效应的本构关系如下：

$$\{\sigma_{ij}\} = [Q_{ij}]\{\varepsilon_{ij}\} - [e_{ij}]^t\{E_j\} \tag{3.331}$$

式中：$\sigma_{ij}$ 和 $\varepsilon_{ij}$ 分别为应力和应变分量；$E_j$ 为电场；$Q_{ij}$ 为弹性常数；$e_{ij}$ 为压电常数，其与 $d_{ij}$ 压电常数的关系为

$$[e_{ij}] = [d_{ij}][Q_{ij}] \tag{3.332}$$

采用正交拉伸型压电材料时，式（3.331）中的一般关系详细形式如下：

$$\begin{Bmatrix}\sigma_{11}\\ \sigma_{22}\\ \sigma_{33}\\ \sigma_{23}\\ \sigma_{13}\\ \sigma_{12}\end{Bmatrix} = \begin{bmatrix}Q_{11} & Q_{12} & Q_{13} & 0 & 0 & 0\\ Q_{21} & Q_{22} & Q_{23} & 0 & 0 & 0\\ Q_{31} & Q_{32} & Q_{33} & 0 & 0 & 0\\ 0 & 0 & 0 & Q_{44} & 0 & 0\\ 0 & 0 & 0 & 0 & Q_{55} & 0\\ 0 & 0 & 0 & 0 & 0 & Q_{66}\end{bmatrix}\begin{Bmatrix}\varepsilon_{11}\\ \varepsilon_{22}\\ \varepsilon_{33}\\ \gamma_{23}\\ \gamma_{13}\\ \gamma_{12}\end{Bmatrix} - \begin{bmatrix}0 & 0 & e_{31}\\ 0 & 0 & e_{32}\\ 0 & 0 & e_{33}\\ 0 & e_{24} & 0\\ e_{15} & 0 & 0\\ 0 & 0 & 0\end{bmatrix}\begin{Bmatrix}E_1\\ E_2\\ E_3\end{Bmatrix}$$

$$\tag{3.333}$$

剪切型压电材料沿轴向（1 或 $x$）极化。拉伸型压电材料绕第 2 方向（2 或 $y$）旋转 90°，再绕第 3 方向（3 或 $z$）旋转 180°，可以得到本构方程。将这些旋转依次应用于式（3.333），得到剪切型压电材料的本构方程如下：

$$\left\{\begin{matrix}\sigma_{11}\\ \sigma_{22}\\ \sigma_{33}\\ \sigma_{23}\\ \sigma_{13}\\ \sigma_{12}\end{matrix}\right\}=\begin{bmatrix}Q_{11} & Q_{12} & Q_{13} & 0 & 0 & 0\\ Q_{21} & Q_{22} & Q_{23} & 0 & 0 & 0\\ Q_{31} & Q_{32} & Q_{33} & 0 & 0 & 0\\ 0 & 0 & 0 & Q_{44} & 0 & 0\\ 0 & 0 & 0 & 0 & Q_{55} & 0\\ 0 & 0 & 0 & 0 & 0 & Q_{66}\end{bmatrix}\left\{\begin{matrix}\varepsilon_{11}\\ \varepsilon_{22}\\ \varepsilon_{33}\\ \gamma_{23}\\ \gamma_{13}\\ \gamma_{12}\end{matrix}\right\}-\begin{bmatrix}e_{11} & 0 & 0\\ e_{12} & 0 & 0\\ e_{13} & 0 & 0\\ 0 & 0 & 0\\ 0 & 0 & e_{35}\\ 0 & e_{26} & 0\end{bmatrix}\left\{\begin{matrix}E_1\\ E_2\\ E_3\end{matrix}\right\}$$

(3.334)

使用厚度较薄的压电片意味着只能施加 $E_3$ 电场。因此,可将伸缩和剪切式压电材料的上述两种本构关系(式(3.333)和式(3.334))压缩写为①

$$\left\{\begin{matrix}\sigma_{11}\\ \sigma_{22}\\ \sigma_{33}\\ \sigma_{23}\\ \sigma_{13}\\ \sigma_{12}\end{matrix}\right\}=\begin{bmatrix}\hat{Q}_{11} & \hat{Q}_{12} & \hat{Q}_{13} & 0 & 0 & 0\\ \hat{Q}_{21} & \hat{Q}_{22} & \hat{Q}_{23} & 0 & 0 & 0\\ \hat{Q}_{31} & \hat{Q}_{32} & \hat{Q}_{33} & 0 & 0 & 0\\ 0 & 0 & 0 & \hat{Q}_{44} & 0 & 0\\ 0 & 0 & 0 & 0 & \hat{Q}_{55} & 0\\ 0 & 0 & 0 & 0 & 0 & \hat{Q}_{66}\end{bmatrix}\left\{\begin{matrix}\varepsilon_{11}\\ \varepsilon_{22}\\ \varepsilon_{33}\\ \gamma_{23}\\ \gamma_{13}\\ \gamma_{12}\end{matrix}\right\}-\begin{bmatrix}e_{11} & 0 & e_{31}\\ e_{12} & 0 & e_{32}\\ e_{13} & 0 & e_{33}\\ 0 & e_{24} & 0\\ e_{15} & 0 & e_{35}\\ 0 & e_{26} & 0\end{bmatrix}\left\{\begin{matrix}E_1\\ E_2\\ E_3\end{matrix}\right\}$$

(3.335)

式中

$$\hat{Q}_{21}=\hat{Q}_{12},\quad \hat{Q}_{31}=\hat{Q}_{31},\quad \hat{Q}_{23}=\hat{Q}_{32}$$

注意:根据所建立的各压电机构的本构方程,当使用延伸型压电致动器时,压电张量的非零分量只有 $e_{31}$、$e_{32}$、$e_{33}$、$e_{24}$ 和 $e_{15}$,而对于剪切型压电致动,压电张量的非零分量只有 $e_{11}$、$e_{12}$、$e_{13}$、$e_{26}$ 和 $e_{35}$。

已知 PZT 陶瓷致动器横向各向同性,因此,$e_{31}=e_{32}$,$e_{15}=e_{35}$[31]。分析薄板时,可以对所有压电和弹性结构层使用假设 $\sigma_{33}\equiv\sigma_z=0$。因此,可以写成

$$\varepsilon_{33}\equiv\varepsilon_z=-\frac{\hat{Q}_{13}\varepsilon_{11}+\hat{Q}_{23}\varepsilon_{22}}{\hat{Q}_{33}}+\frac{e_{33}E_3}{\hat{Q}_{33}} \quad (3.336)$$

---

① $Q_{ij}$ 是一个通用的符号。对于延伸或剪切型机构,其值分别见式(3.333)或式(3.334)。

将式(3.336)代入式(3.335)可得

$$\begin{Bmatrix} \sigma_x \\ \sigma_y \\ \tau_{yz} \\ \tau_{xz} \\ \tau_{xy} \end{Bmatrix} = \begin{bmatrix} \widetilde{Q}_{11} & \widetilde{Q}_{12} & 0 & 0 & 0 \\ \widetilde{Q}_{12} & \widetilde{Q}_{22} & 0 & 0 & 0 \\ 0 & 0 & \widetilde{Q}_{44} & 0 & 0 \\ 0 & 0 & 0 & \widetilde{Q}_{55} & 0 \\ 0 & 0 & 0 & 0 & \widetilde{Q}_{66} \end{bmatrix} \begin{Bmatrix} \varepsilon_x \\ \varepsilon_y \\ \gamma_{yz} \\ \gamma_{xz} \\ \gamma_{xy} \end{Bmatrix} - \begin{bmatrix} \widetilde{P}_1 \\ \widetilde{P}_2 \\ 0 \\ \widetilde{P}_4 \\ 0 \end{bmatrix} E_3 \quad (3.337)$$

式中

$$\sigma_{11} \equiv \sigma_x, \quad \sigma_{22} \equiv \sigma_y, \quad \sigma_{23} \equiv \tau_{yz}, \quad \sigma_{13} \equiv \tau_{xz}, \quad \sigma_{11} \equiv \tau_{xy}$$

$$\varepsilon_{11} \equiv \varepsilon_x, \quad \varepsilon_{22} \equiv \varepsilon_y, \quad \gamma_{23} \equiv \gamma_{yz}, \quad \gamma_{13} \equiv \gamma_{xz}, \quad \gamma_{11} \equiv \gamma_{xy}$$

$$\widetilde{Q}_{11} = \hat{Q}_{11} - \frac{\hat{Q}_{13}^2}{\hat{Q}_{33}}, \quad \widetilde{Q}_{12} = \hat{Q}_{12} - \frac{\hat{Q}_{13}\hat{Q}_{23}}{\hat{Q}_{33}}, \quad \widetilde{Q}_{22} = \hat{Q}_{22} - \frac{\hat{Q}_{23}^2}{\hat{Q}_{33}}$$

$$\widetilde{Q}_{44} = \hat{Q}_{44}, \quad \widetilde{Q}_{55} = \hat{Q}_{55}, \quad \widetilde{Q}_{66} = \hat{Q}_{66}$$

$$\widetilde{P}_1 = \frac{\hat{Q}_{13}}{\hat{Q}_{33}} e_{33} - e_{31}, \quad \widetilde{P}_2 = \frac{\hat{Q}_{23}}{\hat{Q}_{33}} e_{33} - e_{32}, \quad \widetilde{P}_4 = e_{35}$$

在推导中,主材料坐标与所分析问题的坐标重合,延伸型PZT对$x$和$y$方向的应力有贡献,而剪切型PZT只对$x$-$z$方向的剪切应力有贡献。

定义电场强度$E_3$为$E_3 = V/h_k$。其中,$V$为施加在第$k$层上的电压,$h_k$为第$k$层PZT的厚度。

剪切压电致动依赖于剪切变形。因此,对嵌有剪切压电片或连续层的压电层压板的研究必须考虑横向剪切位移。在本节中,两种解的模型(精确的和近似的)都依赖于基于Mindlin理论的位移场(一阶剪切变形理论模型见2.2节)。

假设一阶剪切变形理论模型中的3个位移具有以下形式:

$$\begin{cases} u(x,y,z,t) = u^0(x,y,t) + z\phi_x(x,y,t) \\ v(x,y,z,t) = v^0(x,y,t) + z\phi_y(x,y,t) \\ w(x,y,z,t) = w^0(x,y,t) \end{cases} \quad (3.338)$$

式中:$u^0(x,y,t)$、$v^0(x,y,t)$和$w^0(x,y,t)$分别为中间面上一点在$x$、$y$、$z$方向上的位移,而$\phi_x(x,y,t)$和$\phi_y(x,y,t)$分别为中间面在$x$轴和$y$轴上绕法线的旋转(图3.37)。

如2.2节所述,一阶剪切变形理论模型得到了5个通用的运动控制方程:

$$N_{x,x} + N_{xy,y} + p_x = I_1 \ddot{u}_0(x,y,t) + I_2 \ddot{\phi}_x(x,y,t) \quad (3.339a)$$

$$N_{xy,x}+N_{y,y}+p_y=I_1\ddot{v}_0(x,y,t)+I_2\ddot{\phi}_y(x,y,t) \qquad (3.339\text{b})$$

$$Q_{x,x}+Q_{y,y}+p_z+[\overline{N}_{xx}w_{0,x}+\overline{N}_{xy}w_{0,y}]_{,x}+$$
$$[\overline{N}_{yy}w_{0,y}+\overline{N}_{xy}w_{0,x}]_{,x}=I_1\ddot{w}_0(x,y,z,t) \qquad (3.339\text{c})$$

$$M_{x,x}+M_{xy,y}-Q_x+m_x=I_1\ddot{u}_0(x,y,t)+I_3\ddot{\phi}_x(x,y,t) \qquad (3.339\text{d})$$

$$M_{xy,x}+M_{y,y}-Q_y+m_y=I_1\ddot{V}_0(x,y,t)+I_3\ddot{\phi}_y(x,y,t) \qquad (3.339\text{e})$$

式中

$$I_j=\int_{-h/2}^{h/2}\rho z^{j-1}\mathrm{d}z;\quad j=1,2,3 \qquad (3.340)$$

式中：$N_x$、$N_y$、$N_{xy}$ 为力的结果；$M_x$、$M_y$、$M_{xy}$ 表示力矩结果；$Q_x$、$Q_y$ 表示横向；$p_x$、$p_y$、$p_z$ 和 $m_x$、$m_y$ 分别为外载荷和力矩；$N_{xx}$、$N_{yy}$、$N_{xy}$ 为平面外载荷。力和力矩的结果都取单位长度。像文献中广泛提到的那样，为便于比较[31,43-50]，将研究范围仅限于具有压电诱导力、无外部机械载荷和力矩的平板静态解。对于对称铺层的板，纵向性能（式（3.339a）和式（3.339b））可与弯曲性能（式（3.339c）～式（3.339e））分开求解。基于一阶剪切变形理论模型，这3个运动方程可以用假设的位移（$u_0$, $v_0$, $w_0$）[①] 和转动（$\varphi_x$, $\varphi_y$）表示为

$$\begin{cases}\kappa A_{55}\left[\dfrac{\partial\phi_x(x,y)}{\partial x}+\dfrac{\partial^2 w(x,y)}{\partial x^2}\right]-\dfrac{\partial G_{55}(x,y)}{\partial x}+\kappa A_{44}\left[\dfrac{\partial\phi_y(x,y)}{\partial y}+\dfrac{\partial^2 w(x,y)}{\partial y^2}\right]=0\\[6pt]
D_{66}\left[\dfrac{\partial^2\phi_y(x,y)}{\partial x\partial y}+\dfrac{\partial^2\phi_x(x,y)}{\partial x^2}\right]-\kappa A_{55}\left[\phi_x(x,y)+\dfrac{\partial w(x,y)}{\partial x}\right]+D_{11}\dfrac{\partial^2\phi_x(x,y)}{\partial x^2}\\[6pt]
+D_{12}\dfrac{\partial^2\phi_y(x,y)}{\partial x\partial y}+G_{55}(x,y)-\dfrac{\partial F_{11}(x,y)}{\partial x}=0\\[6pt]
(D_{66}+D_{12})\dfrac{\partial^2\phi_x(x,y)}{\partial x\partial y}-\kappa A_{44}\left[\phi_y(x,y)+\dfrac{\partial w(x,y)}{\partial y}\right]+D_{66}\dfrac{\partial^2\phi_y(x,y)}{\partial x^2}+\\[6pt]
D_{22}\dfrac{\partial^2\phi_y(x,y)}{\partial y^2}-\dfrac{\partial F_{22}(x,y)}{\partial x}=0\end{cases} \qquad (3.341)$$

式中：$D_{ij}$（$i,j=1,2,6$）和 $A_{ii}$（$i=4,5$）是根据层压理论[31]定义的常见弯曲和横向剪切刚度系数。

将压电诱导力 $E_{ii}(x,y)$[②]、力矩 $F_{ii}(x,y)$ 和诱导剪切压电力 $G_{55}(x,y)$ 定义为

---

① 请注意，为了方便起见，在位移表达式中省略了下标0。
② 请注意，$E_{ii}(x,y)$ 并没有出现在式（3.341）中，但是可能出现在式（3.339）a和b中，这取决于压电片电连接的方式（同相或异相）。

$$\begin{cases} E_{ii}(x,y) \equiv \int_{-h/2}^{h/2} \widetilde{P}_i E_3 \mathrm{d}z = \sum_{j=1}^{N} (\widetilde{P}_i E_3)_j, \quad i=1,2 \\ F_{ii}(x,y) \equiv \int_{-h/2}^{h/2} \widetilde{P}_i E_3 z \mathrm{d}z = \sum_{j=1}^{N} (\widetilde{P}_i E_3)_j (z_j - z_{j-1}), \quad i=1,2 \quad (3.342) \\ G_{55}(x,y) \equiv \kappa \int_{-h/2}^{h/2} \widetilde{P}_4 E_3 \mathrm{d}z = \sum_{j=1}^{N} (\widetilde{P}_4 E_3)_j \end{cases}$$

式中：$\kappa$ 为一阶剪切变形理论模型中使用的剪切因子（在大多数情况下，其值为 5/6）；$N$ 为层压板中的压电层数。

基于板的 Navier 或 Lévy 理论[51]，可以求解具有连续压电层的板，其中至少有两个相反的边界条件是简支边。

对于一个沿 $y$ 方向具有相反简支边的板，可以假设板的解形式如下（注意这些函数也必须满足板的边界条件）：

$$\begin{cases} w(x,y) = \overline{w}(x) \sin(\lambda_y y), \quad \phi_x(x,y) = \overline{\phi}_x(x) \sin(\lambda_y y) \\ \phi_y(x,y) = \overline{\phi}_y(x) \cos(\lambda_y y), \quad F_{11}(x,y) = \overline{F}_{11}(x) \sin(\lambda_y y) \quad (3.343) \\ F_{22}(x,y) = \overline{F}_{22}(x) \sin(\lambda_y y), \quad G_{55}(x,y) = \overline{G}_{55}(x) \sin(\lambda_y y) \end{cases}$$

式中

$$\lambda_y \equiv \frac{n\pi}{b}$$

式中：$n$ 为平板的半波数；$b$ 为平板沿 $y$ 方向的长度。

假设压电力沿 $x$ 方向为常数，将建议解（式（3.343））代入 3 个平衡方程（式（3.343）），得到 3 个耦合的正则微分方程组，形式如下：

$$\begin{cases} \kappa A_{55} \left[ \dfrac{\mathrm{d} \overline{\phi}_x(x)}{\mathrm{d}x} + \dfrac{\mathrm{d}^2 \overline{w}(x)}{\mathrm{d}x^2} \right] - \kappa A_{44} [\lambda_y \overline{\phi}_y(x) + \lambda_y^2 \overline{w}(x)] = 0 \\ D_{66} \left[ \lambda_y \dfrac{\mathrm{d} \overline{\phi}_y(x)}{\mathrm{d}x} + \lambda_y^2 \overline{\phi}_x(x) \right] - \kappa A_{55} \left[ \overline{\phi}_x(x) + \dfrac{\mathrm{d} \overline{w}(x)}{\mathrm{d}x} \right] + D_{11} \dfrac{\mathrm{d}^2 \overline{\phi}_x(x)}{\mathrm{d}x^2} \\ - D_{12} \lambda_y \dfrac{\mathrm{d} \overline{\phi}_y(x)}{\mathrm{d}x} + \overline{G}_{55}(x) = 0 \\ (D_{66} + D_{12}) \lambda_y \dfrac{\mathrm{d} \overline{\phi}_x(x)}{\mathrm{d}x} - \kappa A_{44} [\overline{\phi}_y(x) + \lambda_y \overline{w}(x)] + D_{66} \dfrac{\mathrm{d}^2 \overline{\phi}_y(x)}{\mathrm{d}x^2} \\ - D_{22} \lambda_y^2 \overline{\phi}_y(x) = 0 \end{cases} \quad (3.344)$$

3 个耦合方程（式（3.344））可以解耦得到 3 个未知数 $w(x,y)$、$\varphi_x(x,y)$ 和 $\varphi_y(x,y)$[36] 的单个方程，其通用形式如下：

$$\alpha_1 A^6 + \alpha_2 A^4 + \alpha_3 A^2 + \alpha_4 A = 0 \tag{3.345}$$

式中：$A = w$，$\phi_x$ 或 $\phi_y$。

系数 $\alpha_i$ ($i = 1 \sim 4$) 因变量而异。对于变量 $\bar{w}$，式 (3.345) 为

$$\alpha_1 \frac{\mathrm{d}^6 \bar{w}}{\mathrm{d} x^6} + \alpha_2 \frac{\mathrm{d}^4 \bar{w}}{\mathrm{d} x^4} + \alpha_3 \frac{\mathrm{d}^2 \bar{w}}{\mathrm{d} x^2} + \alpha_4 \bar{w} = 0 \tag{3.346}$$

这就得到了下面的特征方程，若代入 $(\bar{w}) = \bar{R} e^{sx}$，则

$$\alpha_1 s^3 + \alpha_2 s^2 + \alpha_3 s + \alpha_4 = 0 \tag{3.347}$$

式中：$S = \bar{S}^2$。

方程 (3.347) 的解得到 3 个实根，其中一个根是负的，另外两个是正的。由此得到根，可将厚度方向上横向位移的通解表示如下：

$$\bar{w} = B_1 \sin(r_1 x) + B_2 \cos(r_1 x) + B_3 \sinh(r_2 x) + B_4 \cosh(r_2 x) + B_5 \sinh(r_3 x) + B_6 \cosh(r_3 x) \tag{3.348}$$

式中

$$r_1 = \sqrt{-s_1}, \quad r_2 = \sqrt{s_2}, \quad r_1 = \sqrt{s_3}$$

6 个常数 $B_1 \sim B_6$ 可通过强化板的边界条件得到。两种弯曲转动 $\varphi_x(x,y)$ 和 $\varphi_y(x,y)$ 的表达式形式相似，但常数不同：

$$\begin{cases} \bar{\phi}_x = \bar{B}_1 \sin(r_1 x) + \bar{B}_2 \cos(r_1 x) + \bar{B}_3 \sinh(r_2 x) + \\ \bar{B}_4 \cosh(r_2 x) + \bar{B}_5 \sinh(r_3 x) + \bar{B}_6 \cosh(r_3 x) \\ \bar{\phi}_y = \bar{\bar{B}}_1 \sin(r_1 x) + \bar{\bar{B}}_2 \cos(r_1 x) + \hat{B}_3 \sinh(r_2 x) + \\ \bar{\bar{B}}_4 \cosh(r_2 x) + \bar{\bar{B}}_5 \sinh(r_3 x) + \bar{\bar{B}}_6 \cosh(r_3 x) \end{cases} \tag{3.349}$$

注意，通过将式 (3.349) 代入耦合运动方程式 (3.344)，常数 $\bar{B}_1 \sim \bar{B}_6$ 和 $\bar{\bar{B}}_1 \sim \bar{\bar{B}}_6$ 可以用常数 $B_1 \sim B_6$ 表示。

对于其他边界条件，要得到板的相对位移和弯曲旋转的解，应使用能量法，如 Rayleigh-Ritz 法。

接下来是 Edery-Azulay 和 Abramovich 的典型结果[36]。首先，介绍了具有连续拉伸或剪切压电层的正方形层压板的性能。层压板的结构见图 3.37。为了简化公式，假设所有层的取向都为 0°。假设正方形板的长度为 10cm，压电材料是层厚 $t$ 为 0.25mm 的 PZT-5H，结构层为石墨环氧树脂复合材料，每层厚度 $t$ 为 0.5mm（板的总厚度为 3mm）。表 3.6 归纳总结了这两种材料的性能。

表3.6 材料属性（力学性能和电性能）

| 力学性能/GPa | 石墨环氧树脂 | PZT-5H | 电性能/(C/m²) | PZT-5H |
|---|---|---|---|---|
| $C_{11}$ | 183.443 | 99.201 | $e_{31}$ | -7.209 |
| $C_{22}$ | 11.662 | 99.201 | $e_{33}$ | 15.118 |
| $C_{33}$ | 11.662 | 86.856 | $e_{24}$ | 12.332 |
| $C_{12}$ | 4.363 | 54.016 | $e_{15}$ | 12.332 |
| $C_{13}$ | 4.363 | 50.778 | | |
| $C_{23}$ | 3.918 | 50.778 | | |
| $C_{44}$ | 2.877 | 21.100 | | |
| $C_{55}$ | 7.170 | 21.100 | | |
| $C_{66}$ | 7.179 | 22.593 | | |

图3.38（a）和（b）描述了由延伸型PZT驱动的全边简支板的俯视图横向位移模式和直线$y=b/2$沿$x$方向的面外位移。这些结果是用现有的精确数学模型（Lévy法）得到的。对于具有连续剪切压电层的板，没有发现横向位移。注意，在压电复合材料梁上也发现了类似的现象[31]。

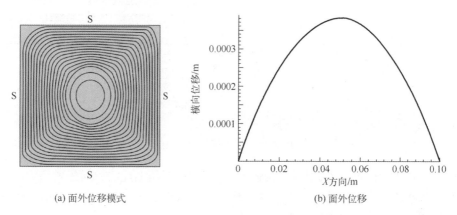

(a) 面外位移模式　　　　　　　(b) 面外位移

图3.38 连续延伸压电层驱动的全边简支板

图3.39（a）和（b）描述了由剪切型PZT驱动的两个反向夹持边板（CCSS）的俯视图横向位移模式和直线$y=b/2$沿$x$方向的面外位移。对于这两种情况，均假定诱导压电力矩恒定（$\overline{F}_{11}=\overline{F}_{22}=1000\mathrm{N}\cdot\mathrm{mm}$），且诱导剪切压电力恒定（$\overline{G}_{55}=1000\mathrm{N}$）。

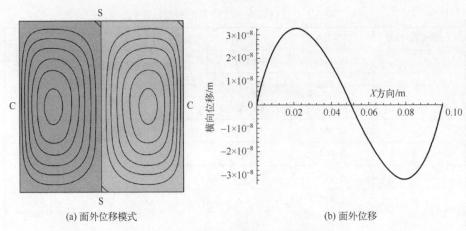

(a) 面外位移模式　　　　　　　(b) 面外位移

图 3.39　连续剪切压电层驱动的 CCSS 板

下面介绍带有延伸或剪切型压电片的板的典型结果[36]。图 3.40 所示为带有压电片的板。

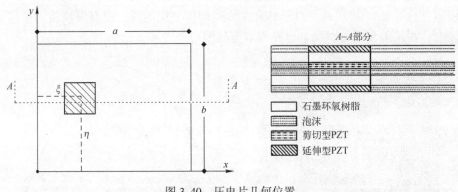

图 3.40　压电片几何位置

压电片的位置由坐标 $\xi$ 和 $\eta$ 确定,它们描述了压电片的中间位置。检测板的表面和芯层均为非结构材料,如泡沫,其材料性能视为零,有一个剪切型压电片和一对延伸型压电片(图 3.40)。

以下结果为一个长和宽为 10cm 的方形板,其中所用的压电片覆盖了板面积的 10%。每个延伸型 PZT 贴片厚度 $t$ 为 0.5mm,而每个剪切型 PZT 贴片的厚度 $t$ 为 0.25mm,且每种结构材料的厚度 $t$ 为 0.5mm。为了解决各种情况,使用了 Rayleigh-Ritz 法。表 3.7 总结了采用 Rayleigh-Ritz 法时,矩形夹芯板在不同边界条件下的假设函数。值得注意的是,假定压片内部的压电力是恒定的。

表 3.7　矩形压电复合材料板的假设函数

| 边界条件 | $\overline{w}$ | $\overline{\phi}_x$ | $\overline{\phi}_y$ |
|---|---|---|---|
| SSSS | $\sum\limits_{n=1}^{\infty}\sum\limits_{m=1}^{\infty} R_{mn} \cdot SX \cdot SY$<br>$\overline{w}(x,0)=0; \overline{w}(x,b)=0$<br>$\overline{w}(0,y)=0; \overline{w}(a,y)=0$ | $\sum\limits_{n=1}^{\infty}\sum\limits_{m=1}^{\infty} T_{mn} \cdot CX \cdot SY$<br>$\overline{\phi}_x(x,0)=0$<br>$\overline{\phi}_x(x,b)=0$ | $\sum\limits_{n=1}^{\infty}\sum\limits_{m=1}^{\infty} G_{mn} \cdot SX \cdot CY$<br>$\overline{\phi}_y(0,y)=0$<br>$\overline{\phi}_y(a,y)=0$ |
| CCCC | $\sum\limits_{n=1}^{\infty}\sum\limits_{m=1}^{\infty} R_{mn} \cdot (1-CX')(1-CY')$<br>$\overline{w}(x,0)=0; \overline{w}(x,b)=0$<br>$\overline{w}(0,y)=0; \overline{w}(a,y)=0$ | $\sum\limits_{n=1}^{\infty}\sum\limits_{m=1}^{\infty} T_{mn} \cdot SX \cdot (1-CY')$<br>$\overline{\phi}_x(x,0)=0$<br>$\overline{\phi}_x(x,b)=0$ | $\sum\limits_{n=1}^{\infty}\sum\limits_{m=1}^{\infty} G_{mn} \cdot SY \cdot (1-CX')$<br>$\overline{\phi}_y(0,y)=0$<br>$\overline{\phi}_y(a,y)=0$ |
| CFCF | $\sum\limits_{\substack{i=1\\j=1}}^{\infty} R_i \cdot x^{2i} \cdot y^{2i}$<br>$\overline{w}(x,0)=0$<br>$\overline{w}(0,y)=0$ | $\sum\limits_{i=1}^{\infty} T_i \cdot x^{2i-1} \cdot y^{2i}$<br>$\overline{\phi}_x(x,0)=0$ | $\sum\limits_{\substack{i=1\\j=1}}^{\infty} G_i \cdot x^{2i} \cdot y^{2i-1}$<br>$\overline{\phi}_y(0,y)=0$ |
| | \multicolumn{3}{c}{$x=0$、$y=0$ 边界处夹紧，$x=a$、$y=b$ 边界处自由} |
| SSCC | $\sum\limits_{n=1}^{\infty}\sum\limits_{m=1}^{\infty} R_{mn} \cdot SX \cdot (1-CY')$<br>$\overline{w}(x,0)=0; \overline{w}(x,b)=0$<br>$\overline{w}(0,y)=0; \overline{w}(a,y)=0$ | $\sum\limits_{n=1}^{\infty}\sum\limits_{m=1}^{\infty} T_{mn} \cdot CX \cdot (1-CY')$<br>— | $\sum\limits_{n=1}^{\infty}\sum\limits_{m=1}^{\infty} G_{mn} \cdot SX \cdot SY$<br>$\overline{\phi}_y(x,0)=0$<br>$\overline{\phi}_y(x,b)=0$ |
| | \multicolumn{3}{c}{$x=0$、$y=b$ 边界处夹紧} |

注：其中，SS 表示简单支撑，C 表示夹持，F 表示自由，且 $SX=\sin\dfrac{m\pi x}{a}$，$CX=\cos\dfrac{m\pi x}{a}$，$SY=\sin\dfrac{n\pi y}{b}$，$CY=\cos\dfrac{n\pi y}{b}$，$CX'=\cos\dfrac{2m\pi x}{a}$，$CY'=\cos\dfrac{2n\pi y}{b}$。

从图 3.41（a）和（b）中可以看到带有中心延伸压电片的全边简支板的结果。对于不同的序列项数（在 $x$ 方向上的序列项数为 $m$），图 3.41（a）中每条线表示沿 $x$ 方向和直线 $y=b/2$ 的中心横向位。在图 3.41（b）中可以看到 $m=13$ 和 $j=1$ 时的横向位移模式。粗线表示 $m=13$ 时最终计算的位移曲线。

图 3.42（a）和（b）描述了在 $\xi=0.2a$、$\eta=0.2b$ 处有延伸型压电片的全边简支板沿 $x$ 方向、直线 $y=0.05b$ 上的横向位移。

剪切型 PZT 压电片的影响如图 3.43（a）和（b）所示。采用位于 $\xi=0.5a$、$\eta=0.5b$ 处的中心剪切压电片驱动的方板进行了结果呈现。图 3.43（a）为沿 $x$ 方向和直线 $y=b/2$ 的横向位移，其中粗线为 $m=12$ 的最终计算位移曲线。图 3.43（b）为 $m=12$、$n=1$ 时整个复合材料板的横向位移模式。值得注意的是，使用剪切型压电片必须使用仅具有偶数项的级数，由于解的特性，第1项（$m=1$）不会导致横向位移，因此至少需要两项来观察横向诱导变形。

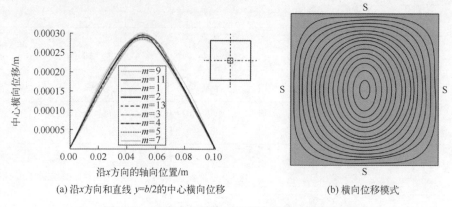

(a) 沿$x$方向和直线$y=b/2$的中心横向位移　　(b) 横向位移模式

图 3.41　具有中心延伸 PZT 压电片的全边简支板

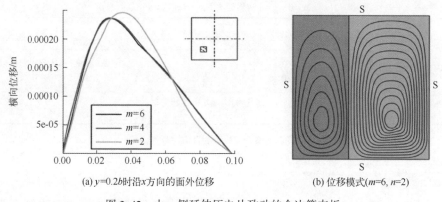

(a) $y=0.2b$时沿$x$方向的面外位移　　(b) 位移模式（$m=6, n=2$）

图 3.42　由一侧延伸压电片致动的全边简支板

本模型的优点在于其鲁棒性，可以在任意位置获得具有无限数量压电片的板的解。

为了检验含有多个压电片的板的解的准确性，本章重新考虑了以前研究过的一种情况：含有一个中心延伸或剪切 PZT 压电片的简支板。将单个压电片分成两个，共同覆盖相同的板区域（它们之间没有重叠），研究该板的弯曲性能。使用两个压电片的板与使用延伸型或剪切型单一压电类型的板得到相同的结果。图 3.44 和图 3.45 分别表示使用一个或两个单独的压电片（延伸型或剪切型）的结果。可以看到一个或两个具有相同面积的压电片的结果具有很好的一致性。

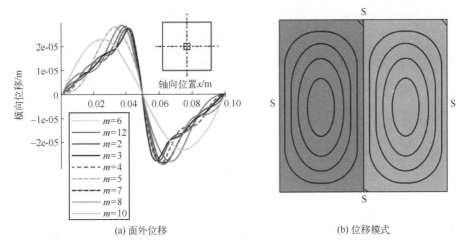

(a) 面外位移　　　　　　　　　(b) 位移模式

图 3.43　由中央剪切型压电片驱动的全边简支板

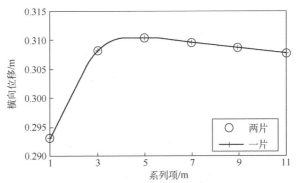

图 3.44　由一个或两个中心延伸压电片驱动的
全边简支板的中心横向位移

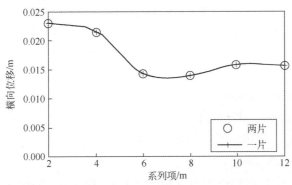

图 3.45　由一个或两个位于中心的剪切型压片驱动的
全边简支板在 $x=a/4$、$y=b/2$ 处的横向位移

## 3.7 附 录 二

### 3.7.1 附录 A

下面将给出并讨论压电耦合系数或机电耦合系数 $k^2$ 的推导。定义压电耦合系数的一种方法见式（3.41），再次呈现如下：

$$k^2 = \frac{(\text{压电能量密度})^2}{\text{电能密度} \cdot \text{机械能密度}} \tag{3.a}$$

文献［52-57］还介绍了使用其他参数的其他类型的定义。Uchino 认为，由于定义的不同，机械能和电能之间的相互转换效率导致机电耦合系数或能量传输系数或效率在某种程度上是混乱的[56]。Uchino[56] 给出的另一个定义具有以下一般形式[57]：

$$\begin{cases} k^2 = \dfrac{\text{存储的机械能}}{\text{输入的电能}} \\ \text{或 } k^2 = \dfrac{\text{存储的电能}}{\text{输入的机械能}} \end{cases} \tag{3.b}$$

单位体积的输入电能可以写成 $0.5\varepsilon_0 \varepsilon^\sigma E^2 = 0.5\bar{\varepsilon}^\sigma E^2$，而单位体积的机械储存机械能为 $0.5 d^2 E^2 / s^E$，代入式（3.b）的第 1 个方程得

$$k^2 = \frac{0.5 d^2 E^2 / s^E}{0.5 \bar{\varepsilon}^\sigma E^2} = \frac{d^2}{\bar{\varepsilon}^\sigma s^E} \tag{3.c}$$

式（3.c）的表达式与式（3.43）相同。对于式（3.b）中的第 2 个表达式，也可以得到与式（3.c）相同的结果①。

文献［54］中提出了一种不同的方法，将压电耦合系数定义为

$$\kappa^2 = \frac{f_{oc}^2 - f_{sc}^2}{f_{oc}^2} \tag{3.d}$$

式中：$f_{oc}$ 为开路电极处的共振频率；$f_{sc}$ 为短路处的共振频率。这一定义为实验确定耦合系数提供了一种方便的方法。

文献［57］中给出了另一种定义，即

---

① 完整推导过程见 Chapter 3 Piezoelectricity, Rupitsch, S. J., *Piezoelectric Sensors and Actuators*, *Topics in Mining, Metallurgy and Materials Engineering*, https://doi.org/10.1007/978-3-662-57534-5_3 © Springer-Verlag GmbH Germany, part of Springer Nature 2019。

$$k^2 = \frac{转换的能量}{供给的能量} \tag{3.e}$$

作者认为，虽然压电耦合因子被认为是评价压电材料效率的有效措施，但式（3.e）所示的因子 $k^2$ 并不总是等于式（3.c）的值。

对于平面应力（图3.a），由式（3.e）可得

$$k_{33}^2 \equiv \frac{转换的能量}{供给的能量} = \frac{0.5 D_3 E_3}{0.5 \varepsilon_3 \sigma_3} = \frac{0.5 \bar{k}_{33}^2 \varepsilon_3^2 / s_{33}^E}{0.5 \varepsilon_3^2 / s_{33}^E} = \bar{k}_{33}^2 \equiv \frac{d_{33}^2}{s_{33}^E \bar{\varepsilon}_{33}^\sigma} \tag{3.f}$$

这表明由式（3.c）的定义可得到式（3.b）的表达式。

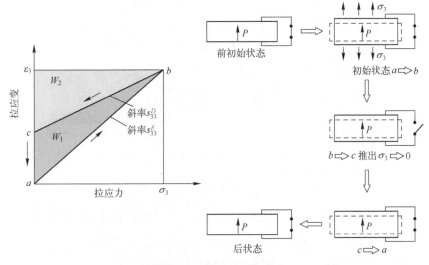

图3.a  准静态应力循环[52]

对于平面应变的情况，Lustig 和 Elata 采用以下本构方程（当 $\varepsilon_1 = \varepsilon_2 = 0$ 时）[57]：

$$\begin{cases} \sigma_3 = C_{33}^E \varepsilon_3 - e_{33} E_3 \\ D_3 = e_{33} \varepsilon_3 + \bar{\varepsilon}_{33}^\varepsilon E_3 \end{cases} \tag{3.g}$$

得到如下关系：

$$k_{33}^2 \equiv \frac{转换的能量}{供给的能量} = \frac{0.5 D_3 E_3}{0.5 \varepsilon_3 \sigma_3} = \frac{0.5 [\kappa_{33}^2/(1+\kappa_{33}^2)] \varepsilon_3^2 C_{33}^E}{0.5 \varepsilon_3^2 C_{33}^E} = \frac{\kappa_{33}^2}{1+\kappa_{33}^2} \tag{3.h}$$

式中

$$\kappa_{33}^2 \equiv \frac{e_{33}^2}{C_{33}^E \bar{\varepsilon}_{33}^\sigma}$$

由式（3.h）可知，转换的能量与供给的能量之比并不一定会得到式（3.c）。

另一个值得讨论的问题是能量传递系数 $\lambda_{max}$[56]。其定义为

$$\begin{cases} \lambda_{max} = \left(\dfrac{输出机械能}{输入电能}\right)_{max} \\ 或 \lambda_{max} = \left(\dfrac{输出电能}{输入机械能}\right)_{max} \end{cases} \quad (3.\text{i})$$

注意式（3.i）与式（3.b）相似，是用式（3.i）中的"输出"代替了式（3.b）中的"存储"。

还要注意效率 $\eta$ 的定义[56]：

$$\begin{cases} \eta = \dfrac{输出的机械能}{消耗的电能} \\ 或 \eta = \dfrac{输出的电能}{消耗的机械能} \end{cases} \quad (3.\text{j})$$

### 3.7.2　附录 B

式（3.205）中的系数形式如下：

$$a_{11}=0, \quad a_{12}=1, \quad a_{13}=0, \quad a_{14}=1 \quad (3.\text{k})$$

$$a_{21}=\sinh(\lambda_1 L), \quad a_{12}=\cosh(\lambda_1 L), \quad a_{13}=\sin(\lambda_2 L), \quad a_{14}=\cos(\lambda_2 L) \quad (3.\text{l})$$

$$a_{31}=k_\theta \lambda_1, \quad a_{32}=\lambda_1^2, \quad a_{33}=k_\theta \lambda_2, \quad a_{34}=-\lambda_2^2 \quad (3.\text{m})$$

$$\begin{cases} a_{41} = \lambda_1^2 \sinh(\lambda_1 L) + k_\theta \lambda_1 \cosh(\lambda_1 L) \\ a_{42} = \lambda_1^2 \cosh(\lambda_1 L) + k_\theta \lambda_1 \sinh(\lambda_1 L) \\ a_{43} = -\lambda_2^2 \sin(\lambda_2 L) + k_\theta \lambda_2 \cos(\lambda_2 L) \\ a_{44} = -\lambda_2^2 \cos(\lambda_2 L) - k_\theta \lambda_2 \sin(\lambda_2 L) \end{cases} \quad (3.\text{n})$$

### 3.7.3　附录 C：表 3.4 中所示的常数

$$a = \frac{q_0 \alpha_0}{6(\alpha_0 - \beta_0^2)}, \quad b = \frac{q_0}{2c_0}, \quad d = \frac{2\alpha_0 \hat{F}_{11} - \alpha_0 q_0}{4(\beta_0^2 - \alpha_0)}, \quad g = \frac{\alpha_0 \hat{F}_{11}}{2(\beta_0^2 - \alpha_0 \beta_0^2)}$$

$$\delta = \frac{\beta_0}{\alpha_0}, \quad \varepsilon = \frac{q_0}{24} + b + \frac{\hat{F}_{11}}{2}, \quad \theta = \frac{b - d - \dfrac{a}{4}}{\dfrac{\alpha_0}{3} \dfrac{\beta_0^2 - \alpha_0}{c_0}}, \quad \theta_1 = Z \frac{d+\theta}{2}$$

$$\theta_2 = \frac{(\beta_0^2 - \alpha_0)}{c_0}\theta, \quad \eta_0 = \frac{q_0}{4} + b + \frac{\hat{F}_{11}}{2} + \frac{a}{4} - \frac{a\delta}{2}$$

$$\eta_1 = \beta_0 - \frac{2(\beta_0^2 - \alpha_0)}{\beta_0} - \frac{2}{\beta_0}\left(\frac{\alpha_0}{3} + \frac{2(\beta_0^2 - \alpha_0)}{c_0}\right), \quad \eta = \frac{\eta_0}{\eta_1}$$

$$\mu = a\delta - \hat{F}_{11} - \frac{q_0}{2} + \frac{(2\alpha_0 - \beta_0^2)}{\beta_0}, \quad \mu_1 = 2\frac{(\alpha_0 - \beta_0^2)}{\beta_0^2 c_0}\eta, \quad \mu_1 = 2\frac{(\alpha_0 - \beta_0^2)}{\beta_0^2 c_0}\eta$$

$$A_0 = \frac{q_0}{\lambda}, \quad B_0 = -\frac{q_0}{2\lambda}, \quad \gamma^2 = \frac{\lambda}{\left(1 - \frac{\beta_0^2}{\alpha_0}\right)\left(1 - \frac{\lambda}{c_0}\right)}, \quad B_1 = \frac{q_0 c_0 (c_0 + \gamma c_0 - \gamma \lambda)}{\gamma \lambda (\lambda - c_0)(c_0 - \gamma c_0 + \gamma \lambda)}$$

$$B_3 = -A_3 = -\gamma\left(1 - \frac{\lambda}{c_0}\right)B_1, \quad B_2 = -B_1 \tan\gamma - \frac{1}{\cos\gamma}\left[\frac{(\lambda + \alpha_0 \hat{F}_{11})}{\lambda \alpha_0} + \frac{q_0}{\left(1 - \frac{\lambda}{c_0}\right)\gamma\lambda}\right]$$

$$B_4 = -B_2, \quad A_1 = \gamma\left(1 - \frac{\lambda}{c_0}\right)B_2, \quad A_2 = -\gamma\left(1 - \frac{\lambda}{c_0}\right)B_1, \quad C_1 = -\gamma\left(1 - \frac{\lambda}{c_0}\right)\frac{\beta_0}{\alpha_0}B_2$$

$$C_2 = \gamma\left(\frac{1-\lambda}{c_0}\right)\frac{\beta_0}{\beta_0}B_1, \quad C_4 = -C_2, \quad C_3 = \frac{\lambda}{\alpha_0 \varphi} - \frac{q_0 \beta_0}{\alpha_0 \lambda}$$

### 3.7.4 附录D：表3.5中所示的常数

注意，那些带有质量块的变量位于无压电片梁。当压电片位于梁的中间时，其他两部分具有相同的性质：

$$A = -\frac{\hat{F}_{11}\alpha_0}{2(\alpha_0 - \beta_0^2)}, \quad A_1 = \frac{\hat{F}_{11}\beta_0}{(\alpha_0 - \beta_0^2)}, \quad B = \frac{\hat{F}_{11}\beta_0(\xi_0 - 1)}{(\xi_0 - 1)(\alpha_0 - \beta_0^2) - (\alpha_0 - \beta_0^2)}$$

$$B_1 = B\frac{\xi_0}{(\xi_0 - 1)}, \quad F = \xi_0(\beta_0 B - \bar{\beta}_0 B_1 + \hat{F}_{11}) + F_1$$

$$F_1 = (1 + \xi_0^2)\frac{\bar{\beta}_0 B_1}{2} - \xi_0^2\frac{\beta_0 B + \hat{F}_{11}}{2}, \quad \delta = \frac{\beta_0}{\alpha_0}, \quad \delta_1 = \frac{\bar{\beta}_0}{\alpha_0}, \quad D_1 = D\left(\frac{\alpha_0 - \beta_0^2}{\bar{\alpha}_0 - \bar{\beta}_0^2}\right)\frac{\bar{\alpha}_0}{\alpha_0}$$

$$C = 2D\frac{\beta_0}{\alpha_0} + A_1, \quad C_1 = A_1\xi_0\frac{\beta_0}{\alpha_0} + A_1\xi_0(2 - \xi_0)\left[\frac{\beta_0}{\alpha_0} - \frac{\bar{\beta}_0}{\bar{\alpha}_0}\frac{(\alpha_0 - \beta_0^2)}{(\bar{\alpha}_0 - \bar{\beta}_0^2)}\right]$$

$$E = 2(A - D), \quad E_1 = 2A\xi_0 + D\xi_0(2 - \xi_0)\left[1 - \frac{\bar{\alpha}_0}{\alpha_0}\frac{(\alpha_0 - \beta_0^2)}{(\bar{\alpha}_0 - \bar{\beta}_0^2)}\right], \quad I = 2D\frac{(\alpha_0 - \beta_0^2)}{\alpha_0 c_0}$$

$$J = -\frac{2}{3}D_1 - I_1, \quad I_1 = 2D_1\frac{(\bar{\alpha}_0 - \bar{\beta}_0^2)}{\alpha_0 c_0} - E_1$$

### 3.7.5 附录 E

矩阵 $A$ 的各项为

$$a_{11}=0, \quad a_{12}=1, \quad a_{13}=0, \quad a_{14}=1, \quad a_{15} \to a_{1,12}=0 \tag{3.o}$$

$$\begin{cases} a_{21}=0, \quad a_{22}=(D_{11})_1 m_{11}\lambda_{11}, \quad a_{23}=0, \quad a_{24}=(D_{11})_1 m_{21}\lambda_{21} \\ a_{25} \to a_{2,12}=0 \end{cases} \tag{3.p}$$

$$\begin{cases} a_{31}=\sinh(\lambda_{11}x_0), \quad a_{32}=\cosh(\lambda_{11}x_0), \quad a_{33}=\sin(\lambda_{21}x_0), \quad a_{34}=\cos(\lambda_{21}x_0) \\ a_{35}=-\sinh(\lambda_{12}x_0), \quad a_{36}=-\cosh(\lambda_{12}x_0), \quad a_{37}=-\sin(\lambda_{22}x_0) \\ a_{38}=-\cos(\lambda_{22}x_0), \quad a_{39} \to a_{3,12}=0 \end{cases}$$

$$\tag{3.q}$$

$$\begin{cases} a_{41}=m_{11}\cosh(\lambda_{11}x_0), \quad a_{42}=m_{11}\sinh(\lambda_{11}x_0), \quad a_{43}=-m_{22}\cos(\lambda_{21}x_0) \\ a_{44}=m_{22}\sin(\lambda_{21}x_0), \quad a_{45}=-m_{12}\cosh(\lambda_{12}x_0), \quad a_{46}=-m_{12}\sinh(\lambda_{12}x_0) \\ a_{47}=+m_{22}\cos(\lambda_{22}x_0), \quad a_{48}=-m_{22}\sin(\lambda_{22}x_0), \quad a_{49} \to a_{4,12}=0 \end{cases}$$

$$\tag{3.r}$$

## 参 考 文 献

[1] Leo, D. J., Kothera, C. and Farinholt, K., Constitutive equations for an induced-strain bending actuator with a variable substrate, Journal of Intelligent Material Systems and Structures 14, November 2003, 707-718.

[2] Crawley, E. F. and De Luis, J., Use of piezoelectric actuators as elements of intelligent structures, AIAA Journal 25 (10), October 1987, 1373-1385.

[3] Crawley, E. F. and Anderson, E. H., Detailed models of piezoceramic actuation of beams, Journal of Intelligent Material Systems and Structures 1 (1), January 1990, 4-25.

[4] Hagood, N., Chung, W. and Von Flotow, A., Modelling of piezoelectric actuator dynamics for active structural control, Journal of Intelligent Material Systems and Structures 1, July 1990, 327-354.

[5] Wang, K., Modeling of piezoelectric generator on a vibrating beam. For completion of Class Project in ME 5984 smart Materials, Virginia Polytechnic Institute and State University, April 2001.

[6] Wang, X., Ehlers, C. and Neitzel, M., An analytical investigation of static models of piezoelectric patches attached to beams and plates, smart Materials and Structures 6, 1997, 204-213.

[7] Wang, L., Ruixiang, B. and Cheng, Y., Interfacial debonding behavior of composite beam/plates with PZT patch, Composite Structures 92 (6), 2010, 1410-1415.

[8] Tylikowski, A., Influence of bonding layer on piezoelectric actuators on a axisymmetric annular plate, Journal of Theoretical and Applied Mechanics 3 (38), 2000, 607-621.

[9] Pietrzakovski, M., Dynamic model of beam-piezoceramic actuator coupling for active vibration control,

Journal of Theoretical and Applied Mechanics 1 (35), 1997, 3-20.

[10] Galichyan, T. A. and Filippov, D. A., The influence of the adhesive bonding on the magnetoelectric effect in bilayer magnetostrictive-piezoelectric structure, Journal of Physics: Conference Series 572, 2014, 012045, 1-6.

[11] Golub, M. V., Buethe, I., Shpak, A. N., Fritzen, C. P., Jung, H. and Moll, J., Analysis of Lamb wave excitation by the partly de-bonded circular piezoelectric wafer active sensors, 11th European Conference on Non-Destructive Testing (ECNDT 2014), October 6-10, 2014, Prague, Czech Republic.

[12] Pohl, J., Willberg, C., Gabbert, U. and Mook, G., Experimental and theoretical analysis of Lamb wave generation by piezoceramic actuators for structural health monitoring, Experimental Mechanics 52, 429.

[13] Chopra, I. and Sirohi, J., smart Structures Theory, Cambridge University Press, 2014, 920.

[14] Crawley, E. F. and Anderson, E. H., Detailed models of piezoceramic actuation of beams, Journal of Intelligent Material Systems and Structures 1 (1), 1990, 4-25.

[15] Wu, K. and Janocha, H., Optimal thickness and depth for embedded piezoelectric actuators, Proceedings of the Third European Conference on Structural Control, 3ECSC, 12-15 July 2004, Vienna University of Technology, Vienna, Austria.

[16] Zehetner, C. and Irschik, H., On the static and dynamic stability of beams with an axial piezoelectric actuation, smart Structures and Systems 4 (1), 2008, 67-84.

[17] Sirohi, J. and Chopra, I., Fundamental understanding of piezoelectric strain sensors, Journal of Intelligent Material Systems and Structures 11 (4), April 2000, 246-257.

[18] Sirohi, J. and Chopra, I., Fundamental behavior of piezoceramic sheet actuators, Journal of Intelligent Material Systems and Structures 11 (1), January 2000, 47-61.

[19] Reddy, J. N., Mechanics of Laminated Composite Plates and Shells: Theory and Analysis, 2nd edn. CRC Press LLC, 2004, 831.

[20] Reddy, J. N., Energy Principles and Variational Methods in Engineering, John Wiley and Sons, August 2002, 608.

[21] Magrab, E. B., Vibrations of Elastic Structural Members, Sijthoff & Noordhoff, 1979, 390.

[22] Abramovich, H. and Livshits, A, Dynamic behavior of cross-ply laminated beams with piezoelectric layers, Composite Structures 25 (1-4), 1993, 371-379.

[23] Abramovich, H. and Livshits, A, Free vibrations of non-symmetric cross-ply laminated composite beams, Jouranl of Sound and Vibration 176 (5), 1994, 597-612.

[24] Miller, S. E. and Abramovich, H., A self-sensing piezolaminated actuator model for shells using a first order shear deformation theory, Journal of Intelligent Material Systems and Structures 6 (5), 1995, 624-638.

[25] Pletner, B. and Abramovich, H, Adaptive suspensions of vehicles using piezoelectric sensors, Journal of Intelligent Material Systems and Structures 6 (6), November 1995, 744-756.

[26] Abramovich, H. and Pletner, B, Actuation and sensing of piezolaminated sandwich type structures, Composite Structures 38, October 1997, 17-27.

[27] Eisenberger, M. and Abramovich, H, Shape control of non-symmetric piezolaminated composite beams, Composite Structures 38, October 1997, 565-571.

[28] Abramovich, H. and Meyer-Piening, H.-R, Induced vibrations of piezolaminated elastic beams, Com-

posite Structures 43, 1998, 47-55.

[29] Abramovich, H. and Livshits, A, Flexural vibrations of piezolaminated slender beams: a balanced model, Journal of Vibration and Control 8 (8), November 2002, 1105-1121.

[30] Abramovich, H, Piezoelectric actuation for smart sandwich structures-closed form solutions, Journal of Sandwich Structures & Materials 5 (4), 2003, 377-396.

[31] Edery-Azulay, L. and Abramovich, H, Piezoelectric actuation and sensing mechanisms-closed form solutions, Composite Structures 64 (3-4), June 2004, 443-453.

[32] Waisman, H. and Abramovich, H, Active stiffening of laminated composite beams using piezoelectric actuators, Composite Structures 58 (1), October 2002, 109-120.

[33] Waisman, H. and Abramovich, H, Variation of natural frequencies of beams using the active stiffening effect, Composites Part B: Engineering 33 (6), September 2002, 415-424.

[34] Abramovich, H, A new insight on vibrations and buckling of a cantilevered beam under a constant piezoelectric actuation, Composite Structures 93, 2011, 1054-1057.

[35] Smart Material Corporation, http://www.smart-material.com.

[36] Edery-Azulay, L. and Abramovich, H., A reliable plain solution for rectangular plates with piezoceramic patches, Journal of Intelligent Material Systems and Structures 18, 2007, 419-433.

[37] Kerr, A. D., An extension of the Kantorovich method, Quarterly of Applied Mathematics 26, 1968, 219-229.

[38] Kerr, A. D. and Alexander, H., An application of the extended Kantorovich method to the stress analysis of a clamped rectangular plate, Acta Mechanica 6, 1968, 180-196.

[39] Yuan, S. and Jin, Y., Computation of elastic buckling loads of rectangular thin plates using the extended Kantorovich method., Composite Structures 66, 1998, 861-867.

[40] Ungbhakorn, V. and Singhatanadgid, P., Buckling analysis of symmetrically laminated composite plates by the extended Kantorovich method, Composite Structures 73, 2006, 120-128.

[41] Aghdam, M. M. and Falahatgar, S. R., Bending analysis of thick laminated plates using extended Kantorovich method, Composite Structures 62, 2003, 279-283.

[42] Yuan, S., Jin, Y. and Williams, F. W., Bending analysis of Mindlin plates by extended Kantorovich method, Journal of Engineering Mechanics 124, 1998, 1339-1345.

[43] Lin, C. C., Hsu, C. Y. and Huang, H. N., Finite element analysis on deflection control of plate with piezoelectric actuators, Composite Structures 35 (4), 1996, 423-433.

[44] Mitchell, J. A. and Reddy, J. N., A refined hybrid plate theory for composite laminates with piezoelectric laminae, International Journal of Solids and Structures 32, 1995, 2345-2367.

[45] Robaldo, A., Carrera, E. and Benjeddou, A., A unified formulation for finite element analysis of piezoelectric adaptive plates. Seventh International Conference on Computational Structures Technology, Lisbon Portugal, 7-9 September, 2004.

[46] Shah, D. K., Joshi, S. P. and Chan, W. S., Static structural response of plates with piezoceramic layers, smart Materials and Structures 2, 1993, 172-180.

[47] Vel, S. S. and Batra, R. C., Exact solution for the cylindrical bending of laminated plates with embedded piezoelectric shear actuators, smart Materials and Structures 10, 2000, 240-251.

[48] Vel, S. S. and Batra, R. C., Exact solution for rectangular sandwich plates with embedded piezoelectric

shear actuators, AIAA Journal 39, 2001, 1363-1373.

[49] Vel, S. S. and Batra, R. C., Analysis of piezoelectric bimorphs and plates with segmented actuators, Thin Walled Structures 39, 2001, 23-44.

[50] Vinson, J. R. and Sierakowski, R. L., Solid Mechanics and its Applications. The Behavior of Structures Composed of Composite Materials, 2nd edn., Kluwer Academic Publishers, 1986, 435.

[51] Zhang, X. D. and Sun, C. T., Analysis of a sandwich plate containing a piezoelectric core, smart Materials and Structures 8, 1999, 31-40.

[52] Wolf, K.-D., Electromechanical energy conversion in asymmetric piezoelectric bending actuators, Ph. D. thesis, D17, Technical University of Darmstadt, Darmstadt, Germany, 2000, 65

[53] Kim, M., Kim, J. and Cao, W., Electromechanical coupling coefficient of an ultrasonic array element, Journal of Applied Physics 99, Paper Id: 074102, 20067. doi: 10.1063/1.2180487.

[54] Neubauer, M., Schwarzendhal, S. M. and Wallaschek, J., A new solution for the determination of the generalized coupling coefficient for piezoelectric systems, Journal of Vibroengineering 14 (1), 2012, 105-110.

[55] Cheng, S. and Arnold, D. P., Defining the coupling coefficient for electrodynamic transducers, The Journal of the Acoustical Society of America 134 (5), 2013, 3561-3572. doi: 10.1121/1.4824347.

[56] Uchino, K., The development of piezoelectric materials and the new perspectives, Chapter 1, In: Advanced Piezoelectric Materials-Science and Technology, 2nd edn., Uchino, K. (ed,), Copyright © 2017 Elsevier Ltd, 93. doi: http://dx.doi.org/10.1016/B978-0-08-102134-4.00001-1.

[57] Lustig, S. and Elata, D., Ambiguous definitions of the piezoelectric coupling factor, Journal of Intelligent Material Systems and Structures 31 (4), 2020, 1689-1696.

# 第 4 章 形状记忆合金

## 4.1 形状记忆合金的基本性能

形状记忆合金是经过适当的热处理和/或机械处理后，能够恢复到之前形状或尺寸的金属材料。通常，这些材料会在较低温度下发生塑性变形，而一旦暴露在较高温度下，就会恢复到变形前的形状。习惯上将那些仅因加热而呈现"形状记忆"的材料称为单向形状记忆合金。其他具有双向形状记忆合金性能的材料在重新冷却后会发生形状变化。这些材料通常会表现出另一种现象，称为超弹性，即去除应力后，表现出较大的弹性应变（6%~7%）并发生弹性恢复，不会留下任何塑性应变。形状记忆和超弹性这两种现象都是由于存在两种冶金物相：在较高温度下的母相奥氏体（Austenite，A）和较低温度下的第 2 相马氏体（Martensite，M）（图 4.1）。

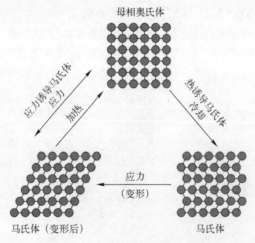

图 4.1 形状记忆效应示意图

从发展历史看，1932 年，瑞典研究人员 Arne Olander 观察到了金-镉合金（Au-Cd）的形状和恢复性能，并指出它实际上产生了运动[1]。1950 年，哥伦

## 第4章 形状记忆合金

比亚大学的Chang和Read利用X射线从微观上观察到了这一不寻常的运动，记录了金-镉合金晶体结构的变化[2]。根据这项研究，铟-钛等其他类似合金被发现。之后的1963年，美国海军武器实验室（Naval Ordinance Laboratory, NOL）的Buehler及其同事发现了镍钛诺合金（镍钛诺）的形状记忆效应[3-4]。表4.1列出了部分形状记忆合金材料及其相关性能[5]。

表4.1 具有形状记忆效应的合金[5]

| 合金 | 组成 | 相变温度范围 | | 相变滞后 | |
|---|---|---|---|---|---|
| | | ℃ | ℉ | Δ℃ | Δ℉ |
| AS-Cd | 44/49 at.%①Cd | −190~−50 | −310~−60 | ≈15 | ≈25 |
| Au-Cd | 46.5/50 at.% Cd | 30~100 | 85~212 | ≈15 | ≈25 |
| Cu-Al-Ni | 14/14.5 wt.%② Al3/4.5 wt.% Ni | −140~100 | −220~212 | ≈35 | ≈65 |
| Cu-Sn | ≈15at.% Sn | −120~30 | −185~85 | — | — |
| Cu-Zn | 38.5/41.5wt.% Zn | −180~−10 | −290~15 | ≈10 | ≈20 |
| Cu-Zn-$X$ ($X$=Si, Sn, Al) | 少量 wt.% X | −180~200 | −290~390 | ≈10 | ≈20 |
| In-Ti | 18/23at.%Ti | 60~100 | 140~212 | ≈4 | ≈7 |
| Ni-Al | 36/38at.% Al | −180~100 | −290~212 | ≈10 | ≈20 |
| Ni-Ti | 49/51at.% Ni | −50~110 | −60~230 | ≈30 | ≈55 |
| Fe-Pt | ≈25at.% Pt | ≈−130 | ≈−200 | ≈4 | ≈7 |
| Mn-Cu | 5/35at.% Cu | −250~180 | −420~355 | ≈25 | ≈45 |
| Fe-Mn-Si | 32wt.% Mn, 6wt.% Si | −200~150 | −330~300 | ≈100 | ≈180 |

① at.%，原子百分比。
② wt.%，重量百分比。

如上所述，形状记忆合金的两相为奥氏体和马氏体，各自有其不同的晶体结构，会产生不同的性能。奥氏体的晶体结构通常为立方晶体，马氏体的晶体结构可为四方晶体、正交晶体或单斜晶体。由于晶格剪切畸变，母相（奥氏体）会转变为马氏体，这种转变也称为马氏体相变。形成的每个马氏体晶体可能具有不同的取向方向，称为变体[6]。马氏体变体的组合通常可能有两种形式：一种是由"自适应"马氏体变体组合成的孪晶马氏体（$M_t$），另一种是某种变体占主导地位的退孪晶（再定向）马氏体（$M_d$）。这种从奥氏体（母相）到马氏体（产物相）或从马氏体到奥氏体的可逆相变使形状记忆合金具备了独特行为。

对于仅冷却材料、不施加机械载荷的简单情况，形状记忆合金晶体结构会从奥氏体转变为马氏体,也称为正向转变。这种转变会形成数种马氏体变体(如镍钛诺高达24种变体),它们的空间排列引起了可忽略不计的平均宏观尺寸变化,产生了孪晶马氏体。从马氏体相加热,使晶体转变回奥氏体相,称为反向转变,同样没有相关的形状变化。

从奥氏体到马氏体的相变及其恢复有4种与相变相关的特征温度。对于正向转变,在没有机械载荷的情况下,材料开始向孪晶马氏体转变的温度称为马氏体起始温度($M_s$),而转变完成的温度称为马氏体结束温度($M_f$)。在此温度及低于该温度时,所有材料均为孪晶马氏体相。当开始进行反向转变并加热时,奥氏体相转变的起始温度为奥氏体起始温度($A_s$),在奥氏体结束温度($A_f$)发生完全转变。图4.2所示为4种转变温度的示意图。

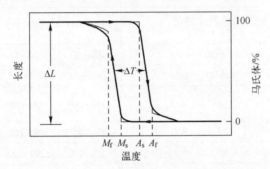

图4.2 形状记忆合金试样的典型相变-温度曲线

目前为止,上述内容讨论了形状记忆合金材料上未施加机械载荷情况下的相变。如果在材料处于孪晶马氏体相时施加载荷,即在低温下,通过重新定向某些变体可能会引入诱导退孪晶过程。与没有晶体形状变化的孪晶过程相比,退孪晶过程会引起宏观形状的变化,释放载荷产生变形。对变形材料不施加任何机械载荷的情况下,将形状记忆合金加热至$A_f$以上的温度,会使其从退孪晶马氏体转变为奥氏体,并呈现完全的形状恢复;冷却到$M_f$以下的温度,会形成没有任何形状变化的孪晶马氏体,这个过程有时称为形状记忆效应。

值得注意的是,启动退孪晶过程所施加的载荷必须高于退孪晶初始所需的最小应力,称为退孪晶启动应力($\sigma_s$)。马氏体退孪晶过程完成时所达到的应力称为退孪晶结束应力($\sigma_f$)。同样,当材料在奥氏体相下施加大于$\sigma_s$的应力时冷却,相变将会产生退孪晶马氏体,进而诱发形变。在应力持续施加的同时,重新加热材料将会出现形状恢复。应注意的是,相变温度将随着施加应力的增加而上升(无论是拉应力还是压应力)。

# 第 4 章 形状记忆合金

另一个有趣的现象是，当材料处于奥氏体相时，仅对材料施加足够大的机械应力就能诱发相变。这种高应力将使奥氏体产生完全孪晶马氏体。如果其温度高于 $A_f$，当卸载到奥氏体相时，可获得完全的形状恢复。这种行为通常称为形状记忆合金的超弹性效应（Pseudoelasticity Effect，PE）（图 4.3）。如图 4.3 所示，马氏体转变启动时加载的应力为 $M_s$ 下的 $\sigma$，而 $M_f$ 下的 $\sigma$ 表示过程的结束。对于卸载过程，$A_s$ 处的 $\sigma$ 是形状记忆合金材料奥氏体相开始时的应力，而 $A_f$ 下的 $\sigma$ 是形状记忆合金材料完成向奥氏体相反向相变时的应力。应注意，对于高于 $M_s$ 温度但低于 $A_f$ 温度的奥氏体相下的材料，预计仅可恢复部分形状。图 4.4 所示为奥氏体、马氏体和超弹性行为下的典型应力-应变曲线，温度分别为 $T_1$、$T_2$ 和 $T_3$，并且 $T_1 > T_3 > T_2$。

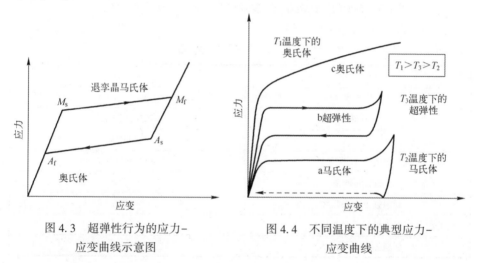

图 4.3 超弹性行为的应力-应变曲线示意图

图 4.4 不同温度下的典型应力-应变曲线

## 4.1.1 形状记忆效应

如上所述，形状记忆合金所展示的第 1 个效应是当它在孪晶马氏体相中变形、然后在 $A_s$ 温度以下卸载时的形状记忆。从该形状开始，加热到 $A_f$ 温度以上时，形状记忆合金将通过转变回母相奥氏体相而恢复其原始形状。形状记忆合金的这一特性如图 4.5 中的三维（3D）曲线所示[6]，该曲线以轴向应力 $\sigma$ 的形式描绘了镍钛诺合金在单轴载荷下的实验行为（镍钛诺合金的特性见表 4.2）。

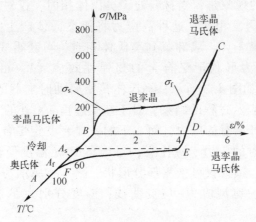

图 4.5 典型的镍钛诺合金实验加载路径显示了其形状记忆效应[6]

表 4.2 镍钛诺合金的性质①

| 密 度 | 6.45g/cm³ |
|---|---|
| 热导率 | 10W/(m·K) |
| 比热 | 322J/(kg·K) |
| 潜热 | 24200J/kg |
| 极限抗拉强度 | 750~900MPa |
| 断裂延伸率 | 15.5% |
| 屈服强度（奥氏体） | 560MPa |
| 杨氏模量（奥氏体） | 75GPa |
| 屈服强度（马氏体） | 100MPa |

① www.tiniaerospace.com。

根据图 4.5，该形状记忆过程从 $A$ 点开始（母相奥氏体）。在无外加机械力的情况下，将材料冷却到 $M_s$ 和 $M_f$ 以下，会形成孪晶马氏体，从而形成图上的 $B$ 点。从 $B$ 点开始，孪晶马氏体承受的应力会超过其起始应力 $\sigma_s$，诱发再取向过程——退孪晶过程。需要注意的是，此时的应力水平远低于马氏体的永久塑性屈服应力。退孪晶过程将在应力 $\sigma_f$ 处完成（位于图 4.5 所示 $\sigma$-$\varepsilon$ 图中的平台末端）。在不加热的情况下，将形状记忆合金的应力从 $C$ 点弹性卸载至零应力 $D$ 点，保持其退孪晶马氏体状态。如果加热开始，在无机械应力的情况下，逆相变由 $A_s$（$E$ 点）开始，并于 $A_f$（$F$ 点）结束。在任意高于 $A_f$ 的温度下，材料将处于奥氏体相，形状记忆合金将恢复其原始形状（$A$ 点）。上述整个过程称为单程形状记忆效应,因为在材料因施加机械应力而被退孪晶后,通过加热实现其形状的恢复。

当形状记忆合金在无机械应力下暴露于循环热载荷时,可能会发生重复的形状变化,这种性能称为双程形状记忆效应或训练。这一过程会引起材料微观结构的变化,导致材料性能的宏观永久性改变。有时需要对形状记忆合金进行训练,以获得滞后响应稳定且非弹性应变饱和的材料。

### 4.1.2 超弹性

形状记忆合金表现出的第 2 个效应是超弹性行为,该行为与应力诱导的转变有关,会导致加载期间产生大的应变,该应变会在高于 $A_f$ 温度下通过应力卸载降低至零。正如 Kumar 和 Lagoudas 所指出的,超弹性过程可以通过图 4.6 中的路径 1($a{\to}b{\to}c{\to}d{\to}e{\to}a$)来描述,或如图 4.6 中的路径 2 描述,该路径在高于 $A_f$ 的标称恒定温度下进行[6]。

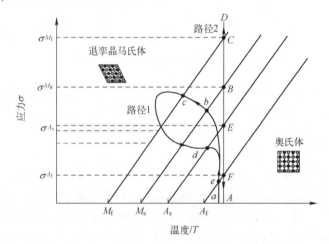

图 4.6 具有两条超弹性加载路径的典型相图[6]

根据图 4.7,超弹性行为过程从 A 点 $A_f$ 以上温度的零应力处开始。当材料处于其母体(奥氏体)相时,施加弹性载荷使曲线到达 B 点。曲线与应力 $\sigma^{M_s}$ 相交,即开始转化为马氏体相图。这个过程伴随着大的非弹性应变的产生(图 4.7)。曲线继续到 C 点,其特征是应力 $\sigma^{M_f}$,此时整个材料处于马氏体相。

在 C 点处,可以观察到与马氏体相弹性载荷相关的 $\sigma$-$\varepsilon$ 曲线斜率的明显变化。机械应力的进一步施加只会导致退孪晶马氏体的弹性变形(从 C 到 D 的路径)。D 点具有最大应力和相关应变的特征。从 D 点开始进入卸载过程,马氏体相所受应力被弹性卸载至 $\sigma^{A_s}$（E 点），于此处开始恢复为奥氏体相。从 D 点到 E 点，应变在没有发生永久塑性应变下可恢复。点 F 以 $\sigma^{A_f}$ 为特征，是曲线与奥氏体相的弹性区域相交的点。将形状记忆合金的应力进一步卸载至零，使材料回到原点 A，此时无任何残余应变。

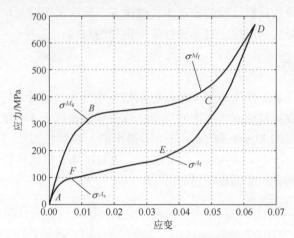

图 4.7 典型形状记忆合金超弹性加载循环[6]

应该注意的是，材料通过一个完整的超弹性循环，将形成一条滞后曲线，代表该循环中消耗的能量。滞后的大小取决于应力水平和测试条件。

### 4.1.3 应用

形状记忆合金的突出性能正吸引着从医学到航空航天应用等许多科学和工程领域研究人员的兴趣。Petrini 和 Migliavacca[7] 围绕形状记忆合金在医学领域的应用撰写了一篇有趣的综述。他们认为，形状记忆合金的形状记忆效应、超弹性、良好的耐腐蚀性和抗弯曲性、生物和磁共振兼容性，可以解释为何过去 20 年形状记忆合金材料在牙科、骨科、血管、神经和外科领域的生物医学设备生产中的巨大市场渗透率。此类应用的典型案例包括形状记忆合金正畸线（图 4.8）、使用镍钛诺制成的钢板治疗骨科问题（图 4.9）以及镍钛诺合金器械在血管领域的各种使用，如支架和过滤器（图 4.10（a）~（c））。读者可以在文献 [7] 中找到形状记忆合金在生物医学领域的更多应用。

尽管以镍钛诺合金为重点的形状记忆合金材料的生物医学应用被认为是该材料的主要用途，但人们也可在机械、航空航天和土木工程等其他工业领域找到涉及形状记忆合金的有趣创新设计。这些应用包括紧固装置、致动器、形状记忆合金致动阀、弹簧、阻尼装置、变形齿（Variable Geometry Chevron，

## 第4章 形状记忆合金

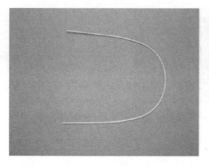
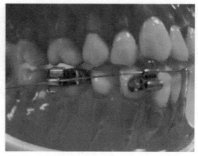

图 4.8 典型形状记忆合金正畸线[7]

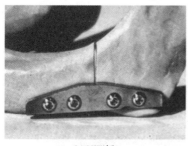
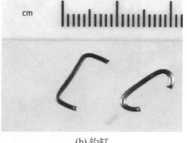

(a) NiTi板　　　　　　　　　(b) 钩钉

图 4.9 用于下颌骨骨折的 NiTi 板[8]和加热前后的钩钉[9]

VGC)、可重构转子叶片、主动铰链销致动器、可变面积喷嘴、锁定机构、热开关,以及许多其他解决实际工业问题的创新设计。图 4.11~图 4.14 重点展示了上述的一些应用。图 4.11 所示为波音公司的 VGC,这是一种旨在降低飞机发动机噪声变形航空航天结构。CDI 航空航天工程公司①展示了一种使用形状记忆合金材料变形机翼的有趣应用(图 4.12)。图 4.13 所示为带有形状记忆合金线的飞机机翼示意图,该形状记忆合金线可以通过施加电压加热来改变形状。这种机械装置可能会取代航空航天工业中现有的传统液压和机电致动器。Pitt 等[11]在 NASA 领导的 SAMPSON 智能进气道项目概述中描述了形状记忆合金线的另一个应用。形状记忆合金线与 Flexskin 或保形模线技术(Conformal Moldline Techndogy,CMT)相结合,这是由波音公司开发的一种旨在提供总体形状变化所需的结构灵活性或符合性的技术,同时也提供结构刚度和光滑的表面。Flexskin 是一种结构杆贯穿其中的弹性板。形状记忆合金线集成的一些细节如图 4.14 所示。

---

① www.continuum-dynamics.com。

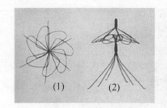  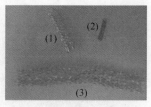

(a)(1)静脉过滤器和(2)Simon过滤器　　(b)室间隔缺损封堵器　　(c)(1)颈动脉支架、(2)冠状动脉支架以及(3)股动脉支架

图 4.10　NiTi 在血管领域的应用[7]

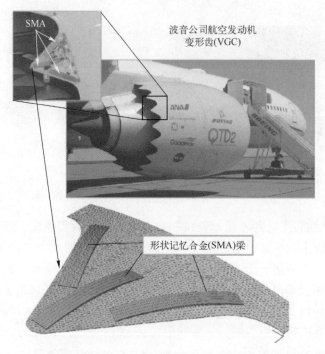

图 4.11　波音公司的变形齿，一种用于喷气式飞机降噪的变形航空结构[10]

  Autosplice 公司①开发了一种带有形状记忆合金线的致动器，该致动器在电子释放之前提供"锁存"控制和锁定。Autosplice 公司声称该致动器在实际应用中也可以提供"突发控制"（图 4.15）。隶属韩国仁和大学的智能结构和系统实验室 $S^3Lab$② 展示了形状记忆合金材料在硬盘驱动器悬架和机器人微夹持

---

① corp. autosplice. com。

② ssslab. com。

器中的有趣应用，如图 4.16（a）和（b）所示。隶属意大利 Pavia 大学土木工程与建筑系的计算力学和先进材料研究组[①]也设计与实现了基于形状记忆合金的装置，如图 4.17 所示。

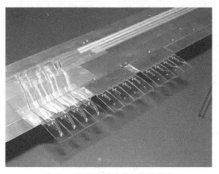
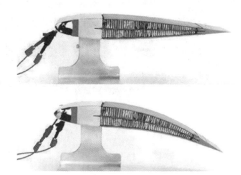

(a) 安装在转子叶片上的CDI形状记忆合金驱动翼片

(b) CDI/洛克希德·马丁公司形状记忆合金控制的连续变形机翼截面

图 4.12　变形机翼

(a) 形状记忆合金致动襟翼　　　　　(b) 传统襟翼

图 4.13　改变机翼形状

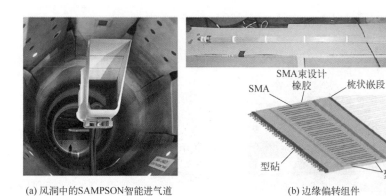

(a) 风洞中的SAMPSON智能进气道　　(b) 边缘偏转组件

图 4.14　集成 CMT/形状记忆合金杆的 NASA SAMPSON 智能进气道项目

---

① www-2.unipv.it/compmech/mat_const_mod.html。

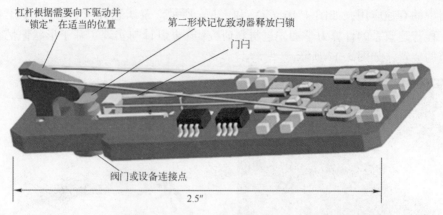

图 4.15 基于形状记忆合金的致动器（引自 Autosplice 公司）

控制输入=0.1A；最大位移=0.1mm
(a) 硬盘驱动器悬挂

加持力=7.5mN；带宽=2Hz
(b) 机器人微夹持器

图 4.16 $S^3Lab$ 的基于形状记忆合金的装置

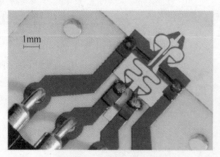

(a) 形状记忆合金微夹持器

(b) 形状记忆合金旋转致动器

图 4.17 计算力学与先进材料研究组的形状记忆合金装置
（资料来源：计算力学与先进材料研究组，www-2.unipv.it/compmech/mat_const_mod_html）

## 第4章 形状记忆合金

隶属德国卡尔斯鲁厄理工学院的微结构技术研究所[①]展示了一种单稳态形状记忆合金基微型阀（图4.18），其外壳由聚合物制成，带有一个集成的流体室、一个薄膜和一个微球致使偏转的形状记忆合金微型致动器。在零电流下，微型阀打开，当供应压力存在时允许流体流过阀门。电加热微型致动器会改变其平面形状记忆状态，从而关闭微型阀。

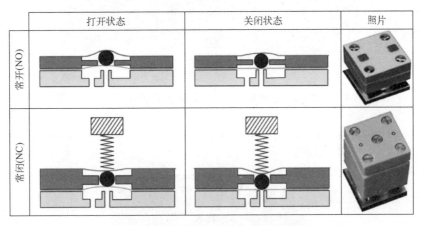

图4.18 德国Karlsruhe理工学院微结构技术研究所的基于单稳态形状记忆合金微型阀

中国制造集团上海棣朱实业有限公司[②]制造了各种形状记忆合金弹簧。在其他合金中，他们正在制造用于汽车和工业用途的双向NiTi形状记忆合金弹簧，其线直径为0.1~10mm（图4.19）。另一个有趣的应用则是在由佐治亚理工学院[③]的教师、研究生和本科生研究的NEESR-RC项目中使用镍钛诺合金弹簧和贝氏垫圈用作阻尼器。

该项目的目的是用形状记忆合金建造的阻尼器（图4.20）改造现有房屋和建筑物，以实现更好的抗震保护。基于形状记忆合金的系统被认为具有延性和能量耗散特性，可有效防止倒塌，并能够显著减少地震发生后的残余变形。一种用于建筑物中支撑构件的拉伸/压缩装置被开发出来，该装置的设计允许在压缩过程中使用镍钛诺合金螺旋弹簧或贝氏垫圈。结果表明，镍钛诺合金螺旋弹簧具有良好的再定中心和阻尼性能，而镍钛诺合金贝氏垫圈具有作为镍钛诺合金阻尼装置基础的良好潜力。

---

[①] www.imt.kit.edu/english/1528.php。
[②] http://zuudee.com/TiNi␣SMAAlloys/shape␣memory␣alloy␣NITI␣ASTM␣F2063-5.html。
[③] neesrcr.gatech.edu。

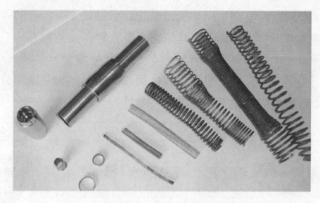

图 4.19 Zuudee 控股集团生产的各种 NiTi 形状记忆合金弹簧

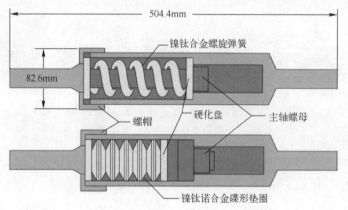

图 4.20 NEESR-RC 项目中设计和制造的两个
基于形状记忆合金的阻尼器

## 4.2 本构方程

如 4.1 节所述,形状记忆合金具有形状记忆效应及超弹性两个特性。自形状记忆效应被 Greninger 和 Mooradian[12]、Buehler 等[3-4]分别在 Cu-Zn 合金、镍钛诺合金中发现以来,镍钛诺合金因其大载荷容量、高可恢复应变(高达 8%)、优异的疲劳性能和相变引起的可变弹性,已被广泛用作智能材料。

许多本构关系被提出来模拟这些特定的性质,如唯象模型[13-15]、微观力学模型以及基于微平面理论的多晶形状记忆合金 3D 模型[16]。

## 第4章 形状记忆合金

在本节中,将介绍形状记忆合金的一维(1D)本构模型,并根据文献中发表的现有经验方法来概述该模型为拟合实验数据而进行的各种调整。最后,将描述一种利用电流加热形状记忆合金材料的方法。

### 4.2.1 形状记忆合金材料的一维本构方程

Tanaka[13-14]证明了Clausius-Duhem不等式①成立的充分条件,即

$$\sigma(\varepsilon, T, \xi) = \rho_0 \frac{\partial \Phi}{\partial \varepsilon} \tag{4.1}$$

式中:应力$\sigma$为应变$\varepsilon$、温度$T$、马氏体分数$\xi$的函数;$\rho_0$和$\Phi$分别为材料的密度和亥姆霍兹自由能②。

通过微分,式(4.1)可以改写为

$$d\sigma = \frac{\partial \sigma}{\partial \varepsilon} d\varepsilon + \frac{\partial \sigma}{\partial T} dT + \frac{\partial \sigma}{\partial \xi} d\xi \equiv E d\varepsilon + \Theta dT + \Omega d\xi \tag{4.2}$$

式中:$E = \rho_0 \frac{\partial^2 \Phi}{\partial \varepsilon^2}$为形状记忆合金的模量;$\Theta = \rho_0 \frac{\partial^2 \Phi}{\partial \varepsilon \partial T}$与热膨胀系数有关;$\Omega = \rho_0 \frac{\partial^2 \Phi}{\partial \varepsilon \partial \xi}$为转变系数。

应该注意的是,根据该模型,仅有的状态变量包括单轴应变$\varepsilon$、温度$T$,以及马氏体相体积分数$\xi$。

通过简化式(4.2)中给出的本构模型,可以得

$$(\sigma - \sigma_0) = E(\xi)(\varepsilon - \varepsilon_0) + \Theta(T - T_0) + \Omega(\xi)(\xi - \xi_0) \tag{4.3}$$

式中:$\sigma_0$、$\varepsilon_0$、$T_0$以及$\xi_0$为材料初始状态的应力、应变、温度和马氏体相体积分数。杨氏模量$E(\xi)$和相变系数$\Omega(\xi)$是关于马氏体体积分数$\xi$的函数,其函数形式可以写成

$$\begin{cases} E(\xi) = E_A + \xi(E_M - E_A) \\ \Omega(\xi) = -\varepsilon_L E(\xi) \end{cases} \tag{4.4}$$

式中:$E_A$和$E_M$分别为奥氏体相和马氏体相的杨氏模量;$\varepsilon_L$为最大可恢复应变。Tanaka关于马氏体体积分数的表达式由耗散势确定,而耗散势则取决于应力和温度,其表达式具有指数形式[13-14]。

---

① Clausius-Duhem不等式是表达连续介质力学中使用的热力学第二定律的一种方法,对确定材料的本构关系在热力学上是否可行时特别有用。
② 在热力学中,亥姆霍兹(Helmholtz)自由能是一种热力学势,它测量在恒温下从封闭的热力学系统获得的"有用"功。

对于从奥氏体到马氏体的转变（即冷却），相体积分数的公式为

$$\xi_{A \to M} = 1 - e^{[\alpha_M(M_S - T) + \beta_M \sigma]} \tag{4.5}$$

而马氏体向奥氏体转变（即加热）时的相体积分数公式为

$$\xi_{M \to A} = e^{[\alpha_A(A_S - T) + \beta_A \sigma]} \tag{4.6}$$

式（4.5）和式（4.6）中出现的材料常数被定义为

$$\begin{cases} \alpha_A = \dfrac{\ln 0.01}{A_S - A_f}, & \beta_A = \dfrac{\alpha_A}{C_A} \\ \alpha_M = \dfrac{\ln 0.01}{M_S - M_f}, & \beta_M = \dfrac{\alpha_M}{C_M} \end{cases} \tag{4.7}$$

每种材料的系数 $E$、$\Theta$、$\Omega$ 以及其他参数，如 $M_s$、$M_f$、$A_s$、$A_f$、$C_A$ 和 $C_M$ 均通过实验测得。

Tanaka 模型的临界应力-温度曲线如图 4.21（a）所示。

Liang 和 Rogers[17]对 Tanaka 提出的一维本构方程进行了进一步的改进。他们用余弦公式给出了马氏体的体积分数。因此，他们关于奥氏体向马氏体转变（即冷却）的方程具有以下形式：

$$\xi_{A \to M} = \frac{1 - \xi_0}{2} \cos[a_M(T - M_f) + b_M \sigma] + \frac{1 + \xi_0}{2} \tag{4.8}$$

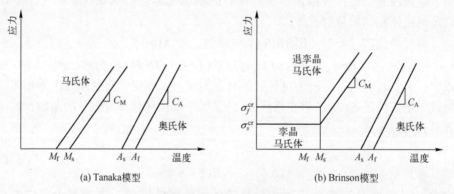

(a) Tanaka模型    (b) Brinson模型

图 4.21　应力-温度曲线

而对于从马氏体到奥氏体的转变（即加热），他们的公式则是

$$\xi_{M \to A} = \frac{\xi_0}{2} \cos[a_A(T - A_s) + b_A \sigma] + \frac{\xi_0}{2} \tag{4.9}$$

式（4.8）和式（4.9）中出现的材料常数定义为

$$a_\mathrm{M} = \frac{\pi}{(M_\mathrm{s}-M_\mathrm{f})}, \quad b_\mathrm{M} = -\frac{a_\mathrm{M}}{C_\mathrm{M}}$$
$$a_\mathrm{A} = \frac{\pi}{(A_\mathrm{f}-A_\mathrm{s})}, \quad b_\mathrm{A} = -\frac{a_\mathrm{A}}{C_\mathrm{A}}$$
(4.10)

式中：$\xi_0$ 为初始马氏体体积分数。需要注意的是，余弦函数的参数值可介于 0 和 π 之间，进而对温度和应力产生限制，即对于从奥氏体相到马氏体相的转变，需要

$$M_\mathrm{f} \leqslant T \leqslant M_\mathrm{s}$$

且 $\quad C_\mathrm{M}(T-M_\mathrm{f}) - \dfrac{\pi}{|b_\mathrm{M}|} \leqslant \sigma \leqslant C_\mathrm{M}(T-M_\mathrm{f})$ (4.11)

而对于马氏体向奥氏体的转变（即加热），其限制条件则是

$$A_\mathrm{s} \leqslant T \leqslant A_\mathrm{f}$$

且 $\quad C_\mathrm{A}(T-A_\mathrm{s}) - \dfrac{\pi}{|b_\mathrm{A}|} \leqslant \sigma \leqslant C_\mathrm{A}(T-A_\mathrm{s})$ (4.12)

需要注意的是，Tanaka 以及 Liang 和 Rogers 提出的模型都只能正确地描述从马氏体到奥氏体的相变及其逆转变，即导致超弹性的应力诱导马氏体相变。低温下的形状记忆效应是由应力诱导马氏体和温度诱导马氏体之间的转换引起的，因此这些模型无法应用于描述马氏体的退孪晶化，而马氏体的退孪晶化正是产生形状记忆现象的原因[18]。

为了解决这一不足，Brinson 开发了一个新模型，将马氏体体积分数分为应力诱导马氏体分数 $\xi_\mathrm{S}$ 和温度诱导马氏体分数 $\xi_\mathrm{T}$，从而得[15,19]

$$\xi = \xi_\mathrm{S} + \xi_\mathrm{T} \tag{4.13}$$

Brinson 模型的原始本构方程是由 Tanaka 的本构方程（式（4.3））略有修改后得到的，并写为

$$\sigma - \sigma_0 = E(\xi)\varepsilon - E(\xi_0)\varepsilon_0 + \Theta(T-T_0) + \Omega(\xi)\xi - \Omega(\xi_0)\xi_0 \tag{4.14}$$

简化该本构方程[19]可得出以下形式：

$$\sigma = E(\xi)(\varepsilon - \varepsilon_\mathrm{L}\xi_\mathrm{S}) + \Theta(T-T_0) \tag{4.15}$$

为了包含温度低于 $M_\mathrm{s}$ 的形状记忆合金，Brinson 模型对 Liang 和 Roger 的转变相公式（式（4.8）和式（4.9））进行了修改，以包含两种类型的马氏体体积分数，即 $\xi_\mathrm{S}$ 和 $\xi_\mathrm{T}$[15]。图 4.21（b）展示了两种体积分数转变的临界应力随温度变化的示意图。退孪晶马氏体的转变现在写为

$$\begin{cases} \xi_S = \dfrac{1-\xi_{S_0}}{2}\cos\left[\dfrac{\pi}{\sigma_s^{cr}-\sigma_f^{cr}}\left[(\sigma-\sigma_f^{cr})-C_M(T-M_s)\right]\right]+\dfrac{1+\xi_{S_0}}{2} \\ \xi_T = \xi_{T_0} - \dfrac{\xi_{T_0}}{1-\xi_{S_0}}(\xi_S-\xi_{S_0}) \end{cases} \quad (4.16)$$

假设温度和应力为

$$\begin{cases} T>M_s \\ \sigma_s^{cr}+C_M(T-M_s)<\sigma<\sigma_f^{cr}+C_M(T-M_s) \end{cases} \quad (4.17)$$

对于

$$\begin{aligned} &T<M_s \\ &\text{且}\quad \sigma_s^{cr}<\sigma<\sigma_f^{cr} \end{aligned} \quad (4.18)$$

可得到以下表达式：

$$\begin{cases} \xi_S = \dfrac{1-\xi_{S_0}}{2}\cos\left[\dfrac{\pi}{\sigma_s^{cr}-\sigma_f^{cr}}(\sigma-\sigma_f^{cr})\right]+\dfrac{1+\xi_{S_0}}{2} \\ \xi_T = \xi_{T_0} - \dfrac{\xi_{T_0}}{1-\xi_{S_0}}(\xi_S-\xi_{S_0})+\Delta T \end{cases} \quad (4.19)$$

$\Delta T_\xi$ 项将具有以下值：

若 $M_f<T<M_s$，且 $T<T_0$

则 $\quad \Delta T_\xi = \dfrac{1-\xi_{T_0}}{2}\{[\cos[a_M(T-M_f)]+1]\} \quad (4.20)$

否则 $\quad \Delta T_\xi = 0$

到母相奥氏体的反向转化由下式给出：

$$\begin{cases} \xi = \dfrac{\xi_0}{2}\left\{\cos\left[a_A\left(T-A_s-\dfrac{\sigma}{C_A}\right)\right]+1\right\} \\ \xi_S = \xi_{S_0} - \dfrac{\xi_{S_0}}{\xi_0}(\xi_0-\xi),\quad \xi_T = \xi_{T_0} - \dfrac{\xi_{T_0}}{\xi_0}(\xi_0-\xi) \end{cases} \quad (4.21)$$

应该注意的是，某些热机械载荷产生 $\xi>1$ 的不可接受马氏体分数，此时 Brinson 模型显示的演化动力学对于这些热机械载荷可能是不正确的。为了解决这一问题，Chung 等[20]将临界应力-温度曲线分为 8 个阶段（图 4.22）：

阶段 1：从奥氏体或温度诱导马氏体向应力诱导马氏体的转变。

阶段 2：温度/应力诱导马氏体和奥氏体的混合物（无转变）。

阶段 3：从马氏体到奥氏体的转变。

阶段4：纯奥氏体（无转变）。

阶段5：从奥氏体或温度诱导马氏体向应力诱导马氏体的转变，以及从奥氏体向温度诱导马氏体的转变。

阶段6：从奥氏体向温度诱导马氏体的转变。

阶段7：从温度诱导马氏体向应力诱导马氏体的转变。

阶段8：应力/温度诱导马氏体的混合物（无转变）。

需要注意的是，Brinson模型在阶段5有一个"弱点"。为了解决这个问题，Chung等在以下5个约束条件下提出了一个修正模型：

（1）在任何情况下，总的马氏体体积分数应符合以下约束条件：$\xi = \xi_S + \xi_T \leq 1$。

（2）当$\sigma = \sigma_f^{cr}(\xi_S = 1)$时，应力诱导马氏体的体积分数必须为1。

（3）当$T = M_f(\xi = \xi_S + \xi_T = 1)$时，总的马氏体体积分数必须为1。

（4）阶段5中的转化动力学必须与阶段1的相连续（图4.22）。

（5）马氏体体积分数的函数都是余弦形式。

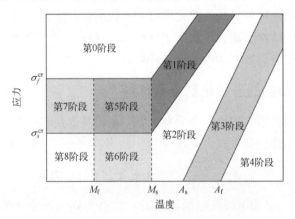

图4.22 分为8个阶段的临界应力-温度曲线[20]

需要注意的是，当只降低温度或只增加应力时，Brinson模型满足上述所有条件。对于同时改变应力和温度的情况，Brinson模型只满足条件（4）和（5）。为了符合上述5个条件，对式（4.18）~式（4.20）进行了修改，并在如下等式[20]中给出了转换方程，前提是临界应力在$M_s$以下为常数：

对于：

$$\begin{cases} T < M_s \\ 且 \quad \sigma_s^{cr} < \sigma < \sigma_f^{cr} \end{cases} \quad (4.22)$$

将得到以下表达式：

$$\begin{cases} \xi_S = \dfrac{1-\xi_{S_0}}{2}\cos\left[\dfrac{\pi}{\sigma_s^{cr}-\sigma_f^{cr}}(\sigma-\sigma_f^{cr})\right]+\dfrac{1+\xi_{S_0}}{2} \\ \xi_T = \Delta T_\xi - \dfrac{\Delta T_\xi}{1-\xi_{S_0}}(\xi_S-\xi_{S_0}) \end{cases} \quad (4.23)$$

$\Delta T_\xi$ 将具有以下值：

若 $M_f < T < M_s$ 且 $T < T_0$

则 $\Delta T_\xi = \dfrac{1-\xi_{S_0}-\xi_{T_0}}{2}\cos\left[\dfrac{\pi}{M_s-M_f}(T-M_f)\right]+\dfrac{1-\xi_{S_0}+\xi_{T_0}}{2}$ (4.24)

否则 $\Delta T_\xi = \xi_{T_0}$

式（4.21）~式（4.24）中的修正解决了上述5个条件，并且可以证明修正后的 Brinson 模型将对所有初始条件给出一致的结果[20]。应该注意的是，早在1998年，Bekker 等就已经对形状记忆合金中一般热机械载荷引起的热弹性相变过程中马氏体体积分数的演化进行了一致的数学描述，得到了一个不允许体积分数超过1的稳定动力学模型[21]。然而，Chung 等在原始 Brinson 模型的基础上添加的模型更好地体现了这种现象。

### 4.2.2 电流加热的形状记忆合金材料

直接加热线状形状记忆合金的方法之一是将其连接到电源上。这种加热类型也称为焦耳加热，其等式由文献［22］给出：

$$(\rho A)c_p \frac{dT(t)}{dt} = I^2 R - h_c A_{circ}[T(t)-T_0] \quad (4.25)$$

式中：$I$ 为电流；$R$ 为单位长度的线电阻；$T(t)$ 为随时间变化的温度；$T_0$ 为环境温度；$\rho A = m$ 表示单位长度线的质量（$\rho$ 表示密度，$A$ 表示线的横截面积）；$c_p$ 为比热；$h_c$ 和 $A_{circ}$ 分别为传热系数和单位长度线的周向面积。假设电流和环境温度均为常数，且不随时间变化，可通过求解式（4.25）得出以下温度随时间变化的解：

$$\begin{cases} T(t)-T_0 = \dfrac{I^2 R}{h_c A_{circ}}(1-e^{-t/t_{ht}})+(T_{start}-T_0)e^{-t/t_{ht}} \\ t_{ht} = \dfrac{(\rho A)c_p}{h_c A_{circ}} \end{cases} \quad (4.26)$$

式中：$T_{start}$ 为加热开始时的温度；$t_{ht}$ 为传热过程的时间常数。对于 $T_{start} = T_0$ 的

情况，得

$$T(t)-T_0=\frac{I^2R}{h_c A_{\text{circ}}}(1-\mathrm{e}^{-t/t_{\text{ht}}}) \qquad (4.27)$$

然后，线上的待测量稳态温度为

$$T_{\text{ss}}=\frac{I^2R}{h_c A_{\text{circ}}}(1-\mathrm{e}^{-t/t_{\text{ht}}})+T_0 \qquad (4.28)$$

通过分析式（4.27）中的指数函数，可以计算当 $t=3t_{\text{ht}}$ 时，温度约为 $T_{\text{ss}}$ 的95%。那么，达到预定温度 $T_{\text{p}}$（从而加热线）所需的时间 $t_{\text{p}}$ 为

$$t_{\text{p}}=-t_{\text{ht}}\ln\frac{T_{\text{ss}}-T_{\text{p}}}{T_{\text{ss}}-T_0} \qquad (4.29)$$

对于线的冷却情况，假设式（4.26）中的 $I=0$，得

$$T(t)-T_0=(T_{\text{start}}-T_0)\mathrm{e}^{-t/t_{\text{ht}}} \qquad (4.30)$$

然后，稳态温度将为环境温度，而达到预定温度的时间将根据以下公式计算：

$$t_{\text{p}}=-t_{\text{ht}}\ln\frac{T_{\text{p}}-T_0}{T_{\text{start}}-T_0} \qquad (4.31)$$

一旦可以估算出加热和冷却的时间，下一个需要解决的问题就是需要向镍钛诺合金线加载什么能量。给定电流 $I(\mathrm{A})$ 和电阻 $R(\Omega)$ 的功率由下式给出：

$$P=I^2R \qquad (4.32)$$

然后，使用以下公式计算电阻 $R$：

$$R=\rho_{\text{r}}\frac{l}{A} \qquad (4.33)$$

式中：$\rho_{\text{r}}$ 为线的电阻率（镍钛诺合金的线电阻值为 $7.6\times10^{-5}\Omega/\mathrm{cm}$）；$l$ 和 $A$ 分别为线的长度和横截面积。将功率乘以时间 $t$ 可得到将给定线从起始温度 $T_{\text{start}}$ 加热到预定温度 $T_{\text{p}}$ 所需的能量 $E$，即

$$E=I^2R\cdot t=c_{\text{p}}\cdot M\cdot(T_{\text{p}}-T_0)+L\cdot M \qquad (4.34)$$

式中：$M$ 是线的质量（g）；$c_{\text{p}}$ 和 $L$ 分别为线的比热和相变潜热①（J/g·C；J/g）。用式（4.33）将电阻表达式代入式（4.34），并求解电流 $I$，得

$$I=\sqrt{\frac{c_{\text{p}}\cdot M\cdot(T_{\text{p}}-T_0)+L\cdot M}{\dfrac{\rho_{\text{r}}l}{A}t}} \qquad (4.35)$$

---

① 镍钛诺的比热 $c_{\text{p}}$ 为 0.8368J/(g·C)，相变潜热 $L$ 为 20J/g。

因此，对于已知长度和横截面的给定质量的线，当材料特性 $c_p$、$L$ 以及 $\rho_r$ 已知时，可以计算在时间 $t$ 内将温度从 $T_0$ 升高到预定温度 $T_p$ 所需的电流 $I$。

## 4.3 文献中的形状记忆合金模型

在本节中，除了 4.2 节展示的 Tanaka、Liang 和 Rogers 以及 Brinson 的模型，将对文献中出现的形状记忆合金模型进行综述。目的是强调形状记忆合金材料的复杂性以及研究人员试图分析和形成分析模型的方式，以尽可能地接近对形状记忆合金材料进行的各种测试。

Cisse 等最近对用于描述形状记忆合金性能的各种模型进行了综述，列出了文献中的大量最新模型[23]。Barbarino 等的另一篇综述在介绍了各种用于描述形状记忆合金材料的本构模型之后，还详细列出了形状记忆合金材料在机翼变形领域中的应用[24]。Paiva 和 Savi 概述了形状记忆合金的本构模型，他们将模型分为 5 类，以此来体现形状记忆合金的一般热机械行为，包括超弹性、由于温度变化而产生的形状记忆效应和相变以及不完全相变[25]。这篇综述包括大量关于本节讨论主题的参考文献。Brocca 等基于静态约束微平面理论开发了一个新模型，得出每个平面只产生 1D 本构关系[26]。Qianhua 等基于超弹性 NiTi 合金的实验结果提出了一个 3D 热机械本构模型，该模型体现了相变硬化、逆转变、奥氏体相和马氏体相之间的弹性失配以及每个相效应的杨氏模量的温度依赖性[27]。Arghavani 等研究了形状记忆合金的 3D 行为，并提出了一个唯象本构模型，该模型将纯再取向机制与纯转变机制完全解耦，通过数值试验再现了成比例和不成比例载荷下形状记忆合金的主要特征，并指出了数值实验与文献中可用的实验结果之间有良好的相关性[28]。

Lagoudas 等开发了一个基于总比吉布斯自由能和耗散势的统一热力模型，因而能够在一个共同的框架下涵盖各种唯象模型[29]。根据他们模型的公式，总应变依赖于马氏体体积分数 $\xi$ 的机械部分 $\varepsilon_{ij}$ 和相变应变张量 $\varepsilon_{ij}^{tr}(\xi)$。因此，广义胡克定律可以写成

$$\sigma_{ij} = C_{ijkl}[\varepsilon_{kl} - \varepsilon_{kl}^{tr} - \alpha_{kl}(T - T_0)] \quad (4.36)$$

式中：$C_{ijkl}$ 为弹性刚度矩阵；$T$ 为温度；$\alpha_{kl}$ 为热膨胀系数。需要注意的是，为了简单起见，他们的统一模型中省略了退孪晶效应[27]。不同模型之间的差异将由转变-硬化函数 $f(\xi, \varepsilon_{ij}^t)$①的具体选择所引起，该函数物理上表示了由于马

---

① $\xi$ 为马氏体体积分数，$\varepsilon_{ij}^t$ 为转化应变矢量。

氏体变体和周围母相之间以及马氏体变体本身之间的相互作用而产生的弹性应变能。在他们的工作中，采用了在有限元中以简单的方式收敛的线性硬化函数，如 Lagoudas 等所描述的，假设硬化函数 $f$ 与转变应变张量 $\varepsilon_{ij}^t$ 是不相关的，则意味着不存在运动转变硬化[29]。那么，函数的以下属性假定保持不变：

（1）如果没有施加外部机械载荷，母相奥氏体是不受应力的。假设 $f(0)=0$，即对于完全奥氏体相，函数值为零，因而满足之前的条件。

（2）因为函数代表了材料中储存的部分弹性应变能，所以它必须为正值。为了满足该条件，应该选择合适的材料常数。

（3）对于所有可能的加载路径，该函数在相变过程中必须是连续的，包括返回点。

基于上述条件，硬化函数 $f(\xi)$ 的选择如下：

$$\begin{cases} f(\xi)=f^M(\xi), & \xi>0 \\ f(\xi)=f^A(\xi), & \xi<0 \end{cases} \tag{4.37}$$

式中：函数 $f^M(\xi)$ 和 $f^A(\xi)$ 被定义为

$$\begin{cases} f^M(\xi)=f^{M_0}(\xi)+\dfrac{1-\xi}{1-\xi^R}[f^A(\xi^R)-f^{M_0}(\xi^R)] \\ f^A(\xi)=f^{A_0}(\xi)+\dfrac{\xi}{\xi^R}[f^M(\xi^R)-f^{A_0}(\xi^R)] \end{cases} \tag{4.38}$$

并且 $\xi^R$ 被定义为返回点的马氏体体积分数，而函数 $f^{M_0}(\xi)$ 和 $f^{A_0}(\xi)$ 必须根据每个选定模型来进行选择。正相变（$\xi>0$）由 $\xi^R \leqslant \xi \leqslant 1$ 给出，而逆相变（$\xi<0$）被定义为 $0 \leqslant \xi \leqslant \xi^R$。

需要注意的是，$f(\xi)$ 的这种形式要满足上述所有（1）~（3）这 3 个条件：

$$\begin{cases} f^{M_0}(0)=0, \ f^{A_0}(0)=0, \ f^{M_0}(\xi) \geqslant 0, \ f^{A_0}(\xi) \geqslant 0, \ \text{其中} \ 0 \leqslant \xi \leqslant 1 \\ \text{且} \ f^{M_0}(1)=f^{A_0}(1) \end{cases} \tag{4.39}$$

对于 Tanaka 的模型[15]（指数模型），必须选择以下 $f^{M_0}(\xi)$ 和 $f^{A_0}(\xi)$ 函数：

$$\begin{cases} f^{M_0}(\xi)=\dfrac{\rho \Delta s_0}{a_e^M}[(1-\xi)\ln(1-\xi)+\xi]+(\mu_1^e+\mu_2^e)\xi \\ f^{A_0}(\xi)=-\dfrac{\rho \Delta s_0}{a_e^A}\xi[\ln(\xi)-1]+(\mu_1^e-\mu_2^e)\xi \end{cases} \tag{4.40}$$

式中：$\rho$ 为密度，且 $\Delta s_0 = s_0^M - s_0^A$，$s_0^M$ 和 $s_0^A$ 分别为马氏体相和奥氏体相的比熵。对于 Liang 和 Rogers 的模型[20]，函数为

$$\begin{cases} f^{M_0}(\xi) = -\int_0^\xi \dfrac{\rho \Delta s_0}{a_c^M}[\pi - \arccos(2\xi - 1)]\mathrm{d}\xi + (\mu_1^c + \mu_2^c)\xi \\ f^{A_0}(\xi) = -\int_0^\xi \dfrac{\rho \Delta s_0}{a_c^A}[\pi - \arccos(2\xi - 1)]\mathrm{d}\xi + (\mu_1^c - \mu_2^c)\xi \end{cases} \quad (4.41)$$

式中：$a_e^M$、$a_e^A$、$a_c^M$、$a_c^A$、$\mu_1^e$、$\mu_1^c$ 均为材料的常数，而参数 $\mu_1^e$、$\mu_1^c$ 可根据连续性条件 $f^{M_0}(1) = f^{A_0}(1)$ 来确定。

若想要用多项式表示直到二次项的 $f(\xi)$，则它们的形式为

$$\begin{cases} f^{M_0}(\xi) = \dfrac{1}{2}\rho b^M \xi^2 + (\mu_1^p + \mu_2^p)\xi \\ f^{A_0}(\xi) = \dfrac{1}{2}\rho b^A \xi^2 + (\mu_1^p - \mu_2^p)\xi \end{cases} \quad (4.42)$$

式中：$b^M$ 和 $b^A$ 分别为正相变和逆相变的线性各向同性硬化模量。引入了参数 $\mu_2$ 来考虑 $\xi=1$ 时的连续性条件。根据所选模型，将得出如下表达式：

$$\begin{cases} \mu_2^e = \dfrac{\rho \Delta s_0}{2}\left(\dfrac{1}{a^A} - \dfrac{1}{a^M}\right): \text{指数模式} \\ \mu_2^c = \dfrac{\pi\rho \Delta s_0}{4}\left(\dfrac{1}{a_c^M} - \dfrac{1}{a_c^A}\right): \text{余弦模式} \\ \mu_2^p = \dfrac{\rho}{2}(b^A - b^M): \text{多项式模式} \end{cases} \quad (4.43)$$

三种不同模型的各种材料常数如表 4.3 所列[29]。

表 4.3 在 $M^{0s}$、$M^{0f}$、$A^{0s}$ 和 $A^{0f}$ 已知的前提下三种模型的材料常数

| 指 数 模 型 | 余 弦 模 型 | 多项式模型 |
|---|---|---|
| $a_e^A = \dfrac{\ln(0.01)}{A^{0s} - A^{0f}}$ | $a_c^A = \dfrac{\pi}{A^{0f} - A^{0s}}$ | $\rho b^A = -\rho \Delta s_o (A^{0f} - A^{0s})$ |
| $a_e^M = \dfrac{\ln(0.01)}{M^{0s} - M^{0f}}$ | $a_c^M = \dfrac{\pi}{M^{0s} - M^{0f}}$ | $\rho b^M = -\rho \Delta s_o (M^{0f} - M^{0s})$ |
| $\gamma^e = \dfrac{\rho \Delta s_o}{2}(M^{0s} + 2A^{0f} - A^{0s})$ | $\gamma^c = \dfrac{\rho \Delta s_o}{2}(M^{0s} + A^{0f})$ | $\gamma^p = \dfrac{\rho \Delta s_o}{2}(M^{0s} + A^{0f})$ |
| $\gamma_e^* = -\dfrac{\rho \Delta s_o}{2}(A^{0s} - M^{0s}) + \dfrac{\rho \Delta s_o}{2\ln(0.01)}(M^{0s} - M^{0f} + A^{0f} - A^{0s})$ | $\gamma_c^* = -\dfrac{\rho \Delta s_o}{2}(A^{0f} - M^{0s}) - \dfrac{\rho \Delta s_o}{4}(M^{0s} - M^{0f} - A^{0f} + A^{0s})$ | $\gamma_p^* = -\dfrac{\rho \Delta s_o}{2}(A^{0f} - M^{0s}) + \dfrac{\rho \Delta s_o}{4}(M^{0s} - M^{0f} - A^{0f} + A^{0s})$ |

文献 [29] 表明,热力学力的基本等式具有如下形式:

$$\bar{\pi} = \sigma_{ij}^{\text{eff}} \Lambda_{ij} + \frac{1}{2}\Delta S_{ijkl}\sigma_{ij}\sigma_{kl} + \Delta\alpha_{ij}\sigma_{ij}\Delta T + \rho\Delta c\left[\Delta T - T\ln\left(\frac{T}{T_0}\right)\right] + \\ \rho\Delta S_0 T - \frac{\partial f(\xi)}{\partial \xi} - \rho\Delta u_0 = \pm Y^* \quad (4.44)$$

式中:$\bar{\pi}$ 是与 $\xi$ 共轭的热力学力,式(4.44)中出现的各种参数被定义为

$$\Delta S_{ijkl} = S_{ijkl}^{\text{M}} - S_{ijkl}^{\text{A}}, \quad \Delta\alpha_{ij} = \alpha_{ij}^{\text{M}} - \alpha_{ij}^{\text{A}}, \Delta c = c^{\text{M}} - c^{\text{A}} \\ \Delta s_0 = s_0^{\text{M}} - s_{i0}^{\text{A}}, \Delta u_0 = u_0^{\text{M}} - u_0^{\text{A}}, \Delta T = T - T_0, Y^* = \sqrt{2Y} \quad (4.45)$$

需要注意的是,在式(4.44)中加号用于正相变,而减号则用于逆相变。此外,式(4.45)可用于分析内部状态变量 $\xi$ 的演化。材料常数 $T^*$ 可以看作相变开始的热力学力 $\bar{\pi}$ 的阈值。为证明统一模型给出结果与文献中现有各种模型相类似,式(4.44)在插入适当的 $f(\xi)$ 函数表达式后可用于 Liang 和 Rogers 的余弦模型,从而得

$$\sigma_{ij}^{\text{eff}} \Lambda_{ij} + \frac{1}{2}\Delta S_{ijkl}\sigma_{ij}\sigma_{kl} + \Delta\alpha_{ij}\sigma_{ij}\Delta T + \rho\Delta c\left[\Delta T - T\ln\left(\frac{T}{T_0}\right)\right] + \rho\Delta s_0 T \\ -\frac{\rho\Delta s_0}{a_c^{\text{M}}}[\arccos(2\xi-1)-\pi] - (\mu_1^c + \mu_2^c) - \frac{f^{\text{A}}(\xi^{\text{R}}) - f^{\text{M}_0}(\xi^{\text{R}})}{1-\xi^{\text{R}}} - \rho\Delta u_0 - Y^* = 0 \quad (4.46)$$

然后,对于一个完整的循环(加载和卸载),有边界条件 $\xi^{\text{R}} = 0$ 和 $f^{\text{A}}(\xi^{\text{R}}) = f^{\text{M}_0}(\xi^{\text{R}}) = 0$。可以根据施加的应力和温度明确地得到 $\xi$ 的表达式:

$$\xi = \frac{1}{2}\left\{\cos\left[\alpha_c^{\text{M}}(T - M^{0f}) - \frac{\alpha_c^{\text{M}}}{C^{\text{M}}H}(\sigma_{ij}^{\text{eff}}\Lambda_{ij} + \frac{1}{2}\Delta S_{ijkl}\sigma_{ij}\sigma_{kl} + \Delta\alpha_{ij}\sigma_{ij}\Delta T)\right] + 1\right\} \quad (4.47)$$

需要注意的是,上述公式由表 4.1 中适当的表达式所获得,而式(4.43)中给出 $\mu_2^c$ 的表达式,并且假设 $\Delta c = 0$。式(4.47)中出现的 $C_{\text{M}}$ 称为马氏体应力影响系数,其形式如下:

$$C_{\text{M}} = -\frac{\rho\Delta s_0}{H} \quad (4.48)$$

若忽略式(4.47)中的 $\frac{1}{2}\Delta S_{ijkl}\sigma_{ij}\sigma_{kl}$ 和 $+\Delta\alpha_{ij}\sigma_{ij}\Delta T$ 项,则得到 Liang 和 Rogers 给出的形式,但下面这项除外[17]:

$$\frac{\sigma_{ij}^{\text{eff}}\Lambda_{ij}^{12}}{H}$$

该项为式（4.47）中的有效驱动应力，而不是施加的外加应力[17]。Tanaka 和 Nagaki 提出的模型可由类似的程序获得[13]。几年后，Lagoudas 等更新了这个统一模型[30]，增加了马氏体相变开始和完成时热力学响应的平滑过渡来改进原始模型[25]。另一个正在处理的特征是机械施加应力的大小对产生有利的马氏体变体的作用，而没有明确考虑马氏体重新取向，从而产生有效的计算工具。第 3 个改进涉及相变概念的临界热力学力的概括，取决于相变的方向（正向或逆向）和施加的机械应力大小。所有这 3 项改进都提供了一个高保真模型，可广泛应用于形状记忆合金材料体系。

在文献［25］中，上述模型以及 Tanaka 和 Nagaki、Liang 和 Rogers 以及 Brinson[15] 模型，均被描述为假定的相变动力学模型，该模型考虑了预先建立的简单数学函数来描述相变动力学。这些模型被认为是目前文献中最流行的模型，因此它们被用于与更多的实验结果进行比较，在形状记忆合金的性能建模中发挥着重要作用。图 4.23 显示了基于表 4.4 的 Tanaka 和 Nagaki、Liang 和 Rogers 以及 Brinson 这 3 个模型对 3 个温度预测的比较。

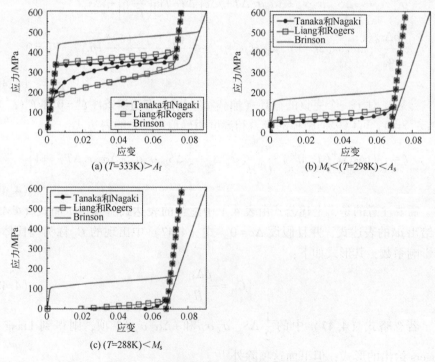

图 4.23　3 种在 3 个代表性温度下假设转化
动力学模型的应变-应力曲线

表 4.4　NiTi 的热机械材料性能[15]

| 材 料 属 性 | 相变温度/K | 模 型 参 数 |
|---|---|---|
| $E_A = 67 \times 10^3 \mathrm{MPa}$ | $M_f = 282$ | $C_M = 8\mathrm{MPa/K}$ |
| $E_M = 26.3 \times 10^3 \mathrm{MPa}$ | $M_s = 291.4$ | $C_A = 13.8\mathrm{MPa/K}$ |
| $\Omega = 0.55 \mathrm{MPa/K}$ | $A_s = 307.5$ | $\sigma_s^{\mathrm{crit}} = 100\mathrm{MPa}$ |
| $\varepsilon_R = 0.067$ | $A = 322$ | $\sigma_f^{\mathrm{crit}} = 170\mathrm{MPa}$ |

有关形状记忆合金的文献也介绍了使用其他假设的研究。Falk 及其同事[31-33]基于 Devonshire 理论[34]提出了一个相对简单的多项式模型①。根据该模型，内部变量或耗散势不是描述超弹性和形状记忆效应的必要条件。因此，应变 $\varepsilon$ 和温度 $T$ 对于该模型是必要条件。相应地，他们提出了自由能势（$\Lambda$）的 6 次多项式模型，即

$$\Lambda(\varepsilon, T) = \frac{a}{2}(T - T_M)\varepsilon^2 - \frac{b}{4}\varepsilon^4 + \frac{b^2}{24a(T_A - T_M)}\varepsilon^6 \tag{4.49}$$

式中：$T_A$ 为奥氏体稳定态时的最低温度；$T_M$ 为马氏体稳定态的最高温度；$a$ 和 $b$ 为正值材料常数。由此产生的本构方程为

$$\sigma(\varepsilon, T) = \frac{\partial \Lambda(\varepsilon, T)}{\partial \varepsilon} = a(T - T_M)\varepsilon - b\varepsilon^3 + \frac{b^2}{4a(T_A - T_M)}\varepsilon^5 \tag{4.50}$$

图 4.24（a）~（c）展示了 Falk 模型在 3 个代表性温度下应力-应变曲线[25]。图 4.24（a）和（b）展示了马氏体退孪晶过程，图 4.24（c）展示了超弹性效应。虽然 Falk 模型没有考虑孪晶效应，但 Falk 多项式以一种定性一致的方式表示了马氏体的退孪晶过程和超弹性。

其他类别的模型包括内部约束的模型。Fremond 的研究即属于这种[35-36]。他的研究考虑了 3 个体积分数、1 个奥氏体分数和 2 个由压缩和拉伸引起的一维情况下的退孪晶变量。三维模型包括应变、温度和总自由能 3 个内部变量。Paiva 等使用 Fremond 模型开发了一种新模型，该模型能够考虑不同的马氏体性质和与孪晶马氏体相关的新体积分数，同时还包括塑性-应变效应和塑性-相变耦合[37]。该模型的方程包括弹性应变、温度以及拉伸退孪晶马氏体、压缩退孪晶马氏体、奥氏体和孪晶马氏体相关的 4 个状态变量。更多详细信息可见文献［37］。

---

① 1949 年，Devonshire 发明了一种现象学理论来描述钛酸钡（$BaTiO_3$）的铁电相变和介电性质的温度依赖性。

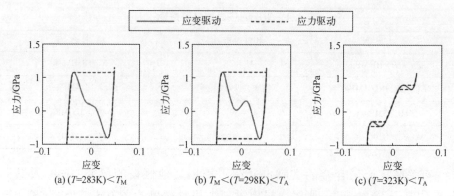

图 4.24　Falk 模型在 3 个代表性温度下应力-应变曲线

# 参 考 文 献

[1] Ölander, A., An electrochemical investigation of solid cadmium-gold alloys, Journal American Chemical Society 54 (10), October, 1932, 3819-3833. doi: 10.1021/ja01349a004.

[2] Chang, L. C. and Read, T. A., Plastic deformation and diffusion less phase changes in metals-The gold-cadmium beta-phase, Trans, AIME 191, 1951, 47-52.

[3] Buehler, W. J., Gilfrich, J. W. and Wiley, R. C., Effects of low-temperature phase changes on the mechanical properties of alloys near omposition TiNi, Journal of Applied Physics 34 (5), 1963, 1475-1477. doi: 10.1063/1.1729603.

[4] Wang, F. E., Buehler, W. J. and Pickart, S. J., Crystal structure and a unique martensitic transition of TiNi, Journal of Applied Physics 36 (10), 1965, 3232-3239. doi: 10.1063/1.1702955.

[5] Shimizu, K. and Tadaki, T., Shape memory alloys, Funakubo, H. (ed.), Gordon and Breach Science Publishers, 1987.

[6] Kumar, P. K. and Lagoudas, D. C., Introduction to shape memory alloys, In: Shape Memory Alloys, Lagoudas, D. C. (ed.), 1, Springer Science + Business Media, Vol. 7, LLC, 2008, 393. doi: 10.1007/978-0-387-47685-8.

[7] Petrini, L. and Migliavacca, F., Biomedical applications of shape memory alloys, Journal of Metallurgy 2011, article ID 501483, Hindawi Publishing Corporation. doi: 10.1155/2011/501483.

[8] Duerig, T. W., Melton, K. N., Stockel, D. and Wayman, C. M., Engineering Aspects of Shape Memory Alloys, London, UK, Butterworth-Heinemann, 1990.

[9] Laster, Z., MacBean, A. D., Ayliffe, P. R. and Newlands, L. C., Fixation of a frontozygomatic fracture with a shape-memory staple, British Journal of Oral and Maxillofacial Surgery 39 (4), 2001, 324-325.

[10] Calkins, F. T., Mabe, J. H. and Butler, G. W., Boeing's variable geometry chevron: morphing aerospace structures for jet noise reduction, SPIE Proceedings Vol. 6171: smart Structures and Materials 2006: Industrial and Commercial Applications of smart Structures Technologies, White, E. V. (ed.).

[11] Pitt, D. M., Dunne, J. P. and White, E. V., SAMPSON smart inlet design overview and wind tunnel

test Part I-Design overview, In: smart Structures and Materials 2002: Industrial and Commercial Applications of smart Structures Technologies, McGowan, A. - M. R. (ed.), Proceedings of SPIE, Vol. 4698, 2002.

[12] Greninger, A. B. and Mooradian, V. G., Strain transformation in metastable beta copper-zinc and beta copper-tin alloys, Transactions of the Metallurgical Society of AIME 128, 1938, 337-368.

[13] Tanaka, K. and Nagaki, S., A thermomechanical description of materials with internal variables in the process of phase transitions, Ingenieur-Archiv 51 (5), 1982, 287-299.

[14] Tanaka, K., A thermomechanical sketch of shape memory effect: one-dimensional tensile behavior, Res Mechanica 18 (3), 1986, 251-263.

[15] Brinson, L. C., One-dimensional constitutive behavior of shape memory alloys: thermomechanical derivation with non-constant material functions and redefined martensite internal variable, Journal of Intelligent Material Systems and Structures 4 (2), 1993, 229-242.

[16] Li, L., Li, Q. and Zhang, F., One-dimensional constitutive model of shape memory alloy with an empirical kinetics equation, Journal of Metallurgy 2011, January 2011, 1-14. article ID 563413, Hindawi Publishing Corporation.

[17] Liang, C. and Rogers, C. A., One dimensional thermomechanical constitutive relations for shape memory material, Journal of Intelligent Material Systems and Structures 1, 1990, 207-234.

[18] Prahlad, H. and Chopra, I., Comparative evaluation of shape memory alloy constitutive models with experimental data, Journal of Intelligent Material Systems and Structures 12, 2001, 383-395.

[19] Brinson, L. C. and Huang, M. S., Simplifications and comparisons of shape memory alloy constitutive models, Journal of Intelligent Material Systems and Structures 7, 1996, 108-114.

[20] Chung, J., Heo, J. and Lee, J., Implementation strategy for the dual transformation region in the Brinson SMA constitutive model, smart Materials and Structure 16 (1), 2007, N1-N5.

[21] Bekker, A. and Brinson, L. C., Phase diagram based description of the hysteresis behavior of shape memory alloys, Acta Mater 46, 1998, 3649-3665.

[22] Song, H., Kubica, E. and Gorbet, R., Resistance modelling of SMA wire actuators, International Workshop on smart Materials, Structures & NDT in Aerospace Conference NDT, 2-4 November 2011, Montreal, Quebec, Canada.

[23] Cisse, C., Zaki, W. and Zineb, T. B., A review of constitutive models and modeling techniques for shape memory alloys, International Journal of Plasticity 76, 2016, 244-284.

[24] Barbarino, S., Saavedra Flores, E. I., Ajaj, R. M., Dayyani, I. and Friswell, M. I., A review on shape memory alloys with applications to morphing aircraft, smart Materials and Structures 23 (6), 2014, article ID 063001.

[25] Paiva, A. and Savi, M. A., An overview of constitutive models for shape memory allows, Hindawi Publishing Corporation, mathematical Problems in Engineering 2006, article ID 56876, 1-30.

[26] Brocca, M., Brinson, L. C. and Bazant, Z. P., Three-dimensional constitutivemodel for shape memory alloys based on microplane model, Journal of the Mechanics and Physics of Solids 50, 2002, 1051-1077.

[27] Qianhua, K., Guozheng, K., Linmao, Q. and Sujuan, G., A temperature-dependent three dimensional super-elastic model considering plasticity for NiTi alloy, International Conference on Experimental Mechanics 2008, Xiaoyuan, H., Xie, H. and Kang, Y. (eds.), Proc. of SPIE Vol. 7375, 73755U, 2009

SPIE.

[28] Arghavani, J., Auricchio, F., Naghdabadi, R., Reali, A. and Sohrabpour, S., A 3-D phenomenological constitutive model for shape memory alloys under multiaxial loadings, International Journal of Plasticity 26, 2010, 976-991.

[29] Lagoudas, D. C., Bo, Z. and Qidwai, M. A., A unified thermodynamic constitutive model for SMA and finite element analysis of active metal matrix composites, Mechanics of Composite Materials and Structures 3 (2), June, 1996, 153-179.

[30] Lagoudas, D. C., Hartl, D., Chemisky, Y., Machado, L. and Popov, P., Constitutive model for the numerical analysis of phase transformation in polycrystalline shape memory alloys, International Journal of Plasticity 32-33, 2012, 155-183.

[31] Falk, F., Model free-energy, mechanics and thermodynamics of shape memory alloys, ACTA Metallurgica 28 (12), 1980, 1773-1780.

[32] Falk, F., One-dimensional model of shape memory alloys, Archives of Mechanics 35 (1), 1983, 63-84.

[33] Falk, F. and Konopka, P., Three-dimensional Landau theory describing the martensitic transformation of shape memory alloys, Journal de Physique 2, 1990, 61-77.

[34] Devonshire, A. F., Theory of barium titanate, Philosophical Magazine 40, 1949, 1040-1063.

[35] Fremond, M., Materiaux a Memoire de Forme, Comptes Redus Mathematique, Academie Sciences, Paris 304 (7), 1987, 239-244.

[36] Fremond, M., Shape Memory Alloy: A Thermomechanical Macroscopic Theory, CISM Courses and Lectures, Vol. 351, New York, Springer, 1996, 3-68.

[37] Paiva, A., Savi, M. A., Braga, M. B. and Pacheco, C. L., A constitutive model for shape memory alloys considering tensile-compressive asymmetry and plasticity, International Journal of Solids and Structures 42 (11-12), 2005, 3439-3457.

# 第 5 章 电流变体与磁流变体

## 5.1 电流变体与磁流变体的基本特性

电流变（ER）和磁流变（MR）材料是两种被视为智能材料的流体。这两种流体都属于非牛顿流体，因为它们不遵守牛顿流体摩擦定律（其均具有可变动力学黏度系数）。Winslow 于 1947 年获得了电流变的专利[1]，其由流体中悬浮的极化粒子组成，其表观黏度在电场作用下会发生变化。当对电流变体施加剪切力时，立即会表现出对外加电场的响应，从而会产生与施加电场强度近似成正比的屈服应力。电流变体的流动性表现为表观黏度的变化，这种特性会随着流体特性和分散粒子的大小与密度的变化而变化[2]。由于具有阻尼性能，电流变体已应用在车辆悬架、减振器和发动机支座，以及离合器、制动器和气门系统上。有关典型见文献 [3-9]。图 5.1 所示为德国达姆施塔特 Fludicon 公司制造的电流变悬架，图 5.2 所示为用于测试电流变体阀门的试验台[8]。

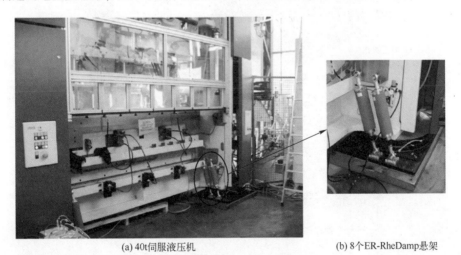

(a) 40t伺服液压机　　(b) 8个ER-RheDamp悬架

图 5.1　40t 伺服液压机的电流变机器悬架

（资料来源：德国 FLUDICON 公司）

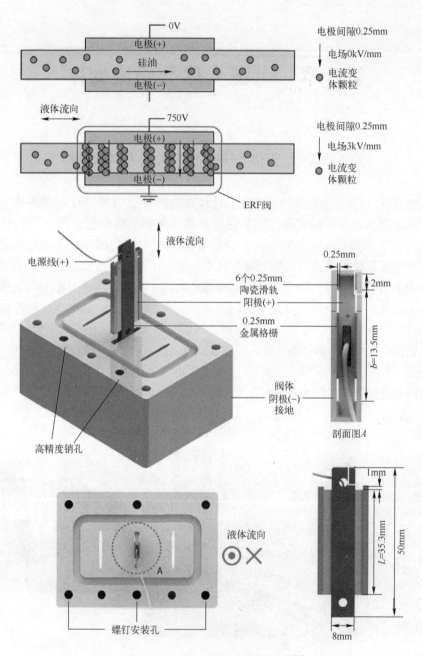

图 5.2 电流变体液压旋转致动器[8]

磁流变体是另一种功能流体，其屈服应力可以通过施加磁场来改变，从而获得比电流变体大 20~50 倍的感应屈服应力。磁流变体也被视为一种智能材料，在外加磁场的作用下能够实现从液体到近固态的可逆且快速（亚毫秒级）变换。应注意，避免将磁流变体与胶态磁流体相混淆，后者的内部颗粒比典型磁流变体内的颗粒小 $1×10^3$ 倍。当施加磁通密度为 1T 数量级的磁场时，磁流变体的表观黏度可能会发生几个数量级的变化。在需要主动控制机械系统的振动或扭矩传输的各种应用场景中，磁流变体的这种特性使其成为一种理想的选择。典型实例包括减振器、制动器、离合器、地震减振器、控制阀和人工关节等[10-16]。图 5.3 所示为一种基于磁流变的车辆悬架减振器①。该减振器设有一个内置磁流变阀，磁流变体会被强制流过该阀。磁流变减振器的活塞就像一个电磁铁，其设计具有适当数量的线圈，用于产生所需的磁场。该减振器有一个贯穿轴，用于防止磁流变体积聚。磁流变离合器如图 5.4 所示。虽然磁流变体可控性强，但其效率有限，至今尚未在汽车动力系统上应用。图 5.4 所示为 Magna 动力总成公司②如何通过将离合器设计、流体开发和磁路优化相结合创造一款高效的磁流变体离合器。

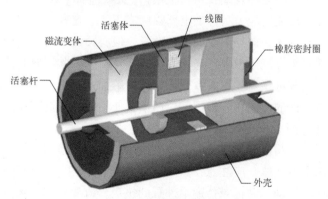

图 5.3　车辆悬架磁流变减振器
（资料来源：美国加利福尼亚大学 ISSL）

---

① 来自美国加利福尼亚大学智能结构与系统实验室（ISSL）。
② www.magna.com/capabilities/powertrain-systems。

图 5.4 磁流变离合器
（资料来源：www.atzonline.com）

为了使人们理解这一现象，Wen 等提出了电流变效应的启发式描述[17]。根据该描述，电流变体由悬浮在液体介质中的固体颗粒构成，因而具有可控的流变性。当对电流变体施加电场时，悬浮颗粒将发生极化并从正极到负电极呈链条状排列。注意，悬浮颗粒的极化是由于固体颗粒与胶体中液体之间存在介电常数差异而形成的①，这种差异会形成有效偶极子。由颗粒形成的链条被外加电场固定在适当的位置，可有效抵抗流动（图 5.5）。由于这种链条结构能够在垂直于外加电场方向上保持剪切力，从而会产生更高的黏度。因此，当流体处于电场作用下时会呈现凝胶状②流变性。文献 [10] 的研究成果表明，该凝胶的强度及其抗流动能力与外加电场强度成正比。这种启发式描述可以解释电场作用下电流变体中链条形成的视觉效应，但实际流变行为复杂很多。由于电流变特性既含有类流体现象，又含有类固体现象，因此需要对传统流变仪进行改进，并且对改进仪器收集的数据进行详细解释[10]。

对于电流变体的性能，应根据其应用模式使用圆柱形流变仪、流量装置或环形泵送工具（图 5.6~图 5.8）进行测量。

图 5.6 所示为圆柱形流变仪以及电流变体的典型试验结果。两个气缸之间的间隙内充满电流变体。应注意，这个间隙很小，为内缸半径的 1/100。使内缸以变化的速度转动，并记录传递给外缸的扭矩。改变外加电场值，重复此步

---

① 胶体是一种均匀的非晶体物质，由一种大分子或超微颗粒的物质分散在另一种物质而成。
② 凝胶（其名称源于明胶，一种半透明、无色、易碎、无味的食物，由各类动物副产品中提取的胶原蛋白制成）是一种固体、果冻状材料，具备从软到硬、从弱到强的性质。凝胶被定义为一种稀释的交联体系，在稳定状态下不流动。

骤。内缸的扭矩与转速之间的关系可转换为剪切应力与剪切应变率之间的关系，如图 5.6 所示。然后，可以通过实验评估这两个重要性质，即给定电场强度、实验曲线斜率下的屈服剪切应力 $\tau_y$ 和电流变体的塑性黏度 $\eta$（本章附录 A）。剪切应力与剪切应变率之间的关系表示如下：

$$\tau = \tau_y + \eta \dot{\gamma} \tag{5.1}$$

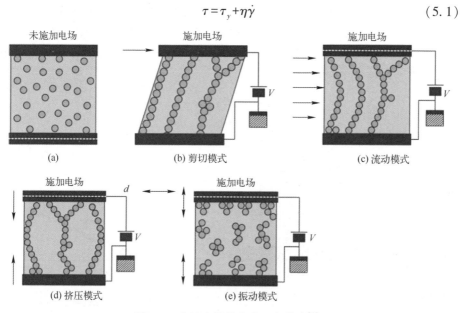

图 5.5  电流变体的各种运行模式[2]

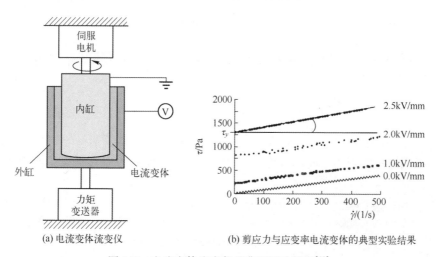

(a) 电流变体流变仪

(b) 剪应力与应变率电流变体的典型实验结果

图 5.6  电流变体流变仪及典型试验结果[10]

式（5.1）类似于具有可控屈服应力的宾汉（Bingham）塑性体①的本构关系[10]。应注意，式（5.1）仅表示电流变体中实际剪切应力和剪切应变率的近似值，但是可为工程应用和基于电流变体的各种结构设计提供有用的信息。

为了确保测量的电流变体符合宾汉塑性方程，可使用流动夹具装置进行测量，如图 5.7 所示。该装置可用于测量随外加电场变化沿通道发生的压降[10]，如图 5.7（b）所示。根据推导该装置运动方程的菲利普（Philips）②研究，沿通道长度的压降 $\Delta p$ 可表示如下：

$$\Delta p = \frac{8\eta Q L}{bh^3} + 2\frac{L}{h}\tau_y \tag{5.2}$$

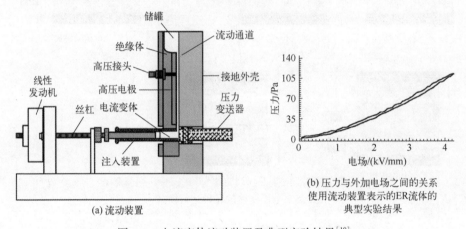

(a) 流动装置

(b) 压力与外加电场之间的关系
使用流动装置表示的ER流体的典型实验结果

图 5.7　电流变体流动装置及典型实验结果[10]

当参数 $\tau^* > 200$ 时，即对于相对较低的流速意味着非常高的屈服应力（$\tau_y$）。$\tau^*$ 的表达式如下：

$$T^* = \frac{bh^2 \tau_y}{12 Q \eta} \tag{5.3}$$

式中：$\eta$ 为流体黏度；$\tau_y$ 为屈服应力；$Q$ 为沿通道的流体流速；$b$ 为通道宽度；$h$ 为高压电极与壳体之间的间隙。对于 $\tau^* < 0.5$，即屈服应力很低或流速很高

---

① 宾汉塑性体是一种黏塑性材料，在低应力下行为类似刚体，但在高应力下则表现为黏性流体。该命名来源于尤金·宾厄姆（Eugene C. Bingham），他在下列文章中提出了数学方程形式：Bingham, E. C., An Investigation of the Laws of Plastic Flow, U. S. Bureau of Standards Bulletin, Vol. 13, 1916, 309-353.

② Phillips, R. W. Engineering Applications of Fluids with a Variable Yield Stress, Ph. D. Dissertation, University of California, Berkeley, 1969.

的情形，压降与屈服应力的关系如下：

$$\Delta p = \frac{12\eta QL}{bh^3} + 3\frac{L}{h}\tau_y \tag{5.4}$$

利用式（5.2）和式（5.4）可以得到黏度和屈服应力的值，二者均为外加电场的函数。若这些结果与使用圆柱形流变仪测得的结果不一致，则流体不是宾汉塑性体，且式（5.1）无效。注意，在大多数测量中，使用两种不同装置获得的结果是一致的，并且式（5.1）的有效性可进行验证。

第3个装置，即环形泵送工具[10]（图5.8），用于在低于屈服应力的条件（称为预屈服区）下获得试验结果。在电流变体预屈服区小应变条件下，使用该装置对复剪切模量 $G'$ 进行测量。该装置含有处于预定电场下的电流变体，并且其内缸以低于屈服应变（$\gamma_y$）的幅度进行振荡。通过改变振荡频率，可得到在不同外加电场下 $G'$ 的变化曲线，如图5.8所示。应注意，随着电场的增强，$G'$ 的值从零（当电流变材料表现为纯流体时）变为接近于柔软橡胶，这清楚地展示了电流变材料从流体到固体的转变过程。

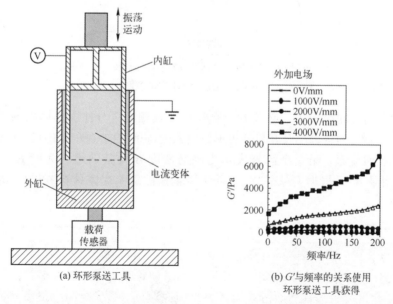

图 5.8 用电流变体环形泵送装置及典型实验结果得到的 $G'$ 值[10]

下面讨论磁流变体的特性。如前所述，磁流变体的特性类似于电流变体，只是两者在激活相关场所需的功率以及流变体（Rheological Fluid，RF）对污染物的稳定性方面存在较大差异（表5.1）。磁流变体由处于液体内的金属颗粒组成。当磁场关闭时，其行为类似于牛顿流体，如图5.9所示。当磁场开启时，磁流变体表现为宾汉流体，这是因为液体中的金属颗粒往往会产生偶极子，从而形成了与磁流变体流动方向相反的约束。在零剪切应变率条件下，宾汉流体必须先克服屈服应力 $\tau_y$ 才能发生流体运动（图5.9）。该屈服应力为外加磁场强度的函数，两者关系如下：

$$\tau = \tau_y(H) + \eta \dot{\gamma} \tag{5.5}$$

式（5.5）与电流变体的式（5.1）相同。

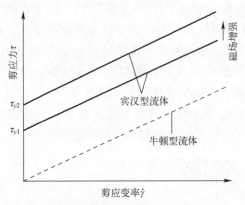

图5.9 剪切应力与剪切应变率的关系——
牛顿流体和宾汉流体的示意图

要获得高屈服应力，必须调节颗粒尺寸及颗粒百分比、外加磁场两个因素。增大颗粒的百分比及其尺寸可形成更强的链条结构（图5.10），从而导致更高的屈服应力。第2个因素表示为磁通密度 **B** 及其随磁场强度 **H** 的变化（本章附录B）。根据文献［18］，基于羰基铁的磁流变体具有100kPa的工作屈服应力。

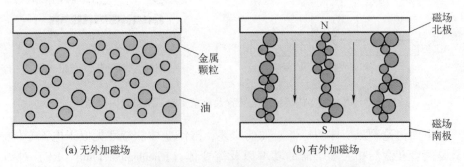

(a) 无外加磁场　　　　　(b) 有外加磁场

图5.10 磁流变体-有/无施加磁场时的行为示意图
（资料来源：www.intechopen.com）

建议使用本章附录 B 中式（5.i）所示的具有 $H=f(B)$ 关系的线性截面，这样可以大大减少迟滞问题，并获得一种简单的设计方法。图 5.11 所示为磁流变体的 3 种工作模式：流量（阀）模式、剪切模式和挤压模式。

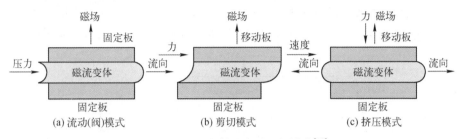

图 5.11　磁流变体的各种运行模式[18]

与上文对电流变体的描述相似，磁流变体也由提供润滑（添加剂组合）并含有金属颗粒和悬浮颗粒的液体形成。磁流变体内的液体可以是碳氢化合物、矿物油或硅基油。羰基铁、铁粉或铁/钴合金为悬浮颗粒的理想材料，其尺寸为微米级，因为这种材料在磁流变体中具有高饱和磁化强度[19]。它们体积可达液体和颗粒总体积的 50%。第 3 种物质由添加剂组成，其中包括表面活性剂和稳定剂[20]。需要使用添加剂来控制流体黏度、金属颗粒沉降速率以及颗粒间摩擦，以防止经过预定次数的使用循环后出现增稠效应。添加剂包含悬浮剂、摩擦改性剂、触变①液体和耐磨（防腐）成分，包括润滑脂、环烷酸亚铁（$C_{22}H_{14}O_4Fe$）或油酸亚铁（$C_{36}H_{66}FeO_4$）等分散剂，以及硬脂酸锂或硬脂酸钠等触变添加剂。

上述电流变体的各种装置也适用于相应的磁流变体（图 5.6 和图 5.7）。使用磁流变体构建的装置可以通过上述 3 种运行模式（图 5.11）实现特定用途。减振器或阻尼器采用了流量（阀）运行模式（图 5.11）。向减振器内部线圈施加电流（图 5.12）会产生磁场。当活塞移动时会产生一种力，进而在活塞末端形成位移或速度。

制动器和离合器使用剪切模式（图 5.11）进行操作。图 5.13 所示为各种常用制动器设计的示意图②。

第 3 种模式，即挤压模式（图 5.11），可用于阻滞低速运动和高作用力应用中的振动。虽然这种应用在文献中并不常见，但一些研究者已经设计了一些实现方法。Alghamdi 和 Olabi 在其研究论文中介绍了一种基于磁流变体挤压模

---

① 触变是流体随时间变化的剪切变稀特性。
② 来自比利时布鲁塞尔自由大学主动结构实验室，http://scmero.ulb.ac.be。

式设计的车辆减振器（图 5.14）[21]。他们使用了由 Lord 公司[22]生产的磁流变体 F-140CG，其属性如表 5.1 所列。该减振器仅需很小的功率（2~24V，1~2A，功率为 2~48W）即可提供磁场，因此已经成为该类减振器的理想选择。通过在活塞两侧施加拉伸和压缩力可产生很高的净力。研究结果表明，增加磁线圈的电流可以产生很宽的变阻尼力，因此可以在较大范围内对所需阻尼进行控制。

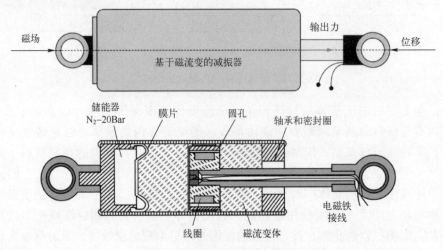

图 5.12 基于磁流变的减振器
（资料来源：www.intechopen.com）

表 5.1 文献 [20] 中使用的磁流变体特性

| 特 性 | 取值或限值 |
| --- | --- |
| 颜色 | 深灰色 |
| 磁流变体型号 | 磁流变 F-140CG |
| 基础液 | 碳氢化合物 |
| 黏度（Pa·s）（40℃时） | 0.0280±0.070 |
| 密度/(g/cm$^3$) | 3.54~3.74 |
| 固体重量含量/% | 85.44 |
| 闪点①/℃ | >150 |
| 工作温度/℃ | -40~130 |
| 颗粒类型 | 羰基铁 |
| 粒度/μm | 0.88~4.03 |

① 化学物质的闪点为流体蒸发形成可燃浓度气体的最低温度。

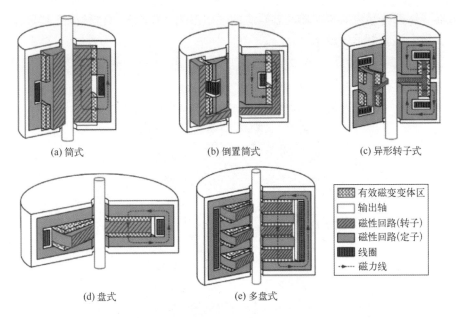

图 5.13　各种磁流变体制动器设计

(资料来源：www.scmero.ulb.be)

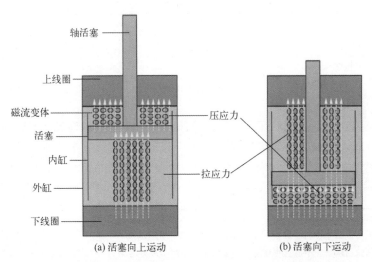

图 5.14　磁流变体减振器在挤压模式下的工作原理[21]

近期的一项研究[17]提出了另一个概念，即基于磁流变体挤压模式设计的减振器，如图 5.15 所示。该研究主要对减振器进行了有限元分析，揭示了减振器导体部分涡流对减振器动态响应的强烈影响。作者建议继续推进其研究工作，结合实验揭示该现象。

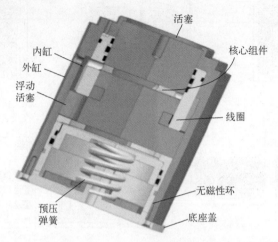

图 5.15　基于磁流变体挤压模式的减振器 CAD 模型[17]

表 5.2 列出了电流变体和磁流变体的各种特性，包括各种常用数据。

表 5.2　电流变体和磁流变体的典型特性[23]

| 特　性 | 电 流 变 体 | 磁 流 变 体 |
| --- | --- | --- |
| 颗粒材料 | 聚合物、沸石①等 | 铁磁②、亚铁磁③等 |
| 典型粒度/μm | 0.1~10 | 0.1~10 |
| 载液 | 油、介电凝胶及其他聚合物 | 水、合成油、非极性和极性液体等 |
| 密度/(g/cm³) | 1~2 | 3~5 |
| 关闭黏度/(Pa·s)（25℃时） | 0.1~0.3 | 0.1~0.3 |
| 所需磁场 | ~3kV/mm | ~3kOe④ |
| 屈服应力 $\tau_y$/kPa | 10 | 100 |
| 激活类型 | 高电压 | 电磁体或永磁体 |

① 沸石是一种微孔硅铝酸盐矿物质，通常用作商业吸附剂和催化剂。
② 铁磁性是某些材料（如铁）形成永磁体或被磁体吸引的基本机制。
③ 亚铁磁材料的原子群具有相反的不等磁矩，可产生自发磁化。
④ kOe=千奥斯特，CGI 系统中辅助磁场强度（$H$）的单位；1kOe=1 达因/麦克斯韦。

根据表 5.2 所列的磁流变体和电流变体数据及其特性，可得出以下结论：

（1）由于磁流变体的屈服应力 $\tau_y$ 比电流变体约高 10 倍，使用磁流变体能够实现最高的能量吸收水平。

(2) 磁流变体的活化时间很短（约 10ms），但活化时间会受组件几何尺寸的影响。

(3) 磁流变体对污染物和杂质不敏感（5.2 节），而电流变体对这些因素较为敏感。使用添加剂可以解决沉淀问题，从而使磁流变体的使用更加可靠和持久。

(4) 使用线圈感应永磁系统可使磁流变系统更加可靠并且降低沉降的敏感性，从而形成一种故障安全操作机制。

(5) 电流变体和磁流变体装置所需的功率差不多（接近 50W），但两者在所需电压和电流方面的固有差异使得磁流变器件更具吸引力，因为磁流变器件可以直接通过普通低压电源供电。

(6) 磁流变体远比电流变体有效。但是，必须对激活流体所需的外围设备予以考虑。

(7) 通过对固体和液体组分的密度匹配或使用纳米颗粒，可以解决电流变体悬浮液沉降时间这一主要问题。

(8) 电流变体的另一个问题是空气的击穿电压，即约 3kV/mm，该值接近其运行所需的电场强度。

## 5.2 电流变体和磁流变体建模

从工程应用角度看，电流变体的建模基于以下假设：电流变体的行为符合宾汉塑性模型（5.1 节），该模型将流体中因施加电场而产生的剪切应力与流体的黏度和屈服应力进行了关联[17,24-35]。此模型可表示如下：

$$\tau = \tau_y + \eta_{pl} \dot{\gamma} \tag{5.6}$$

式中：$\tau$ 为流体中的剪切应力；$\tau_y$ 为给定电场强度下的屈服剪切应力；$\eta_{pl}$ 为电流变体的塑性黏度；$\dot{\gamma}$ 为剪切应变率。电流变体行为的另一种表现方法是将宾汉塑性模型（pl）划分为前屈服区和后屈服区，即

$$\begin{aligned} \tau &= \eta \dot{\gamma}, & |\tau| < \tau_c \\ \tau &= \tau_y + \eta_{pl} \dot{\gamma}, & |\tau| > \tau_c \end{aligned} \tag{5.7}$$

式中：$\eta$ 为预屈服的黏度；$\tau_c$ 为标志着从预屈服区向后屈服区过渡的剪切应力阈值。应注意，除了宾汉模型，一些研究人员[36]还使用 Herschel-Bulkley 模型来描述电流变体（本章附录 C）。

式（5.6）和式（5.7）中给出的模型表明，在剪切应力作用下，在克服屈服应力之前不会发生位移。在低于屈服应力的条件下，电流变悬浮液为类固

态；但在高于屈服应力时，电流变悬浮液则表现为黏度极高的液体。

应注意，宾汉（Bingham）模型适用于均匀、稳定的流动情况，在这种情况下，瞬态或启动效应可以忽略。当瞬态行为变为主导行为或流体处于动态载荷下时（具有冲击应力且阻尼问题不能忽略），宾汉模型的表现则不会特别理想[17]。

电流变体承受正常压缩应力的能力，即在其挤压模式下工作（图5.5（d）），是工程师们非常感兴趣的另一个问题。根据 Monkman 的研究，电流变体在压缩应力作用下的特性与牛顿流体相似[25]。但是，当外加电场密度达到一定值时，宾汉塑性模型效应会突然发生。这种现象可以通过增大施加电压或施加力来减小间隙宽度（图5.16，在这种情况下，电流变体层在两个电极之间会受到压缩，仅允许其垂直运动），这两种方式均会导致电极之间电场强度增大。Monkman 对各种类型的电流变体进行了大量测试。图5.17所示为在氯化石蜡（50LV 系列）中使用 30%聚甲基丙烯酸锂悬浮液时，在 4 种不同的施加电压下，其间隙与轴向压缩应力 $\sigma$ 之间的典型关系曲线[27]。应注意，图中仅展现了在压缩应力作用下形成的可测量硬化区域（曲线转折点后方）。可以看出，随着间隙的缩小，稳态区域表示从牛顿流体到表观固态的相变完成。宾汉塑性现象开始后，位移为塑性且不可逆，而电场保持不变。电流变体将返回其原始液体状态，然后解除外加电场。

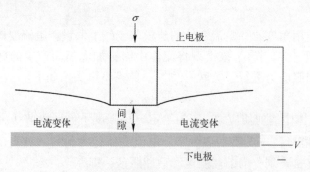

图 5.16　承受压缩应力的电流变体[25]

通过将 $y$ 轴的值除以初始间隙 $g_0$，重新绘制图 5.17 所示的图形，可以获得塑性模量曲线的线性部分对应的电流变体硬化状况，即

$$E = \frac{\sigma}{\varepsilon} \tag{5.8}$$

式中

$$\varepsilon = \ln\left(\frac{g}{g_0}\right)$$

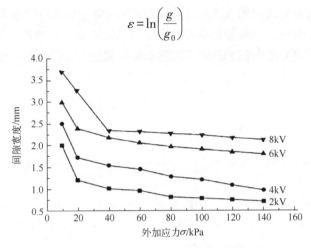

图 5.17　30%聚甲基丙烯酸锂悬浮液在氯化石蜡中的电流变体压缩特性[25]

应变 $\varepsilon$ 定义为在间隙宽度内相对较大位移引起的真实应变。当电流变体从液体状态转化为塑性固体时，式（5.8）中定义的模量至少增大了一个数量级[25]。研究发现，当对间隙施加 2000V 电压时，典型电流变体的硬化悬浮液能够承受超过 1GPa[24-26]的法向应力。

需要注意的是，当向流变体施加电场时，其硬化悬浮液的杨氏模量有所增加，如下关系：

$$E_i = -\frac{\mathrm{d}\sigma_i}{\mathrm{d}\varepsilon_i} \tag{5.9}$$

式中：$i=x, y, z$，$\sigma_i$ 为 $x$、$y$ 或 $z$ 方向的法向应力，$\varepsilon_i$ 为 $x$、$y$ 或 $z$ 方向的应变。研究表明[17]，流变体中的 $E_i$ 对于外加电场方向具有很强的依赖性，杨氏模量仅对于 1 自由度系统有效。

流变体的体积杨氏模量定义了硬化悬浮液在恒定压力下的刚度 $p$（3 自由度系统），可表示为以下关系：

$$E_{\mathrm{bulk}} = \frac{\mathrm{d}p}{\mathrm{d}\varepsilon_V} \tag{5.10}$$

式中：$\varepsilon_V = \dfrac{\mathrm{d}V}{V}$；$p$ 为施加压力；$V$ 为初始体积。研究发现[17]，零外加电场条件下的 $E_{\mathrm{bulk}}$ 初始值近似于流变体中液相的 $E_{\mathrm{bulk}}$。

除了上述处理压缩应力问题的挤压模式（在较低速运动与高作用力的应用），电流变体还有两种工作模式：第 1 种为剪切模式（图 5.5（b）），第 2 种为流动模式或阀模式（图5.5（c））。剪切模式至少存在一个移动电极，而阀模式使用固定电极。典型剪切模式的应用包括离合器、致动器、卡盘和锁定装置，而阀门、减振器和阻尼器则采用流动运行模式。

文献［17］给出了在剪切模式下产生的力，即

$$F = F_v + F_{rh} = \frac{\eta S A}{h} + \tau_y A \tag{5.11}$$

式中：$\eta$ 为流体黏度；$\tau_y$ 为屈服应力；$A = L \times b$ 为有效剪切区面积（$L$ 和 $b$ 分别表示有效长度和有效宽度）；$h$ 为有效厚度（即电极之间的间隙）；$S$ 为电极之间的相对速度。

基于压力驱动流动模式（或阀模式）的应用中产生的压降，可用5.1节中流动夹具装置（图5.7）的式（5.2）~式（5.4）进行计算，即

$$\begin{cases} \Delta p = \frac{8\eta Q L}{b h^3} + c \frac{L}{h} \tau_y, & c = 2, \quad T^* > 200 \\ \text{或} \quad \Delta p = \frac{12\eta Q L}{b h^3} + c \frac{L}{h} \tau_y, & c = 3, \quad T^* < 0.5 \end{cases} \tag{5.12}$$

在 $T^*$ 的表达式中，描述式（5.12）右侧第2项与第1项之间比例的参数由下式给出：

$$T^* = \frac{b h^2 \tau_y}{12 Q \eta} \tag{5.13}$$

式中：$Q$ 为流体流速。

磁流变体的行为类似于相应的电流变体。它们以下列形式遵循宾汉塑性模型：

$$\tau = \tau_y(H) + \eta \dot{\gamma} \tag{5.14}$$

式中：$\tau_y(H)$ 为屈服剪切应力，其随外加磁场 **H** 的变化而改变；$\eta$ 为流体的黏度；$\dot{\gamma}$ 为剪切应变率。在低于屈服应力时（显示 $10^{-3}$ 范围内的应变），该流体将表现出黏弹性，即

$$\tau = G\gamma, \quad \tau < \tau_y(H) \tag{5.15}$$

式中：$G$ 为复合材料模量。图5.18示意了磁流变体的磁滞、黏性等复杂特性。

Bossis 等建立了一种常用的屈服应力预测模型[37-39]。根据该模型可知，其适用于 $\alpha = (\mu_p / \mu_f) \geqslant 1$ 的情形①，即假设由颗粒组成的无限链条沿施加电场方向排列，如图5.19（a）所示。在应变作用下，这些链条将发生与应变一致的

---

① $\mu_p$ 是颗粒的磁导率，$\mu_f$ 是流体的磁导率。

变形，从而形成角 $\theta$ 与剪切应变 $\gamma$ 之间的关系：

$$\gamma = \tan\theta \qquad (5.16)$$

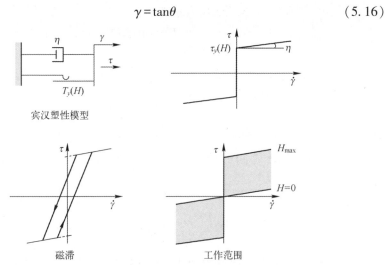

图 5.18　磁流变体的特性示意图

(资料来源：比利时布鲁塞尔自由大学（ULB）活性结构实验室；http://scmero.ulb.ac.be)

屈服应力的场依赖性可以用幂律 $\tau_y \sim H^n(1<n<2)$ 进行表示。代入 $n=2$，可得线性磁性材料情况，这种情况出现在低磁场中或者低磁导率颗粒中[39]。

用一个适用于电流变体和磁流变体的简单模型可描述饱和强度对屈服应力的影响。该模型假设：由于粒子的高磁导率，可以忽略 $H_i$ 粒子内部的磁场①，从而确定两个粒子间隙内的两个域（图 5.19（b））：

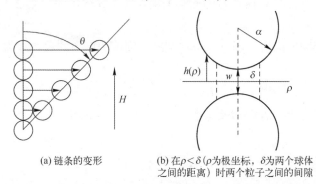

图 5.19　Bossis 等的模型[39]（$H_g = M_s$）

---

① 注意磁场的各种符号：$H_0$ 为外部磁场，$H_f$ 为磁流变液相内部的磁场，$H_i$ 为磁流变固相（颗粒）内的磁场，$M_s$ 的单位为 kA/m。

(1) 磁极域：$\rho<\delta$（$\rho$ 为极坐标）且磁场由饱和磁化量 $H_g = M_s$ 给出。

(2) 第 2 个域：$\rho>\delta$ 且磁场表示如下：

$$H_g = \frac{H(a+0.5w)}{h(\rho)}$$

式中：$H$ 为悬浮物中的平均磁场值；$a$ 为粒子半径；项 $h(\rho)$ 为对称平面与球体表面之间的距离，由以下公式给出：

$$h(\rho) = \frac{1}{2}\left(w + \frac{\rho^2}{a}\right) \tag{5.17}$$

式中：$w$ 为两个球体之间的最小间隙。当 $\rho = \delta$ 时，要求 $H_g = M_s$，则可得距离 $\delta$。

对于情形

$$\frac{H}{M_s} \ll 1$$

可通过对分开两个颗粒的平面磁场积分，得到两个球体之间的径向力：

$$F_r = \frac{\mu_0}{2}\int_0^a a_0 (H_g - H)^2 \times 2\pi\rho \mathrm{d}\rho = \pi a^2 \mu_0 M_s^2 \left(\frac{H}{M_s} - \frac{\varepsilon}{w}\right) + \pi a^2 \mu_0 M_s^2 \frac{H}{M_s} \tag{5.18}$$

式中

$$\varepsilon = \frac{w}{a}$$

应注意，方程（5.18）右侧的第 1 项对应于 $\rho<\delta$ 区域，第 2 项对应于 $\rho>\delta$ 区域。

剪切应变取近似值，即 $\gamma = \tan\theta \approx \sin\theta$，则剪切应力的表达式为[39]

$$\tau(\gamma) = F_r \frac{N}{L^2}\sin\theta\cos^2\theta = \frac{3}{2}\Phi \frac{F_r}{\pi a^2}\frac{\gamma}{1+\gamma^2} \tag{5.19}$$

式中：$N/L^2$ 项为单位表面的链条数；$\Phi$ 为固体颗粒的体积百分比①。用剪切应变表示 $\varepsilon$，即

$$\varepsilon = 2(\sqrt{1+\gamma^2} - 1) \tag{5.20}$$

同时，根据剪切模量（在 $\gamma \ll 1$ 的情况下有效）的式（5.18）和式（5.19），屈服应力可表示为

---

① 磁场平方在连接两个相邻球体的单位矢量上的投影导致 $\cos^2\theta$ 项，而 $\sin\theta$ 项来自剪切方向上径向力的投影。

$$\begin{cases} G = 3\mu_0 \Phi H M_s \\ \tau_y = 2.31\Phi\mu_0 M_s^{0.5} H^{1.5} \end{cases} \quad (5.21)$$

注意，$G$ 的表达式可根据剪切模量的斜率与 $\gamma=0$ 处的剪切应变获得，屈服应力可从 $\gamma=\gamma_c$ 的最大值得出。

$$\gamma_c = \sqrt{\frac{2H}{1.5M_s + 6H}} \quad (5.22)$$

Lemaire 等使用了一个与图 5.20 中所示模型类似的模型[40]。

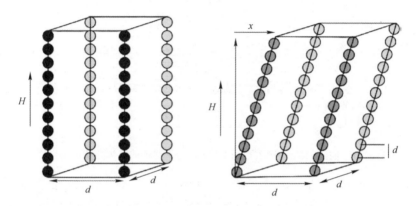

图 5.20　Lemaire 等的模型中所假设的链条示意图[40]

根据该模型，当在 $x$ 方向（垂直于磁场）产生剪切应变时，剪切应力可表示为

$$\tau = \frac{F_{\text{shear}}}{S} = -h\frac{\delta W}{\delta x} = -h\frac{\delta \mu_{zz}}{\delta x}H^2 \quad (5.23)$$

式中：$W = \frac{1}{2}\mu_{zz}H^2$ 为静磁能；$H$ 为介质内部的磁场强度平均值；$h$ 为链条内两个粒子之间沿 $z$ 的初始距离；$\mu_{zz}$ 为沿磁导率张量场方向的对角分量。有关该模型的更多详细信息见文献 [40]。

许多研究人员还开发出一些类似的电流变体模型。Rusika 开发了一个模型，并用该模型描述了由固体颗粒和载体油液组成悬浮液的反应过程，从而定义了电流变体。此外，还提出了一些解决方案，并对其误差进行了评估[30]。

根据电流变体的工作方式和装置，在对电流变体的大量研究中提出了相应的本构方程。例如，Nilsson 和 Ohlson 推导出了其构建模型的本构方程，其中将力表示为挤压模式下夹住电流变体的两个平行板的位移和速度的函数（运动仅限于两个平行板的上下运动）[41]。两个圆形板仅能进行垂直方向移动，从而实现电流变体的径向流动（图 5.21）。当施加电场时，悬浮液中的颗粒（硅油中的塑性球）会在两个平行板之间形成链条。Nilsson 和 Ohlson 将粒子链条建模为四方柱体，其中心之间有边 $a$ 和距离 $b$。他们将体积分数命名为 $f$。研究发现，两个平行板之间的力可表示为

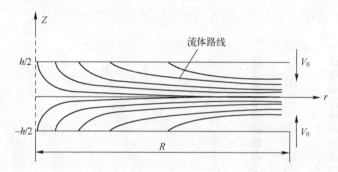

图 5.21 假定的两个平面之间以速度 $v_0$ 上下移动的流体路线示意图[41]

$$F = F_\mathrm{p} + F_\mathrm{e} = k \cdot f \frac{\pi R^2}{h}\left(3\eta \frac{R^2}{a^2}v_0 + E \cdot \delta\right) \tag{5.24}$$

式中：$F_\mathrm{p}$ 和 $F_\mathrm{e}$ 分别为孔压和柱传递的力引起的垂直力；$k$ 为常数，$0<k<1$，其中 $k=0$ 表示零电场，$k=1$ 表示纯液相[41]；$R$ 为圆板半径；$h$ 为圆板之间的距离；$E$ 为柱状材料的杨氏模量；$\delta$ 为轴向位移；$\eta$ 为流体黏度①。

据称，与在剪切模式下运行的同样装置相比，以此方式构建的减振器可提供高 1 个数量级的力（有关该运行模式的定义，见 5.1 节）。

文献 [32] 使用了一种基于链状聚集体的模型。链状聚集体因在电流变悬浮液颗粒中产生的感应偶极子之间相互作用而形成，这与 Bossis 在文献 [39] 中提出的磁流变体模型相似。由此可以计算各种流体与电场情况下的剪切应力，且计算结果与实验结果之间具有较好的定性一致性。

Wineman 和 Rajagopal 利用连续力学方法描述了一些流体的实验行为，推导出了电流变体的三维本构方程[33]。当流体从静止状态开始变形时，表现出类似固体的响应；在正弦剪切应力作用下，表现出类黏弹性；在剪切应变率或电场的突然变化下表现出随时间变化的响应。

---

① 文献 [31] 的黏度 $\eta$ 值在 $k=0$ 处是 $25\times10^{-2}\mathrm{m}^2/\mathrm{s}$，在 $k=1$ 处降至 $7\times10^{-2}\mathrm{m}^2/\mathrm{s}$。

为了进行设计,必须确定流变体材料的其他参数。如表 5.3 所列[17],电流变体和磁流变体需要的运行功率最高可达 50W。电流变体装置需要最高 5kV 的电压和最低 10mA 的电流,而磁流变体装置则需要高达 25V 的低压和高达 2A 的高电流。文献 [17] 给出了电流变体和磁流变体所需最小功率的半经验公式(所需最小功率为最小有效体积(Minimum Active Volume,MAV)的函数),即

$$\begin{cases} 电流变体:P_{\min} = \dfrac{0.001\text{MAV}}{\Delta t} \\ 磁流变体:P_{\min} = \dfrac{0.1\text{MAV}}{\Delta t} \end{cases} \tag{5.25}$$

式中:$\Delta t$ 为所需的切换时间(s)。

表 5.3 流变流体的典型特性[17]

| 特　　性 | 电流变体 | 磁流变体 |
| --- | --- | --- |
| 最大屈服应力 $\tau_{y\max}$/(kPa) | 2~5 | 50~100 |
| 最大电场 | ~4kV/mm(受击穿限制) | ~250kA/m(受饱和限制) |
| 黏度 $\eta$/(Pa·s) | 0.1~1.0 | 0.1~1.0 |
| 工作温度范围/℃ | +10°~+90°(电离,DC)<br>-10°~+125°(不电离,AC) | -40°~+150°(受载液限制) |
| 稳定性 | 不耐受杂质 | 不受大多数杂质影响 |
| 响应时间 | <ms | <ms |
| 密度 $\rho$/(g/cm³) | 1~2 | 3~4 |
| $H_p/\tau_y^2$/(s/Pa) | $10^{-7} \sim 10^{-8}$ | $10^{-10} \sim 10^{-11}$ |
| 最大能量密度/(J/cm³) | 0.001 | 0.1 |
| 常用电源/W | 2~50(2~5kV@1~10mA) | 2~50(2~25V@1~2A) |
| 辅助材料 | 任何导电表面 | 铁或钢 |
| 运行所需的流变体量 | 高 | 低 |

表 5.3 中,磁流变体受污染与杂质的影响远低于电流变体,而电流变体对大气冷凝水的敏感度更高。因此,电流变体在重载应用条件下的寿命有限,而磁流变体可能会在悬浮液中表现出固相特性。应注意,流变体的物理性质与温度有密切的关系,因此在设计基于这些流体的装置时应予以考虑。

通过比较这两种流变体的性能,可以明显看出,电流变体的最大屈服应力比磁流变体的最大屈服应力(50~100kPa)小 1 个数量级(2~5kPa),见

表5.3。因此，在电流变体设计中需要兆安培伏级电压，比磁流变体大2个数量级，存在以下关系[17]：

$$\text{MAV} \approx \begin{cases} \alpha P \left[ \dfrac{F_{on}}{F_{off}} \right] \times 10^{-2}, & \text{电流变体} \\ \alpha P \left[ \dfrac{F_{on}}{F_{off}} \right] \times 10^{-4}, & \text{磁流变体} \end{cases} \quad (5.26)$$

式中：$P$ 为所需功率（W）；$F_{on}$ 和 $F_{off}$ 分别为"开启状态"与"关闭状态"的最小力（N），剪切模式下旋转应用的常数 $\alpha$ 值为1，阀模式下线性应用的常数 $\alpha$ 值为2。

## 5.3 电流变体和磁流变体的阻尼

如前两节所述，电流变体和磁流变体都显示出相对较高的阻尼系数，这使其可能成为各种系统中减振器的理想选择。本节旨在为读者提供相关计算数据和公式，以评估电流变体和磁流变体的阻尼系数，供工程应用参考。图 5.22 所示为在 ER-RheDamp 装置中对电流变体施加 4kV 电场导致垂直加速度大幅下降的情况（图 5.2）。当然，加速度从初始值下降了近 7 倍，这一事实表明电流变体具备有效衰减振动的高级功能。

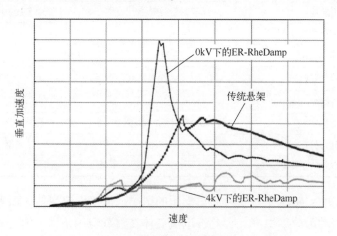

图 5.22 电流变减振器 ER-RheDamp 的性能
（资料来源：由德国达姆施塔特 FLUDICAN 公司设计制造）

Kohl 和 Tichy 给出了两种含电流变体的黏滞减振器的阻尼系数表达式：第

1种减振器的基本结构为两个固定平行板之间的流体,第2种减振器为两个静止同心圆柱之间的环形流体[42]。电场从间隙的两侧施加,流体在间隙内流动。图5.23所示为第1个组件的各种尺寸。Kohl和Tichy提出了沿通道的压力梯度(为施加到活塞上的力$F_{plunge}$的函数)表达式(图5.23),即

$$\frac{dp}{dx} = -\frac{F_{plunge}}{WH_pL} \tag{5.27}$$

式中:$W$为通道宽度;$L$为通道长度。

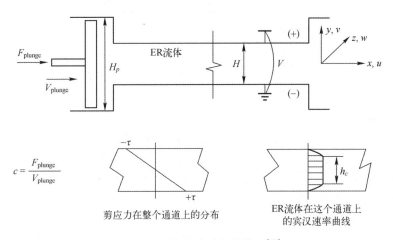

图5.23 平行板式减振器模型[42]

将作用力与活塞速度$F_{plunge}$联系起来,可得到阻尼系数$c$,即

$$F_{plunge} = cV_{plunge} = (c_N c^*)V_{plunge} = \left[\left(\frac{12\eta WL}{H}H^{*2}\right)c^*\right]V_{plunge} \tag{5.28}$$

式中:$c_N$为阻尼系数的牛顿部分(当不施加电压时);$c^*$为修正阻尼因子,其定义如下:

$$c^* = \frac{2}{3}\left(1 + \frac{\tau_y^*}{4H^{*2}}\right)\cos\left(\frac{a}{3}\cos A\right) + \frac{1}{3}\left(1 + \frac{\tau_y^*}{4H^{*2}}\right) \tag{5.29}$$

式中

$$A = 1 - \frac{\tau_y^*}{36H^{*6}\left(1+\frac{\tau_y^*}{4H^{*2}}\right)^3}, \quad \tau_y^* = \left|\frac{\tau_y H_p}{\eta V_{plunge}}\right|, \quad H^* = \frac{H_p}{H} \tag{5.30}$$

式中：$\eta$ 为流体的黏度。因无量纲屈服应力 $\tau_y^*$ 与电场呈线性关系①，修正阻尼系数 $c^*$ 随着施加在电流变体上的电场的增强而增加[42]。

第 2 个装置[42]为同心圆筒式减振器，如图 5.24 所示。关于小间隙近似的主要假设如下：$x$ 方向（纵向）的速度剖面与两个圆柱的曲率无关。由此可使用前一个装置的速度分布和 $h$ 值。

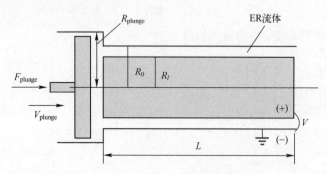

图 5.24　同心圆筒式减振器模型[42]

对于该类装置，式（5.27）~式（5.30）的具体形式如下：

沿同心圆柱的压力梯度表达式为施加于活塞上的力 $F_{\text{plunge}}$ 的函数（图 5.24），即

$$\frac{\mathrm{d}p}{\mathrm{d}x} = -\frac{F_{\text{plunge}}}{\pi R_{\text{plunge}}^2 L} \tag{5.31}$$

若导致的阻尼系数为 $c$，活塞的力-速度关系如下：

$$F_{\text{plunge}} = c V_{\text{plunge}} = (c_N c^*) V_{\text{plunge}} = \left[\left(\frac{6\pi\eta R_p^{*4}}{(1-\kappa)^3}\right) c^*\right] V_{\text{plunge}} \tag{5.32}$$

式中：$c_N$ 为阻尼系数的牛顿部分（当不施加电压时）；$c$ 为修正阻尼因子，其定义如下：

$$c^* = \frac{2}{3}\left(1 + \frac{(1-\kappa)^2 \tau_y^*}{2R_p^{*2}}\right)\cos\left(\frac{a}{3}\cos A\right) + \frac{1}{3}\left(1 + \frac{(1-\kappa)^2 \tau_y^*}{2R_p^{*2}}\right) \tag{5.33}$$

式中

---

① 结果表明，式（5.29）中出现的 $a/3$ 可通过以下条件获得：在不施加电压的情况下（即 $\tau_y^* = 0$ 或 $\tau_y = 0$），校正系数为 $c^* = 1$（且 $A = 1$）。

$$A = 1 - \frac{(1-\kappa)^6 \tau_y^{*3}}{4R_p^{*6}\left(1+\frac{(1-\kappa)^2 \tau_y^*}{2R_p^{*2}}\right)^3}, \quad \tau_y^* = \left|\frac{\tau_y R_{\text{plunge}}}{\eta V_{\text{plunge}}}\right|, \quad R_p^* = \frac{R_{\text{plunge}}}{R_0}, \quad \kappa = \frac{R_i}{R_0}$$

(5.34)

注意,式(5.33)中的 a/3 项是以与式(5.29)相同的方式确定的(见上页脚注①)。与前述一样,因无量纲屈服应力 $\tau_y^*$ 与电场呈线性关系,修正阻尼因子 $c*$ 随施加在电流变上的电场强度的增强而增大[42]。

一些文献中详细介绍了磁流变型减振器的阻尼性能[43-54]。虽然没有关于磁流变体减振器阻尼系数的显式表达式,但是可以尝试从这些文献的各种详细公式中推导出该阻尼系数。

Nishiyama 等对磁流变体中平板在低磁场条件下的特性进行了研究[45]。当磁流变体中平板处于低磁场条件时,阻尼系数可表示如下:

$$\begin{cases} c_{\text{eq.}} = \sqrt{\frac{c^2}{2} + \frac{m_A^2 \omega^2}{2}} \\ m_A = \frac{k}{\omega^2}\left(1 - \frac{\cos\phi}{a}\right) - m; \quad a \equiv \frac{\Delta z}{\Delta z_g} \end{cases}$$

(5.35)

式中:$m_A$ 为添加的质量;$m$ 为平板的质量;$k$ 为连接到板的弹簧常数;$z$ 为板的位移;$z_g$ 为励磁机的位移;$\omega = 2\pi f$,其中 $f$ 为频率。注意,振幅比 $a$ 和相位差 $\phi$ 通过实验进行确定。有关更多详细信息,见文献[45]。

文献[43]给出了阻尼系数的另一个表达式,并设计了一个用于低频应用场景的磁流变减振器。将羰基铁粒子浸于硅油中,该固相的体积分数在 20%~35%。阻尼系数 $\zeta$ 的表达式在文献[43]中给出:

$$\zeta = \frac{1}{\sqrt{2}} \times \frac{\sqrt{[(a-1^2)(a-\sqrt{a^4-a^2})]}}{(a^2-1)}$$

(5.36)

式中:$a$ 为峰值振幅比,文献[43]中报道在 $f$ 为 9.24Hz 时 $\zeta$ 为 0.2814,干摩擦情况下 $\zeta$ 为 0.2139,然而在加入剪切效应时的平均阻尼系数 $\zeta_{\text{av}}$ 为 0.277。

当磁流变阻尼器作为可控半主动组件嵌入控制系统时,要求所选的数值模型能够描述包括迟滞在内的非线性行为。文献[47]列出了适用于磁流变体(有时也适用于电流变体)的数学模型。在表 5.4 所述各种模型中均有提供,并展示它们的示意图和运动方程,这里仅提供最常见的模型。第 1 个模型为 Stanway 等提出的理想化机械宾汉模型,其包括一个与库仑摩擦元件平行的黏性减振器,如图 5.25 所示[55]。

表 5.4 磁流变模型的分类

| 建 模 技 术 | 磁流变阻尼器模型 |
|---|---|
| 宾汉模型 | 原始宾汉模型[55-72] |
| | 改进的宾汉模型[57] |
| | Gamota 和 Filisko 模型[58]① |
| | Occhiuzzi 等更新的宾汉模型[59] |
| | 三元模型[50]② |
| 双黏性模型 | 非线性双黏性模型[55] |
| | 非线性迟滞双黏性模型[73] |
| | 非线性迟滞反正切模型[74] |
| | 集总参数双黏性模型[75] |
| 黏弹性塑性模型 | 通用黏弹性塑性模型[76] |
| | Li 等的黏弹性塑性模型[77] |
| 刚度黏度弹性滑动模型 | 刚度黏度弹性滑动（SVES）模型[62] |
| 流体力学模型 | 流体力学模型[78] |
| 麦克斯韦模型 | Makris 等的 BingMax 模型[61]③ |
| | 麦克斯韦非线性滑块模型[79] |
| Bouc④-Wen⑤模型 | 简单的 Bouc-Wen 模型[80] |
| | 修正的 Bouc-Wen 模型[81] |
| | 剪切模式阻尼器的 Bouc-Wen 模型[82] |
| | 大尺度模式阻尼器的 Bouc-Wen 模型[83] |
| | 电流相关的 Bouc-Wen 模型[84] |
| | 电流-频率-幅度相关的 Bouc-Wen 模型[85] |
| | 非对称 Bouc-Wen 模型[86] |
| Dahl 模型⑥ | 修正的 Dahl 模型[72] |
| | 黏性 Dahl 模型[66] |
| LuGre 模型⑦ | Jimenez 和 Alvarez 修正的 LuGre 模型[63] |
| | Sakai 等修正的 LuGre 模型[64-65] |
| 双曲正切模型 Sigmoid 模型 | Kwok 等的双曲正切模型[71] |
| | Ma 和 Wang 等的 Sigmoid 模型[69-70] |
| 等效模型 | Oh 的等效模型[68] |

续表

| 建 模 技 术 | 磁流变阻尼器模型 |
|---|---|
| 相变模型 | 相变模型[67] |

① 该模型最初是为电流变体开发的。
② 该模型最初是为电流变体开发的。
③ 该模型最初是为电流变体开发的。
④ Bouc, R., "Modèle mathématique d'hystérésis: application aux systèmes à un degré de liberté. Acustica (in French) Vol. 24, 16–25, 1971.
⑤ Wen, Y. K., Method for random vibration of hysteretic systems, Journal of Engineering Mechanics (American Society of Civil Engineers), Vol. 102, No. 2, 249–263, 1976.
⑥ Dahl, P. R., A solid friction model. Technical report, The Aerospace Corporation, El Secundo, CA, 1968.
⑦ LuGre 模型是法国 Lund 和 Grenoble 之间的合作成果。

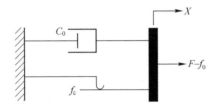

图 5.25 磁流变减振器的宾汉原理模型

该减振器产生的输出力 $F$ 可表示为

$$F = f_c \text{sign}\dot{x} + c_0 \dot{x} + f_0 \tag{5.37}$$

式中：$c_0$ 为黏性阻尼系数；$f_c$ 为流体屈服应力引起的摩擦力；$f_0$ 为一种偏移力（考虑测量力中由流体蓄能器引起的非零平均值）；$\dot{x}$ 为速度[47]。

另一个模型为扩展宾汉模型[55,58-72,81]，如图 5.26 所示。该黏弹性-塑性模型由宾汉模型和线性固体（Zener 电流变单元）的 3 参数单元串联而成[53]。该模型的运动方程如下：

$$\begin{cases} F(t) = k_1(x_2 - x_1) + c_1(\dot{x}_2 - \dot{x}_1) + f_0 \\ \quad = f_c \text{sign}(\dot{x}) + c_0 \dot{x} + f_0 = k_2(x_3 - x_2) + f_0, \ |F(t)| > f_c \\ F(t) = k_1(x_2 - x_1) + c_1 x_2 f_0 = k_2(x_3 - x_2) + f_0, \ |F(t)| < f_c \end{cases} \tag{5.38}$$

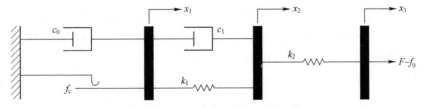

图 5.26 磁流变减振器的扩展宾汉模型

式中：宾汉模型使用的 $c_0$ 和摩擦力 $f_c$ 在前文已定义；场常数 $c_1$、$k_1$ 和 $k_2$ 与流体在预屈服区的弹性特性相关[53]。

一个常用的模型是 Bouc-Wen 模型（表5.4），最初由 Bouc 提出，后来 Wen 对其进行了扩展。该模型用于模拟迟滞系统对随机激励的响应[53]。根据图 5.27，力输出可以表示如下：

$$F = c_0 \dot{x} + k_0(x - x_0) + \alpha z \tag{5.39}$$

式中

$$\dot{z} = -\gamma |\dot{x}| z |z|^{n-1} - \beta \dot{x} |z|^n + \delta \dot{x}$$

式中：$z$ 为迟滞分量；$x_0$ 为被引入有累积器存在的模型；$\alpha$、$\beta$、$\gamma$、$\delta$ 和 $n$ 为控制力-速度曲线的形状参数，这些参数是电流、振幅和频率的函数。

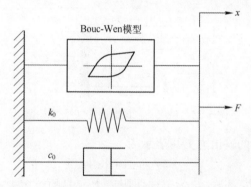

图 5.27 一种磁流变体减振器的 Bouc-Wen 原理模型

如文献 [48] 所述，为确定 Bouc-Wen 特性参数，以模拟磁流变体减振器迟滞响应。Kwok 等提出了非对称 Bouc-Wen 模型（表5.4），其修改表达式如下[86]：

$$F = c_0[\dot{x} - \mu \text{sign}(\dot{z})] + k_0(x - x_0) + \alpha z \tag{5.40}$$

式中

$$\dot{z} = \{-[\gamma \text{sign}(z\dot{x}) + \beta]|z|^n + \delta\}\dot{x}$$

式中：$\mu$ 为调整速度的比例因子。

为了更好地预测磁流变减振器在应力屈服点附近的性能，Spencer 等对 Bouc-Wen 模型进行了修正，增加了两个附加机械部件（弹簧和缓冲器），如图 5.28 所示[81]。此模型的输出力 $F$ 可表示如下：

$$F = \alpha z + c_0(\dot{x} - \dot{y}) + k_0(x - y) + k_1(x - x_0) = c_1\dot{y} + k_1(x - x_0) \tag{5.41}$$

式中

$$\dot{z} = -\gamma|\dot{x} - \dot{y}|z|z|^{n-1} - \beta(\dot{x} - \dot{y})|z|^n + \delta(\dot{x} - \dot{y})$$

且

$$\dot{\chi} = \frac{\alpha z + c_0 \dot{\chi} + k_0(x-y)}{c_0 + c_1}$$

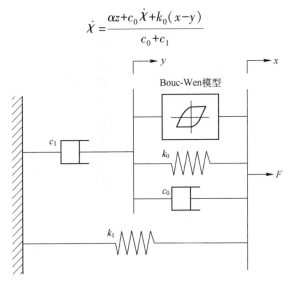

图 5.28  修改的 Bouc-Wen 磁流变减振器原理模型

注意，如文献 [53] 所述，迟滞部件再次用 $z$、弹簧 $k_1$ 和初始位移 $x_0$ 表示，允许存在附加刚度和储能器。假设各种参数与电压 $v$ 线性相关，并将这些参数应用于电流驱动器，可知：

$$\begin{cases} \alpha = \alpha(u) = \alpha_a + \alpha_b u \\ c_1 = c_1(u) = c_{1a} + c_{1b} u \\ c_0 = c_0(u) = c_{0a} + c_{0b} u \\ \dot{u} = -\eta(u-v) \end{cases} \quad (5.42)$$

式中：$u$ 为实际输出信号；$\eta$ 为时间常数。

Kwok 等[71]提出了另一个类似于 Bouc-Wen 的模型[48]。该模型插入了一个迟滞模型，用以预测磁流变体减振器的阻尼力（图 5.29）。该模型可表示如下：

$$F = c\dot{x} + kx + \alpha z + f_0 \quad (5.43)$$

式中

$$z = \tanh[\beta \dot{x} + \delta \mathrm{sign}(x)]$$

其中：$c$ 和 $k$ 分别为黏性系数和刚度系数；$\alpha$ 为迟滞标度因子；$z$ 为迟滞变量；$f_0$ 为因储能器存在产生的减振器力偏移；$\beta$ 和 $\delta$ 为待识别模型参数。文献 [48] 提供了宾汉模型、Bouc-Wen 模型、迟滞模型和实验数据之间的比较，如图 5.30 所示。实验模型与数值模型之间存在明显的偏差。

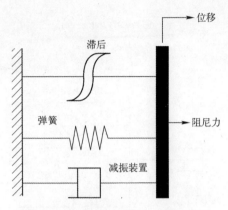

图 5.29 磁流变减振器迟滞原理模型

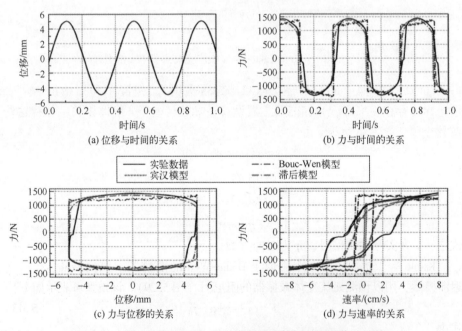

图 5.30 磁流变体减振器-多个模型试验数据和预测值的比较，激励信号采用振幅 5mm、电流 1.5A 的 2.5Hz 正弦激励[48]（见彩插）

另一个模型是改进的 Dahl 模型，据称该模型成功地再现了低速区的力-速度关系[43]。该模型由模拟摩擦的简化 Dahl 模型组成，能够反映迟滞和零滑移位移，但不能描述 Stribeck 效应①或黏性行为。其典型迟滞如图 5.31 所示。

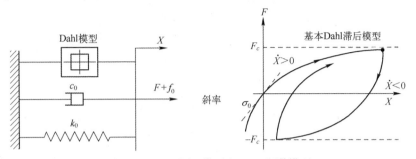

图 5.31 Dahl 原理模型和 Dahl 迟滞模型

于是，可以将模型表示如下：

$$\frac{\mathrm{d}F}{\mathrm{d}x} = \sigma_0 \left[ 1 - \frac{F}{F_c} \mathrm{sign}(\dot{x}) \right] \quad (5.44)$$

式中：$\sigma_0$ 和 $F_c$ 分别为刚度和库仑摩擦。注意，图 5.31 展示了增加和缩小位移的不同行为[43]。

基于 Dahl 改进模型的磁流变减振器产生的力可表示如下：

$$F = K_0 x + C_0 \dot{x} + F_d z - f_0 \quad (5.45)$$

式中

$$\dot{z} = \sigma \dot{x} [1 - \mathrm{sign}(\dot{x})]$$

迟滞环形状是由 $\sigma$ 确定的[43]。与改进的 Bouc-Wen 模型（5.42）情况相似，模型参数与外加磁场的关系可以通过假设线性关系获得，即

$$\begin{cases} C_0 = C_0(u) = C_{0s} + C_{0d} u \\ F_d = F_d(u) = F_{ds} + F_{dd} u \end{cases} \quad (5.46)$$

而且

$$\dot{u} = -\eta(u - v)$$

与前文所述相同，式中 $v$ 为在电流驱动器上施加的电压，$u$ 为实际信号输出，$\eta$ 为时间常数。$C_{0s}$ 和 $F_{ds}$ 分别为磁流变减振器在 0V 时的阻尼系数和库仑力，$C_{0d}$ 和 $F_{dd}$ 为实验确定的常数。

Wereley 等[73]采用分段线性函数，引入了两个阻尼系数，提出了以非线性双黏性模型构建的迟滞环路（图 5.32），一个用于预屈服条件，另一个用于后屈服条件[87]。这些函数的形式表示如下：

---

① Stribeck 效应或 Stribeck 曲线清楚地显示了摩擦的最小值，作为全液膜润滑和一些固体微凸体相互作用之间的分界线。

$$f_h = \begin{cases} c_{po}\dot{x} - f_y, & x \leq -\dot{x}_1, \dot{x} > 0 \\ c_{pr}(\dot{x} + v_h), & -\dot{x}_1 \leq \dot{x} \leq \dot{x}_2, \dot{x} > 0 \\ c_{po}\dot{x} + f_y, & \dot{x}_2 \leq \dot{x}, \dot{x} > 0 \\ c_{po}\dot{x} + f_y, & \dot{x}_1 \leq \dot{x}, \dot{x} > 0 \\ c_{pr}(\dot{x} + v_h), & -\dot{x}_2 \leq \dot{x} \leq \dot{x}_1, \dot{x} > 0 \\ c_{po}\dot{x} - f_y, & \dot{x} \leq \dot{x}_2, \dot{x} > 0 \end{cases} \quad (5.47)$$

式中：$f_y$ 为从零速度下后屈服分支投影导出的常数 ($\dot{x}=0$)；$v_y$ 为迟滞环路的宽度，且

$$\begin{cases} \dot{x}_1 = \dfrac{f_y - c_{pr} v_h}{C_{pr} - c_{po}} \\ \dot{x}_2 = \dfrac{f_y + C_{pr} v_h}{C_{pr} - C_{po}} \end{cases} \quad (5.48)$$

式中：各参数通过实验确定。

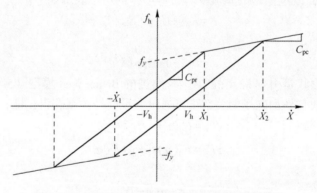

图 5.32　磁流变减振器的非线性迟滞双黏性模型

基于上述各种模型，可以将磁流变或电流变提供的力带入包含这类减振器的系统运动方程中，从而得到位移/速度随时间变化的数据。

## 5.4　附　录　三

### 5.4.1　附录 A

讨论材料黏度时，必须区分动态黏度（$\eta$）和运动黏度（$v$）。表达式

如下：

$$\eta = \frac{\tau}{\dot{\gamma}} \tag{5.a}$$

$$v = \frac{\eta}{\rho} \tag{5.b}$$

式中：$\tau$ 为剪切应力（$N/m^2$ 或 Pa）；$\dot{\gamma}$ 为剪切应变率（1/s）；$\rho$ 为密度（$kg/m^3$）。

对于国际单位制（SI）系统，动态黏度 $\eta$ 具有以下单位：

$$1\text{Pl} = 1\text{Pa} \cdot \text{s} = 1\frac{\text{N} \cdot \text{s}}{\text{m}^2} = 1\frac{\text{kg}}{\text{s} \cdot \text{m}} \tag{5.c}$$

而对于厘米克秒单位制 CGS 系统，则为

$$1\text{P} = \frac{1}{10}\text{Pa} \cdot \text{s} = \frac{1}{10}\frac{\text{N} \cdot \text{s}}{\text{m}^2} = 1\frac{\text{g}}{\text{s} \cdot \text{cm}} = 1\frac{\text{dyne} \cdot \text{s}}{\text{cm}^2} \tag{5.d}$$

将 P 除以 100 将得到一个更小的单位，即厘泊（cP），其中，

$$1\text{P} = 100\text{cP}$$

$$1\text{cP} = \frac{1}{100}\text{P} = \frac{1}{1000}\text{Pa} \cdot \text{s} = \frac{1}{1000}\frac{\text{Ns}}{\text{m}^2} = \frac{1}{100}\frac{\text{g}}{\text{s} \cdot \text{cm}} = \frac{1}{100}\frac{\text{dyne} \cdot \text{s}}{\text{cm}^2} \tag{5.e}$$

对于 SI 系统，运动黏度 $v$ 也可以用斯（St）表示，其中，

$$1\frac{\text{m}^2}{\text{s}} = 10^4 \text{St} = 10^4 \frac{\text{cm}^2}{\text{s}} \tag{5.f}$$

一个较小的单位是由斯除以 100 获得的厘斯（cSt），由此可得

$$1\text{St} = 100\text{cSt}$$

$$1\frac{\text{m}^2}{\text{s}} = 10^6 \text{cSt} = 10^6 \frac{\text{mm}^2}{\text{s}} \tag{5.g}$$

表 5.a 给出了动态（或绝对）黏度 $\eta$ 的典型值。

表 5.a  各种流体的典型动态黏度值

| 流　体 | 室温下的动力学黏度 $\eta/(\text{Pa} \cdot \text{s})$ |
|---|---|
| 水 | $1 \times 10^{-3}$ |
| 橄榄油 | $\sim 1 \times 10^{-1}$ |
| 甘油 | $\sim 1 \times 10^0$ |
| 蜂蜜（液态） | $\sim 1 \times 10^{+1}$ |
| 玻璃（液态） | $\sim 1 \times 10^{+40}$ |

此外，还应注意，黏度与温度有关，其依据是以下公式：

$$\eta(T) = A e^{\left(\frac{b}{T+273}\right)} \tag{5.h}$$

式中：常数 $A$ 和 $b$ 通过实验确定。

### 5.4.2 附录 B

$H$（磁场强度）与 $B$（磁场密度）的关系由下式给出：

$$H \equiv \frac{B}{\mu_0} - M \tag{5.i}$$

式中：$\mu_0$ 为真空磁导率，其值如下：

$$\mu_0 = 4\pi \times 10^{-7} \mathrm{V \cdot s/(A \cdot m)} \tag{5.j}$$

$M$ 为磁化矢量场，其定义为单位体积的净磁偶极矩。

在 SI 系统中，$B$ 和 $H$ 具有以下单位：

$$\begin{aligned} 1B(\mathrm{T}) &= 1\Phi\left(\frac{\mathrm{Wb}}{\mathrm{m}^2}\right) \\ H&\left(\frac{A}{\mathrm{m}}\right) \end{aligned} \tag{5.k}$$

式中：T 为特斯拉；$\Phi$ 为磁通量。

在 CGS 系统中，$B$ 和 $H$ 具有以下单位：

$$\begin{cases} B(\mathrm{G}) \\ H(\mathrm{Oe}) \end{cases} \tag{5.l}$$

式中，G 为高斯，且 1T=10000G；Oe 为奥斯特。

### 5.4.3 附录 C

Herschel-Bulkley 流体模型[88]是一个非线性非牛顿流体模型，试图描述流体的剪切变稀（如普通油漆）和剪切增稠（如悬浮在水中的玉米淀粉）效应。其本构方程的形式如下：

$$\tau = \tau_y + k\dot{\gamma}^n \tag{5.m}$$

式中：$\tau$ 为剪切应力；$\tau_y$ 为屈服剪切应力；$k$ 为一致性指数；$\dot{\gamma}$ 为剪切应变率；$n$ 为流动指数。当 $\tau<\tau_y$ 时，Herschel-Bulkley 流体的行为类似于固体，而在 $\tau_y$ 上方，其行为表现为流体。当 $n>1$ 时，流体表现为剪切增稠；而 $n<1$ 时，则表现为剪切变稀。如果 $n=1$ 且 $\tau_y=0$，Herschel-Bulkley 流体会降级为牛顿（Newton）流体（图 5.a）。

有效黏度可表示为

$$\eta_{\mathrm{eff}} = \begin{cases} \eta_0, & |\dot{\gamma}| \leqslant \dot{\gamma}_y \\ k|\dot{\gamma}|^{n-1} + \tau_y|\dot{\gamma}|^{-1}, & |\dot{\gamma}| \geqslant \dot{\gamma}_y \end{cases} \tag{5.n}$$

此处，选择满足以下方程式的极限黏度 $\eta_0$：

$$\eta_0 = k\,\dot{\gamma}^{n-1} + \tau_y\,\dot{\gamma}_y^{-1} \tag{5.o}$$

在后屈服区，Herschel-Bulkley 流体显示出的表观黏度关系为

$$\frac{d\tau}{d\gamma} \equiv \eta_{app} = nk\,\dot{\gamma}^{n-1} \tag{5.p}$$

应注意，对于 $n=1$，可得

$$k = \eta_0 - \frac{\tau_y}{\dot{\gamma}_y} \equiv \eta_{pl} \tag{5.q}$$

替换式（5.m）中的 $k$ 值，当 $n=1$ 时可得到宾汉流体方程，即

$$\tau = \tau_y + \eta_{pl}\dot{\gamma} \tag{5.r}$$

这与 5.2 节中的式（5.6）类似。

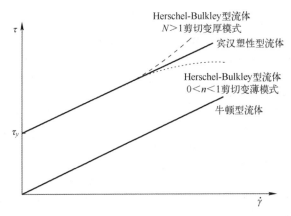

图 5.a　牛顿流体、Herschel-Bulkley 流体和宾汉塑性流体模型的剪切应力-剪切应变率示意图

# 参 考 文 献

[1] Winslow, W. M., Method and means for translating electrical impulses into mechanical forces, US Patent Specification 2417850, 1947.
[2] Ahn, Y. K., Yang, B. S. and Morishita, S., Directionally controllable squeeze film damper using electro-rheological fluid, Journal of Vibration and Acoustics 124, January 2002, 105-109.
[3] Stanway, R., Sproston, J. L. and EL-Wahed, A. K., Applications of electro-rheological fluids in vibration control: a survey, Smart Materials and Structures 5, 1996, 464-482.
[4] Morishita, S. and Ura, T., ER fluid applications to vibration control devices and an adaptive neural-net controller, Journal of Intelligent Material Systems and Structures 4, 1993, 366-372.

[5] Peel, D. J., Stanway, R. and Bullough, W. A., Dynamic modeling of an ER vibration damper for vehicle suspension applications, Smart Materials and Structures 5, 1996, 591-606.

[6] Duclos, T. G., Carlson, J. D., Chrzan, M. J. and Coulter, J. R., Electrorheological fluids - materials and applications, In: Intelligent Structural Systems, Tzou, H. S. and Anderson, G. L. (eds.), Kluwer Academic Publishers, 1992, 213-241.

[7] Nguyen, Q.-A., Jorgensen, S. J., Ho, J. and Sentis, L., Characterization and testing of an electrorheological fluid valve for control of ERF actuators, Actuators 4, 2015, 135-155.

[8] Dyke, S. J., Spencer, B. F. Jr., Sain, M. K. and Carlson, J. D., Modeling and control of magnetorheological dampers for seismic response reduction, Smart Materials and Stnutures 5, 1996, 565-575.

[9] Choi, H. J. and Jhonb, M. S., Electrorheology of polymers and nanocomposites, Soft Matter, The Royal Society of Chemistry 5, 2009, 1562-1567.

[10] Carlson, J. D. and Spencer, B. F. Jr., Magneto-rheological fluid dampers for semi-active seismic control, Proceedings of the 3rd International Conference on Motion and Vibration Control, Vol. 3, 1996, 35-40, Chiba, Japan.

[11] Aslam, M., Liang, Y. X. and Chao, D. Z., Review of magnetorheological (MR) fluids and its applications in vibration control, Journal of Marine Science and Applications 5 (3), September 2006, 17-29.

[12] Baranwal, D. and Deshmukh, T. S., MR-fluids technology and its application- a review, International Journal of Emerging Technology and Advanced Engineering 2 (12), December 2012, 563-569.

[13] De Vicente, J., Klingenbergb, D. J. and Alvareza, R. H., Magnetorheological fluids: a review, Soft Matter, The Royal Society of Chemistry 7, 2011, 3701-3710. doi: 10.1039/cosmo1221a.

[14] Zhu, X., Jing, X. and Cheng, L., Magnetorheological fluid dampers: a review on structure design and analysis, Journal of Intelligent Material Systems and Structures 28 (8), 2012, 839 - 873. doi: 10.1177/1045389x12436735.

[15] Kciuk, S., Turczyn, R. and Kciuk, M., Experimental and numerical studies of MR damper with prototype magnetorheological fluid, Journal of Achievements in Materials and Manufacturing Engineering 39 (1), 2010, 52-59.

[16] Wen, W., Huang, X. and Sheng, P., Electrorheological fluids: structures and mechanisms, Soft Matter, The Royal Society of Chemistry 4, 2008, 200-210. doi: 10.1039/b710948m.

[17] Szary, M. L., The phenomena of electrorheological fluid behavior between two barriers under alternative voltage, Archives of Acoustics 29 (2), 2004, 243-258.

[18] Olabi, A. G. and Grunwald, A., Design and application of magneto-rheological fluid, Materials and Design 28, 2007, 2658-2664.

[19] Bin Mazlan, S. A., The behavior of magnetorheological fluids in squeeze mode, Ph. D. Thesis submitted to School of Mechanical and Manufacturing Engineering, Faculty of Engineering and Computing, Dublin City University, North Ireland, August 2008.

[20] Alghamdi, A. A. and Olabi, A. G., Novel design concept of magneto rheological damper in squeeze mode, 15th International Conference on Experimental Mechanics, ICEM15, Paper No. 2607, Porto, Portugal, 22-27 July 2012.

[21] Lord Corporation company, Lord technical data sheet for MRF-140CG, http://www.lord.com/products-and-solutions/magneto-rheological- (mr) /product.xml.1646/2, 2012.

[22] Sapinski, B. and Goldasz, J., FE Simulation of a Magnetic Circuit in a MR Squeeze-Mode Damper, 22nd International Congress on Sound and Vibration, Florence, Italy, 12-16 July, 2015.

[23] Huang, J., Zhang, J. Q., Yang, Y. and Wei, Y. Q., Analysis and design of a cylindrical magnetorheological fluid brake, Journal Material Process Technology 129, 2002, 559-562.

[24] Klingenberg, D. J., Dierking, D. and Zukoski, C. F., Stress-transfer mechanisms in lectrorheological suspensions, Journal of Chemical Society Faraday Transactions 87 (3), 1991, 425-430.

[25] Monkman, G. J., The electrorheological effect under compressive stress, Journal of Physics D: applied Physics 28, 1995, 588-593.

[26] Davis, L. C. and Ginder, J. M., Electrostatic forces in electrorheological fluids, In: Progress in Electrorheology, Havelka, K. O. and Filisko, F. E. (eds.), New York, Plenum Press, 1995, 107-114.

[27] Brooks, D. A., A practical High Speed ER actuator, Actuator 1992, 3rd International Conference on New Actuators, Germany, Bremen, 1992, 110-115.

[28] Hoppe, R. H. W. and Litvinov, W. G., Problems on electrorheological fluid flows, Communications on Pure and Applied Analysis (CPAA) 3, 2004, 809-848.

[29] Hoppe, R. H. W., Litvinov, W. G. and Rahman, T., Problems of stationary flow of electrorheological fluids in a cylindrical coordinate system, SIAM Journal of Applied Mathematics 65 (5), 2005, 1633-1656.

[30] Ruzicka, M., Modeling, mathematical and numerical analysis ofelectrorheological fluids, Applications of Mathematics 49 (6), 2004, 565-609. http://dml.cz/dmlcz/134585.

[31] Ursescu, A., Channel flow of electrorheological fluids under an inhomogeneous electric field, Ph. D. Thesis, submitted to Institute of Mechanics, Darmstadt University of Technology (DUT), Darmstadt, Germany, 21st January, 2005.

[32] See, H., Constitutive equation for electrorheological fluids based on the chain model, Journal of Physics D: Applied Physics 32, 2000, 1625-1633.

[33] Wineman, A. S. and Rajagopal, K. R., On constitutive equations for electrorheological materials, Continuum Mechanics Thermodynamics 7, 1995, 1-22.

[34] Gavin, H. P., Hanson, R. D. and Mc-Clamroch, N. H., Control of structures using electrorheological dampers, Paper # 272, 11th World Conference on Earthquake Engineering, Acapulco, Mexico, 23-28 June, 1996, Elsevier Science Ltd.

[35] Prusa, V. and Rajagopal, K. R., Flow of an electrorheological fluid between eccentric rotating cylinders, Preprint No. 2010-025, Necas Center for Mathematical Modeling, Research team 1, Mathematical Institute of the Charles University, Sokolovska 83, 18675, Praha 8, Czech Republic, 23rd of July, 2010. http://ncmm.karlin.mff.cuni.cz/.

[36] Lee, D. Y. and Wereley, N. M., Quasi-steady Herschel-Bulkley analysis of electro- and magneto-rheological flow mode dampers, Journal of Intelligent Material Systems and Structures 10, 1999, 761-769.

[37] Delivorias, R., The potential of magnetorheological fluid in crashworthiness design, MSc. Thesis, submitted to Automotive Engineering - Vehicle Safety, Department of Mechanical Engineering, Eindhoven Univerity of Technology (EUT), Eindhoven, The Netherlands, Decemberer 2005.

[38] Bossis, G., Lemaire, E., Volkova, O. and Clercx, H., Yield stress in magnetorheological and electro-rheological fluids: a comparison between microscopic and macroscopic structural models, Journal of Rheolo-

gy 41, 1997, 687-704.

[39] Bossis, G., Volkova, O., Lacis, S. and Meunier, A., Magnetorheology: fluids, structures and rheology, In: Ferrofluids, Odenbach, S. (ed.), Berlin, Springer, 2002.

[40] Lemaire, E., Meunier, A., Bossis, G., Liu, J., Felt, D., Bashtovoi, P. and Matoussevitch, N., Influence of the particle size on the rheology of magnetorheological fluids, Technical report, Universite de Nice, Laboratoire de Physique Matiere Condensee, France, June 1995.

[41] Nillson, M. and Ohlson, N. G., An electrorheological fluid in squeeze mode, Journal of Intelligent Material Systems and Structures 11, July 2000, 545-554. doi: 10.1106/MB24-94JR-T6LX-648L.

[42] Kohl, J. G. and Tichy, J. A., Expressions for coefficients of electrorheological fluid dampers, Lubrication Science 10 (2), February 1998, 135-143.

[43] Prabhu, S. R. B., Harisha, S. R. and Gangadharan, K. V., Design, synthesis and fabrication of magneto-rheological fluid damper for low frequencies application, Journal of Mechanical and Civil Engineering (IOSR-JMCE), e-ISSN: 2278-1684, p-ISSN: 2320-334X, International Conference on Advances in Engineering & Technology, 2014 (ICAET-2014), 60-63.

[44] Liao, W. H. and Lai, C. Y., Harmonic analysis of a magnetorheological damper for vibration control, Smart Materials and Structures 11, 2002, 288-296.

[45] Nishiyama, H., Oyama, T. and Fujita, T., Damping characteristics of MR fluids in low magnetic fields, International Journal of Modern Physics B 15 (6&7), 2001, 829-836.

[46] Kelso, S., Denoyer, K., Blankinship, R., Potter, K. and Lindler, J., Experimental validation of a novel stictionless magnetorheological fluid isolator, SPIE conference on Smart Structures and Materials, Paper #5052-24, San Diego, CA, USA, March 2-6, 2003.

[47] Braz-Cesar, M. T. and Barros, R. C., Experimental behavior and numerical analysis of MR dampers, Proceeding of the 15th World Conference on Earthquake Engineering (15WCEE), Lisbon, Portugal, 24-28 September, 2012.

[48] Truong, D. Q. and Ahn, K. K., MR fluid damper and its application to force sensorless damping control system, Chapter 15, © 2012 Truong and Ahn, licensee InTech. This is an open access chapter distributed under the terms of the Creative Commons Attribution License. (http://creativecommons.org/licenses/by/3.0), http://dx.doi.org/10.5772/51391.

[49] Ambhore, N. H., Hivarale, S. D. and Pangavhane, D. R., A comparative study of parametric models of magnetorheological fluid suspension dampers, International Journal of Mechanical Engineering and Technology (IJMET), ISSN 0976-6340 (Print), ISSN 0976-6359 (Online) 4 (1), January-February 2013, © IAEME, 222-232.

[50] Dimock, G. A., Yoo, J. H. and Wereley, N. M., Quasi-steady Bingham biplastic analysis of electrorheological and magnetorheological dampers, Journal of Intelligent Material Systems and Structures 13 (9), September 2002, 549-559. doi: 10.1106/104538902030906.

[51] Yang, G., Spencer, B. F. Jr., Carlson, J. D. and Sain, M. K., Large-scale MR fluid dampers: modeling and dynamic performance considerations, Engineering Structures 24, 2002, 309-323.

[52] Sapinski, B., Linearized characterization of a magnetorheological fluid damper, Mechanics 24 (2), 2005, 144-149.

[53] Butz, T. and Von Stryk, O., Modelling and simulation of electro-and magnetorheological fluid dampers,

Journal of Applied Mathematics and Mechanics – ZAMM (Zeitschrift fur Angewandte Mathhematik und Mechanik) 82 (1), 2002, 3-20.

[54] Ambhore, N. H., Hivarale, S. D. and Pangavhane, D. R., A study of Bouc–Wen model of magnetorheological fluid damper for vibration control, International Journal of Engineering Research and Technology (IJerT), ISSN 2279-0181, 2 (2), February 2013, 1-6.

[55] Stanway, R., Sproston, J. L. and Stevens, N. G., Non-linear modelling of an electrorheological vibration damper, Journal of Electrostatics 20 (2), 1987, 167-184.

[56] Bingham, E. C., An investigation of the laws of plastic flow, U.S. Bureau of Standards Bulletin 13, 1916, 309-353.

[57] Nakamura, M. and Sawada, T., Numerical study on the laminar pulsating flow of slurry, The Journal of Non-Newtonian Fluid Mechanics 22 (2), 1987, 191-206.

[58] Gamota, D. R. and Filisko, F. E., Dynamic mechanical studies of electrorheological materials: moderate frequencies, Journal of Rheology 35 (3), 1991, 399-425.

[59] Occhiuzzi, A., Spizzuoco, M. and Serino, G., Experimental analysis of magnetorheological dampers for structural control, Smart Materials and Structures 12, 2003, 703-711.

[60] Powel, J. A., Modelling the oscillatory response of an electrorheological fluid, Smart Materials and Structures 3, 1994, 416-438.

[61] Makris, N., Burton, S. A. and Taylor, D. P., Electrorheological damper with annular ducts for seismic protection applications, Smart Materials and Structures 5, 1996, 551-564.

[62] Madhaven, V., Wereley, N. M. and Kamath, G. M., Hysteresis modelling of semi-active magnetorheological helicopter dampers, Journal of Intelligent Material Systems and Structures 10 (8), 1999, 624-633.

[63] Jimenez, R. and Alvarez, L., Real time identification of structures with magnetorheological dampers, Proc. 41st IEEE Conference on Decision and Control 1, 2002, 1017-1022.

[64] Sakai, C., Ohmori, H. and Sano, A., Modeling of MR damper with hysteresis for adaptive vibration control, Proc. 42nd IEEE Conference on Decision and Control 4, 2003, 3840-3845.

[65] Terasawa, T., Sakai, C., Ohmori, H. and Sano, A., Adaptive identification of MR damper for vibration control CDC, Proc. 43rd IEEE Conference on Decision and Control 3, 2004, 2297-2303.

[66] Ikhouane, F. and Dyke, S. J., Modelling and identification of a shear magnetorheological damper, Smart Materials and Structures 16 (3), 2007, 605-616.

[67] Wang, L. X. and Kamath, H., Modelling hysteretic behavior in magnetorheological fluids and dampers using phase-transition theory, Smart Materials and Structures 16 (6), 2006, 1725-1733.

[68] Oh, H. U., Experimental demonstration of an improved magnetorheological fluid damper for suppression of vibration of a space flexible structure, Smart Materials and Structures 13 (5), 2004, 1238-1244.

[69] Ma, X. Q., Rakheja, S. and Su, C. Y., Relative assessments of current dependent models for magnetorheological fluid dampers, ICNSC'06, Proc. IEEE International Conference on Networking, Sensing and Control, 23-26 April 2006, Ft. Lauderdale, FL, USA, 510-515.

[70] Wang, E. R., Ma, X. Q., Rakheja, S. and Su, C. Y., Modelling the hysteretic characteristics of a magnetorheological fluid damper, The Proceedings of the Institution of Mechanical Engineers 217, 2003, 537-550.

[71] Kwok, N. M., Ha, Q. P., Nguyen, T. H., Li, J. and Samali, B., A novel hysteretic model for magnetorheological fluid dampers and parametric identification using particle swarm optimization, Sensors and Actuators 132 (2), 2006, 441-451.

[72] Bastien, J., Michon, G., Manin, L. and Dufour, R., An analysis of the modified Dahl and Masing models: application to a belt tensioner, Journal of sound and vibration 302 (4-5), 2007, 841-864. ISSN 0022-460X.

[73] Wereley, N. M., Kamath, G. M. and Pang, L., Idealized hysteresis modelling of er and MR dampers, Journal of Intelligent Material Systems and Structures 9 (8), 1998, 642-649.

[74] Li, W. H., Zhang, P. Q., Gong, X. L. and Kosasih, P. B., Characterization and Modeling a MR Damper Under Sinusoidal Loading, in Electrorheological Fluids and Magnetorheological Suspensions (ERMR 2004), Lu, K., Shen, R. and Liu, J. (eds.), World Scientific, 14 June 2005, 769-775.

[75] Sims, N. D., Holmes, N. J. and Stanway, R., A unified modelling and model updating procedure for electrorheological and magnetorheological vibration dampers, Smart Materials and Structures 13 (1), 15 December 2003, 100-121.

[76] Katona, M., A visco-elastic-plastic constitutive model with a finite element solution methodology, Technical Report, R866, Civil Engineering Laboratory, Naval Construction Battalion Center, Port Hueneme, California, 93043, June 1978, 159.

[77] Li, W. H., Yao, G. Z., Chen, G., Yeo, S. H. and Yap, F. F., Testing and steady state modeling of a linear MR damper under sinusoidal loading, Smart Materials and Structures 9, 2000, 95-102.

[78] Hong, S. R., Choi, S. B., Choi, Y. T. and Wereley, N. M., A hydro-mechanical model for hysteretic damping force prediction of er damper: experimental verification, Journal of Sound and Vibration 285 (4-5), August 2005, 1180-1188.

[79] Chae, Y., Ricles, J. M. and Sause, R., Maxwell nonlinear slider model for seismic response prediction of semi-active controlled magnetorheological dampers, COMPDYN 2011, III ECCOMAS Thematic Conference on Computational Methods in Structural Dynamics and Earthquake Engineering, Papadrakis, M. and Fragiadakis, M. and Plevris (eds.), Corfu, Greece, 26-28 May 2011.

[80] Wen, Y. K., Method for random vibration of hysteretic systems, Journal of Engineering Mechanics (American Society of Civil Engineers) 102 (2), 1976, 249-263.

[81] Spencer, B. F. Jr., Dyke, S. J., Sain, M. K. and Carlson, J. D., Phenomenological model of a magnetorheological damper, Journal of Engineering Mechanics 123, 1997, 230-238.

[82] Tse, T. and Chang, C., Shear-mode rotary magnetorheological damper for small-scale structural control experiments, Journal of Structural Engineering 130 (6), 2004, 904-911.

[83] Rodriguez, A., Iwata, N., Ikhouane, F. and Rodellar, J., Model identification of a large-scale magnetorheological fluid damper, Smart Materials and Structures 18 (1), January 2009, 1-12.

[84] Atabay, E. and Ozkol, I., Application of a magnetorheological damper modeled using the current-dependent Bouc-Wen model for shimmy suppression in a torsional nose landing gear with and without freeplay, Journal of Vibration and Control 20, August 2014, 1622-1644.

[85] Dominguez, A., Sedaghati, R. and Stiharu, I., A new dynamic hysteresis model for magnetorheological dampers, Smart Materials and Structures 15 (5), 2006, 1179-1189.

[86] Kwok, N. M., Ha, Q. P., Nguyen, M. T., Li, J. and Samali, B., Bouc-Wen model parameter iden-

tification for a MR fluid damper using computationally efficient genetic algorithms, ISA Transactions 46 (2), 2007, 167-179.

[87] Rakheja, S., Ma, X. Q. and Su, C. Y., Development and relative assessments of model characterizing the current dependent hysteresis properties of magnetorheological fluid dampers, Journal of Intelligent Materials and Structures 18 (5), 2007, 487-502.

[88] Herschel, W. H. and Bulkley, R., Konsistenzmessungen von Gummi-Benzollösungen, Kolloid Zeitschrift 39, 1926, 291-300. doi: 10.1007/BFO 1432034.

# 第6章 磁致伸缩与电致伸缩材料

## 6.1 磁致伸缩材料的特性

众所周知，铁磁①类材料在磁化时形状或尺寸会发生改变（$10^{-6}$数量级的体积变化），这一特性称为磁致伸缩。为了理解这一特性，可以将铁磁材料的内部结构划分为多个磁畴，其中每个磁畴都具有均匀的磁极化。通过施加磁场可引起磁畴之间的边界移动和旋转，进而导致材料尺寸的变化。这些磁畴移动和旋转的能力可归因于材料晶体结构的各向异性。这种各向异性在易磁化方向产生最小的系统自由能。由于晶向与不同长度相关联，在铁磁材料内诱发了应变。磁致伸缩材料能够将磁能转化为机械能，反之亦然，从而可用于制造致动器和传感器。在致动器模式下（图6.1中的典型致动器），线圈因通电而产生磁场，引发磁致伸缩材料在易磁化方向的尺寸变化。然而，在传感器模式下，施加到磁致伸缩材料上的机械力会导致其磁场发生变化，从而在线圈中产生可测量的电流（图6.2中的典型传感器）。

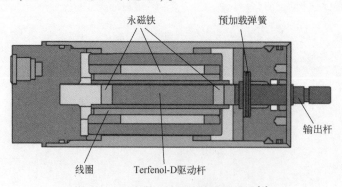

图6.1　一种典型的磁致伸缩致动器[1]

---

① 铁磁性：某些材料（如铁、镍、钴及其合金）形成永久磁铁或被磁铁吸引的特性。

# 第6章 磁致伸缩与电致伸缩材料

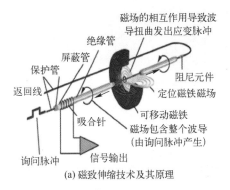
(a) 磁致伸缩技术及其原理

(b) 分体式磁致伸缩传感器

图 6.2　一种典型的磁致伸缩传感器（iTarget 传感器公司）

磁致伸缩系数 λ（可以是正数或负数）定义如下：

$$\lambda = \frac{\Delta L}{L}(H 饱和时) \tag{6.1}$$

式中：$L$ 为原始长度；$\Delta L$ 为长度的变化；$H$ 为单位测得的磁场（A/m）。注意，饱和是指当材料的磁化强度不再随着外加磁场 $H$ 的增加而增加，进而导致总磁通密度 $B$ 达到平稳状态。应该区分开 $H$ 和 $B$ 这两个符号①，前者的测量单位为 A/m，后者的测量单位为 T 或 N/（A·m）（另见本章末尾的附录 A；$H \equiv (B/\mu_0) - M$）。图 6.3 所示为正 λ（膨胀）和负 λ（收缩）的含义。众所周知，由于磁致伸缩系数 λ 具有各向异性，Lee 推导出了与各向异性磁致伸缩一致的最简易方程[2]，该方程仅包含 $\lambda_{100}$ 和 $\lambda_{111}$ 两个常数，代表当晶体从理想退磁状态沿 [100] 和 [111] 晶向磁化至饱和时的总应变。该等式如下：

$$\lambda_s = \frac{3}{2}\lambda_{100}\left(\alpha_1^2\beta_1^2 + \alpha_2^2\beta_2^2 + \alpha_3^2\beta_3^2 - \frac{1}{3}\right) + 3\lambda_{111}(\alpha_1\alpha_2\beta_1\beta_2 + \alpha_2\alpha_3\beta_2\beta_3 + \alpha_3\alpha_1\beta_3\beta_1) \tag{6.2}$$

式中：$\alpha_1$、$\alpha_2$、$\alpha_3$ 为磁化畴相对于参考坐标系（表示为 1、2、3）的方向余弦；$\beta_1$、$\beta_2$、$\beta_3$ 为磁致伸缩应变相对于参考坐标系的方向余弦。例如，对于多晶的、未织构化的立方材料，饱和磁致伸缩系数 $\lambda_s$ 的简化表达式如下：

$$\lambda_s = \frac{2}{5}\lambda_{100} + \frac{3}{5}\lambda_{111} \tag{6.3}$$

式中：符号 100 和 111 为结晶方向。表 6.1 列出了一些常见磁致伸缩材料的典型 λ 值[3]。

---

① $M$ 为磁化矢量场，$\mu_0$ 为磁性常数，见 5.4.2 节。

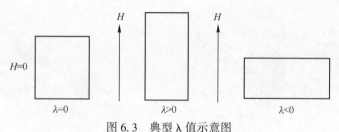

图 6.3　典型 λ 值示意图

表 6.1　常见铁磁材料的磁致伸缩系数

| 材　料 | 磁致伸缩系数/×10⁻⁶ | | | |
|---|---|---|---|---|
| | $\lambda_{100}$ | $\lambda_{111}$ | $\lambda_s$（计算值） | $\lambda_s$（计算值） |
| 铁 | 19.5 | −18.8 | −3.5 | −7 |
| 镍 | −45.9 | −25.3 | −32.9 | −34 |
| 钴 | 多晶材料 | | | −55 |
| Terfenol-D[①] | 90 | 1640 | 1020 | 2000 |

① Terfenol-D（$Tb_xDy_{1-x}Fe_2$），其中 Ter 为铽，Fe 为铁，NOL 表示海军军械实验室，D 为镝。

　　Terfenol-D 为致动器最常用的磁致伸缩合金之一，其产生的应变比传统磁致伸缩材料大 100 倍，比传统压电陶瓷大 2~5 倍。另一种有趣的磁致伸缩材料为 Galfenol[①]（$Fe_{100-x}Ga_x$），这是一种铁镓合金，由美国海军研究人员于 1999 年发现[4]，可通过改变 $x$ 实现所需的磁性和力学性能。表 6.2 列出了这两种合金的典型特性，而图 6.4 所示为美国 Etrema Products 公司制造的 Galfenol 和 Terfenol-D 样件[②]。根据文献 [5]，Galfenol 合金在非常低的磁场中可表现出适当的磁致伸缩性能，具有非常低的磁滞，并在 −20~80℃ 范围内显示具有高拉伸强度（相比于压电材料）和有限的磁力学性能变化[4]。一般来说，这些合金（对于 Ga 含量小于 20%）是可加工的、可延展的且可焊接的，具有高居里温度且耐腐蚀。因此，Galfenol 合金也可用在致动器和传感器上。

表 6.2　典型 Terfenol-D 和 Galfenol 的物理特性[①]

| 特　性 | 值 | |
|---|---|---|
| | Terfenol-D | Galfenol |
| 标准成分 | $Tb_{0.3}Dy_{0.7}Fe_{1.92}$ | $Fe_{81.6}Ga_{18.4}$ |
| 密度/(kg/m³) | 9200~9300 | 7800 |

① Galfenol（[EQN3117]），其中 Gal 为 gallium（镓）的缩写，Fe 为 iron（铁）的缩写，NOL 是美国海军军械实验室。

② www.etrma.com/。

续表

| 特　　性 | 值 | |
|---|---|---|
| | Terfenol-D | Galfenol |
| 硬杨氏模量/GPa | 50~90 | 60~80 |
| 软杨氏模量/GPa | 18~55 | 40~60 |
| 体积模量/GPa | 90 | 125 |
| 声速/(m/s) | 1395~2444 | 2265~2775 |
| 抗拉强度/MPa | 28~40 | 350 |
| 抗压强度/MPa | 300~880 | — |
| 疲劳强度/MPa @完全扭转（$R=-1$） | — | 75 |
| 维氏硬度/(HV)[2] | 650 | 227 |
| 最小层压厚度/mm | 1.0 | 0.25 |
| 热膨胀系数（CTE）（$10^{-6}$/℃）@25℃ | 11 | 11 |
| 比热/(kJ/(kg·K)) | 0.33 | — |
| 热导率（W/(mK)）@25℃ | 13.5 | 15~20[3] |
| 熔点/℃ | 1240 | 1450 |
| 电阻率/(Ω/m) | $60\times10^{-3}$ | $85\times10^{-3}$ |
| 居里温度/℃ | 380 | 670 |
| 应变（估计线性值）/ppm | 800~1200 | 200~250 |
| 能量密度/(kJ/m³) | 4.9~25 | 0.3~0.6 |
| 压磁常数 $d_{33}$/(nm/A) | 6~10 | 20~30 |
| 耦合系数 | 0.7~0.8 | 0.6~0.7 |
| 相对磁导率 $\mu_r$ | 2~10 | 75~100[4] |
| 饱和磁通密度/T | 1 | 1.5~1.6 |

注：1. 美国 ETREMA Products 公司，www.etrema.com。

2. 维氏硬度是一种材料硬度的度量，是根据材料在菱形金刚石压头负载下产生的压痕尺寸计算得出的。[HV]=维氏菱形金刚石编号。

3. 基于低碳钢的估算值。

4. 高度依赖于 Galfenol 的压力状态。在接近 0~13 千磅/平方英寸压力下的测量值范围为 300~20。

本章附录 B 提供了 Terfenol-D 和 Galfenol 超磁致伸缩材料的其他数据。

除了 Terfenol-D 和 Galfenol 合金，另一种常见的磁致伸缩复合材料是非晶合金 $Fe_{81}Si_{3.5}B_{13.5}C_2$，商品名为 Metglas 2605SC[①]，其具有约 20 微应变或更大的

---

① www.metglas.com/products/magnetic_materials/2605sa1.asp。

高饱和磁致伸缩常数 $\lambda$，以及小于 1kA/m 的低磁各向异性场强（以达到磁饱和）。图 6.5 和表 6.3 给出了该材料的特性。

图 6.4　有钻孔和无钻孔的 Galfenol 杆和 Terfenol-D 件

（资料来源：www.etrema.com）

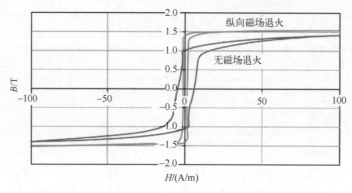

图 6.5　Metglas 合金 2605SA1——一种典型的磁滞回线（见彩插）

（资料来源：www.metglas.com/products/magnetic_materials/2605sa1.asp）

表 6.3　典型 Metglas ® 2605SA1 和 2605HB1M 磁性合金的特性

| 特　性 | 值 |
| --- | --- |
| 铸态饱和感应/T | 1.56 |
| 退火后的最大直流磁导率/$\mu$ | 600000 |
| 铸态最大直流磁导率/$\mu$ | 45000 |
| 饱和磁致伸缩/ppm | 27 |
| 电阻率/$\mu\Omega\cdot$cm | 130 |
| 居里温度/℃ | 395 |
| 厚度/$\mu$m | 23 |
| 标准有效宽度 | — |

# 第6章 磁致伸缩与电致伸缩材料

续表

| 特　性 | 值 |
|---|---|
| 最小/mm | 5 |
| 最大/mm | 213 |
| 密度/(g/cm³) | 7.18 |
| 维氏硬度（50g 载荷） | 900 |
| 抗拉强度/GPa | 1~2 |
| 弹性模量/GPa | 100~110 |
| 层压系数/% | 84 |
| 热膨胀/(ppm/℃) | 7.6 |
| 结晶温度/℃ | 510 |
| 持续使用温度/℃ | 150 |

磁致伸缩材料最显著的现象之一为迟滞行为，如图6.6所示。这种磁滞回线的形状称为"蝴蝶回线"，可以使用 Jiles-Atherton 模型[6]获得。虽然该模型最初是为各向同性材料而开发的，但后来进行了扩展[7-8]，使其可用于各向异性磁性材料的建模。

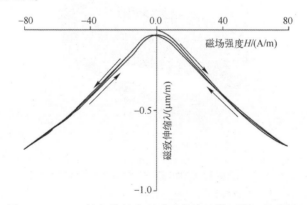

图6.6　Mn-Zn 铁氧体的典型磁致伸缩迟滞回线（见彩插）

(资料来源：https://commons.wikimedia.org/wiki/File:Magnetostrictive_hysteresis_loop_of_Mn-Zn_ferrite.png)

目前，磁致伸缩器件已应用于以下领域：超声波清洗机、大功率直线电机、自适应光学定位器、主动振动/噪声控制系统、医疗和工业超声波、泵和声纳。此外，还开发出了磁致伸缩线性马达、反应物致动器和调谐减振器，以及多种超声磁致伸缩换能器，用于外科手术工具、水下声纳以及化学和材料处理等领域[1,9-12]。磁致伸缩致动器受限于其温度依赖性，会表现出小的位移，当过热时可能引起一些问题。然而，磁致伸缩致动器的优点包括对低电压的响应、不随时间衰减、较少的迟滞、耐磨损、耐撕裂以及共振时的高动态应变。图 6.7 展示了使用 Terfenol-D 致动器改变机翼形状的概念实现变形。

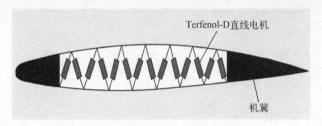

图 6.7　用作 Gulfstream Ⅲ 飞机双梁机翼桁架翼肋的 Terfenol-D 直线电机
(资料来源：www.machinedesign.com)

Cedrat Technologies[①] 利用 Terfenol-D 超磁致伸缩材料制造出了磁致伸缩致动器，其设计在静态或动态应用中可以低工作电压（<12V）下产生高作用力（>20kN）和大蠕变（>200μm）（图 6.8）。

图 6.8　CEDRAT 磁致伸缩致动器
(资料来源：www.cedrat-technologies.com)

磁致伸缩位置传感器本质上是一种声波传感装置。高分辨率时钟可测量声波在固定参考点和移动磁铁之间的传播时间。根据已知的声波速度和运行时间来计算磁铁的绝对位置。此外，其磁铁不与波导管接触，所以不存在零件磨损现象。基本的磁致伸缩位置传感器由 4 个基本部件组成：定位磁铁、波导管、拾音器（声波转换器）和驱动器以及信号调节电子设备。导电"波导"线

---

① www.cedrat-technologies.com/en/technologies/actuators/magnetic-actuators-motors.html。

(通常由镍基磁致伸缩合金制成）载有一个短脉冲电流——询问脉冲。当该脉冲信号沿着波导传播时，会产生一个沿轴向围绕波导的同心磁场。当波导磁场与定位磁铁的永久磁场发生交互时，磁致伸缩效应会导致波导上产生应变，进而形成压力波并沿着波导在远离定位磁体的两个方向上以声速传播（约2850m/s）。一束波在波导管的远端被阻尼机构吸收，有助于防止来自波导末端引起干扰的反射；而另一束波传输到拾音器。完整的波导、阻尼模块和拾取组件通常称为传感元件（Sensing Element，SE）。典型的MTS（美国Mechanical Testing & Simulation公司）磁致伸缩传感器如图6.9所示。

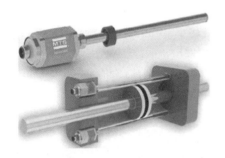
(a) MTS连续位置反馈传感器(MTS提供)

(b) 线性位置传感器/非接触式/绝对磁致伸缩/带SSI接口-50-2500mm，100g,200V/m, SIL2

图6.9　MTS磁致伸缩传感器

## 6.2　磁致伸缩材料的本构方程

忽略热效应的影响，磁致伸缩材料的本构方程可表示为如下张量表达式[13-14]：

$$S_{ij} = s^H_{ijkl}T_{kl} + d_{kij}H_k + m_{klij}H_kH_l$$
$$B_j = d^*_{jkl}T_{kl} + \mu^T_{jk}H_k \quad (6.4)$$

式中：$S$和$T$分别为机械应变和应力；$d$为磁致伸缩常数（从$S$-$H$回线线性部分的斜率获得）；$B$和$H$分别为磁通密度和磁场强度；$s^H$为恒定磁场下的弹性柔度；$\mu^T$为恒定应力下的磁导率；$m_{klij}$为场磁致伸缩模量张量，物理上表示单位外磁场产生的磁致伸缩应变（其量纲为$m^2/A^2$）。对于低激励水平，使用线性方程[15]来产生类似于压电材料的形式（见第3章）：

$$S = s^H T + dH \quad (6.5a)$$
$$B = d^* T + \mu^T H \quad (6.5b)$$

式（6.5a）通常称为逆向效应，式（6.5b）称为直接效应，类似于压电方程。这些等式传统上用于感应和致动应用。

两个磁致伸缩常数 $d$ 和 $d^*$ 定义如下：

$$\begin{cases} d = \dfrac{\partial S}{\partial H}_{@\,T=\text{const}} \\ d^* = \dfrac{\partial B}{\partial T}_{@\,H=\text{const}} \end{cases} \quad (6.6)$$

对于小应变，可以假定 $d = d^*$。改变式（6.5）的边界会得出以下矩阵形式：

$$\begin{Bmatrix} T \\ B \end{Bmatrix} = \begin{bmatrix} E^H & -e \\ e^* & \mu^S \end{bmatrix} \begin{Bmatrix} S \\ H \end{Bmatrix} \quad (6.7)$$

式中：$E^H$ 为恒定磁场下磁致伸缩材料的杨氏模量；$\mu^S$ 为恒定应变下的磁导率；常数 $e$ 和 $e^*$ 定义如下：

$$\begin{aligned} e &= E^H d \\ e^* &= E^H d^* \end{aligned} \quad (6.8)$$

对于轴 $x_3$ 作为磁极化方向并处于应力下的多晶铁磁材料，式（6.5）具有的矩阵表达式如下：

$$\begin{Bmatrix} S_1 \\ S_2 \\ S_3 \\ S_4 \\ S_5 \\ S_6 \end{Bmatrix} = \begin{bmatrix} s_{11}^H & s_{12}^H & s_{13}^H & 0 & 0 & 0 \\ s_{12}^H & s_{11}^H & s_{13}^H & 0 & 0 & 0 \\ s_{13}^H & s_{13}^H & s_{33}^H & 0 & 0 & 0 \\ 0 & 0 & 0 & s_{44}^H & 0 & 0 \\ 0 & 0 & 0 & 0 & s_{44}^H & 0 \\ 0 & 0 & 0 & 0 & 0 & s_{66}^H \end{bmatrix} \begin{Bmatrix} T_1 \\ T_2 \\ T_3 \\ T_4 \\ T_5 \\ T_6 \end{Bmatrix} + \begin{bmatrix} 0 & 0 & d_{31} \\ 0 & 0 & d_{31} \\ 0 & 0 & d_{33} \\ 0 & d_{15} & 0 \\ d_{15} & 0 & 0 \\ 0 & 0 & 0 \end{bmatrix} \begin{Bmatrix} H_1 \\ H_2 \\ H_3 \end{Bmatrix} \quad (6.9)$$

$$\begin{Bmatrix} B_1 \\ B_2 \\ B_3 \end{Bmatrix} = \begin{bmatrix} 0 & 0 & 0 & 0 & d_{15} & 0 \\ 0 & 0 & 0 & d_{15} & 0 & 0 \\ d_{31} & d_{31} & d_{33} & 0 & 0 & 0 \end{bmatrix} \begin{Bmatrix} T_1 \\ T_2 \\ T_3 \\ T_4 \\ T_5 \\ T_6 \end{Bmatrix} + \begin{bmatrix} \mu_{11}^T & 0 & 0 \\ 0 & \mu_{11}^T & 0 \\ 0 & 0 & \mu_{33}^T \end{bmatrix} \begin{Bmatrix} H_1 \\ H_2 \\ H_3 \end{Bmatrix} \quad (6.10)$$

式中：对于压电材料，应变和应力矢量的定义相类似，即

$$\begin{Bmatrix} \varepsilon_1 \\ \varepsilon_2 \\ \varepsilon_3 \\ \gamma_{23} \\ \gamma_{13} \\ \gamma_{12} \end{Bmatrix} \equiv \begin{Bmatrix} S_1 \\ S_2 \\ S_3 \\ S_4 \\ S_5 \\ S_6 \end{Bmatrix}; \begin{Bmatrix} \sigma_1 \\ \sigma_2 \\ \sigma_3 \\ \tau_{23} \\ \tau_{13} \\ \tau_{12} \end{Bmatrix} \equiv \begin{Bmatrix} T_1 \\ T_2 \\ T_3 \\ T_4 \\ T_5 \\ T_6 \end{Bmatrix} \quad (6.11)$$

式中：$\varepsilon_i$ 和 $\gamma_{ij}$ 分别为法向应变和剪切应变；$\sigma_i$ 和 $\tau_{ij}$ 分别为法向应力和剪切应力。假设 **B** 和 **H** 在磁性一侧呈线性关系，而 $S$ 和 $T$ 在弹性一侧呈线性关系，则可以得到内能[13]如下：

$$U = \frac{S_i T_i}{2} + \frac{H_m B_m}{2} = \frac{T_i S_{ij} T_j}{2} + \frac{T_i d_{im} H_m}{2} + \frac{H_m d_{mi} T_i}{2} + \frac{H_m \mu_{mk} H_k}{2} \quad (6.12)$$
$$= U_e + U_{em} + U_{me} + U_m = U_e + 2U_{em} + U_m$$

式中：$U_e$ 和 $U_m$ 分别为系统的纯弹性能和磁能，而 $U_{em} = U_{me}$ 为相互磁弹性能。

耦合系数 $k$ 为一个重要的品质因数，定义如下：

$$k = \frac{U_{me}}{\sqrt{U_e \cdot U_m}} \quad \text{或} \quad k^2 = \frac{U_{me}^2}{U_e \cdot U_m} \quad (6.13)$$

例如，假设磁场和应力仅存在于 $x_3$ 的方向上，即

$$\begin{cases} H_1 = H_2 = 0, \quad H_3 \neq 0 \\ T_1 = T_2 = T_4 = T_5 = T_6 = 0, \quad T_3 \neq 0 \end{cases} \quad (6.14)$$

则由式 (6.9) 和式 (6.10) 可得

$$\begin{cases} B_1 = B_2 = 0, \quad B_3 \neq 0 \\ S_4 = S_5 = S_6 = 0, \quad S_1 = S_2 \neq 0, \quad S_3 \neq 0 \end{cases} \quad (6.15)$$

但是，由于 $T_1 = T_2 = 0$，$S_1$ 和 $S_2$ 对系统的弹性能量没有贡献，那么对于这种特殊情况的耦合系数 $k^2$（式 (6.13)）可表示如下：

$$k_{33}^2 = \frac{d_{33}^2}{\mu_{33}^T \cdot S_{33}^H} \quad (6.16)$$

如文献 [16] 所述，对于纵向耦合情况，假设在 $x_3$ 方向上，所有未知量、应变、应力和磁场均与其平行。在这种情况下，可以省略下标，即

$$S = s^H T + d \cdot \boldsymbol{H} \quad (6.17\text{a})$$
$$\boldsymbol{B} = d \cdot T + \mu^T \boldsymbol{H} \quad (6.17\text{b})$$

将式 (6.17a) 表示为 $T$ 和 $\boldsymbol{B}$ 的函数，并将式 (6.17b) 表示为 $S$ 和 $\boldsymbol{H}$ 的函数，则

$$S = s^B T + \frac{d}{\mu^T} B \tag{6.18a}$$

$$B = \frac{d}{S^H} S + \mu^S H \tag{6.18b}$$

式中：两个新常数 $s^B$ 和 $\mu^S$ 定义如下：

$$s^B \equiv \frac{\partial S}{\partial T}\bigg|_{@B=\text{const.}} = s^H \left(1 - \frac{d^2}{s^H \mu^T}\right) = s^H (1 - k^2) \tag{6.19a}$$

$$\mu^S \equiv \frac{\partial B}{\partial H}\bigg|_{@S=\text{const.}} = \mu^T \left(1 - \frac{d^2}{S^H \mu^T}\right) = \mu^T (1 - k^2) \tag{6.19b}$$

$$\mu^S S^H = \mu^T S^B \tag{6.19c}$$

式中：$k^2$ 为之前 $x_3$ 方向定义的耦合系数，见式（6.13）。

磁致伸缩材料的线性本构方程的另一种表现形式使用了以下变量（称为恒定应力下的磁阻率）：

$$\begin{cases} g = \dfrac{d}{\mu^T} \\ v^T \equiv \dfrac{1}{\mu^T} \end{cases} \tag{6.20}$$

则本构方程可表示为

$$S = s^B T + g \cdot B \tag{6.21a}$$
$$H = -g \cdot T + v^T B \tag{6.21b}$$

不同的耦合常数 $\kappa^2$ 定义如下：

$$\kappa^2 = -\frac{g^2}{s^B \cdot v^T} = -\frac{k^2}{(1-k^2)} \tag{6.22}$$

由于其假定属性，所以很少使用。

本构方程（式（6.21a）和式（6.21b））可以进一步修改为以下公式：

$$T = \frac{S}{S^B} - \lambda \cdot B = S \cdot c^B - \lambda \cdot B, \quad c^B = \frac{1}{s^B} \tag{6.23a}$$

$$H = -\lambda \cdot S + v^S \cdot B \tag{6.23b}$$

式中：$c^B$ 为恒定磁场下的刚度矩阵。经典的磁致伸缩常数 $\lambda$ 可表示为（通过将式（6.23a）和式（6.23b）等效于式（6.18a）和式（6.18b））

$$\begin{cases} \lambda = \dfrac{d}{\mu^T \cdot s^B} = \dfrac{d}{\mu^S \cdot s^H} \\ v^S = \dfrac{1}{\mu^S} \end{cases} \tag{6.24}$$

# 第6章 磁致伸缩与电致伸缩材料

为了获得耦合系数 $k^2$，必须首先计算式（6.23a）和式（6.23b）右侧系数的交叉乘积比例，得出 $\lambda^2 \mu^S s^B$。然后可以得出以下表达式，另见式（6.16）：

$$\lambda^2 \cdot \mu^S \cdot s^B = \left(\frac{d^2}{\mu^T \cdot s^B \cdot \mu^S \cdot s^H}\right) \mu^S \cdot s^B = \frac{d^2}{\mu^T \cdot s^H} = k^2 \tag{6.25}$$

最后，可以推导出 $d$ 和 $\lambda$ 之间的关系如下[16]：

$$d = \frac{1}{\lambda} \frac{k^2}{(1-k^2)} \Rightarrow k^2 = \frac{\lambda \cdot d}{(1+\lambda d)} \tag{6.26}$$

下面介绍磁致伸缩材料的一维模型。假设一个根据磁致伸缩材料建立的一维模型，如图6.10所示。杆的长度为 $L$，横截面积为 $A$，杨氏模量为 $E$。拉力 $P$ 作用在杆上，产生应变 $S$ 和应力 $T=P/A$。然后计算图中系统的各个能量贡献。由于存在应变 $S$ 和应力 $T$，杆内累积的应变能由下式给出（应变-应力关系见式（6.7））：

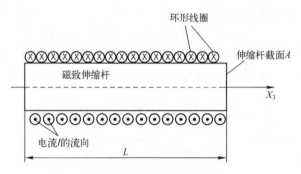

图6.10 一维示意图模型

$$\begin{aligned} U_{\text{strain}} &= \frac{1}{2}\int_v S \cdot T \cdot \mathrm{d}v = \frac{1}{2}\int_v S \cdot (E^H \cdot S - eH) \cdot \mathrm{d}v \\ &= \frac{1}{2}\int_v S \cdot E^H \cdot S \cdot \mathrm{d}v - \frac{1}{2}\int_v S \cdot e \cdot H \cdot \mathrm{d}v = \frac{A \cdot L}{2}E^H \cdot S^2 - \frac{A \cdot L}{2}S \cdot e \cdot H \end{aligned} \tag{6.27}$$

式中：$\mathrm{d}v$ 为杆的微分体积。磁致伸缩杆中的磁能计算如下（$\boldsymbol{B}$-$\boldsymbol{H}$ 关系见式（6.7））：

$$\begin{aligned} U_{\text{magn.}} &= \frac{1}{2}\int_v \boldsymbol{B} \cdot \boldsymbol{H} \cdot \mathrm{d}v = \frac{1}{2}\int(e^* \cdot S + \mu^S H) \cdot \boldsymbol{H} \cdot \mathrm{d}v \\ &= \frac{1}{2}\int_v e^* \cdot S \cdot \boldsymbol{H} \cdot \mathrm{d}v + \frac{1}{2}\int_v \boldsymbol{H} \cdot \mu^S \cdot \boldsymbol{H} \cdot \mathrm{d}v = \frac{A \cdot L}{2}e^* \cdot S \cdot \boldsymbol{H} + \frac{A \cdot L}{2}\mu^S \cdot \boldsymbol{H}^2 \end{aligned} \tag{6.28}$$

具有 $N$ 匝和电流 $I$ 的线圈所做的外部磁功由下式给出：

$$W_{\text{coil}} = I \cdot N \cdot \mu^T \cdot H \cdot A \tag{6.29}$$

最终贡献为力 $P$ 所做的机械功：

$$W_{\text{mech}} = P \cdot \Delta X_3 = P \cdot S \cdot L \tag{6.30}$$

总势能为上述 4 种贡献的总和，即

$$\pi = -U_{\text{strain}} + U_{\text{magn.}} - W_{\text{coil}} + W_{\text{mech}} \tag{6.31}$$

将各表达式（式（6.27）~式（6.30））代入式（6.31），可得

$$\pi = -\frac{1}{2} A \cdot L \cdot E^H \cdot S^2 + A \cdot L \cdot e \cdot S \cdot H$$
$$+ \frac{1}{2} A \cdot L \cdot \mu^S \cdot H^2 - I \cdot N \cdot \mu^T \cdot H \cdot A + P \cdot S \cdot L \tag{6.32}$$

使用具有两个变量 $S$ 和 $H$ 的变分原理（如哈密顿（Hamilton）原理[①]），可给出两个线性方程如下：

$$-A \cdot L \cdot E^H S + A \cdot L \cdot e \cdot H + P \cdot L = 0 \tag{6.33}$$

$$A \cdot L \cdot e \cdot S + A \cdot L \cdot \mu^S \cdot H - I \cdot N \cdot \mu^T \cdot A = 0 \tag{6.34}$$

式（6.32）和式（6.33）除以 $A \cdot L$，可得

$$E^H S - e \cdot H = \frac{P}{A} \tag{6.35}$$

$$e \cdot S + \mu^S \cdot H = \frac{I \cdot N \cdot \mu^T}{L} \tag{6.36}$$

求解两个未知数 $S$ 和 $H$，同时使用关系式 $\mu^S = \mu^T - d \cdot E^H \cdot d^*$ 和 $e = E^H \cdot d$，得

$$S = \frac{I \cdot N \cdot \mu^T \cdot A \cdot e + \mu^S \cdot P \cdot L}{A \cdot L \cdot \mu^T \cdot E^H} \tag{6.37}$$

$$H = \frac{I \cdot N}{L} - \frac{P}{A \cdot e}\left(1 - \frac{\mu^S}{\mu^T}\right) \tag{6.38}$$

式（6.37）可以简化如下：

$$S = \lambda + S_T \tag{6.39}$$

式中：磁致伸缩 $\lambda$ 定义如下：

$$\lambda = \frac{I \cdot N \cdot \mu^T \cdot A \cdot e}{A \cdot L \cdot \mu^T \cdot E^H} = \frac{I \cdot N \cdot d}{L} \tag{6.40}$$

---

[①] $\delta\left(\int^\pi \mathrm{d}t\right) = 0$，其中 $t$ 表示时间。

机械应变$S_T$定义如下：

$$S_T = \frac{\mu^S \cdot P}{A \cdot \mu^T \cdot E^H} = \frac{P}{A} \cdot \frac{1}{\overline{E^H}} \qquad (6.41)$$

式中

$$\overline{E^H} \equiv E^H \left(\frac{\mu^T}{\mu^S}\right)$$

式中：$\overline{E^H}$为修正后的杨氏模量。由于存在关系$\mu^S = \mu^T - d \cdot E^H \cdot d^*$，$\overline{E^H}$也可以表示为

$$\overline{E^H} \equiv E^H + \frac{e^2}{\mu^S} \qquad (6.42)$$

当$(\mu^T/\mu^S)$比值等于1时，修正的杨氏模量$\overline{E^H}$等于$E^H$。对于所有其他情况（包括Terfenol-D），这一比值不等于1且杨氏模量值也不同（更大）。类似于压电性，可以将"阻尼力"定义为达到的最高力。其表达式如下：

$$F_{\text{blooking}} \equiv \frac{E^H \cdot A}{L} \Delta L = E^H \cdot A \cdot S_{\max} = E^H \cdot A(\lambda + S_T)_{\max} \approx E^H \cdot A \cdot \lambda_{\max} \qquad (6.43)$$

除了本章所述的内容，还应关注其他用于描述磁致伸缩材料本构方程的模型（见文献［17］中的综述）。此类模型包括微观激励模型、每个单晶转换机制的近似、基于材料唯象描述的宏观本构模型（主要为Preisach模型[18]，其最初是为描述铁磁体的磁化而开发的；可以通过正确选择模型的各种参数来处理宽范围的迟滞回线），以及基于热力学原理的模型。

## 6.3 电致伸缩材料的特性

电致伸缩是所有介电材料中存在的一种共有特性，类似于磁致伸缩材料，当施加外部电场时，在其晶格中显示出轻微的离子位移，导致在施加电场的方向上产生整体应变。

由此产生的应变与极化的平方成正比。改变电场的极性（从+到-）不会改变应变的方向。在数学层面，这种关系可以表示如下：

$$S_{ij} = Q_{ijkl} \cdot P_k \cdot P_l \qquad (6.44)$$

式中：$S_{ij}$为二阶应变张量；$Q_{ijkl}$为电致伸缩系数的四阶张量；$P_k$、$P_l$为一阶极化张量。应注意，与压电材料的线性关系相比，电致伸缩材料的极化为二次关系（图6.11）。尽管所有电介质均表现出某种电致伸缩行为，但只有弛豫铁电体具有非常高的电致伸缩常数，如铌镁酸铅（PMN）、铌镁酸铅-钛酸铅（PMN-PT）

和锆钛酸镧铅（PLZT）。应注意，由于式（6.44）所示的关系，施加机械应力不会产生电荷，意味着电致伸缩材料与压电材料不同，不能用作传感器。通过比较压电材料和电致伸缩材料发现，锆钛酸铅（PZT）具有低应变（约为0.06%）和显著的迟滞（为15%~20%），而电致伸缩PMN材料则表现出更高的应变（约为0.1%）和更低的迟滞（为1%~4%）。但是，由于工作温度的限制，PMN需要进行隔热。例如，TRS公司①开发了3种用于声纳的PMN-PT复合材料，即PMN-15、PMN-38和PMN-85，3种材料的工作温度范围分别为0~30℃、10~50℃、75~95℃。在2℃下，PMN-15在30kV/cm的电场下表现出约0.14%的应变；而在22℃和相同的电场下，其应变下降到0.115%。

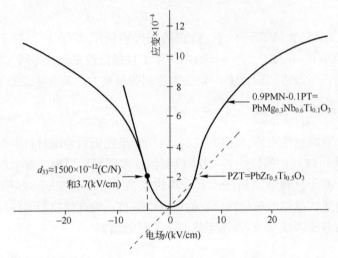

图6.11　电致伸缩（PMN-PT）和压电（PZT）陶瓷中的机电耦合[19]

表6.4列出了钛酸铅PT EC-97②、铌镁酸铅PMN EC-98以及同一公司制造的软质PZT这3种电致伸缩陶瓷的典型机电性能。与压电现象的研究相比，电致伸缩材料的研究更为有限[19-27]。图6.12所示为德国Piezomechanik Dr. Lutz Pickelmann GmbH公司制造的致动器和堆栈，表明通过简单的电压控制可减少蠕变并精确复位到零点，并声称这些影响至少比压电元件小1个数量级。基于电致伸缩的堆栈能够提供6~25μm的蠕变和50~12N/μm的刚度，或者当施加电压增至120V时，能够提供6~40μm的更高蠕变和120~20N/μm的刚

---

① www.trstechnologies.com/Materials/Electrostrictive-Ceramics。
② 来源于EXELIS Inc.电子陶瓷产品和材料规格目录（原始数据来自EDO Corp.，该公司目前是EXELIS Inc.的子公司），www.exelisinc.com。

度。图 6.13 显示了两种电致伸缩致动器（来源于 Newport 公司①）：第 1 种为 AD-30，具有 30μm 的蠕变；第 2 种为 AD-100，具有 100μm 的蠕变，分辨力 0.04μm，可提供 45N 的力。

表 6.4  电致伸缩陶瓷与软质 PZT 的典型机电特性（@25℃）

| 特　性 | PT EC-97 | PMN EC-98 | 软质 PZT EC-65 |
|---|---|---|---|
| 密度 $\rho/(kg/m^3)$ | 6700 | 7850 | 7500 |
| 杨氏模量/GPa | 128 | 61 | 66 |
| 居里温度/℃ | 240 | 170 | 350 |
| 机械强度（薄盘） | 950 | 70 | 100 |
| 介电常数@1kHz | 270 | 5500 | 1725 |
| 耗散常数@1kHz/% | 0.9 | 2.0 | 2.0 |
| $k_{31}$ | 0.01 | 0.35 | 0.36 |
| $k_p$ | 0.01 | 0.61 | 0.62 |
| $k_{33}$ | 0.53 | 0.72 | 0.72 |
| $k_{15}$ | 0.35 | 0.67 | 0.69 |
| $d_{31}/(\times 10^{-12} m/V)$ | -3.0 | -312 | -173 |
| $d_{33}/(\times 10^{-12} m/V)$ | 68.0 | 730 | 380 |
| $d_{15}/(\times 10^{-12} m/V)$ | 67.0 | 825 | 584 |
| $g_{31}/(\times 10^{-3} V\, m/N)$ | -1.7 | -6.4 | -11.5 |
| $g_{33}/(\times 10^{-3} V\, m/N)$ | 32.0 | 15.6 | 25.0 |
| $g_{15}/(\times 10^{-3} V\, m/N)$ | 33.5 | 17.0 | 38.2 |
| $\varepsilon_{11}^E/(\times 10^{-12} m^2/N)$ | — | 16.3 | 15 |
| $\varepsilon_{12}^E/(\times 10^{-12} m^2/N)$ | — | -5.6 | -5.3 |
| $\varepsilon_{33}^E/(\times 10^{-12} m^2/N)$ | 7.7 | 21.1 | 18.3 |
| $S_{11}^D/(\times 10^{-12} m^2/N)$ | — | 14.3 | 13.2 |
| $S_{12}^D/(\times 10^{-12} m^2/N)$ | — | -7.6 | -7.3 |
| $S_{33}^D/(\times 10^{-12} m^2/N)$ | — | 10.2 | 8.8 |
| 泊松比 $v$ | — | 0.34 | 0.31 |
| 老化率（每 10 年变化）/% | | | |
| 介电常数 | -0.3 | -1.5 | -0.8 |
| 耦合常数 | -0.4 | -0.4 | -0.3 |
| 共振频率 | 0.05 | 0.4 | 0.2 |

① www.newport.com。

续表

| 特　性 | PT EC-97 | PMN EC-98 | 软质 PZT EC-65 |
|---|---|---|---|
| 电场依赖性 | | | |
| 最大正电场/(V/mm) | — | 900（仅限 DC） | 600 |
| 最大负电场/(V/mm) | — | 450（仅限 DC） | 300 |
| 应用场（25℃） | 79 | 79 | 79 |
| 介电常数增加/% | 1.5 | 22.5 | 12.0 |
| 耗散系数 | 0.8 | 6.2 | 7.3 |

注：假设横向各向同性，即 $(\cdots)_{23}=(\cdots)_{13}$。

 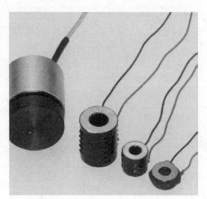

(a) 致动器　　　　　　　　　(b) 预加载外壳内的堆栈

图 6.12　电致伸缩致动器

（资料来源：www.piezomechanick.com）

(a) Newport AD-30 超高分辨力　　(b) Newport AD-100 超高分辨力
　　电致伸缩致动器　　　　　　　　　电致伸缩致动器

图 6.13　Newport 电致伸缩致动器

最后，AOA Xinetics①制造了铌镁酸铅（PMN：RE）电致伸缩多层致动器，其中 PMN 层厚在 0.1~0.15μm（图 6.14）。典型的性能规格如下：
(1) 最佳温度范围：10~30℃。
(2) 典型工作电压：0~100V。
(3) 0~100V 之间的典型迟滞：<5%。
(4) 室温下的光学蠕变：λ/2000@ 0.63μm。
(5) 分辨力：μm。
(6) 工作频率：取决于驱动。
(7) 蠕变范围：在 100V 和 25℃下为 3.1~30μm。

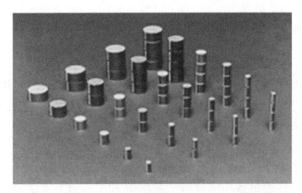

图 6.14　AOA Xinetics 电致伸缩多层 PMN 致动器

## 6.4　电致伸缩材料的本构方程

如 6.3 节所述，电致伸缩效应表现为应变（应力）和极化 $P$（或外加电场）的二次关系，如图 6.15 所示。任何介电材料，无论是陶瓷还是聚合物，有无对称中心，均会显示电致伸缩效应。

应变 $S_{ij}$ 和极化张量 $P_k$（或 $P_l$）之间的关系可表示如下（假设没有施加机械应力）[21,28-36]：

$$S_{ij} = Q_{ijkl} \cdot P_k \cdot P_l \tag{6.45}$$

式中：$Q_{ijkl}$ 为与电荷相关的电致伸缩系数（$m^4/C^2$）。对于各向同性的电致伸缩材料（如聚合物），式（6.45）可简化如下：

---

① http://www.northropgrumman.com/BusinessVentures/AOAXinetics/IntelligentOptics/Products/Pages/Actuators.aspx。

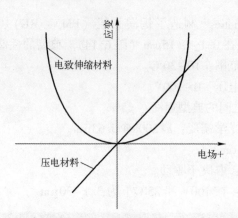

图 6.15 电致伸缩和压电材料的应变与外加电场关系示意图

$$S_1 = Q_{13} \cdot P^2, \quad S_2 = Q_{23} \cdot P^2, \quad S_3 = Q_{33} \cdot P^2 \qquad (6.46)$$

式中：$S_1$、$S_2$（$S_1 = S_2$ 平面各向同性）为横向应变（垂直于极化轴），$S_3$ 为纵向应变（平行于极化轴）。另一个式（图 6.15）可将应变和外加电场（无机械应力）关联起来，即

$$\boldsymbol{S}_{ij} = \boldsymbol{M}_{ijkl} \cdot E_k \cdot E_l \qquad (6.47)$$

式中：$M_{ijkl}$ 为电致伸缩系数（张量符号）（m²/V²）；$S_{ij}$ 为应变；$E_k$（或 $E_l$）为施加的场。对于线性电致伸缩材料，系数 $M$ 和 $Q$ 存在如下关系：

$$\boldsymbol{M}_{ijkl} = \boldsymbol{Q}_{mnkl} \cdot \chi_{mi} \cdot \chi_{nj} \qquad (6.48)$$

式中：$\chi_{mi}$（或 $\chi_{nj}$）为电介质极化率。

应注意，极化 $P_m$ 可通过如下关系式与电场 $E_m$ 关联起来：

$$P_m = (\varepsilon - \varepsilon_0) E_m = D_m - \varepsilon_0 \cdot E_m \qquad (6.49)$$

式中：$\varepsilon_0$ 为电介质真空介电常数（= 8.85×10$^{-12}$ F/m）；$\varepsilon$ 为其他位置的介电常数；$D_m$ 为电位移。

另一种表示极化和电场之间关系的公式如下：

$$P_k = \chi_{kl} \cdot E_l \qquad (6.50)$$

式中，如前所述，$\chi_{kl}$ 为电介质极化率张量[36]。

根据希望定义的独立变量，使用唯象描述，如势能可表示如下：

$$\pi(T, E) = -\frac{1}{2} s_{ijkl} T_{ij} T_{kl} - M_{ijkl} T_{ij} E_k E_l - \frac{1}{2} \varepsilon_{ij} E_i E_j \qquad (6.51)$$

导致电致伸缩的非线性耦合机电本构方程如下：

$$\begin{cases} \boldsymbol{S}_{ij} = S_{ijkl} \cdot T_{kl} + \boldsymbol{M}_{ijkl} \cdot E_k \cdot E_l \\ D_i = \boldsymbol{\varepsilon}_{ij} \cdot E_j + 2 M_{klij} \cdot T_{kl} \cdot E_j \end{cases} \qquad (6.52)$$

式中：$S_{ij}$和$T_{ij}$分别为应变和应力；$s_{ijkl}$、$M_{ijkl}$和$\varepsilon_{ij}$为弹性柔量、电致伸缩系数和介电常数矩阵。类似于压电材料的情况，$D_i$和$E_j$分别为电位移和电场。

对于将应力和极化（$P$）作为独立变量的吉布斯自由能表达式[①]，忽略高阶项时，可得

$$\Delta G(T,P) = -\frac{1}{2} s_{ijkl} \cdot T_{ij} \cdot T_{kl} - Q_{mnij} \cdot P_m \cdot P_n \cdot T_{ij} + \frac{1}{2}\gamma_{mn} \cdot P_m \cdot P_n \quad (6.53)$$

式中

$$S_{ij} \equiv -\left(\frac{\partial \Delta G}{\partial T_{ij}}\right)_{@P,\text{Temp. const}}$$

$$E_m \equiv \left(\frac{\partial \Delta G}{\partial P_m}\right)_{@T,\text{Temp.}=\text{const}}$$

由此其本构方程可表示如下：

$$\begin{aligned} S_{ij} &= s_{ijkl} \cdot T_{kl} + Q_{mnij} \cdot P_i \cdot P_j \\ P_m &= \chi_{mn} \cdot E_n + 2M_{mnij} \cdot E_n \cdot T_{ij} \end{aligned} \quad (6.54)$$

式中：$\gamma_{mn}$为电介质极化率的线性倒数。再次使用吉布斯自由能的表达式定义如下：

$$\begin{cases} S_{ij} \equiv -\left(\dfrac{\partial \Delta G}{\partial T_{ij}}\right)_{@E,\text{Temp.}=\text{const}} \\ P_m \equiv -\left(\dfrac{\partial \Delta G}{\partial E_m}\right)_{@T,\text{Temp.}=\text{const}} \end{cases} \quad (6.55)$$

该本构方程可表示为

$$\begin{cases} S_{ij} = S_{ijkl} \cdot T_{kl} + M_{mnij} \cdot E_m \cdot E_n \\ P_m = \chi_{mn} \cdot E_n + 2M_{mnij} \cdot E_n \cdot T_{ij} \end{cases} \quad (6.56)$$

应注意，在没有外部机械应力（$T_{kl}=0$）的情况下，由式（6.52）、式（6.54）和式（6.56）可得

$$S_{ij} = M_{ijkl} \cdot E_k \cdot E_l \quad (6.57)$$

或者

$$S_{ij} = Q_{mnij} \cdot P_i \cdot P_j$$

这正是本节前述的式（6.47）和式（6.45）。

电致伸缩材料的可逆效应可表示为

---

[①] 吉布斯自由能、吉布斯能量、吉布斯函数或自由焓是一种热力学势，用于在恒温热力学系统测定其可能执行的最大或可逆功。

$$\begin{cases} Q_{mnij} = -\frac{1}{2}\frac{\partial \gamma_{mn}}{\partial T_{ij}} \\ M_{mnij} = \frac{1}{2}\frac{\partial \chi_{mn}}{\partial T_{ij}} \end{cases} \quad (6.58)$$

另一个可逆效应涉及电场对压电系数 $d_{mij}$ 的依赖性，即

$$M_{mnij} = \frac{1}{2}\left[\frac{\partial d_{mij}}{\partial E_n}\right] \quad (6.59)$$

并且

$$Q_{mnij} = \frac{1}{2}\left[\frac{\partial d_{mij}}{\partial P_n}\right]$$

可得出如下关系：

$$\begin{cases} d_{ijm} \equiv \left[\dfrac{\mathrm{d}S_{ij}}{\mathrm{d}E_m}\right] = 2M_{ijmn} \cdot E_n \\ g_{ijm} \equiv \left[\dfrac{\mathrm{d}S_{ij}}{\mathrm{d}P_m}\right] = 2Q_{ijmn} \cdot P_n \\ d_{ijm} = \chi_{mk} \cdot g_{ijm} = 2\chi_{mk} \cdot Q_{ijmn} \cdot P_n \end{cases} \quad (6.60)$$

式 (6.45) 的显式表达式（对于具有点阵 $m3m$ 的 PMN 晶体）如下：

$$\begin{Bmatrix} S_1 \\ S_2 \\ S_3 \\ S_4 \\ S_5 \\ S_6 \end{Bmatrix} = \begin{bmatrix} Q_{11} & Q_{12} & Q_{12} & 0 & 0 & 0 \\ Q_{12} & Q_{11} & Q_{12} & 0 & 0 & 0 \\ Q_{12} & Q_{12} & Q_{11} & 0 & 0 & 0 \\ 0 & 0 & 0 & Q_{44} & 0 & 0 \\ 0 & 0 & 0 & 0 & Q_{44} & 0 \\ 0 & 0 & 0 & 0 & 0 & Q_{44} \end{bmatrix} \begin{Bmatrix} P_1^2 \\ P_2^2 \\ P_3^2 \\ P_2 P_3 \\ P_3 P_1 \\ P_1 P_2 \end{Bmatrix} \quad (6.61)$$

且 $Q_{44} = 4Q_{1212} = 2(Q_{11} - Q_{12})$。

文献 [31-35] 中提出了此本构方程的另一种形式。假设电场不饱和，则需要增加高阶项，该本构方程可表示如下：

$$\begin{cases} \{S\} = [S^D]\{T\} + [g(Q,D)]\{D\} \\ \{E\} = -2[g(Q,D)]\{T\} + [\beta^T]\{D\} \end{cases} \quad (6.62)$$

式中：$[\beta^T] = [\varepsilon^T]^{-1}$ 为恒定应力下的电介质不渗透性矩阵；$\{S\}$、$\{T\}$ 为凝聚应变和应力张量；$\{E\}$、$\{D\}$ 为电场和电位移矢量；$[S^D]$ 为恒定电位移下的弹性柔度矩阵；$[g(Q,D)]$ 定义如下：

$$[g] = \begin{bmatrix} Q_{11}D_1 & Q_{12}D_1 & Q_{12}D_1 & 0 & (Q_{11}-Q_{12})D_3 & (Q_{11}-Q_{12})D_2 \\ Q_{12}D_2 & Q_{11}D_2 & Q_{12}D_2 & (Q_{11}-Q_{12})D_3 & 0 & (Q_{11}-Q_{12})D_1 \\ Q_{12}D_3 & Q_{12}D_3 & Q_{11}D_3 & (Q_{11}-Q_{12})D_2 & (Q_{11}-Q_{12})D_1 & 0 \end{bmatrix}$$
(6.63)

发现图 6.15 中所示的电致伸缩现象与实验中所发生的现象并不完全相同：实验中高于某一电场时，应变回线变平，显示电场饱和。图 6.16 显示了这种行为，其中 $E_s$ 为饱和电场。

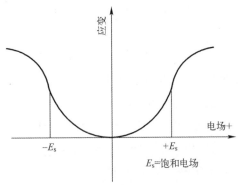

图 6.16　具有饱和电场（$E_s$）的电致伸缩材料的应变与外加电场关系示意图

在这种情况下，极化与电场的关系如文献 [30-37] 中所示如下：

$$P = (\varepsilon - \varepsilon_0) E_S \tanh\left[\frac{E}{E_S}\right] = \chi E_S \tanh\left[\frac{E}{E_S}\right] \quad (6.64)$$

使用方程（6.45），可将饱和情况下的应变与极化的关系表示为

$$\{S\} = [Q]\{P\}^2 = [Q](\varepsilon-\varepsilon_0)^2 E_S^2 \tanh^2\left[\frac{E}{E_S}\right] = [Q]\chi^2 E_S^2 \tanh^2\left|\frac{E}{E_S}\right| \quad (6.65)$$

因此，电场的范围被分成如下两个区域：

$$\begin{cases} 对于 \ E \leq E_S \Rightarrow P \approx \chi E & 且 \quad S = ME^2 \\ 对于 \ E \geq E_S \Rightarrow P \approx \chi E_S & 且 \quad S = ME_S^2 = \text{const} \end{cases} \quad (6.66)$$

对于饱和情况，该本构方程可用 Voight 表达式①表示如下[30-34]：

$$\begin{cases} T_i = c_{ij}^P \cdot S_j - c_{ij} \cdot Q_{jmn} \cdot P_m \cdot P_n \\ E_m = -2c_{ij}^P \cdot Q_{jmn} \cdot S_i \cdot P_n + \chi_m^{-1} \cdot P_m^S \operatorname{arctanh}\left(\frac{P_m}{P_m^S}\right) + 2c_{ij}^P \cdot Q_{jmn} \cdot Q_{ikl} \cdot P_n \cdot P_k \cdot P_l \end{cases}$$
(6.67)

---

① Voigt 符号：$\{\sigma_{xx}, \sigma_{yy}, \sigma_{zz}, \tau_{yz}, \tau_{xz}, \tau_{xy}\} \equiv \{T_1, T_2, T_3, T_4, T_5, T_6\}$。

式中：$c_{ij}^P$ 为在恒定极化 $P$ 下的刚度矩阵；$P_m^S$ 为饱和极化。对于各向同性材料的简单情况，其刚度矩阵表示如下：

$$[c_{ij}^P] = \frac{E_{\text{Young's}}}{(1+v)(1-2v)} \begin{bmatrix} 1-v & v & v & 0 & 0 & 0 \\ v & 1-v & v & 0 & 0 & 0 \\ v & v & 1-v & 0 & 0 & 0 \\ 0 & 0 & 0 & \frac{1-2v}{2} & 0 & 0 \\ 0 & 0 & 0 & 0 & \frac{1-2v}{2} & 0 \\ 0 & 0 & 0 & 0 & 0 & \frac{1-2v}{2} \end{bmatrix} \quad (6.68)$$

然而电致伸缩系数均为 0，以下情况除外：

$$\begin{cases} Q_{111} = Q_{222} = Q_{333} \\ Q_{123} = Q_{133} = Q_{211} = Q_{233} = Q_{311} = Q_{322} \\ 2(Q_{111} - Q_{222}) = Q_{412} = Q_{523} = Q_{613} \end{cases} \quad (6.69)$$

对于一维分析，式（6.47）可表示为

$$S = ME^2 \quad (6.70)$$

二次关系（式（6.70））可能会对致动器的实现造成一些问题。因此，为了便于使用电致伸缩材料，希望将二次关系转换成线性关系，这可以通过叠加直流电（Direct Current，DC）场和交流电（Alternating Current，AC）场来实现，即

$$E = E_{\text{dc}} + E_{\text{ac}} \quad (6.71)$$

将式（6.71）带入式（6.70）可得

$$S = M(E_{\text{dc}} + E_{\text{ac}})^2 = M(E_{\text{dc}}^2 + 2E_{\text{dc}}E_{\text{ac}} + E_{\text{ac}}^2) \quad (6.72)$$

注意，$ME_{\text{dc}}^2 = \text{const} \equiv S_{\text{dc}}$，可将式（6.72）表示为

$$S - S_{\text{dc}} \approx M(2E_{\text{dc}}E_{\text{ac}} + E_{\text{ac}}^2) \quad (6.73)$$

假设叠加的交变场比直流场和交流场的乘积小得多。即 $E_{\text{ac}}^2 \ll 2E_{\text{dc}}E_{\text{ac}}$，可得出应变和交流场之间的线性关系如下：

$$S - S_{\text{dc}} \approx M(2E_{\text{dc}}E_{\text{ac}}) \quad (6.74)$$

定义项 $2ME_{\text{dc}} \equiv d_{\text{eff}}$，式中 $d_{\text{eff}}$ 为有效压电应变系数，可得

$$S - S_{\text{dc}} \approx d_{\text{eff}} E_{\text{ac}} \quad (6.75)$$

为了比较电致伸缩材料和压电材料的性能，需要比较在相同方向上作用的压电材料的 $d_{\text{eff}}$ 和 $d$。

最后，对于图6.17所示的电致伸缩杆的情况，用式（6.62）可导出一个简单的一维解[31]。杆的尺寸为长度$L$、宽度$b$和厚度$t$，其中$t \ll L$且$b \ll L$。唯一非0的应力为$T_3$（$T_1 = T_2 = T_4 = T_5 = T_6 = 0$）。唯一的电位移为$D_3$，而$D_1 = D_2 = 0$（因为电极垂直于$x_3$轴）。对于一维情况，式（6.62）可简化如下：

$$\begin{cases} S_3 = S_{33}^D T_3 + Q_{11} D_3^2 \\ E_3 = -2Q_{11} T_3 D_3 + \beta_{33}^T D_3 \end{cases} \quad (6.76)$$

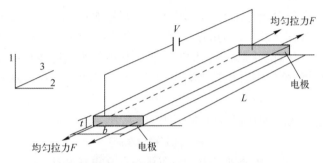

图6.17 具有几何尺寸的电致伸缩杆示意图

从沿着$x_3$轴的力的平衡过程中，可得出在垂直于$x_3$轴所有横截面上的应力$T_3$是恒定的，即

$$T_3 = \frac{F}{b \cdot t} = \text{const} \quad (6.77)$$

应用高斯定理（$(\partial D_3 = \partial x_3) = 0$），可得$D_3$为沿着电致伸缩杆的常数。因此，根据式（6.76）中的第2个方程，电场也是恒定的且

$$E_3 = -\frac{V}{L} \quad (6.78)$$

然后求解$D_3$，可得

$$D_3 = -\frac{V}{L \left( \beta_{33}^T - \dfrac{2Q_{11} F}{b \cdot t} \right)} \quad (6.79)$$

将$D_3$的表达式代入式（6.76）的第1个方程，可得到应变$S_3$，即

$$S_3 = s_{33}^D \frac{F}{b \cdot t} + \frac{Q_{11} \cdot V^2}{L^2 \left( \beta_{33}^T - \dfrac{2Q_{11} F}{b \cdot t} \right)^2} \quad (6.80)$$

## 6.5 附录四

### 6.5.1 附录A

为了理解磁致伸缩现象，需要一些磁学和电路方面的基本背景知识，这些知识在本附录中进行介绍。

磁场是一个矢量场，用两个密切相关的符号 $B$ 和 $H$ 表示。$B$ 称为磁通量密度或磁感应强度，其单位为国际单位制中的特斯拉（T）或牛顿/安培·米（N/A·m），而 $H$ 称为磁场或磁化场强度，其单位为国际单位制中的安培/米（A/m）。在 CGS 单位中，$B$ 以高斯（G）[①]为测量单位，$H$ 以奥斯特（Oe）[②]为测量单位。

$B$ 通常定义为施加在运动电荷上的洛伦兹力，即

$$F=q(E+V\times B) \tag{6.a}$$

式中：$F$, $q$, $E$, $V$, $B$ 分别是力、粒子的电荷、电场和磁场，×表示叉乘。相应地，1C 的电荷以 1m/s 的速度通过 1T 的磁场时，会受到 1N 的力。因此，以下关系成立：

$$1T = 1\,\frac{V \cdot s}{m^2} = 1\,\frac{N}{A \cdot m} = 1\,\frac{J}{A \cdot m^2} = 1\,\frac{H \cdot A}{m^2} = 1\,\frac{Wb}{m^2} = 1\,\frac{kg}{C \cdot s} = 1\,\frac{N \cdot s}{C \cdot m} = 1\,\frac{kg}{A \cdot s^2} \tag{6.b}$$

式中：A 表示安培；C 表示库仑；H 表示亨利（电感单位），Wb 表示韦伯（磁通量单位，$\Phi_B$）。

$B$ 的另一种定义方式是使用 Biot-Savart 定律，该定律定义了长度为 $dl$ 导电元件上电流 $I$ 产生的磁场，即

$$B(r) = \frac{\mu_0}{4\pi}\int_C \frac{Idl \times \hat{r}}{r^2} \tag{6.c}$$

式中：$\mu_0$ 为真空渗透率，定义如下：

$$\mu_0 = 4\pi \times 10^{-7}\,\frac{N}{A^2} \approx 1.25666370614\cdots\times 10^{-6}\,\frac{H}{m}\,\text{或}\,\frac{T \cdot m}{A}\,\text{或}\,\frac{Wb}{A \cdot m}\,\text{或}\,\frac{V \cdot s}{A \cdot m} \tag{6.d}$$

对于无限长直导体（图 6.a），电流为 $I$ 时，与其垂直距离为 $d$ 处的磁场为

---

[①] 1T=10000G。
[②] 致敬丹麦物理学家 Hans Christian Ørsted。他观察到载流线圈中产生的磁场使安培表（测量电流的仪器）在开关时发生了偏转，从而发现了磁场和电流之间的关系。

$$B = \frac{\mu_0 I}{2\pi d} \tag{6.e}$$

再次参考图6.a,对于半径为 $R$ 的单线圈,其中心的磁场可表示如下:

$$B = \frac{\mu_0 I}{2R} \tag{6.f}$$

而对于具有 $N$ 个线圈的螺线管(图6.a),则表示如下:

$$B = \frac{\mu_0 N I}{L} \equiv \mu_0 n I, \quad n = \frac{N}{L} \tag{6.g}$$

式中:$n$ 为单位长度的线圈数;$N$ 为线圈数;$L$ 为线圈长度。

$H$ 和 $B$ 之间的一般关系可以表示为

$$H \equiv \frac{B}{\mu_0} - M \tag{6.h}$$

式中:$M$ 为材料区域磁化强度的磁化矢量场,其定义为该区域单位体积的净磁偶极矩。对于感应磁通量大的材料,该关系可简化如下(图6.b):

$$B \equiv \mu H \tag{6.i}$$

式中:$\mu$ 为材料的磁导率(H/m 或 N/A$^2$)。相对磁导率定义如下:

$$\mu_r \equiv \frac{\mu}{\mu_0} \tag{6.j}$$

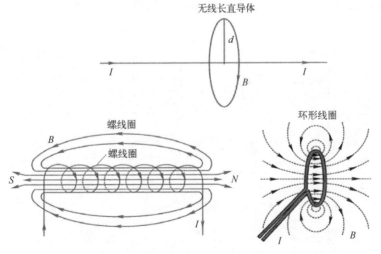

图6.a 定义磁场 $B$ 的各种配置

(资料来源:www.thunderbolts.info)

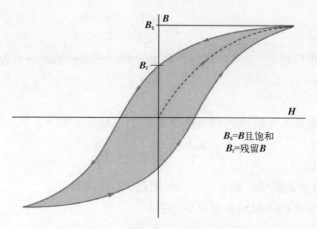

图 6.b 一种典型的 $B$-$H$ 迟滞回线

式中：$\mu_0 = 4\pi \times 10^{-7} \dfrac{N}{A^2}$。另一项为磁化率，其定义如下：

$$\chi_m \equiv \mu_r - 1 \tag{6.k}$$

### 6.5.2 附录 B

图 6.c~图 6.m 所示为最著名的公司之一——Etrema Products 公司生产的两种超磁致伸缩材料 Terfenol-D 和 Galfenol 的数据。各种影响表示为不同参数的函数。

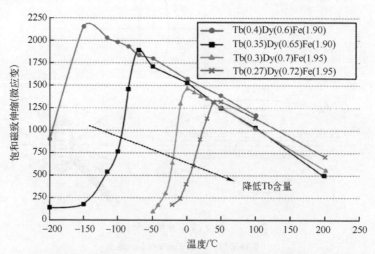

图 6.c 典型 Terfenol-D 合金：温度影响（见彩插）
（资料来源：ETREMA 专有资料）

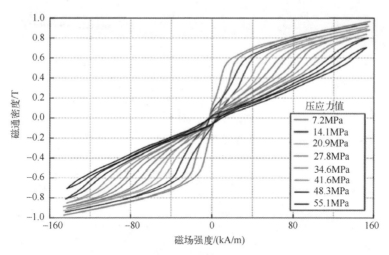

图 6.d 典型 Terfenol-D 合金：不同压应力下磁通密度与磁场强度之间的关系（见彩插）
（资料来源：ETREMA 专有资料）

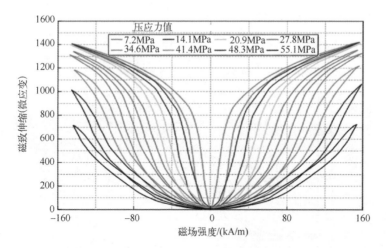

图 6.e 典型 Terfenol-D 合金：不同压应力下磁致伸缩与磁场之间的关系（见彩插）
（资料来源：ETREMA 专有资料）

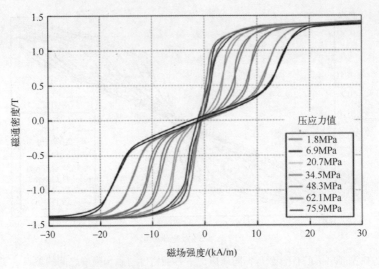

图 6.f 典型 Galfenol 合金（生长状态，无应力退火 BH）：
不同压应力下磁通量密度与磁场强度之间的关系（见彩插）
（资料来源：ETREMA 专有资料）

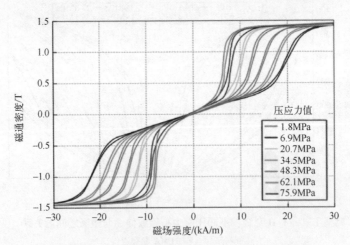

图 6.g 典型 Galfenol 合金（应力退火 BH）：
不同压应力下磁通密度与磁场强度之间的关系（见彩插）
（资料来源：ETREMA 专有资料）

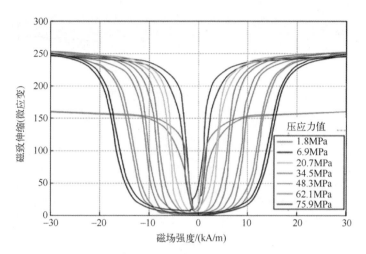

图 6.h 典型 Galfenol 合金（见彩插）（生长状态，无应力退火 BH）：
不同压应力下磁致伸缩与磁场强度之间的关系

（资料来源：ETREMA 专有资料）

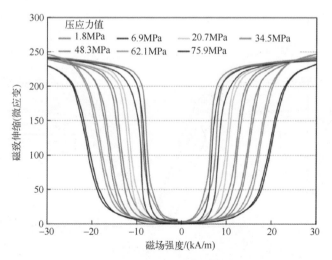

图 6.i 典型 Galfenol 合金（应力退火 BH）：
不同压应力下磁致伸缩与磁场强度之间的关系（见彩插）

（资料来源：ETREMA 专有资料）

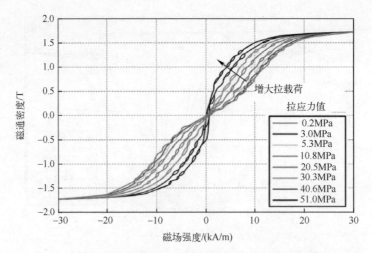

图 6.j 典型 Galfenol 合金（生长状态，无应力退火 BH）：
不同拉应力下磁致伸缩与磁场强度之间的关系（见彩插）
（资料来源：ETREMA 专有资料）

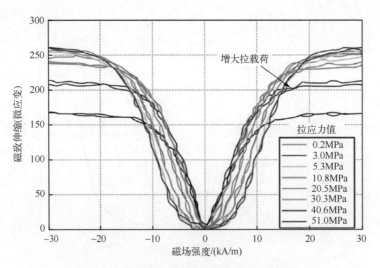

图 6.k 典型 Galfenol 合金（应力退火 BH）：
不同拉应力下磁致伸缩与磁场强度之间的关系（见彩插）
（资料来源：ETREMA 专有资料）

# 第 6 章 磁致伸缩与电致伸缩材料

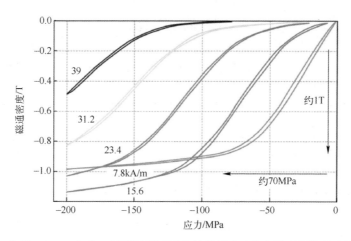

图 6.1 典型 Galfenol 合金：不同场强下通量密度相对于应力强度的降低（见彩插）

（资料来源：ETREMA 专有资料）

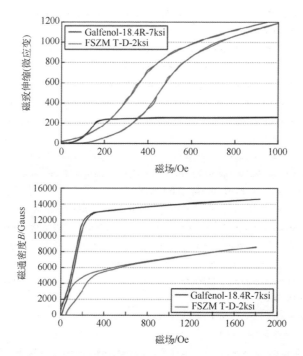

图 6.m 典型 Terfenol-D 和 Galfenol 合金的比较（见彩插）

（资料来源：ETREMA 专有资料）

# 参 考 文 献

[1] Olabi, A. G. and Grunwald, A., Design and application of magnetostrictive materials, Materials & Design 29 (2), 2008, 469-483.

[2] Lee, E. W., Magnetostriction and magnetomechanical effects, Reports on Progress in Physics 18, 1955, 185-229.

[3] Gosh, A. K., Introduction to Transducers, Delhi, PHI Learning Private Limited, 2015, 323.

[4] Clark, A. E., Wun-Fogle, M., Restorff, J. B. and Lograsso, T. A., Magnetic and magnetostrictive properties of Galfenol alloys under large compressive stresses PRICM-4: Int. Symp. on Smart Materials - Fundamentals and System Applications, Pacific Rim Conf. on Advanced Materials and Processing (Honolulu, Hawaii), 2001.

[5] Atulasimha, J. and Flatau, A. B., A review of magnetostrictive iron-gallium alloys, Smart Materials and Structures 20, 2011, 1-15.

[6] Jiles, D. C. and Atherton, D. L., Theory of ferromagnetic hysteresis, Journal of Applied Physics 55 (5), March 1984, 2115-2120.

[7] Ramesh, A., Jiles, D. C. and Roderick, J. M., A model of anisotropic anhysteretic magnetization, IEEE Transactions on Magnetics 32 (5), 1996, 4234-4236.

[8] Szewczyk, R., Validation of the anhysteretic magnetization model for soft magnetic materials with perpendicular anisotropy, Materials 7 (7), 2014, 5109-5116.

[9] Chowdhury, H. A., A finite element approach for the implementation of magnetostrictive material Terfenol-D in automotive Cng Fuel injection actuation, A master thesis in Engineering, submitted to the School of Mechanical and Manufacturing Engineering, Faculty of Engineering and Computing, Ireland, Dublin City University, July 2008, 167.

[10] Belahcen, A., Magnetoelasticity, magnetic forces and magnetostriction in electrical machines, Ph. D. Thesis, Submitted to Department of Electrical and Communications Engineering, Helsinki University of Technology, Helsinki, Finland, August, 2004, 115.

[11] Pons, J. L., A comparative analysis of piezoelectric and magnetostrictive actuators in smart structures, Boletin de la Sociedad Espaniola de Ceramica y Vidrio, In English 44 (3), 2005, 146-154.

[12] Poeppelman, C., Characterization of magnetostrictive iron-gallium alloys under dynamic conditions, Undergraduate honors thesis, The Ohio State University, 2010, 62.

[13] Du Trâemolet De Lacheisserie, E., Magnetostriction: Theory and Applications of Magnetoelasticity, Boca Raton, FL: CRC Press, 1993.

[14] Wang, L. and Yuan, F. L., Vibration energy harvesting by magnetostrictive material, Smart Materials and Structures 17 (4), 2008, 1-14.

[15] IEEE standard on magnetostrictive materials: Piezomagnetic nomenclature IEEE STD 319-1990, 1991.

[16] Engdahl, G., Modeling of giant magnetostrictive materials, In: Handbook of Giant Magnetostrictive Materials, Engdahl, G. (ed.), Ch. 2, Elsevier, 20 October 1999.

[17] Linnemann, K., Klinkel, S. and Wagner, W., A constitutive model for magnetostrictive and piezoelectric

materials, Mitteilung 2 (1008), Institut für Baustatik, Universität Karlsruhe, 76128 Karlsruhe, Germany, 40.

[18] Preisach, F., Über die magnetische Nachwirkung, Zeitschrift für Physik A, Hadrons and Nuclei 94 (5), 1935, 277–302.

[19] Kay, H. F., Electrostriction, Reports on Progress in Physics 18, 1955, 230–250.

[20] Cross, L. E., Piezoelectric and electrostrictive sensors and actuators for adaptive structures and smart materials, Proc. AME 110th Annual Meeting., San Francisco, USA, December, 1989.

[21] Zhang, Q., Pan, W., Bhalla, A. and Cross, L. E., Electrostrictive and dielectric response in lead magnesium niobate–lead titanate (0.9PMN, 0.1PT) and lead lanthanum zirconate titanate (PLZT 9.5/65/35) under variation of temperature and electric field, Journal American Ceramic Society 72 (4), 1989, 599–604.

[22] Newnham, R. E., Xu, Q. C., Kumar, S. and Cross, L. E., Smart ceramics, Ferroelectrics 102, 1990, 77–89.

[23] Cross, L. E., Newnham, R. E., Bhalla, A. S., Dougherty, J. P., Adair, J. H., Varadan, V. K. and Varadan, V. V., Piezoelectric and electrostrictive materials for transducers applications, FINAL REPORT ad-a250 889, Vol. 1, Office of Naval Research, PennState, The Materials Research Laboratory, University Park, PA, USA, June 3, 1992.

[24] Sherrit, S. and Mukherjee, B. K., Electrostrictive materials: characterization and application for ultrasound, Proc. of the SPIE Medical Imaging Conference, Vol. 3341, San Diego, California, USA, February 1998.

[25] Waechter, D. F., Liufu, D., Camirand, M., Blacow, R. and Prasad, S. E., Development of highstrain low hysteresis actuators using electrostrictive lead magnesium niobate (PMN), Proc. 3rd CanSmart Workshop on Smart Materials and Structures, 28–29 September 2000, St Hubert, Quebec, Canada, 31.

[26] Giurgiutiu, V., Pomirleanu, R. and Rogers, C. A., Energy-based comparison of solid-state actuators, Report # USC-ME-LAMSS-200-102, Laboratory for Adaptive Materials and Smart Structures, University of South Carolina, Columbia, SC 299208, USA, March 1, 2000.

[27] Ursic, H., Zarnik, M. S. and Kosec, M., ($Pb(Mg_{1/3}Nb_{2/3})O_3 - PbTiO_3$ PMN-PT) material for actuator applications, Smart Material Research 2011, article ID 452901.

[28] Damjanovic, D. and Newnham, R. E., Electrostrictive and piezoelectric materials for actuator applications, Journal of Intelligent Materials Systems and Structures 3, April 1992, 190–208.

[29] Sundar, V. and Newnham, R. E., Electrostriction and polarization, Ferroelectrics 135, 1992, 431–446.

[30] Hom, C. L. and Shankar, N., A fully coupled constitutive model for electrostrictive materials, Journal of Intelligent materials Systems and Structures 5, 1994, 795–801.

[31] Debus, J.-C., Dubus, B., McCollum, M. and Black, S., Finite element modeling of PMN electrostrictive materials, Paper presented at the 3rd ICIM/ECSSM '96, Lyon, France, 1996.

[32] Newnham, R. E., Sundar, V., Yimnirun, R., Su, J. and Zhang, Q. M., Electrostriction: nonlinear electromechanical coupling in solid dielectrics. Journal of Physical Chemistry B 101, 1997, 10.141–10.150.

[33] Coutte, J., Debus, J.-C., Dubus, B., Bossut, R., Granger, C. and Haw, G., Finite element model-

ing of PMN electrostrictive materials and application to the design of transducers, Proc. of the 1998 IEEE International Frequency Control Symposium, IEEE catalog No. 98CH36165, Ritz-Carlton Hotel, Pasadena, California, USA, 27-29 May, 1998.

[34] Pablo, F. and Petitjean, B., Characterization of 0.9PMN-0.1PT patches for active vibration control of plate host structures, Journal of Intelligent materials Systems and Structures 11 (11), 2000, 857-867.

[35] Coutte, J., Dubus, B., Debus, J.-C., Granger, C. and Jones, D., Design, production and testing of PMN-PT electrostrictive transducers, Ultrasonics 40, 2002, 883-888.

[36] Li, J. Y. and Rao, N., Micromechanics of ferroelectric polymer-based electrostrictive composites, Journal of the Mechanics and Physics of Solids 52, 2004, 591-615.

[37] Lallart, M., Capsal, J.-F., Kanda, M., Galineau, J., Guyomar, D., Yuse, K. and Guiffard, B., Modeling of thickness effect and polarization saturation in electrostrictive polymers, Sensors and Actuators B: Chemical 171-172, 2012, 739-746.

# 第 7 章 智能材料在结构中的应用

## 7.1 航空航天领域

过去的几年里,通用智能结构领域正在进行的大型且雄心勃勃的项目之一是欧洲智能飞机结构项目(Smart Intelligent Aircraft Structures,SARISTU)。该项目是一个旨在降低飞机重量和运营成本、改善飞行剖面特定气动性能的 2 级集成项目,该项目重点关注 3 个不同技术领域的整合活动:①机翼保形变形;②自我感知;③采用纳米增强树脂开发多功能结构。

SARISTU 项目组的协调工作由空客公司负责,会集了来自 16 个国家的 64 家合作伙伴。该项目总预算为 5100 万欧元,部分资金由欧盟委员会根据 FP7-AAT-2011-RTD-1(拨款协议编号 284562)提供。该项目于 2011 年 9 月 1 日开始,于 2015 年 8 月 31 日完成。最后一次会议于 5 月在俄罗斯莫斯科举行,并对会议演讲稿(共 55 篇论文)进行了汇编和出版[1]。

SARISTU① 的目标是通过整合各种技术来降低航空旅行的成本。据称层流机翼的不同保形变形概念可将飞行阻力降低 6%,从而改善飞机的性能、实现更高的燃料经济性并降低飞行所需的燃料载荷。据称可将机身产生的噪声降低到 6dB,从而减少机场附近空中交通噪声的影响并增加航空飞行的频次。SARISTU 的另一个重要目标是对制造产业链中结构健康监测(Structural Health Monitoring,SHM)进行集成,使检查成本降低 1%。与目前的蒙皮/桁条/框架系统相比,在航空树脂中添加碳纳米管可降低 3% 的重量,这些技术的组合有望使电气结构网络的安装成本降低 15%,还有改善组件级损伤容限和导电性等优点。

SARISTU 项目如图 7.1 所示,其具有保形变形、综合传感和多功能概念三大基础[1]。该项目的分工如下:

---

① www.saristu.eu。

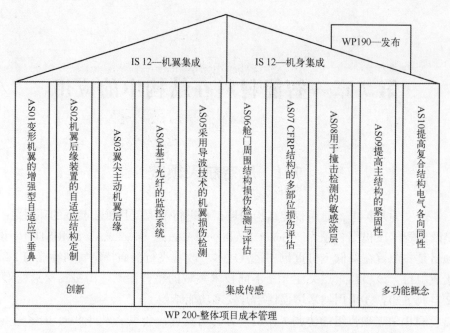

图 7.1　SARISTU 项目结构[1]

（1）AS01 变形机翼的增强型自适应下垂鼻（Enhanced Adaptive Droop Nose，EADN），由 DLR① 负责。根据结构和风洞试验，设计、计算、制造并集成了一个 IS12 变形机翼的 3D 大尺寸自适应下垂鼻，用于机翼的确认和验证。结果表明，变形和无间隙下垂鼻装置能够提供层流效应，进而减小起飞、巡航、着陆过程中的阻力，并降低整体噪声的产生。EADN 装置中采用了其他一些技术（图 7.2）来提供防鸟撞、除冰、表面和雷电保护。此外，该装置还实现了低复杂性致动系统、轻质结构、先进制造解决方案、功能集成和抗疲劳性能。

（2）AS02 机翼后缘装置的自适应结构定制，由意大利航空航天研究中心（the Ztalian Aerospace Research Centre，CIRA）② 负责。对机翼后缘装置进行了计算、设计、制造和测试，通过弦向致动使 $L/D$ 比提高了 10%，并通过展向升力再分配使根部弯矩（Root Bending Moment，RBM）降低了 10%。自适应机翼后缘装置（Adaptive Trailing Edge Device，ATED）还可能降低巡航油耗（约 3%）和总重量，因为可以减少产生的阻力和 RBM。基于多肋 SDOF 系统的

---

① www.dlr.de。
② www.cira.it/en。

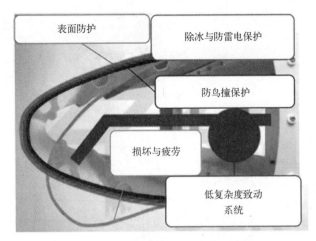

图7.2 增强型自适应的下垂鼻(ETAD)及其要求示意图

ATED架构,其均配备有用于进行激活的杆状承重致动器。肋拱由硬杆连接的刚性部件制成,其中任意一个部件均可作为独立的致动器工作(每个肋拱一个)。保形或差动致动器可产生弧度(机翼极线)或后掠角(升力分布)的变化。其无缝蒙皮采用了聚合物薄层和泡沫。ATED示意图如图7.3所示。

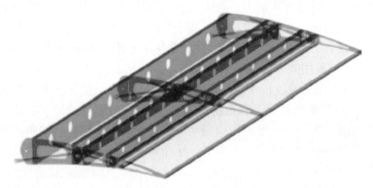

图7.3 ATED示意图

(3) AS03翼尖主动机翼后缘,由EADS创新工厂[①]负责。实现了主动小翼(图7.4)的工程解决方案,该主动小翼称为WATE—翼尖主动机翼后缘。这一翼尖主动机翼后缘能够降低临界飞行条件下的机翼载荷,进而降低机翼重量。根据飞机结构和空气动力/气动弹性的要求,进行了WAET结构件和系统的实验室设计、集成和试验。

---

① www.inmaproject.eu。

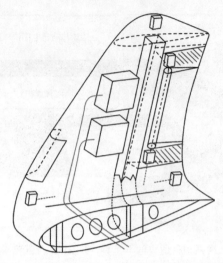

图 7.4 带有活动标签和变形（阴影）过渡区域的 WATE 示意图

（4）AS04 基于光纤的监控系统——概念的评估、价值、风险和开发，由 INASCO Hellas[①] 负责。在该工作分支内，开发了基于光纤的监控系统，监测复合增强型的飞机结构[②]。该系统能够感知机翼和机身处的应变；该系统的应变读数使用机翼后缘形状监测作为驱动，从而感知保形变形结构的形状并检测加强件的脱离情况。在制造过程中，基于光纤的监控系统在复合材料结构上执行，并监控组件在整个寿命周期内的状态。光纤布拉格光栅（Fiber Bragg Grid，FBG）的复用能力可提供以上应变读数，从而使监控系统更小更轻。

（5）AS05 采用导波技术的机翼损伤检测，由意大利那不勒斯费德里克二世大学（DIAS）航空航天工程系负责[③]。该工作分支旨在设计、制造和实施基于超声导波测量的复合机翼损伤检测系统。通过理论建模，将超声导波的传播特性与复合材料的弹性特性联系起来，从而对其进行无损表征。弹性波与缺陷的相互作用分析可用于提高超声 NDE 对复合结构中隐藏缺陷的检测有效性，并有助于开发有效的 SHM 系统。高频超声传感器阵列位于给定结构的关键区域，以分析受控源传播的导波特性。该系统应能够检测复合机翼加强蒙皮上几乎看不见的损伤（Barely Visible Damage，BVID）和可见的损伤（Visible Damage，VID）。在制造阶段，传感器集成在结构的子部件内，因而可减少维护时间与检查间隔，从而极大地减少了相关部件所需的检查时间，真正降低了与维

---

① www.inasco.com。
② 应用场景（类似于其他项目中的工作包）。
③ www.dias.unina.it。

护和检查相关的寿命周期成本。

（6）AS06 采用集成超声传感器的撞击损伤评估，由 Martin Bach（德国 EADS 有限公司）负责①。该工作分支旨在通过在结构内永久性嵌入压电换能器来为飞机机身提供自我感知功能，通过一键模式下的导波致动和感应对损伤进行检测与评估（图 7.5）。该工作对有关损伤检测、传感器集成、质量保证、接触和数据传输的自感知能力进行了评估。

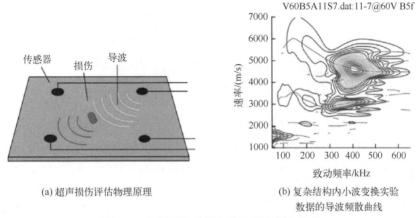

(a) 超声损伤评估物理原理　　(b) 复杂结构内小波变换实验
　　　　　　　　　　　　　　　　数据的导波频散曲线

图 7.5　使用嵌入式压电换能器的损伤评估

（资料来源：Marti Bach②）

（7）AS07 CFRP 结构的多部位损伤评估，由空客-D 负责②。AS07 工作分支旨在开发一种用于评估存在多处损伤的航空结构（典型的机身壁板）适航性的预测方法。对采用有限元建模进行的虚拟测试与物理测试的结果进行了比较，并将这些结果集成到多点损伤预测方法中。该预测方法通过结合结构健康监测系统信息与简化的预测方法模型，开发出一种能够提供决策支持的损伤评估工具。该工作采用了光纤和 PZT 传感器。最后，将选定的结构健康监测技术与用于演示任务的多现场冲击损伤试验件评估工具进行集成和验证。

（8）AS08 用于撞击检测的敏感涂层，由法国 EADS 负责③。众所周知，当金属结构受到撞击时，可以通过视觉检测到局部的形状变形。然而在复合结构上，撞击可能意味着表面没有可见迹象的内部损伤。在组装或维护任务中，这些意外撞击可能出现在飞机机身的内部。这就是为什么目前的复合材料制造

---

① martin.bach@eads.net。
② www.airbusgroup.com/int/en/group-vision/global-presence/germany.html。
③ EADS-欧洲航空防务和航天公司（European Aeronautic Defence and Space Company NV），是 2014 年前空客集团 SE 的名称（www.airbusgroup.com）。

的飞机被设计为有结构的厚度裕度,并能承受被忽视的潜在损伤。因此,在本AS中,主要的目标是通过使用智能涂层提高不可见冲击损伤(BVID)的阈值,从定义的能量阈值揭示发生的可见撞击,进而便于复合材料结构的撞击检测。降低 BVID 阈值可以减少厚度裕度,从而减轻重量(图7.6)。

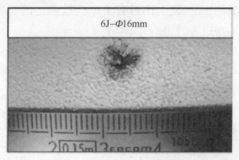

图 7.6　AS08-涂料中包含的微胶囊显示的撞击

(9) AS09 通过改善损伤容限来提高主结构的坚固性,由 Tecnalia 创新与应用研究院负责[①]。该方案通过添加如纳米材料和各类聚合物夹层材料的增韧剂来评估复合材料的增韧效果。该方案与 AS10 一起交付多功能复合材料和潜在的改进黏合剂。该工作分支旨在提升对损伤容限至关重要的材料特性,这些特性主要包括模式 1 和模式 2 断裂韧性以及撞击后压痕和层间剪切强度(图7.7)。提高材料的损伤容限可以减少部件的厚度和重量。这些优点反过来可以降低燃料消耗,因此有益于环保和节省成本(图7.8)。

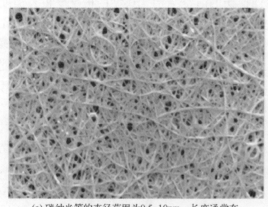

(a) 碳纳米管的直径范围为0.5~10nm,长度通常在几纳米到几十微米之间(资料来源:Unidym)　　(b) 不溶性纤维

图 7.7　AS09-使用的碳纳米管和不可溶掩饰物

---

① www.tecnalia.com/en/2。

# 第7章 智能材料在结构中的应用

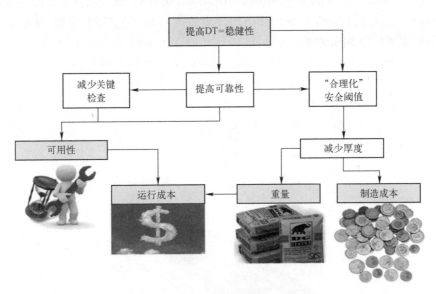

图7.8　AS09-可靠性提升的框图

（10）AS10 复合结构电气各向同性的提高，由 Tecnalia 创新与应用研究院负责。本应用场景的构建旨在评估可集成到电气结构网络中的技术，从而降低复合结构的重量并提高耐久性。这些评估涉及从纳米材料到金属带和涂层的各种技术与材料（图7.9）。将 AS09 中关于损伤容限的研究纳入本评估中，并对可提升损伤容限和电气耐久性的多功能材料可制造性进行了验证。

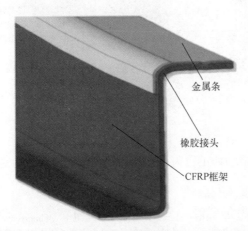

图7.9　一种经 EADS 创新工厂特别许可使用的电力结构网络（ESN）用金属带

(11) IS12 机翼装配集成和试验,由 Alenia 负责①。该集成方案(Integration Scenario,IS)旨在将保形变形和结构健康监测(基于光纤和 PZT)集成到机翼演示机上,并测试和验证耐用结构尺寸的各种方法(图 7.10 和图 7.11),开发了一款演示机,可用于计算跨度 4.5m 的全尺寸机翼模型,包括带有创新的三维致动器的小翼、前缘和后缘(图 7.10)。全尺寸保形变形装置具有结构和致动功能,机翼后缘和小翼是连续的,而前缘的保形变形则是非自动化的、机械的且离散的。对实施的结构进行了低速风洞和数值验证。图 7.12 总结了该工作分支的各种技术。

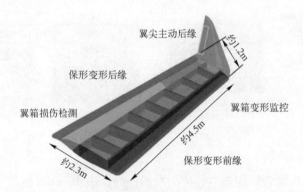

图 7.10　IS12 内 SARISTU 机翼上开发和实施的某些技术

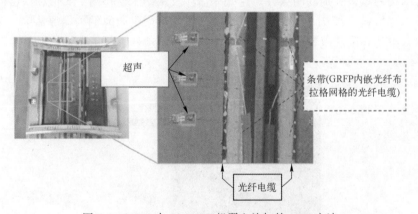

图 7.11　IS12 内 SARISTU 机翼上施加的 SHM 方法

---

① 阿莱尼亚·马基航空公司(Alenia Aermacchi)是芬梅卡尼卡(Finmeccanica S. P. A.)的子公司,活跃于航空领域。自 2016 年 1 月 1 日起,阿莱尼亚·马基航空公司的业务并入芬梅卡尼卡航空部门的飞机和航空结构部门。www.finmeccanica.com。

## 第7章 智能材料在结构中的应用

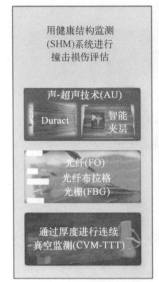

图 7.12　IS12 内 SARISTU 中正在研究的某些技术[1]

（12）IS13 机身装配、集成、测试和验证，由 Ben Newman 为代表的空客运营股份有限公司①和 TECCON 咨询与工程股份有限公司②负责。该工作分支的主要目标是依据应用场景中开发和设计的技术，提供有代表性的实际飞机结构，对这些技术的潜力进行评估，旨在实现以下 SARISTU 目标：①多位置损伤检测能力，可将结构检查成本降低 1%；②提升损伤表征能力，通过先进的算法将结构检查成本降低 1%；③实现结构健康监测与纳米增强树脂结合，以提高损伤容限并将重量降低 3% 以上；④将电气结构网络安装成本降低 15%。

通过使用以下 4 种极具前景的技术来实现这些目标（图 7.13）：①声学-超声结构健康监测的门环绕结构和迷你门；②损伤容限、纳米增强树脂的低电导率、低成本电气结构网络（Electrical Structure Network，ESN）安装及有限结构健康监测的低面板；③用于中等电导率、低成本高性能雷击防护（Lightning Strike Protection，LSP）的低成本 ESN 的侧面板；④多种结构健康监测性能的上面板，测试结构为#1。

---

① 原名空客德国有限公司，http://www.airbus.com/。
② 截至 2013 年 4 月，TECCON Consulting & Engineering GmbH 更名为 ALTRAN Aviation Engineering GmbH，http://www.altran.de/。

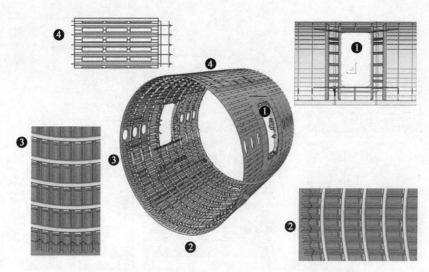

图 7.13 IS13 内用于评估最有应用前景的 4 种通用机身面板

（13）WP 190-宣传，由 EASN-TIS 领导的宣传机构①。该工作分支涉及各种宣传活动，旨在重点宣传 SARISTU 项目的业绩和成果。

（14）WP 200-整体项目管理，由德国空客公司负责，TECCON 咨询工程股份有限公司提供支持。其任务是对 SARISTU 项目进行管理，包括各个时间阶段项目的业绩监控和目标实现。

在展示了项目开展的各种活动后，接下来介绍一些与其智能结构性质相关的亮点。所有这些活动均在文献［1］中进行了很好的展示。

在 SARISTU 项目中，设计了一种先进的变形机翼后缘装置，计算了颤振、致动强度和气动性能，并最终完成了制造和测试[2]。由机电致动器驱动的紧凑轻量杠杆可实现机翼后缘（Trailing Edge，TE）的变形，而 TE 的翼肋由无缝蒙皮[3-13]覆盖（图 7.14）。图 7.15 所示为先进机翼后缘构造示意图。

SARISTU 项目的最大成就之一是采用聚合物层和泡沫制成的无缝蒙皮[3]，该蒙皮使得机翼后缘能以变形的方式运动，进而向上和向下偏转（图 7.16 和图 7.17）。

---

① 欧洲航空科学网络-技术创新服务（EASN-TIS）是一家总部位于比利时的中小企业，主要参与航空和航空运输相关的研究项目，http://easn-tis.com/。

# 第7章 智能材料在结构中的应用

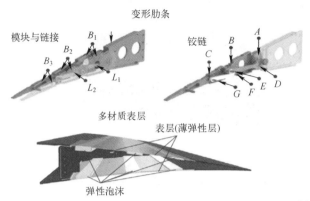

图 7.14 自适应机翼后缘装置（ATED）的基本组件示意图[2]

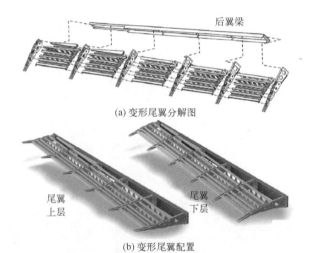

图 7.15 变形机翼后缘构型[2]

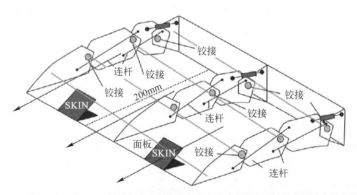

(a) SARISTU示意方案：在横跨方向上支持负载，通过弦向运动获得变形

313

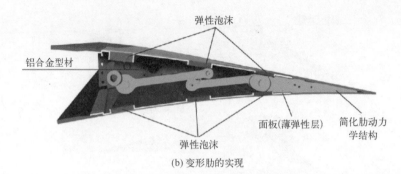

(b) 变形肋的实现

图 7.16　SARISTU 变形尾翼示意图[3]

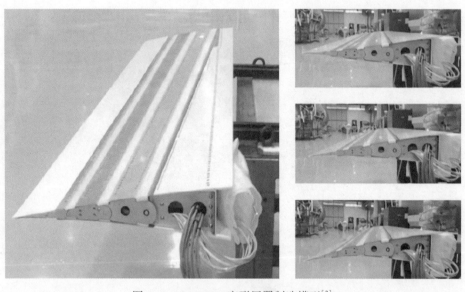

图 7.17　SARISTU 变形尾翼制造模型[3]

EADN[4]还涉及一些技术，包括表面保护、除冰程序、闪电和鸟撞保护以及获取形状变化的简单运动学。Invent①制造了一款4m长的机翼前缘样品，用于进行风洞和地面试验、鸟撞试验，而两个小模型则装备有钛箔。EADN典型的试验剖面如图 7.18 所示。

设计和制造了一款用于研究机翼弯曲与前缘展开组合[5]的试验验证机，对

---

①　德国布伦瑞克市 INVENT GmbH，http://www.invent-gmbh.de/。

EADN 部件进行了进一步评估（图 7.19），以验证不同技术集成的技术成熟度水平（Readiness Level，RL）。

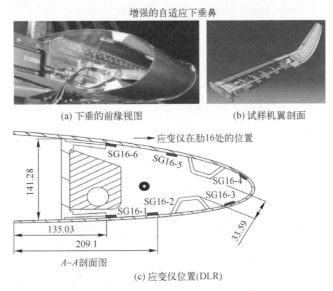

图 7.18　EADN 典型的试验剖面[4]

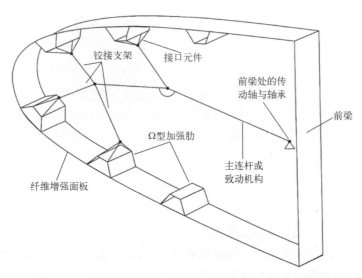

图 7.19　使用常规蒙皮材料的整体变形机翼前缘示意图

本书开发了一种揭示复合材料所制备结构损伤的先进方法，并利用在可见光和紫外光下有明显颜色变化的微胶囊对该方法进行了测试[6]。该方法采用了隐色染料①显示剂系统。在进行任意撞击之前，其两个部件相互分离，且没有任何着色。当微胶囊在撞击作用下发生破裂时，会和涂层内的显示剂相结合。因此，在撞击位置会出现一个彩色污点。SARISTU项目中开发的使用微胶囊的概念、典型结果以及多层概念如图7.20所示。

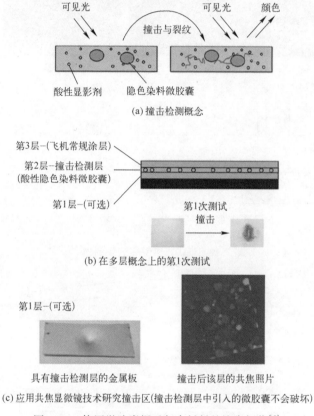

图7.20 使用微胶囊揭示复合材料的撞击损伤[6]

使用光纤和应变计对施加在翼梁上的载荷进行监控研究的工作已经完成[7]。将光纤嵌入带状物中，并黏合到翼梁上。第2款监测系统包括传统的应变计，其设计用于测量风洞试验期间机翼上的载荷，并可与气动载荷模型进行交互（图7.21），开发了用于识别机翼内部载荷和剖面载荷的相关算法。这些结果已经应用于先进航空航天结构健康与使用监控系统内。

---

① 隐色染料（希腊语中leukos为白色）是一种可以在两种化学形式之间切换的染料：一种是被封装的无色试剂，另一种是以呈弱酸化合物形式的显色剂直接引入微胶囊外部的涂层黏合剂中。

# 第 7 章　智能材料在结构中的应用

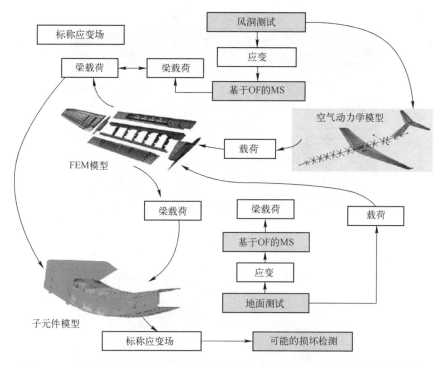

图 7.21　有限元模型、气动弹性模型、子部件模型与监控系统相互作用框图[7]

文献［8］研究了另一种先进结构健康监测技术方法，该方法基于超声（Acoustic Ultrasonic，AU）原理，通过在门四周结构中永久嵌入压电换能器网络来实现。超声导波可提供关于被测结构完整性的信息（图 7.22）。文献［8］的工作包括验证方法的推导，该验证方法从在结构中引入损伤的方式开始，并以无损检测（Non-Destructive Inspection，NDI）作为参考。然后，对换能器网络的查询及相关数据进行评估，并将其集成到图形用户界面（Graphical User Interface，GUI）中。

文献［9］采用了一种类似方法，其通过嵌入结构中的压电致动器和传感器网络来激发与接收兰姆波，从而实现结构健康监测（图 7.23）。Smart Layer®[10]和 DuraAct™型[11-12]压电换能器可用于产生与接收兰姆波（图 7.24）。该文献描述了全尺寸复合材料机身门周围结构的各个开发和制造阶段，该结构配备了基于压电换能器的集成结构健康监测（SHM）系统（图 7.25 和图 7.26）。应特别注意用于制造结构（图 7.25）和先进传感器（图 7.26）的先进设备，即自动纤维铺放机器人。

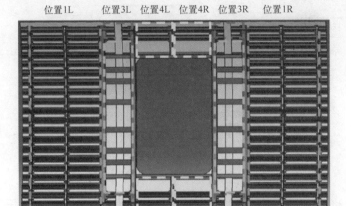

图 7.22　含有永久嵌入式超声压电换能器的门环绕结构的撞击位置[8]

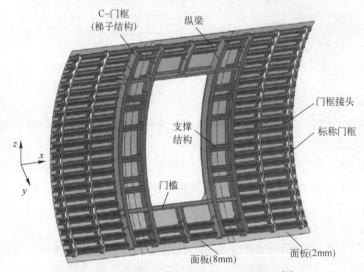

图 7.23　无 SHM 网络的完整门环绕结构[9]

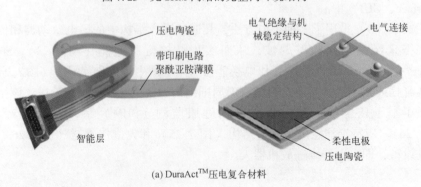

(a) DuraAct$^{TM}$压电复合材料

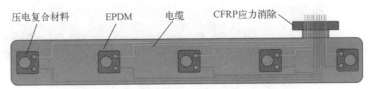

(b) 基于DuraAct™压电复合材料技术的传感器阵列

图 7.24　用于兰姆波基 SHM 的压电换能器[9]

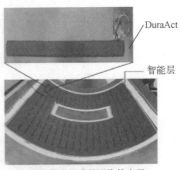

(a) 使用自动纤维布设机器人进行表层铺装　　(b) 具有集成传感器网络的表层

图 7.25　门环绕结构的自动化制造过程和使用 SHM 的集成压电换能器[9]

图 7.26　制造带有集成压电换能器的门环绕结构，用于形成 SHM 网络[9]

　　SARISTU 项目中 AERNNOVA[14]开发了另一种有趣的 SHM 方法，使用基于多个超声波的相控阵监测增强寿命评估（Phased Array Monitoring for Enhanced Life Assessment，PAMELA）Ⅲ SHM™ 系统（图 7.27）进行无线控制。该系统集成了所有必需的硬件、固件和软件，以使给定结构内嵌入的集成压电相控阵（Phased Array，PhA）换能器产生激励信号，从而获取响应信号、将数据发送到中央主机和/或进行高级信号处理，最终获得 SHM 图。

　　文献 [14] 总结了损伤测试结果及解释和识别金属与复合材料二者结构损伤的系统有效性。图 7.28~图 7.30 展示了整个项目获得的典型成果。

图 7.27　PAMELA Ⅲ SHMTM 系统[14]
（其复合盒内的所有电子元件放置在相阵列换能器适配器上，并与结构黏合，以进行 SHM 测试）

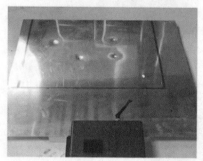

(a) 4处撞击损伤的铝合金面板　　(b) 用PAMELA SHM系统获得的图(a)损伤图像

图 7.28　在受损铝制平板上使用 PAMELA Ⅲ SHMTM 系统的损伤结果图像[14]

(a)　　　　　　　　　　　　　　(b)

图 7.29　PAMELA Ⅲ SHMTM 系统在 3 舱翼盒验证机上的安装[14]

## 第7章 智能材料在结构中的应用

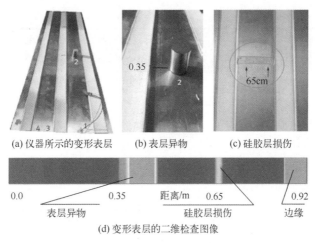

(a) 仪器所示的变形表层　　(b) 表层异物　　(c) 硅胶层损伤

(d) 变形表层的二维检查图像

图 7.30　PAMELA Ⅲ SHMTM 系统对变形蒙皮上施加的损伤结果[14]

## 7.2　医疗领域

医疗领域是使用智能材料相对较多的行业之一,主要涉及镍钛诺合金和压电材料[15-22]。

由于形状记忆合金与人体的生物相容性,其在医疗领域的应用被广泛认可并接受。镍钛诺合金通常是一种二元化学体系(约50%镍和50%钛),但是如 $Ni_{0.53}Ti_{0.47}$ 此类镍含量稍高的合金会产生形状记忆现象,因此已在医疗领域获得广泛应用[15-17]。2012年,全球镍钛诺合金医疗器械半成品市场的估值为15亿美元,预计到2019年会增长到25亿美元;而2012年全球镍钛诺合金医疗器械成品医疗部件市场价值82亿美元,预计到2019年底将达到173亿美元①。表7.1列出了一些知名的相关公司。

表 7.1　全球镍钛诺合金医疗器械公司

| 公司 | 总部所在国家 | 电子邮件地址 |
| --- | --- | --- |
| Abbott 实验室有限责任公司 | 美国 | www.abbott.com |
| 波士顿科技公司 | 美国 | www.bostonscientific.com |
| Cook 医疗有限责任公司 | 美国 | www.cookmedical.com |

---

① 根据标题为"Nitinol Medical Devices Market——Global Industry Analysis, Size, Share, Growth, Trends and Forecast, 2013-2019"的一份报告,http://www.transparencymarketresearch.com。

| 公司 | 总部所在国家 | 电子邮件地址 |
|---|---|---|
| Covidien 股份有限公司 | 美国① | www.medtronic.com/covidien |
| C. R. Bard 有限责任公司 | 美国 | www.crbard.com |
| Custom Wire 技术有限责任公司 | 美国 | www.customwiretech.com |
| ENDOSMART 股份有限公司 | 德国 | www.endosmart.com |
| Medtronic 公司 | 美国 | www.medtronic.com |
| 镍钛诺合金器械及组件有限责任公司 | 美国 | www.nitinol.co |
| Terumo 公司 | 日本 | www.terumomedical.com |

① Covidien 股份有限公司是一家爱尔兰公司（总部位于爱尔兰共和国都柏林市），2014年6月并入美国美敦力公司。

镍钛诺合金广泛应用于心脏病学、神经病学、骨科和介入放射学领域的仪器和专用设备，成功拓展了其医疗仪器市场。

由于镍钛诺合金丝线具有较高的可恢复应变、良好的抗扭结性、可操纵性和可扭转性，已应用在血管造影术领域。镍钛诺合金的另一个广泛应用领域是镍钛诺合金基支架①，用于支撑食道和胆管等管状通道（内腔）的内周，或为冠状动脉、颈动脉、髂动脉、主动脉和股动脉等血管提供内部支撑。冠状动脉内的支架及各种其他支架的示意图如图7.31和图7.32所示。

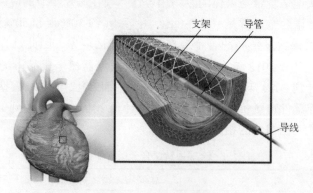

图 7.31  冠状动脉支架
（资料来源：www.wikipedia.org/wiki/Stent）

① "支架"一词来源于 Charles Thomas Stent 博士（1807—1885年），他是伦敦的一位牙医，开发了一种牙科设备来辅助形成牙齿印模。

## 第 7 章 智能材料在结构中的应用

(a) Express™ SD肾脏与胆道预置支架系统

(b) ALIMAZZ-ES™食管支架

(c) 医疗器材-血管

图 7.32　镍钛诺合金制备的各种支架

（资料来源：www.bostonscientific.com；www.endotek.merit.com）

支架为球囊扩张型，即血管成形术球囊用于打开阻塞的血管并扩张支架，或者为自扩张型，即支架被推出导管后立即打开来支持已经扩张的管腔，这是由于它的形状记忆特性。大多数镍钛诺合金支架为自扩张型。支架在其输送导管内会被压缩。当导管到达相应位置时，将支架推出，并进行扩张，以支撑血管壁的原始形状。其他医疗应用包括肾脏取石器（图 7.33）和植入物（图 7.34），其中镍钛诺合金制成的 Conventus DRS™[①]由于具有超弹性和形状记忆效应，可折叠成小尺寸并通过小切口植入体内。到达体内相应位置后，即可释放，并使其扩张至原始形状。

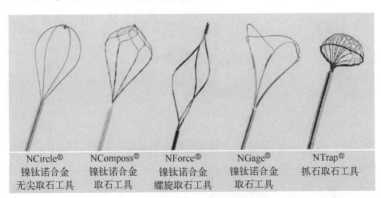

图 7.33　镍钛诺合金取石工具

（资料来源：www.cookmedical.com）

图 7.35 展示了用于骨骼固定的各种镍钛诺合金器械，而图 7.36 展示了镍钛诺合金的典型牙科应用以及镍钛诺合金植入物的固定螺钉。

---

① www.conventusortho.com。

(a) Conventus DRSTM植入物

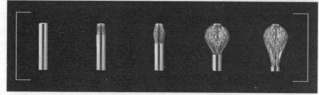

(b) 镍钛诺合金植入物的各阶段

图 7.34　用于手腕的 Conventus DRS™ 植入物

(资料来源：www.conventusortho.com/patients/our-solution-patients/)

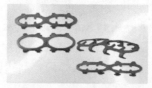

(a) 镍钛诺合金便携式接骨板　　(b) 镍钛诺合金四肢骨折内固定器械-Ⅰ型

(c) 镍钛诺合金四肢骨折　　(d) 镍钛诺合金四肢骨折　　(e) 适用于髌骨骨折内固定的
　　内固定器械-Ⅱ型　　　　　内固定器械-Ⅲ型　　　　　镍钛诺合金器械

图 7.35　中国江苏 IAWA 生物技术工程有限公司的镍钛诺合金接骨器械

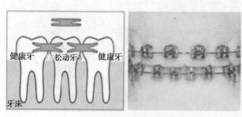 

(a) 牙科镍钛诺合金器械　　　　　　　　　(b) 骨折植入物固定螺钉

图 7.36　镍钛诺合金植入物的牙科应用和骨折固定螺钉

(资料来源：www.totalmateria.com/page.aspx?ID=CheckArticle&site=ktn& NM=212；www.mxortho.com/company/)

Shabalovskaya 等对镍钛诺合金与人体的相容性进行了评述[18]。据称，目前开发的表面厚度可能从几个纳米到微米不等，若在应变下保持表面完整性，并且在镍钛合金表面上没有富镍亚层，则可有效防止镍释放到人体内。

最近的一项应用是压电骨手术，允许选择性切割矿化组织，同时保留软组织。与洁牙器类似，可将 25~35kHz 的高频振动传输给金属尖端。

然而，压电外科器械的功率比洁牙器高 3~6 倍。该技术的主要优点包括高精度，增加曲线截骨术便利性的设计、对软组织创伤更小、可保留神经和血管结构、减少出血、骨骼热损伤最小以及整体改善愈合过程。与旋转仪器或振荡锯不同，压电设备需要轻微的压力和尖端的持续运动。掌握这项技术需要进行专业培训（图 7.37~图 7.39）。

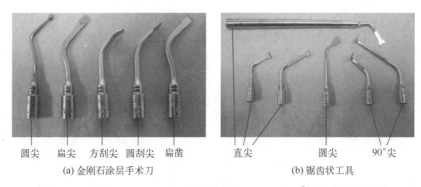

图 7.37　法国 Acteon Satalec 公司制造的 Piezotome 2® 系统的不同尖端

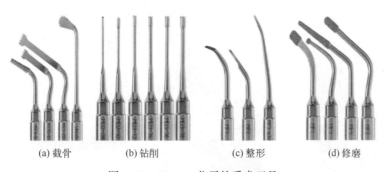

图 7.38　Mectron 公司的手术刀具

（资料来源：www.dental.mectron.com/products/piezosurgery/）

当使用手术钻或锯时，切割效率与施加在骨骼上的压力相关。对于压电单元，切割由仪器尖端的高频振动驱动；压力过大会阻碍振动、降低效率并产生

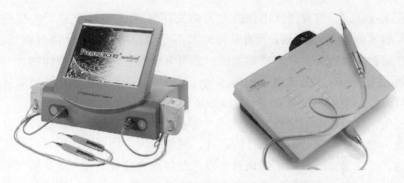

图7.39 Mectron公司的压电外科手术系统
(资料来源：www.dental.mectron.com/products/piezosurgery/)

摩擦热。仪器手柄用改进的笔夹握持。施加适度的力（1.5~3N）使尖端振动，该力相当于接近1N的手写轴向力。已经证明，冷却冲洗水最小流速为30mL/min，工作压力为1.5~2.0N，不会对骨骼温度造成负面影响[19-20]。因此，建议皮质骨骼的切割载荷为1.5N。

最后，压电换能器已应用于医疗领域的高频超声成像设备[21-22]。医疗超声用换能器由厚度模式谐振器组成。对压电元件进行激励时，其振动格栅即会发射超声波（图7.40和图7.41）。例如，PZT中的声波传播速度约为4350m/s，在5MHz时的标称厚度（$\lambda/2$）为435μm，此外，其他因素也会影响换能器谐振[22]。如何在谐振频率与元件尺寸之间权衡是该阵列设计的主要挑战。为了增加可用材料的选择余地，同时还开发出了一些压电复合材料（图7.41）。

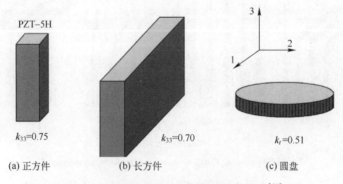

(a) 正方件　　　　　(b) 长方件　　　　　(c) 圆盘

图7.40 几何尺寸对机电耦合系数的影响[21]

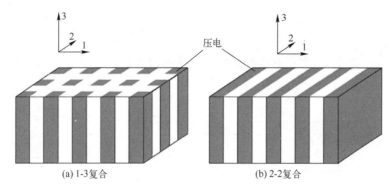

(a) 1-3复合  (b) 2-2复合

图 7.41 使用压电复合材料作为超声波源（嵌入在低密度聚合物中的小 PZT 棒）[21]

## 7.3 压电发动机

发动机是一种可产生连续直线或旋转运动的装置。相比于传统励磁发动机，利用逆压电效应产生运动的压电发动机具有以下优点：不需要传统励磁发动机的强磁场，可以小型化、低功率，且更可靠。微型发动机能够满足精密定位需求，如集成电路技术中的掩模对准、光纤对准、医疗导管放置、移动电话相机中的自动聚焦和光学变焦、药品处理等。此类发动机可以小型化至 4mm 以下，并且可提供高达 0.1μm 或更高的位置精度。它们在需要精密运动和避免近磁场干扰的工程和医疗领域中有着广泛的应用。压电发动机主要包括线性发动机和旋转发动机两类。在这些发动机中，可将因逆压电效应引起压电材料的微小位移转换为连续的平移运动或旋转运动。

当电能转化为机械能时，逆压电效应[24]可驱动发动机工作。在电场的作用下，材料内会产生应变。由电压 $V$ 引发的在平行膨胀与收缩、横向膨胀与收缩、平行剪切与弯曲（串联和并联）模式下的发动机工作如图 7.42 所示。压电块在施加电压之前的尺寸为厚度 $t$、宽度 $w$ 和长度 $L$，$P$ 代表极化矢量。由于施加电压 $V$ 所导致的尺寸变化由如下表达式给出：

$$\Delta t = V \cdot d_{33}, \quad \text{对于平行伸出(回缩)} \tag{7.1}$$

$$\frac{\Delta L}{L} = \frac{\Delta w}{w} = \frac{V \cdot d_{31}}{t}, \quad \text{对于横向伸出(回缩)} \tag{7.2}$$

$$\Delta x = V \cdot d_{15}, \quad \text{对于平行剪切} \tag{7.3}$$

$$\Delta x = \frac{3 \cdot L^2 \cdot V \cdot d_{31}}{2 \cdot t^2}, \quad \text{对于串联弯曲} \tag{7.4}$$

$$\Delta x = \frac{3 \cdot L^2 \cdot V \cdot d_{31}}{t^2}, \quad 对于并联弯曲 \tag{7.5}$$

式中：$d_{33}$、$d_{31}$ 和 $d_{15}$ 为压电应变系数（见本书第 2 章）。

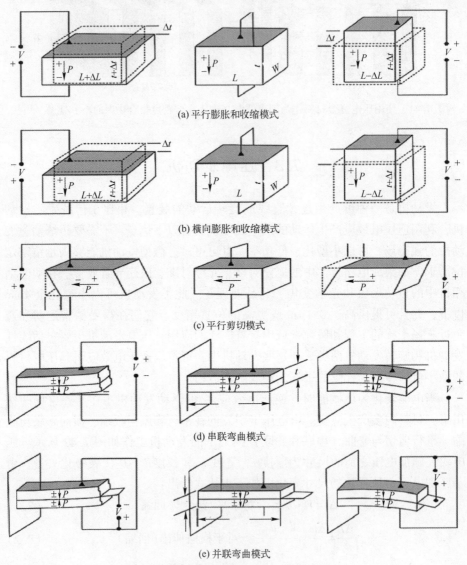

(a) 平行膨胀和收缩模式

(b) 横向膨胀和收缩模式

(c) 平行剪切模式

(d) 串联弯曲模式

(e) 并联弯曲模式

图 7.42　压电材料的发动机工作
（资料来源：美国 PIEZO SYSTEMS 公司）

## 7.3.1 线性压电发动机

已知有两类线性压电发动机：一类在低频下工作，另一类在超声频率下工作。

一类低频线性压电发动机为装夹型，其可产生像蜗杆一样的滑块线性运动。这种压电发动机称为"英制蜗杆"发动机，由 Burleigh 仪器公司首次设计并获得专利。该发动机的工作原理如图 7.43 所示。该英寸蜗杆发动机由 3 组压电致动器组成：两组为夹紧致动器（1 和 2），一组为驱动致动器（3）。致动器按顺序操作，以实现可移动滑块的运动。

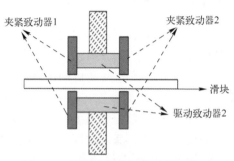

图 7.43　压电英制蜗杆发动机示意图

致动器 1 和 2 为交替支撑滑块的夹持致动器。致动器 3 为在滑块运动方向上伸展或收缩的驱动致动器。起初，所有致动器均与滑块断开连接。当一个夹持致动器夹持滑块时，另一个夹持致动器断开连接。换言之，在任何情况下只有一个夹持致动器夹持滑块。驱动致动器在滑块运动方向上横向膨胀或收缩。操作步骤如图 7.44 所示。

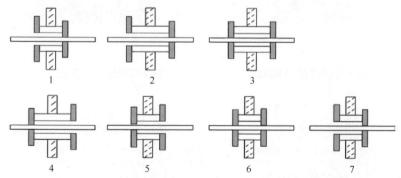

图 7.44　英制蜗杆发动机工作的 7 个步骤（一个循环）

起初，所有致动器均打开且未运转。
步骤 1：关闭夹持致动器 2。

步骤2：膨胀驱动致动器。
步骤3：关闭夹持致动器1。
步骤4：打开夹持致动器2。
步骤5：收缩驱动致动器。
步骤6：关闭夹持致动器2。
步骤7：打开夹持致动器1。

重复这些步骤，即可使线性发动机工作。

### 7.3.2 旋转压电发动机

与励磁微型发动机相比，压电旋转发动机更便于按比例缩小；与静电发动机相比，压电旋转发动机具有更大的能量密度，所以人们非常看好压电旋转发动机的应用前景。其中，一种旋转压电发动机的工作基于 Piezo LEGS 行走原理①，该原理为非共振型，即驱动腿的位置在任何给定时刻均已知。这确保了在整个速度范围内能够很好地控制发动机的运动。Piezo LEGS 发动机的性能在几个方面与直流或步进发动机不同。基于摩擦的 Piezo LEGS 发动机运动是通过驱动腿和驱动盘之间的接触摩擦进行传递的。对于每个波形周期，Piezo LEGS 发动机可行进一整步，称为一个波形（wfm）步（在菱形波形空载下约为 1.5mrad）。完整的一步如图 7.45 所示。驱动轴的转速等于 wfm 步角乘以波形频率（1.5mrad×2kHz＝3rad/s＝170（°）/s）。

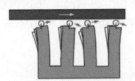
1.启动时，4个腿全部拉长并弯曲，压在发动机电枢上。

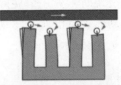
2.一对腿缩回离开电枢并向左移动，而另一对腿向右弯曲，将电枢推向所示方向。

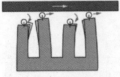
3.最初缩回的那对腿现在伸出来推动电枢，而推动电枢向右移动的那对腿现在缩回。

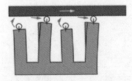
4.第二对向右弯曲继续向所示方向推动电枢，而原来的那对腿现在向左移动，准备开始下一个行走循环。

图 7.45 Piezo LEGS 行走原理示意图（一个循环）

---

① 美国佛罗里达州克利尔沃特常青大道14881号，http://www.micromo.com/。

还有一些其他结构的旋转压电发动机。其中，一种结构基于前进波（图7.46）。其定子由一个下面粘有压电陶瓷的弹性环组成。当施加高频交流电压（AC）时，压电陶瓷会发生振动。在与定子固有频率相当的激励频率（即共振频率）下，可达到 $2\sim3\mu m$ 的机械振幅。

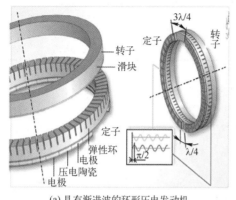
(a) 具有渐进波的环形压电发动机

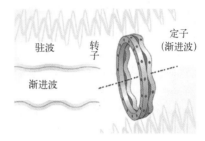
(b) 示意图

图 7.46　基于前进波的旋转压电发动机

（资料来源：www.pierretoscani.com/echo_shortpress.html）

## 7.3.3　超声压电发动机的特性和分类

相比于基于励磁效应的传统发动机，基于压电材料的超声发动机具有许多优势，自问世以来有望实现广泛应用[25-44]。

**1. 超声发动机的特性**

超声发动机的优点包括：

（1）结构紧凑、设计灵活、比扭矩（扭矩/重量比）大。因为压电元件可激发不同类型的振动，包括纵向、弯曲和扭转振动，所以超声发动机具有结构紧凑和设计灵活的优点。如表7.2所列，其转矩密度为传统发动机的3~5倍。

表 7.2　电磁发动机和超声发动机之间的比较[38]

| 发动机分类 | 制造商 | 堵转扭矩/(N·m) | 空载转速/(r/min) | 重量/g | 扭矩密度/(N·m/g) | 效率/% |
|---|---|---|---|---|---|---|
| 有刷励磁直流型 | MICROMO① | 0.00332 | 13500 | 11 | 0.0302 | 71 |
| 有刷励磁直流型 | MAXON② | 0.0127 | 5200 | 38 | 0.0334 | 70 |
| 有刷励磁直流型 | MABICHI MOTOR③ | 0.0153 | 14500 | 36 | 0.0425 | 53 |
| 有刷励磁直流型 | Aeroflex④ | 0.00988 | 4000 | 256 | 0.00386 | 20 |

续表

| 发动机分类 | 制造商 | 堵转扭矩/(N·m) | 空载转速/(r/min) | 重量/g | 扭矩密度/(N·m/g) | 效率/% |
|---|---|---|---|---|---|---|
| 三相励磁交流型 | Astro[5] | 0.0755 | 11500 | 340 | 0.0222 | 20 |
| 纵扭驻波超声型 | Kumada[6] | 1.334 | 120 | 150 | 0.889 | 80 |
| 行波超声型，$\Phi 60$ | Shinsei[7] | 1.0 | 150 | 260 | 0.385 | 35 |
| 行波超声型，$\Phi 60$ | PDLab[8] | 1.2 | 180 | 250 | 0.522 | 30 |

① 美国 PIEZOSYSTEMS 公司。

② http://www.maxonmotor.com/maxon/view/content/index。

③ http://www.mabuchi-motor.co.jp/en_US/technic/t_0401.html。

④ http://ams.aeroflex.com/motion/motion-motors-brushless.cfm。

⑤ www.astroflight.com。

⑥ Kumada, A.；Piezotech 公司，日本东京[45]。

⑦ http://www.shinsei-motor.com。

⑧ http://jiangsuusm.en.made-in-china.com/company-Jiangsu-TransUSM-Co-Ltd-.html。

(2) 低速下的高扭矩无须齿轮即可直接驱动负载，定位精度和响应速度大大提高。基于这一优点，减少了因齿轮箱、振动和噪声、能量损失以及由传动引起的位置误差导致的额外体积和重量。

(3) 发动机的运动部件（转子）具有惯性小、响应快速（微秒级）、自锁和保持转矩高等优点。超声发动机可在几毫秒内达到稳定的速度，而由于定子和转子之间的摩擦，制动速度更快。

(4) 位置/速度的良好可控性和位移的高分辨率。因为定子的工作频率很高，而转子或滑块较轻，所以超声发动机在伺服系统中可达到微米级甚至纳米级控制精度。超声发动机的响应非常快，其位移分辨率很高。

(5) 无电磁干扰。超声发动机的工作方式与传统发动机不同；在工作过程中不会产生磁场，且不会受到电磁干扰。

(6) 噪声低。超声发动机的工作频带通常大于 20kHz，超出了人类听觉的阈值。此外，因为发动机可直接驱动负载，所以避免了齿轮箱减速产生的噪声。

(7) 可在极端环境条件下工作。压电和摩擦材料的合理设计和适当选择，可使超声发动机在极端环境条件下（真空或高/低温）正常工作。

超声发动机的缺点包括：

(1) 功率输出小、效率低。超声发动机有两个能量转换过程。第 1 个过程通过逆压电效应将电能转化为机械能。第 2 个过程通过定子和转子间的摩擦将定子的振动转变为转子的宏观单向运动。这两个过程都会产生能量损失,尤其是后者。因此,超声发动机的效率较低。目前,行波超声发动机的效率约为30%,输出功率小于50W。

(2) 使用寿命短,不适合持续工作。在摩擦传动过程中,定子和转子间存在摩擦和磨损问题。此外,高频振动会导致转子和压电材料的疲劳损伤,尤其是在大输出功率和高环境温度的条件下。因此,长时间连续工作后,使用寿命会缩短,性能会降低。

(3) 对驱动信号的特殊要求。为了激发定子的谐振,发动机对激励信号的振幅、频率和相位有特殊要求。当发动机温度变化时,压电元件激励信号的频率需要适当调整,以保持输出性能的稳定。因此,超声发动机驱动器的电路有时会比较复杂。

**2. 超声发动机的分类**

设计灵活、结构多样的超声发动机没有统一的分类方法。表7.3列出了一些按不同角度的分类。

表7.3 超声发动机分类

| 视　　角 | 类　　型 |
| --- | --- |
| 波传播方式 | 行波;驻波 |
| 动作输出方式 | 直线;旋转 |
| 定转子接触状态 | 接触;无接触 |
| 压电元件对定子的励磁条件 | 谐振;无谐振 |
| 转子的自由度数 | 单一度数 |
| 某方向上工作模式的位移 | 平面外;平面内 |
| 定子的几何形状 | 盘;环;棒;壳 |
| 旋转方向 | 单向;双向 |

超声发动机作为利用振动的典型产品,根据振动形式的分类基本上能够反映这些发动机的特性。基于此工作原理,超声发动机可分为如下5类:

1) 第1类:基于纵向振动

基于纵向振动模式的发动机属于驻波发动机。该类发动机使用具有电能-机械能高转换效率的Langevin振子。Langevin谐振器(图7.47)[46]是一种预应力夹层换能器:一个圆盘或一对压电陶瓷圆盘夹在金属端部之间,并通过金属上的螺栓将其置于压缩偏置下。这种预应力显著降低了压电陶瓷工作过程中的拉伸应力。

333

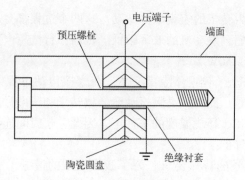

图 7.47 Langevin 振动器示意图

超声发动机接触面上摩擦材料的磨损是科学家们必须解决的问题之一。Sashida 提出的第 1 台驻波超声发动机就属于此类发动机[47]。定子在一个方向上的纵向振动可通过柔性片的变形转换为转子的旋转运动。1989 年，Kurosawa 等提出了一种使用两个纵向振动复合模式的线性超声发动机，该发动机具有更高的驱动效率[48]。实验证明，其可实现 76W/kg 的大功率密度，最大输出力为 51N，最高速度为 0.55m/s。

2) 第 2 类：基于纵向-弯曲振动复合模式

1989 年，Tomikawa 等基于矩形板的纵向-弯曲振动复合模式设计了一种线性发动机，如图 7.48 所示[49]。该发动机利用矩形板的第 1 纵向振动模式和第 4 弯曲振动模式来实现驱动腿的椭圆运动，实验测得样机效率为 20.8%。之后，Ueha 和 Tomikawa 也提出了一种使用第 1 纵向振动模式和第 8 弯曲振动模式的扁平线性超声发动机[50]。实验结果表明，该样机的最大空载速度为 0.7m/s，最大推力为 4N。该发动机具有结构简单、外形扁平和速度快的特点，

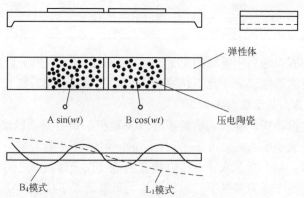

图 7.48 Tomikawa 的板式纵向/弯曲发动机

特别适合于纸张、卡片等轻薄物体的传动。注意，早在 1995 年，Nikon 和 NEC 株式会社就已经分别生产出该类发动机产品。

1992 年，Onishi 和 Naito 设计了一种两腿结构的 IT 形线性超声发动机，如图 7.49 所示[51]。偏转布局中两个堆叠的压电陶瓷激发了 IT 形弹性体腿部的纵向和横向弯曲模式，进而合成了驱动导轨的椭圆运动。其激励频率约为 90kHz，施加电压的相位差为 90°。该发动机的空载速度为 30cm/s，最大推力为 10N。1993 年，SUNSYN 公司制造并商业化了这种用于 X-Y 定位系统的 IT 形线性超声发动机，实现了线性超声发动机的首次应用。

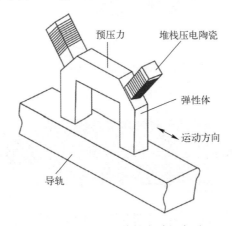

图 7.49　PI 形的线性发动机定子

3）第 3 类：基于纵向-扭转振动复合模式

如图 7.50 所示，Kurosawa 和 Ueha 开发了一种纵向-扭转振动复合发动机，该发动机的独特之处在于层叠式压电振子产生的纵向振动[52]。这种振动在低电压和非共振条件下可以具有更大的振幅。转子直径为 50mm，总长度为 82mm。当转子上施加 90N 的预压力、在扭转激振器上施加 34V（以 $V_{rms}$ 来衡量，其中 rms 为均方根）电压时，测得该发动机的空载转速为 100r/min，最大扭矩为 0.7N·m，最大效率为 33%。图 7.51 展示了该实验室开发的有刷纵向-扭转振动复合超声发动机。在发动机内部，纵向与扭转压电陶瓷分别布置在定子和转子内。这种设计可以单独调整定子和转子的结构参数，以保持纵向-扭转振动的模态频率尽可能接近。发动机结构的合理设计可以增加运行过程中接触面上的预压力，进而提升输出性能和工作效率。因为扭转振动压电陶瓷布置在转子内，所以可通过电刷向转子提供电能。这种发动机称为有刷纵向-扭转振动复合超声发动机。该发动机的直径为 45mm，长度为 210mm，最大输出扭矩为 2.5N·m。

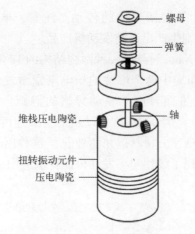

图 7.50 Kurosawa 和 Ueha 的纵向-扭转发动机

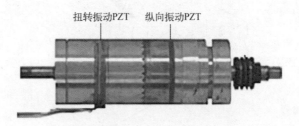

图 7.51 电刷式纵向-扭转发动机

4）第 4 类：基于弯曲振动

根据定子的结构，基于弯曲振动模式的超声发动机可以分为杆型、环型和盘型三类，三者均属于行波超声发动机。近年来，基于面外弯曲振动模式的棒型定子的 USM 由于其结构简单、便于制造、低成本等优点，已成为微型致动器研究领域中的一个热点。某些棒型定子超声发动机已应用在微型透镜聚焦系统和医疗内窥镜系统上。

5）第 5 类：基于面内振动

面内振动发动机有伸缩型、弯曲型和扭转型三种类型。1989 年，Takano 开发了一种基于面内伸缩和弯曲振动模式的超声发动机，如图 7.52 所示。当相位差为 π/2 的交流电压信号分别施加在两个区域的圆形压电陶瓷上时，驱动点 $A\ (A')$ 的径向和切向运动合成了用于驱动转子的椭圆运动。转子直径为 10mm，厚度为 2mm，工作频率为 43.3kHz，最大输出扭矩为 40mN·m，效率为 3.5%。2005 年，Zhou 等开发了一种使用了空心圆柱体面内伸缩振动模式的线性超声发动机，如图 7.53 所示[53]。

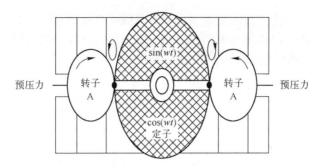

图 7.52　基于面内模式的超声发动机

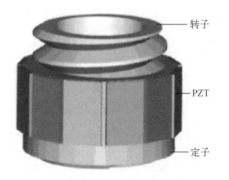

图 7.53　基于面内伸缩模式的线性超声发动机

注意，除了上述 5 种类型，还有一些基于其他模式的超声发动机，如扭转-弯曲振动复合模式、纵向-剪切振动复合模式。

## 7.3.4　超声压电发动机的工作原理

在输入和输出连杆间的接触区域产生椭圆运动的压电发动机是最受欢迎的一种。为达到这一目标，利用了对输出链路或行波的斜向冲击。在压电发动机中，通过利用斜向冲击产生的摩擦力在输入和输出链接间传递运动与能量。这可以通过接触区域具有相位差的两个振荡运动（法向和切向分量）$u_y$ 和 $u_x$ 来实现，该相位差被用于改变输出连杆的运动方向。这两种运动均可以通过一个或两个主动连杆的共振来实现。不同的振荡模式提供了开发不同类型压电发动机的可能性，包括纵向、横向、剪切和扭转等模式。采用斜向冲击的压电发动机具有很宽的频率范围。其下限位于更低的超声频率（用于消除声学作用），即 16~20kHz；其上限为几个兆赫。基于弹性体内行波运动和输出连杆间摩擦作用的行波运动压电发动机，其工作原理类似于谐波牵引传动。沿着输入连杆表面传播的瑞利波在接触区域形成椭圆运动。瑞利波是纵向波和剪切波的耦合波。因此,弹性介质中的每个表面点均沿着椭圆轨迹运动。压电发动机利用了弯曲波、剪切波、扭转波和纵向波。压电陶瓷中的行波由电场激发。行波运动压电发动机的特性[54]如表7.4所列。

表7.4 一些行波压电发动机的特性

| 发 动 机 | 单 位 | USR60 | USR45 | USR30 |
|---|---|---|---|---|
| 工作频率 | kHz | 40 | 43 | 42 |
| 工作电压 | $V_{rms}$ | 100 | 100 | 100 |
| 额定扭矩 | N·m | 0.38 | 0.15 | 0.04 |
| 额定输出功率 | W | 4.0 | 2.3 | 1.0 |
| 额定转速 | r/min | 100 | 150 | 250 |
| 机械时间常数 | ms | 1 | 1 | 1 |
| 重量 | g | 175 | 69 | 33 |
| 旋转误差 | % | 2 | 2 | 2 |
| 使用寿命 | h | 1000 | 1000 | 100 |
| 工作温度范围 | ℃ | -10~+50 | -10~+50 | -10~+50 |

# 参 考 文 献

[1] Wolcken, P. C. and Papadopoulos, M., (eds.), Smart Intelligent Aircraft Structures (SARISTU), Proceedings of the Final Project Conference, 19-21 of May 2015, Moscow, Russia, Springer International Publishing AG Switzerland 2016, 1039.

[2] Dimino, I., Ciminello, M., Concilio, A., Pecora, R., Amoroso, F., Magnifivo, M., Schueller, M., Gratias, A., Volovick, A. and Zivan, L., Distributed actuation and control of a morphing wing trailing edge, Smart Intelligent Aircraft Structures (SARISTU), Proceedings of the Final Project Conference, Springer International Publishing AG Switzerland, 2016, 171-186.

[3] Schorsch, O., Luhring, A. and Nagel, C., Elastomer-based skin for seamless morphing of adaptive wings, Smart Intelligent Aircraft Structures (SARISTU), Proceedings of the Final Project Conference, Springer International Publishing AG Switzerland, 2016, 187-197.

[4] Snop, V. and Horak, V., Testing overview of the EADN samples, Smart Intelligent Aircraft Structures (SARISTU), Proceedings of the Final Project Conference, Springer International Publishing AG Switzerland, 2016, 85-96.

[5] Kintscher, M., Kirn, J., Storm, S. and Peter, F., Assessment of the SARISTU enhanced adaptive droop nose, Smart Intelligent Aircraft Structures (SARISTU), Proceedings of the Final Project Conference, Springer International Publishing AG Switzerland, 2016, 113-140.

[6] Monier, L., Le Jeune, K., Kondolff, I. and Vilaca, G., Coating for detecting damage with a manifest col-

or change, Smart Intelligent Aircraft Structures (SARISTU), Proceedings of the Final Project Conference, Springer International Publishing AG Switzerland, 2016, 735-743.

[7] Airoldi, A., Sala, G., Evenblij, R., Koimtzoglou, C., Loutas, T., Carossa, G. M., Mastromauro, P. and Kanakis, T., Load monitoring by means of optical fibers and strain gages, Smart Intelligent Aircraft Structures (SARISTU), Proceedings of the Final Project Conference, Springer International Publishing AG Switzerland, 2016, 433-469.

[8] Bach, M., Dobmann, N. and Bonet, M. M., Damage introduction, detection and assessment of CFRP door surrounding panel, Smart Intelligent Aircraft Structures (SARISTU), Proceedings of the Final Project Conference, Springer International Publishing AG Switzerland, 2016, 947-957.

[9] Schmidt, D., Kolbe, A., Kaps, R., Wierach, P., Linke, S., Steeger, S., Dungern, F., Tauchner, J., Breu, C. and Newman, B., Development of a door surround structure with integrated structural health monitoring system, Smart Intelligent Aircraft Structures (SARISTU), Proceedings of the Final Project Conference, Springer International Publishing AG Switzerland, 2016, 935-945.

[10] Acellent Technologies Inc., Smart Layer®, http://www.acellent.com, May 2015.

[11] Wierach, P., Elektromechanisches Funktionmodul, German Patent # DE10051784C1, 2002.

[12] Wierach, P., Development of piezocomposites for adaptive systems, DLR-Forschungsbericht DLR-FB 2010-23, Technical University Braunschweig, Braunschweig, Ph. D. Thesis, 2010.

[13] Schorsch, O., Luhring, A., Nagel, C., Pecora, R. and Dimino, I., Polymer based morphing skin for adaptive wings, Proceedings of the 7th ECCOMAS Thematic Conference on Smart Structures and Materials, SMART 2015, Azores, Portugal, 3-6 June, 2015.

[14] Alcaide, A. and Martin, F., PAMELA SHM system implementation on composite wing panels, Smart Intelligent Aircraft Structures (SARISTU), Proceedings of the Final Project Conference, Springer International Publishing AG Switzerland, 2016, 545-555.

[15] Duerig, T., Pelton, A. and Stockel, D., An overview of Nitinol medical applications, Materials Science and Engineering A 273-275, 1999, 149-160.

[16] Stockel, D., Nitinol medical devices and implants, Minimal Invasive Therapy & Allied Technology 2 (2), 2000, 81-88.

[17] Morgan, N. B., Medical shape memory alloy applications-the market and its products, Materials Science and Engineering A 378, 2004, 16-23.

[18] Shabalovskaya, S., Anderegg, J. and Van Humbeeck, J., Critical overview of Nitinol surfaces and their modifications for medical applications, Acta Biomaterialia 4, 2008, 447-467.

[19] Abella, F., de Ribot, J., Doria, G., Sindreu, D. S. and Roig, M., Applications of piezoelectric surgery in endodontic surgery: a literature review, JOE 40 (3), March 2014, 325-332. doi: http://dx.doi.org/10.1016/j.joen.2013.11.014.

[20] Hennet, P., Piezoelectric bone surgery: a review of the literature and potential applications in veterinary oromaxillofacial surgery, Frontiers in Veterinary Science 2 (8), May 2015, 6. doi: http://dx.doi.org/10.3389/fvets.2015.00008.

[21] Shung, K. K., Cannata, J. M. and Zhou, Q. F., Piezoelectric materials for high frequency imaging applications: a review, Journal of Electroceramics 19, 2007, 139-145.

[22] Martin, K. H., Lindsey, B. D., Ma, J., Lee, M., Li, S., Foster, F. S., Jiang, X. and Dayton, P.

A., Dual frequency piezoelectric transducers for contrast enhanced ultrasound imaging, Sensors 14, 2014. doi: 10.3390/s141120825, 20.825–20.842.

[23] Hagood, N. W. and McFarland, A. J., Modeling of a piezoelectric rotary ultrasonic motor, IEEE Transactions on Ultrasonics, Ferroelectrics, and Frequency Control 42 (2), March 1995, 210–224.

[24] Bansevicius, R. and Tolocka, R. T., Piezoelectric actuator s, Chapter 20.3, The Mechatronics Handbook, Bishop, R. H. – Editor-in-Chief, CRC Press LLC, 2002.

[25] Sharp, S. L., Design of a linear ultrasonic piezoelectric motor, Master of Science Thesis, Department of Mechanical Engineering, Brigham Young University, Provo, UT, USA, August 2006, 189 pp.

[26] Yoon, M.-S., Khansur, N. H., Lee, K.-S. and Park, Y. M., Compact size ultrasonic linear motor using a dome shaped piezoelectric actuator, Journal of Electroceramics 28 (2), 2012, 123–131.

[27] Pirrotta, S., Sinatra, R. and Meschini, A., A novel simulation model for ring type ultrasonic motor, Meccanica 42 (2), April 2007, 127–139.

[28] Liu, Y., Chen, W., Liu, J. and Shi, S., A rotary ultrasonic motor using bending vibration transducers, IEEE Transactions on Ultrasonics, Ferroelectrics, and Frequency Control 57 (10), October 2010, 2360–2364.

[29] Flynn, A. M., Piezoelectric ultrasonic micromotors, MIT Artificial Intelligence Laboratory, Computer Science and Artificial Intelligence Lab (CSAIL) Artificial Intelligence Lab Publications, AI Technical reports (1964–2004), June 1995, http://hdl.handle.net/1721.1/7086.

[30] Kanda, T., Makino, A. and Oomori, Y., A cylindrical micro ultrasonic motor using micromachined piezoelectric vibrator, Okayama University, Japan, Ultrasonics Symposium, 2005 IEEE (Vol.1), 18–21 September 2005.

[31] Uchino, K., Cagatay, S., Koc, B., Dong, S., Bouchilloux, P. and Strauss, M., Micro piezoelectric ultrasonic motors, ADM001697, ARO – 44924.1 – EG – CF, International Conference on Intelligent Materials (5th) (Smart Systems & Nanotechnology), State College, PA, 14–17 June 2003.

[32] Watson, B., Friend, J. and Yeo, L., Piezoelectric ultrasonic resonant motor with stator diameter less than 250μm: the Proteus motor, Journal of Micromechanics and Microengineering 19, 2009, 1–5.

[33] Hirata, H. and Ueha, S., Characteristics estimation of a traveling wave type ultrasonic motor, IEEE Transactions on Ultrasonics, Ferroelectrics, and Frequency Control 40 (4), July 1993, 402–406.

[34] Uchino, K., Piezoelectric ultrasonic motors: overview, Smart Materials and Structures 7 (3), 1998, 273–285.

[35] Dong, S., Yan, L., Wang, N. and Viehland, D., A small, linear, piezoelectric ultrasonic cryomotor, Applied Physics Letters 86, 2005, article ID 053501, 3 pp.

[36] Wallaschek, J., Contact mechanics of piezoelectric ultrasonic motors, Smart Materials and Structures 7 (3), 1998, 369–381.

[37] Zhang, H., Dong, S.-X., Zhang, S.-Y., Wang, T.-H., Zhang, Z.-N. and Fan, L., Ultrasonic micromotor using miniature piezoelectric tube with diameter of 1.0 mm, Ultrasonics 44, 2006, e603–e606.

[38] Zhao, C. S., Ultrasonic Motors Technologies and Applications, Beijing, Science Press, China, 2010.

[39] Xiaolong, L., Junhui, H., Lin, Y. and Chunsheng, Z., A novel in-plane rotary ultrasonic motor, Chinese Journal of Aeronautics 27 (2), 2014, 420–424.

[40] Newton, D., Garcia, E. and Horner, G. C., A linear piezoelectric motor, Smart Materials and Structures 7 (3), 1998, 295–305.

[41] Ho, S.-T. and Shin, Y.-J., Analysis of a linear piezoelectric motor driven by a single-phase signal, 2013 IEEE International Ultrasonics Symposium (IUS), 21–25 July 2013, 481–484, Prague, The Czech Republic.

[42] Bauer, M. G., Design of a linear high precision ultrasonic piezoelectric motor, Ph. D. Thesis, Mechanical Engineering Department, Raleigh, North Carolina State University, 2001.

[43] Lopez, J. F., Modeling and optimizationof ultrasonic linear motors, Ph. D. Thesis, Faculte des Sciences et Techniques de L'ingenieur, Ecole Polytechnique Federale de Lausanne, Switzerland, Thesis No. 3662, 10th November 2006, 281.

[44] El Ghouti, N., Hybrid Modelling of a Traveling Wave Piezoelectric Motor, Ph. D. Thesis, Department of Control Engineering, Aalborg University, DK-9220 Aalborg Ø, Denmark, Doc. no. D-00-4383, May 2000.

[45] Kumada, A., Piezoelectric revolving motors applicable for future purpose, IEEE 7th Intenational Symposium on Applications of Ferroelectrics, 213–219, Urbana-Champaign, IL, USA, 6–8 June 1990.

[46] Carotenuto, R., Iula, A. and Pappalardo, M., A displacement amplifier using mechanical demodulation, Applied Physics Letters 73 (18), November 1998, 2573–2575.

[47] Sashida, T., Motor device using ultrasonic oscillation, US patent #4562374, 16th of May 1984.

[48] Kurosawa, M., Nakamura, K., Okomoto, T. and Ueha, S., An ultrasonic motor using bending vibrations of a short cylinder, IEEE Transactions on Ultrasonics, Ferroelectrics, and Frequency Control 36 (5), September 1989, 517–521.

[49] Tomikawa, Y., Takano, T. and Umeda, H., Thin rotary and linear ultrasonic motors using a double-mode piezoelectric vibrator of the first longitudinal and second bending mode, Japan Journal of Applied Physics, 31, 1992, 3073–3076.

[50] Ueha, S. and Tomikawa, Y., Ultrasonic Motors, Theory and Applications, Oxford Science Publication, Clarendon Press, 1993.

[51] Onishi, K. and Naito, K., Ultrasonic linear motor, US patent #5134334 A, July 28th, 1992.

[52] Kurosawa, M. and Ueha, S., Hybrid transducer type ultrasonic motor, IEEE Transactions on Ultrasonics, Ferroelectrics, and Frequency Control 38 (3), 1991, 89–92.

[53] Zhou, T., Zhang, K., Chen, Y., Wang, H., Wu, J., Jiang, K. and Xue, P., A cylindrical rod ultrasonic motor with 1 mm diameter and its application in endoscopic OCT, Chinese Science Bulletin 50 (8), 2005, 826–830.

[54] Bishop, R. H., The Mechatronics Handbook, CRC Press, 2002, 1290.

# 第8章 基于智能材料的能量收集

## 8.1 压电能量收集

能量收集,有时也称为能量采集,是指从环境的各种能量源中获取能量的一种方式,如直接从太阳或人造太阳光中获取的以环境光、射频等形式存在的无线电射频、各种热源和机械振动源,并将其转化为电能提供给不同用户。能量收集可通过以下一个或多个原理得以实现[1]:光伏、塞贝克效应(Seebeck effect)①、RF 和机械源的电磁、静电以及压电效应。近年来,随着对人体运动或风致振动等新能源的研究和分析,致力于能量采集的研究急剧增加,其目标是发现周围环境中可用能量提取和储存的新方法。因此,机械能的来源包括运输领域(如汽车、卡车、火车甚至飞机)、基础设施(如桥梁、道路、隧道、农场和房屋)、工业(如电机、泵、压缩机、振动、噪声)、环境(如风、洋流、波浪)以及人体活动和运动(如呼吸、行走、跳跃、慢跑、呼气以及手臂和/或手指运动)。本章致力于讨论环境振动中的压电式能量收集,其中压电材料使用 $d_{31}$ 工作模式,并使用 $d_{33}$ 工作模式直接施加应力。Roundy、Lopes 和 Gallo 罗列了一系列不同环境中振动源,并通过压电式采集器进行收集(表 8.1)[2-3]。

表 8.1 环境振动源

| 振 动 源 | 峰值加速度/(m/s²) | 相关频率/Hz |
|---|---|---|
| 3 轴机床底座 | 10.0 | 70 |
| 厨房的搅拌器套管 | 6.4 | 121 |
| 衣服烘干机 | 3.5 | 121 |
| 洗衣机 | 0.5 | 109 |

---

① 塞贝克效应(Seebeck effect)又称为第一热电效应,是指由两种不同导体或半导体的温度差异而引起两种物质间电压差的热电现象。

# 第8章 基于智能材料的能量收集

续表

| 振 动 源 | 峰值加速度/(m/s$^2$) | 相关频率/Hz |
|---|---|---|
| 冰箱 | 0.1 | 240 |
| 读取 CD 时的笔记本电脑 | 0.6 | 75 |
| 小型微波炉 | 2.25 | 121 |
| 关门时的门框 | 3.0 | 125 |
| 办公楼的空调通风口 | 0.2~1.5 | 60 |
| 面包机 | 1.03 | 121 |
| 行人走动的甲板 | 1.3 | 385 |
| 汽车发动机舱 | 12.0 | 200 |
| 木结构办公楼的 2 层 | 0.2 | 100 |
| 靠近繁华街道的外部窗户（2 英尺×3 英尺） | 0.7 | 100 |
| 快速拍脚后跟 | 3.0 | 1 |
| 汽车仪表盘 | 3.0 | 13 |

图 8.1 展示了使用收集器将能量从其来源转化为电能的示意图。根据收集器的类型，力和速度被转化为了电压和电流。

图 8.1 能量收集系统原理图

本章将讨论基于压电材料的能量收集装置以及其将环境机械振动或直接施加应力转化为电能（或能量）的性能。图 8.2 展示了两种最常见的由机械能产生电能的模式示意图。

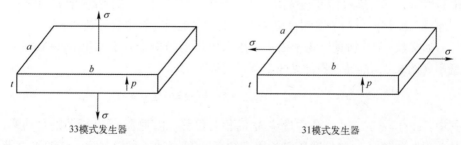

图 8.2 两种产生电能的模式示意图

在图 8.2 中，$P$ 表示极性方向，$\sigma$ 表示施加在尺寸为 $a \times b \times t$（宽×长×厚）压电板上的机械应力。根据不同的工作模式，$d_{33}$ 工作模式的机械、电与几何尺寸之间的关系由式（8.1）给出：

$$\begin{cases} Q_{@V=0} = d_{33} \cdot \sigma \cdot (a \cdot b) \\ V_{@Q=0} = g_{33} \cdot \sigma \cdot t \end{cases} \tag{8.1}$$

式中：$Q$ 为压电电极上的感应电荷；$V$ 为感应电压；$d_{33}$ 和 $g_{33}$ 分别为压电电荷常数和压电电压常数。

对于 $d_{31}$ 工作模式，得

$$\begin{cases} Q_{@V=0} = d_{31} \cdot \sigma \cdot (a \cdot b) \\ V_{@Q=0} = g_{31} \cdot \sigma \cdot t \end{cases} \tag{8.2}$$

为了获得两种发电模式的电能，必须将电荷乘以电压，再除以 2，得到 $d_{33}$ 工作模式如下：

$$U = \frac{1}{2} Q \cdot V = \frac{1}{2} g_{33} \cdot d_{33} \cdot \sigma_{33}^2 \cdot (a \cdot b \cdot t) = \frac{1}{2} g_{33} \cdot d_{33} \cdot \sigma_{33}^2 \cdot \text{Vol} \tag{8.3}$$

式中：Vol 表示压电材料的体积。相应地，对于 $d_{31}$ 工作模式，能量将被定义为

$$U = \frac{1}{2} Q \cdot V = \frac{1}{2} g_{31} \cdot d_{31} \cdot \sigma_{11}^2 \cdot (a \cdot b \cdot t) = \frac{1}{2} g_{31} \cdot d_{31} \cdot \sigma_{11}^2 \cdot \text{Vol} \tag{8.4}$$

为了实现振动能量收集，需要一种称为采集器的设备，其通常由一个悬臂梁组成，该梁配备有粘在承载梁一侧或两侧的压电贴片或压电层。为了将压电梁的固有频率降低到环境频率，在压电梁的自由端加装一个质量块，称为末端质量。压电梁的底座与振动物体相连接，然后利用合理的电路将机械能从中收集转化为电能。由于 PZT 材料的脆性（具有陶瓷材料的特性）或 PVDF 材料的柔韧性，作为压电材料时，还需要一个承载梁。图 8.3 展示了一个典型的压电悬臂梁，其一维自由度系统由一个质量块、弹簧和一个阻尼器组成，压电装置通过二极管桥与调节器和存储系统（电容电池或超级电容）进行连接。

如文献［4］所述，用于描述一维系统运动的模型属于 Williams-Yates 模型[5]（图 8.3，推导过程详见 8.2 节）可以写为

$$M \cdot \ddot{z}(t) + C_d \cdot \dot{z}(t) + K \cdot z(t) = -M \cdot \ddot{y}(t) \tag{8.5}$$

式中：$z(t) = x(t) - y(t)$ 是质量块 $M$ 的相对位移。阻尼器 $C_d$ 上耗散的总功率，是关于振动频率 $\omega$ 和悬臂压电梁固有频率 $\omega_n$ 的函数，可以写为（参见 8.2 节中的详细推导）

# 第8章 基于智能材料的能量收集

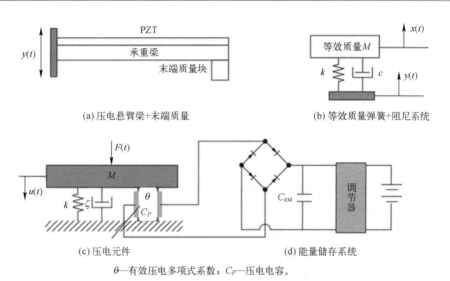

(a) 压电悬臂梁+末端质量　　　　　(b) 等效质量弹簧+阻尼系统

(c) 压电元件　　　　　　　　　　(d) 能量储存系统

$\theta$—有效压电多项式系数；$C_P$—压电电容。

图8.3　典型的压电悬臂梁储能系统示意图[4]

$$p_\text{d}(\omega) = \frac{2\zeta \cdot M \cdot \left(\dfrac{\omega}{\omega_\text{n}}\right)^3 \cdot \omega^3 \cdot Y_0^2}{\left[1-\left(\dfrac{\omega}{\omega_\text{n}}\right)^2\right]^2 + \left[2\zeta \cdot \left(\dfrac{\omega}{\omega_\text{n}}\right)\right]^2} \tag{8.6}$$

式中：$Y_0$ 为惯性质量块 $M$ 的位移；$\zeta$ 为阻尼比。根据以下定义：

$$\omega_\text{n}^2 \equiv \frac{K}{M}, \quad \frac{C_d}{M} \equiv 2\zeta \cdot \omega_\text{n} \Rightarrow \zeta = \frac{C_d}{2\sqrt{M \cdot K}} \tag{8.7}$$

将压电悬臂梁的频率调节到环境频率，即在共振状态下工作，$\omega = \omega_\text{n}$，可以得到最大可用功率为

$$p_\text{d}(\omega_\text{n})_\text{max} = \frac{2\zeta \cdot M \cdot \omega_\text{n}^3 \cdot Y_0^2}{4\zeta^2} = \frac{M \cdot \omega_\text{n}^3 \cdot Y_0^2}{2\zeta} \tag{8.8}$$

可以用运动的加速度 $\overline{a}$ 重写式（8.8），而 $\overline{a} = \omega_\text{n}^2 Y_0$，得

$$p_\text{d}(\omega_\text{n})_\text{max} = \frac{M \cdot \overline{a}^2}{2\omega_\text{n} \cdot \zeta} \tag{8.9}$$

虽然 Williams-Yates 模型最初是针对具有黏滞阻尼的电磁转换器推导出来的，但是通过式（8.8）和式（8.9）计算的功率通常适用于任何线性惯性转换器。根据文献［4］，鉴于压电转换机理的复杂性，在分析压电采集器时，需要使用另一个模型（Ertuk-Inman 模型[6-8]）来进行分析。采用图 8.4 Ertuk 和 Inman 串联的伯努利-欧拉梁理论编写了以下梁运动方程式[6]：

$$(\overline{EI})\frac{\partial^4 w_{\rm rel}^{\rm s}(x,t)}{\partial x^4}+c_{\rm S}I\frac{\partial^5 w_{\rm rel}^{\rm s}(x,t)}{\partial x^4 \partial t}+c_{\rm a}\frac{\partial w_{\rm rel}^{\rm s}(x,t)}{\partial t}+m\frac{\partial^2 w_{\rm rel}^{\rm s}(x,t)}{\partial t^2}$$
$$+\overline{V}_{\rm s}v_{\rm s}(t)\left[\frac{{\rm d}\delta(x)}{{\rm d}x}-\frac{{\rm d}\delta(x-L)}{{\rm d}x}\right]=-\left[m+M_{\rm end}\delta(x-L)\right]\frac{\partial^2 w_{\rm base}(x,t)}{\partial t^2} \qquad (8.10)$$

式中：串联的压电耦合项 $\overline{V}_{\rm s}$ 由下式给出：

$$\overline{V}_{\rm s}=\frac{\overline{e}_{31}b}{2h_{\rm p}^{\sim}}\left[\frac{h_{\rm s}^{\sim 2}}{4}-\left(h_{\rm p}^{\sim}+\frac{h_{\rm s}^{\sim}}{2}\right)^2\right] \qquad (8.11)$$

式中：$w_{\rm base}$ 为具有平动和小旋转分量的有效基底位移；$w_{\rm rel}^{\rm s}$ 为串联处的相对横向位移；$v_{\rm s}(t)$ 为穿过电阻负载（串联时）的电压响应；$(\overline{EI})$ 为带有转动惯量 $I$ 的弯曲刚度；$m$ 为单位长度梁的质量；$c_{\rm a}$ 为外部黏滞阻尼（空气或其他周围流体）；$c_{\rm S}I$ 为内部应变速率（或 Kelvin-Voight 阻尼）；$M_{\rm end}$ 为位于悬臂梁自由端的末端质量；$\overline{e}_{31}$ 为压电常数；$\delta(x)$ 为 Dirac $\delta$ 函数。对于并联情况，运动方程式为

$$(\overline{EI})\frac{\partial^4 w_{\rm rel}^{\rm p}(x,t)}{\partial x^4}+c_{\rm S}I\frac{\partial^5 w_{\rm rel}^{\rm p}(x,t)}{\partial x^4 \partial t}+c_{\rm a}\frac{\partial w_{\rm rel}^{\rm p}(x,t)}{\partial t}+m\frac{\partial^2 w_{\rm rel}^{\rm p}(x,t)}{\partial t^2}+$$
$$\overline{V}_{\rm p}v_{\rm p}(t)\left[\frac{{\rm d}\delta(x)}{{\rm d}x}-\frac{{\rm d}\delta(x-L)}{{\rm d}x}\right]=-\left[m+M_{\rm end}\delta(x-L)\right]\frac{\partial^2 w_{\rm base}(x,t)}{\partial t^2} \qquad (8.12)$$

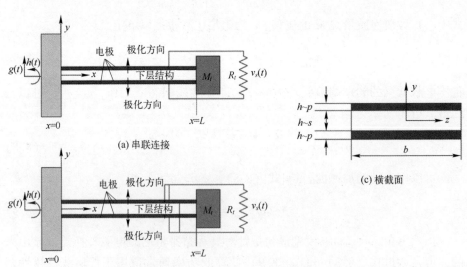

图 8.4 双晶片悬臂结构[6]

并联的反向耦合项 $\bar{V}_p$ 为

$$\bar{V}_p = 2\bar{V}_s = \frac{\bar{e}_{31}b}{h_{\tilde{p}}}\left[\frac{h_{\tilde{s}}^2}{4} - \left(h_{\tilde{p}} + \frac{h_{\tilde{s}}}{2}\right)^2\right] \tag{8.13}$$

而质量 $m$ 和弯曲刚度则可以写为

$$\begin{cases} m = b(\rho_{\tilde{s}}h_{\tilde{s}} + 2\rho_{\tilde{p}}h_{\tilde{p}}) \\ (\overline{EI}) = \frac{2b}{3}\left\{\bar{E}_{\tilde{s}}\frac{h_{\tilde{s}}^3}{8} + \bar{c}_{11}^E\left[\left(h_{\tilde{p}} + \frac{h_{\tilde{s}}}{2}\right)^3 - \frac{h_{\tilde{s}}^3}{8}\right]\right\} \end{cases} \tag{8.14}$$

式中：$\rho_{\tilde{s}}$ 和 $\rho_{\tilde{p}}$ 分别为承载结构和压电材料的质量密度；$\bar{E}_{\tilde{s}}$ 为承载结构的杨氏模量；$\bar{c}_{11}^E$ 为恒定电场下的弹性刚度。由于唯一的机械应变是由轴向应变引起的，电位移可以表示为

$$D_s = \bar{e}_{31}S_1^{\tilde{p}} + \bar{\varepsilon}_{33}^S E_3 \tag{8.15}$$

式中：$\bar{\varepsilon}_{33}^S = \varepsilon_{33}^T - \frac{d_{31}^2}{S_{11}^E}$，而 $\bar{\varepsilon}_{33}^S$ 为恒定应变下的介电常数分量（假设为平面应力情况）；$\varepsilon_{33}^T$ 为恒定应力下的介电常数。通过高斯定律计算电位移，并假设压电层在仅有电阻负载 $R_{load}$ 的电路中工作，可以得到如下的电路方程式[5-26]：

$$\frac{\bar{\varepsilon}_{33}^S \cdot b \cdot L}{h_{\tilde{p}}}\frac{\mathrm{d}v(t)}{\mathrm{d}t} + \frac{v(t)}{R_{load}} = -\bar{e}_{31}h_{\tilde{p}_c}b\int_0^L \frac{\partial^3 w_{rel}(x,t)}{\partial x^2 \partial t}\mathrm{d}x \tag{8.16}$$

式中：$b$、$h_{\tilde{p}}$ 和 $L$ 分别为压电层的宽度、厚度和长度；$h_{\tilde{p}_c}$ 为中性轴和压电层中心之间的距离。将 $w_{rel}(x,t)$ 如下的模态表达式代入式 (8.16)，模态表达式为

$$w_{rel}(x,t) = \sum \phi_r(x)\varphi_r(t) \tag{8.17}$$

得

$$\frac{\bar{\varepsilon}_{33}^S \cdot b \cdot L}{h_{\tilde{p}}}\frac{\mathrm{d}v(t)}{\mathrm{d}t} + \frac{v(t)}{R_{load}} = \sum_{i=1}^{\infty}\kappa_r\frac{\mathrm{d}\varphi_r(t)}{\mathrm{d}t} \tag{8.18}$$

式中：$\kappa_r$ 为电路中的模态耦合项，定义为

$$\kappa_r = -\bar{e}_{31}h_{\tilde{p}_c}b\int_0^L \frac{\mathrm{d}^2\phi_r(x)}{\partial x^2}\mathrm{d}x = -\bar{e}_{31}h_{\tilde{p}_c}b\frac{\mathrm{d}\phi_r(x)}{\partial x}\bigg|_{x=L} \tag{8.19}$$

根据文献 [5-6]，对稳态谐波基激励响应的耦合电压可以写为

$$v_S(t) = V_S \mathrm{e}^{\mathrm{j}\omega t} = \frac{\sum_{r=1}^{\infty}\frac{\mathrm{j}\omega\kappa_r F_r}{\omega_r^2 - \omega^2 + 2\zeta_r\omega_r\omega}}{\frac{1}{R_{load}} + \mathrm{j}\omega\frac{C_p}{2} + \sum_{r=1}^{\infty}\frac{\mathrm{j}\omega\kappa_r\chi_r}{\omega_r^2 - \omega^2 + 2\zeta_r\omega_r\omega}}\mathrm{e}^{\mathrm{j}\omega t} \tag{8.20}$$

式中：$\omega$ 为激励频率；$C_p = \varepsilon_{33}^s \cdot b \cdot L/h_p$ 为压电层的电容；$\chi_r$ 为反向模态耦合

项；$\omega_r$、$\zeta_r$ 和 $F_r$ 分别为无阻尼固有频率、机械阻尼比和第 $r$ 模式下的模态机械力。相应地，可以定义 $|v(t)|$ 为峰电压振幅、$|v(t)|^2/R_{load}$ 为峰功率振幅、$|v(t)|^2/(2R_{load})$ 为平均功率振幅。对于 $\omega \approx \omega_r$ 的情况，式（8.20）可以简化得出单模电压响应及其相关峰功率振幅"精简"表达式[5]。

图 8.5~图 8.9 展示了典型的基于振动弯曲的商用能量收集器。

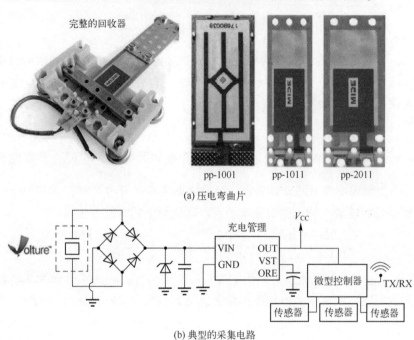

图 8.5 Mide 公司典型的 Volture™ 振动收集器

（资料来源：http://www.mide.com/collections/vibration-energy-harvesting-with-protected-peizos）

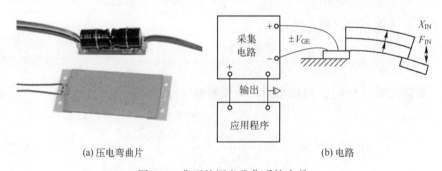

图 8.6 典型的压电收集系统产品

（资料来源：https://www.piezo.com/prodproto4Ekit.html）

# 第 8 章　基于智能材料的能量收集

图 8.7　PI 公司的压电弯曲能量收集器
（资料来源：https://www.piceramic.com/applications/piezo-energy-harvesting.html）

(a) 压电俘能器　　　　　(b) 测试装置

图 8.8　用于交通振动的压电收集器及实验室测试装置[17]

图 8.9　基于日本 Ceratec 工程公司（http://www.ceratec-e.com）压电收集器的冲击电池

在介绍了振动能量收集器之后，接下来介绍其他类型的从直接应力中收集能量的表达式[25-26]。直接施加应力的压电发电机在 $d_{33}$ 工作模式下通常不会发生共振，因此发电机的作用类似于电容器，即

$$C_\mathrm{p} = \frac{\varepsilon_{33}^T A}{h} \tag{8.21}$$

式中：$\varepsilon_{33}^T$ 为恒定应力下的介电常数；$A$ 和 $h$ 分别为电容器的面积和厚度。因

此，由应力 $\sigma_{33}$ 获得的能量可以写成

$$\begin{cases} U = \dfrac{1}{2} C_p \cdot V_{out}^2 = \dfrac{1}{2} C_p \cdot g_{33}^2 \cdot \sigma_{33}^2 \cdot h^2 \\ V_{out} = -g_{33} \cdot \sigma_{33} \cdot h = -g_{33} \cdot \dfrac{F_{33}}{A} \cdot h \end{cases} \quad (8.22)$$

式中：$A$ 为压电发电机的面积；$F_{33}$ 为垂直于该面积上施加的力；$g_{33}$ 为压电电压常数。

根据电容的定义（式 (8.21)），将 $g_{33}$ 和 $d_{33}$ 之间的关系、压电常数（$\varepsilon_{33}^T \cdot g_{33} = d_{33}$），以及压电材料的体积（Vol=$A \cdot h$）代入式 (8.22) 得

$$U = \dfrac{1}{2} C_p \cdot V_{out}^2 = \dfrac{1}{2} (g_{33} \cdot d_{33}) \cdot \sigma_{33}^2 \cdot \text{Vol} \quad (8.23)$$

而功率可通过以下表达式进行计算：

$$\text{Power} = U \cdot f = \dfrac{1}{2} (g_{33} \cdot d_{33}) \cdot \sigma_{33}^2 \cdot f \cdot \text{Vol} = \dfrac{1}{2} \dfrac{d_{33}^2}{\varepsilon_{33}^T} \cdot \sigma_{33}^2 \cdot f \cdot \text{Vol} \quad (8.24)$$

为了使功率最大化，需要施加尽可能大的应力，同时使用大量压电材料，并且选择一种具有高$(g_{33} \cdot d_{33})$值的压电材料。因此，要达到最大能量，就必须使用与最大能量相同的值加上更高的频率。

文献中发表了大量涉及能量收集的相关研究。下面将介绍这些研究中的典型例子。在文献 [1] 中，对收集问题和材料选择进行了扩展综述。该综述基于功率密度和带宽，回顾了从共振到非共振设备包括旋转解决方案在内的工作模式和设备配置。Roundy 展示了一项供后来许多研究人员参考的基础研究，在该研究中他评估了基于振动能量收集设备的有效性[2]。根据他的计算，在 50~350Hz 时，加速度 1~10m/s² 振动产生的可用理论功率密度范围为 0.5~100mW/cm³。他推导出的有效性规则考虑了耦合系数、器件的品质因数、质量以及外部电负载最大化电功率传输方式。该方法可应用于电磁、压电、磁致伸缩、静电等多种收集方式。Lopes 和 Gallo 对 2014 年之前利用压电器件从环境中获取能量的测试和应用进行了综述[3]。Beeby 等对微系统应用中的能量收集振动源进行了综述，该综述除了用于冲击耦合、共振和人体运动的压电基收集器，还包括基于电磁和静电的发电机[5]。

Sebald 等利用非线性模型对具有压电式机电耦合的 Duffing 振荡器进行了实验研究，结果表明其功率带宽增大了 5.45 倍，而输出功率降低为原来的 5/12[9]。Vatansever 等展示了同时利用 PVDF 和压电纤维复合结构从不同的风速与水滴中收集能量的实验结果，表明基于 PVDF 的设备能更有效地从这两种能源中收集能量[10]。Pozzi 和 Zhu 设计并测试了拉伸压电双晶片在人行走时收集膝关节能量的性能，结果显示，模型预测和实验结果之间具有良好的相关性[11]。

## 第 8 章 基于智能材料的能量收集

Ling 等综述了从人类活动中收集能量方面的研究,如鞋装式、脉冲激励式和冲击驱动式收集器,具有旋转膝关节的关节运动收集器、旋转收集器和柔性可穿戴收集器[12]。以桥梁为例,展示了民用基础设施和交通运输的能源及其获取方式。而其他的能量来源,如使用柔性压电薄膜作为收集器的风的流动,树木在风中的摇摆运动,由于水流而产生的流致振动的获取以及来自雨滴的能量也包括在该综述中。Wekin 发表了一篇关于如何通过表征和比较各种用作采集器的压电材料,进而从热声发动机中获取能量的硕士论文[13]。Bowen 等进行了另一项广泛的综述,总结了基于压电器件作为潜在的能量收集器从振动中收集机械能的应用,并提出了复杂压电设备的各种优化方法[14]。他们介绍了利用铁电和多铁材料将光转化为化学能或电能,并讨论了如何将其应用于收集器的方法。热波动收集也使用热电基采集器解决。Wang 等研究了基于线性振动收集器中应用的耦合因子,该收集器产生的频域耦合较弱[16]。Dhingra 等在介绍了各种目前使用不同类型的压电收集器将环境中可用能量转化为电能的项目[18]。Sodano 和 Inman 通过比较 3 种类型的折弯机[20]:压电系统公司的黏接在 0.0635mm 厚铝板上的 PSI-5H4E(PZT);$8.255 \times 5.715 cm^2$ 黏接在相同铝板上的压电纤维复合材料(Macro Fiber Composite,MFC)① 贴片;以及像悬臂梁一样工作的快速包装致动器②(约为 $10.16 \times 2.54 cm^2$),充实了文献中可用的实验数据库。3 种器件的效率分别在谐振、啁啾 0~500Hz 和随机 0~500Hz 频率下进行了测试,展示了根据激励类型对电池充电的不同效率。文献 [21] 对商用压电陶瓷进行了综述,推导并给出了不同的描述典型负载情况和商用压电陶瓷的最优值,最后,该文章推导出了改进发电机解决方案的设计规则。文献 [23] 中介绍了一个有趣的应用,其中压电收集器提供了直升机桨叶内部传感器所需能量。其测试了两个设想:第 1 个是在细长的低耦合梁状结构中放置压电贴片;第 2 个是在直升机旋翼的滞后阻尼器中引入压电叠层,这需要对滞后阻尼器进行调整。根据所进行的研究,功率是在单一位置大量产生的,并可以使用基本解析方程式进行分析。

Kim 等进行了利用圆形"钹"型压电换能器(图 8.10)获取机械振动并将其转化为电能的研究[24]。当金属盖受到外部轴向载荷并引发振动时,压电圆板会产生径向和轴向振动。与多层堆叠换能器相比,压电圆板承受轴向外载荷时产生径向和轴向振动,能够产生高的转换系数 $d_{eff}$(有效压电电场常数)和 $g_{eff}$(有效压电电压常数)。这是由钹的特殊结构所造成的,其包含了允许金属端盖用作入射轴向应力机械放大成相反符号的径向应力的空腔[24]。有效压

---

① 来自智能材料公司,www.smart-material.com/MFC-product-main.html。
② 来自前 ACX 公司,现在归于 Midé 公司。

电电场常数可通过下式计算：

$$d_{\text{eff}} = d_{33} + A d_{31} \qquad (8.25)$$

式中：$A$ 为放大因子。依赖盖子的设计，放大因子可为 10~100。记住 $d_{31}$ 有一个负值，而 $A$ 为一个较高的值，很明显钹的转换率非常高。在直径为 29mm、厚度为 1mm 的"钹"上进行的实验结果显示，在 100Hz 的频率下施加 7.8N 的力，在 400kΩ 的电阻器上测得的产生功率为 39mW。报告指出使用专门设计的 DC-DC 转换器，可以在占空比 2%、开关频率 1kHz 下将 30mW 的功率转换为 5kΩ 的阻抗。文献［25-26］中描述了另一种潜在的应用，其中压电圆盘承受频率为 5Hz 的轴向压缩应力。结果表明，尽管当增加轴向压缩应力时，输出功率会增加，但当经受数千次循环时，圆盘无法承受太高应力，因此必须降低应力，从而产生较低的功率。这一问题可以通过添加更多的圆盘来解决，从而增加压电材料的体积，收集机械能并将其转化为电能。表 8.2 总结了文献中各种压电基收集器及其功率输出的数据。对于基于振动的能量收集器，其功率范围通常为毫瓦，而只有文献［25-26］中给出的结果展示了功率为瓦的收集器（由于转换中涉及体积相对较大的压电材料）。

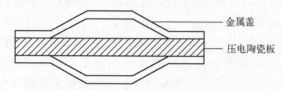

图 8.10 "钹"型压电换能器截面示意图

表 8.2 典型的收集器及其功率输出

| 能量源 | 设备类型 | 尺寸 | 平均收集功率 | 参考文献 |
|---|---|---|---|---|
| 人行走 | 鞋装收集器：安装在鞋垫下面的 PVDF 壁 | 两个 8 层 2μm 厚的 PVDF 夹在 2mm 厚的环氧树脂柔性基板上 | 在 250kΩ 负载和 0.9Hz 频率下为 1.3mW | ［12］ |
| 人行走时脚跟撞击 | 鞋装收集器：PZT 插在鞋脚跟下 | 两个 5cm×5cm×0.0381cm 的 PZT 传感器黏接在一块预应力 5cm×8.5cm 的弯曲弹簧钢片上 | 在 500kΩ 负载和 0.9Hz 频率下为 8.4mW | ［12］ |
| 人体运动 | 脉冲激励收集器：一个标准的圆柱形质量块，通过磁吸引来驱动压电双形态梁阵列 | 8 根 72mm×5mm×0.5mm 的横梁，上下铺 0.2mm 厚的 PZT 507① | 在 2.7m/s² 和 2Hz 频率下为 2.1mW | ［12］ |

## 第8章 基于智能材料的能量收集

续表

| 能量源 | 设备类型 | 尺寸 | 平均收集功率 | 参考文献 |
|---|---|---|---|---|
| 人体运动 | 冲击驱动收集器：一个移动的质量冲击压电弯曲器 | 体积 $25cm^3$，重量60g | 振幅为10cm，0.5Hz频率下0.047mW，10Hz频率下0.6mW | [12] |
| 人体运动 | 旋转膝关节收集器 | 由130μm厚金属垫片夹在2层125μm厚聚合物制成的双晶片，串联状态 | 在 $7.1m/s^2$，10kΩ 负载和320Hz频率下为57.6mW，在 $7.1m/s^2$，2kΩ负载和300Hz频率下为3.646μW | [11-12] |
| 人体缓慢运动 | 弹性可穿戴PVDF壳型结构收集器 | 壳结构由127μm×30mm×5mm 聚酯膜+PVDF 膜制成，110μm×20mm×2mm | 在90Ω 负载和3.3Hz频率下为0.87mW | [12] |
| 风流 | PVDF压电薄膜 | 压电薄膜面积 $7.44cm^2$，厚度64μm | 150nF 电容，风速12.3m/s，收集51.66μJ | [12] |
| 水流 | 使用PVDF或PZT-5H薄膜进行流动诱导振动 | 压电薄膜包括24μmPVDF+2个28μm 电极层+125μm 聚酯层 | 当压力差为 1.790～2.392kPa 时，输出峰值电压为1.77~2.30V | [12] |
| 雨滴 | PVDF薄板 | 25μm 厚 $d_{31}=2J/N$ 单拉伸PVDF薄片 | 接近1μW | [12] |
| 汽车振动 | 压电薄膜元件层压在聚酯薄膜（聚酯膜）+末端质量 | 12.19mm×30mm×1.57mm 薄片 | 在 $8.25m/s^2$ 时接近770mV | [19] |
| 振动 | 压电陶瓷（PSI-5H4E②） | 压电陶瓷片粘在一个40mm×80mm×1.016mm 铝悬臂板上 | 最大瞬时功率为1.5~2mW，平均功率为0.14~0.2mW | [15] |
| 振动 | 双晶片作为悬臂梁 | 整体尺寸（包括末端质量）= $1cm^3$ | 在 $2.25m/s^2$，250kΩ 负载和85Hz 频率下为0.2mW；在 $2.25m/s^2$，250kΩ 负载和60Hz 频率下为0.38mW | [2] |
| 立交桥振动 | 悬臂式压电梁，端部质量（12g），共振频率接近14.5Hz | 两个双压电晶片贴片QP20W（购于Mide公司）夹着一个 40mm×220mm×0.8mm 钢板 | 在15Hz 激励下0.03mW | [17]（图8.8） |
| 时间相关的力 | 一个装有PVDF和摩擦电层的拱门③ | 拱门尺寸 7cm×3cm | PVDF 单独产生 80V 7.62μA/$cm^2$，摩擦电层单独产生 380V 4.3μA/$cm^2$ | [22] |
| 随时间变化力产生的振动 | 压电"钹"类换能器 | 尺寸：直径29mm，厚度1mm | 在 100Hz 400kΩ 接近 7.8N 39mW | [24]（图8.10） |

353

续表

| 能量源 | 设备类型 | 尺　　寸 | 平均收集功率 | 参考文献 |
|---|---|---|---|---|
| 时间相关的力 | 压缩的分层圆盘 | 12个直径10mm,厚度4mm的圆盘;8个直径15mm,厚度4mm的圆盘 | 75MPa、5Hz、2.8~3MΩ下0.14~0.15W;在100MPa、5Hz、2.8MΩ下0.35W | [25-26] |

① 来自 Morgan Electro Ceramics。www.morgantechnicalceramics.com/products。
② 来自压电系统有限责任公司。
③ 摩擦发电机是由两种具有不同摩擦电特性的材料制成的聚合物薄片叠加而成。一旦受到机械位移,两层之间的摩擦由于纳米尺度的表面粗糙度产生相等的电荷,但在两边有相反的符号。

## 8.2　电磁能量收集

原则上,电磁能量收集是在以给定频率振动的结构上进行的。发电机被放置在这些振动结构上,而磁铁则在线圈内振动。磁铁的相对运动被机电换能器转换为电能。最著名的单自由度模型是 Williams 等开发的,将在下面进行介绍[27]。如图 8.11 所示,一个质量块 $m$、一个弹簧 $k$ 和一个阻尼器 $C_d$ 串联在一起,注意,上述提到的换能器由阻尼器 $C_d$ 表示。该系统位于一个具有位移为 $y(t)$ 的箱体内,而质量块的位移(相对于外部参考系统)为 $x(t)$。质量块的相对位移为 $z(t)$,其表达式如下:

$$z(t)=x(t)-y(t)\Rightarrow x(t)=z(t)+y(t) \tag{8.26}$$

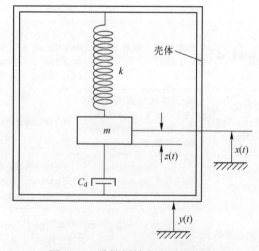

图 8.11　线性惯性发电机的原理

如文献[27]所述，质量块、弹簧和阻尼器的单一运动方程式可以写成

$$m \cdot \ddot{x}(t) + C_d[\dot{x}(t) - \dot{y}(t)] + k[x(t) - y(t)] = 0 \quad (8.27)$$

或者使用式（8.26），可以得

$$m \cdot \ddot{z}(t) + C_d \cdot \dot{z}(t) + k \cdot z(t) = -m \cdot \ddot{y}(t) \quad (8.28)$$

应用拉普拉斯变换①，得

$$-m \cdot s^2 \cdot Y(s) = s \cdot Z(s)\left[m \cdot s + C_d + \frac{k}{s}\right] \quad (8.29)$$

文献[27]提供了上述机械系统的等效电路，即

$$-I(s) = V(s)\left[C_d \cdot s + \frac{1}{R} + \frac{1}{s \cdot L}\right] \quad (8.30)$$

式中：$I(s)$和$V(s)$为输入电流和感应电压；$C_d$为电容；$R$为电阻；$L$为电感。通过比较式（8.29）和式（8.30），在引入常数$k_e$（转换器电磁常数）时，得到了电气和机械之间的关系为

$$\begin{cases} I(s) = \dfrac{m}{k_e^2} s^2 \cdot Y(s), \quad V(s) = k_e \cdot s \cdot Z(s) \\ C = \dfrac{m}{k_e^2}, \quad R = \dfrac{k_e^2}{C_d}, \quad L = \dfrac{k_e^2}{k} \end{cases} \quad (8.31)$$

基于电气参数得到的等效电路如图8.12所示。注意，阻尼器$C_d$由电阻器$R$表示。因此，阻尼器中消耗的功率具有以下表达式：

$$P_d(s) = \frac{V^2(s)}{R} = \frac{\dfrac{k_e^2 \cdot m^2 \cdot s^6 \cdot Y^2(s)}{[m \cdot s^2 + C_d \cdot s + k]^2}}{\dfrac{k_e^2}{C_d}} = = \frac{C_d \cdot m^2 \cdot s^6 \cdot Y^2(s)}{[m \cdot s^2 + C_d \cdot s + k]^2} \quad (8.32)$$

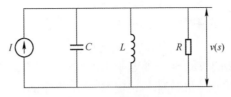

图8.12　等效电路

---

① 拉普拉斯变换是一个积分变换，它的函数是一个正实变量$t$（时间）变成复变量$s$（频率）的函数。

假设谐波振动具有以下表达式，$Y(t)=Y_0 \cdot \cos(\omega t)$、$\omega=2\pi \cdot f$，式中 $f$ 是频率，代入式（8.32），得到频域中的耗散功率，即

$$p_d(\omega)=\frac{C_d \cdot m^2 \cdot \omega^6 \cdot Y_0^2}{(k-m \cdot \omega^2)^2+(\omega \cdot C_d)^2} \tag{8.33}$$

参考质量块、弹簧和阻尼器系统的固有频率，共振频率 $\omega_n$ 和相关阻尼比 $\zeta$ 的定义如下：

$$\omega_n^2 \equiv \frac{k}{m}=\frac{1}{L \cdot C}, \quad \frac{C_d}{m} \equiv 2\zeta \cdot \omega_n \Rightarrow \zeta = \frac{C_d}{2\sqrt{m \cdot k}}=\frac{1}{2R}\sqrt{\frac{L}{C}} \tag{8.34}$$

利用固有频率和阻尼比，式（8.33）可被修改为

$$p_d(\omega)=\frac{C_d \cdot \left(\frac{\omega}{\omega_n}\right)^3 \cdot \omega^3 \cdot Y_0^2}{\omega_n\left\{\left[1-\left(\frac{\omega}{\omega_n}\right)^2\right]^2+\left[2\zeta \cdot \left(\frac{\omega}{\omega_n}\right)\right]^2\right\}}=\frac{2\zeta \cdot m \cdot \left(\frac{\omega}{\omega_n}\right)^3 \cdot \omega^3 \cdot Y_0^2}{\left[1-\left(\frac{\omega}{\omega_n}\right)^2\right]^2+\left[2\zeta \cdot \left(\frac{\omega}{\omega_n}\right)\right]^2} \tag{8.35}$$

应当注意的是，无论机电换能器的类型如何，式（8.35）适用于任何线性惯性发电机。如果将发电机的固有频率设计为激励频率，即 $\omega=\omega_n$，并代入式（8.35），可得到谐振频率下的耗散功率为

$$p_d(\omega_n)=\frac{2\zeta \cdot m \cdot \omega_n^3 \cdot Y_0^2}{4\zeta^2}=\frac{m \cdot \omega_n^3 \cdot Y_0^2}{2\zeta} \tag{8.36}$$

为了使耗散功率最大化，必须增加质量，提高固有频率和位移，同时尽量减小阻尼比。

为了实现电磁收集器，在振动质量块 $m$ 上附着一个永磁体，同时在圆形壳体周围缠绕线圈。由于线圈内部质量块 $z$ 的相对位移，线圈中将感应出电压 $V_c(t)$，其表达式为

$$V_c(t)=k_e \cdot \dot{z}(t) \tag{8.37}$$

因此，电流 $I_c(t)$ 作用在振动磁铁上的力可表示为

$$F_c(t)=k_e \cdot I_c(t) \tag{8.38}$$

假设连接到电路的线圈有一个电阻 $R_c$、一个自感 $L_c$ 和一个电气负载 $R_L$，得到一个由力输入部分、机械部分、电气部分和外部电气负载4部分构成的"新"等效电路，如图8.13所示。

应注意，机械部件中的代表电阻 $R_m$ 定义为

$$R_m=\frac{k_e^2}{C_m} \tag{8.39}$$

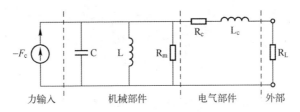

图 8.13　等效电路以及线圈的电气部分和外部电气负载示意图

式中：$C_m$ 既代表机械阻尼，又代表电磁铁和涡流损耗。值得注意的是，在共振时（$\omega = \omega_n$），并联阻抗 $C$ 和 $L$ 变成无穷大[27]。因此，共振时系统的 Thévenin5① 等效电路简化为图 8.14 所示的等效电路。根据文献 [27]，电压 $V$（图 8.14 中）由下式给出：

$$V = -\frac{k_e \cdot m \cdot \ddot{y}(t)}{C_m} = \frac{k_e \cdot m \cdot \omega_n^2 \cdot Y_0}{C_m} \qquad (8.40)$$

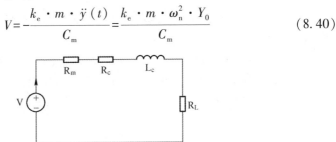

图 8.14　Thévenin 系统在共振时的等效电路

为了评估外部电阻上的功耗，$R_L$ 上的电流计算为

$$I \cdot Z_N = V \Rightarrow I = \frac{k_e \cdot m \cdot \omega_n^2 \cdot Y_0}{C_m \cdot Z_N} = \frac{k_e \cdot m \cdot \omega_n^2 \cdot Y_0}{C_m \cdot \sqrt{(R_c + R_L + R_m)^2 + (\omega_n L_c)^2}} \qquad (8.41)$$

那么，共振时的功率可以写成

$$p_L(\omega_n) = I^2 \cdot R_L = \frac{k_e^2 \cdot m^2 \cdot \omega_n^4 \cdot Y_0^2}{C_m^2 \cdot [(R_c + R_L + R_m)^2 + (\omega_n L_c)^2]} R_L \qquad (8.42)$$

为了使功率 $p_L(\omega_n)$ 最大化，必须使用外部负载，使得 $R_L \gg R_c$ 且 $R_L \gg \omega_n \cdot L_c$。同时，通过将电阻抗与机械阻尼阻抗相匹配，即 $k_e = \sqrt{R_L \cdot C_m}$，将式（8.42）改写为

---

① Thévenin 的定理表明，任何线性电路，无论多么复杂，都有可能简化为具有单个电压源和串联电阻连接到电负载的等效电路。

$$p_L(\omega_n)_{max} = \frac{m^2 \cdot \omega_n^4 \cdot Y_0^2}{4C_m} \tag{8.43}$$

由 $2\zeta_m \cdot \omega_n = \dfrac{C_m}{m}$,式(8.43)转化为

$$p_L(\omega_n)_{max} = \frac{m \cdot \omega_n^3 \cdot Y_0^2}{8\zeta_m} \tag{8.44}$$

式中:$\zeta_m$ 为机械阻尼比。根据式(8.44),可以得出使电磁线性惯性发电机功率最大化的设计原则[1]:

(1) 根据所设计的系统,惯性质量 $m$ 应具有很高的值。

(2) 一旦选定了质量的值,弹簧系数 $k$ 应调整为 $\omega_n = \omega_{base}$($\omega_{base}$ 是激励频率)。

(3) 发电机是在共振时工作的,因此必须确保箱体的内部尺寸容许振动质量的最大位移。

(4) 将线圈阻抗最小化,使其比电气负载($R_L$)小1个数量级。

(5) 为了降低阻尼比,机械阻尼和杂散①损耗应尽可能小。

(6) 必须选择电磁铁因子 $k_e$,使其满足式(8.39)中给出的关系,即

$$R_m = \frac{k_e^2}{C_m}$$

需要注意的是,实际中并不是所有上述设计规则都可以实现,而应该折中以获得最佳设计。例如,如果 $k_e$ 选择一个较大的值,而 $R_c$ 保持为一个较小的值,这就需要增加线圈的匝数;如果线圈的空间有限,这反过来又会增加 $R_c$ 的值。

基于上述模型,Williams 等设计、制造并测试了一个演示器(一个总直径为 5mm、总高度为 700μm 的圆柱体),在 4MHz 的频率下产生了 0.3μW 的功率[27]。

其他研究人员也给出了类似的结果。Roundy 利用电磁、压电、磁致伸缩和静电转换器技术估算了获取功率[2]。据称,根据材料类型和操作参数的不同,每种技术都可能有较高的耦合系数(0.6~0.8)。因此,需要根据操作环境和设计问题的条件来决定最合适的技术。基于作者测量的一系列常见振动,在 50~350Hz 范围内、加速度为 1~10m/s² 的振动,最大理论功率密度范围为 0.5~100mW/cm³。Beeby 等回顾了无线自供电微系统的振动能量收集技术现

---

① 杂散损失是所有不能准确确定的损失。对于电磁铁发电机,包括由于不确定的磁场、谐波磁链脉动和绕组中的涡流造成的损失。

## 第8章 基于智能材料的能量收集

状[28]。考虑了由惯性质量块、弹簧和阻尼器组成的振动发电机，同时考虑了压电、电磁和静电3种类型能量转换机制。对现有的压电发电机进行了全面综述，包括冲击耦合、谐振和基于人体的器件。还对电磁发电机进行了综述，包括大型分立器件和晶圆级集成设备。另外，还展示了静电发电机①，其可分为面内重叠变化、面内间隙闭合和面外间隙闭合。Arnold 对紧凑型磁力发电系统（小于几个立方厘米）进行了综述，同时介绍了磁力发电机小型化的方法，评估了设备的设计和性能，同时集成了高性能硬磁材料、微型芯叠片、低摩擦轴承、高速转子动力学和紧凑、高效的电源转换器[29]。Beeby 等的另一篇文章展示了一种紧凑型电磁发电机（部件体积为 $0.1cm^3$，实际体积为 $0.15cm^3$），在52Hz 低谐振频率、$0.59m/s^2$ 加速度下，在 $4k\Omega$ 的电阻负载上收集了 $46\mu W$ 的能量[30]。在 2300 匝线圈的微型发电机中显示出 428mV（rms）的电压，这一电压足以用于后续的整流和升压电路。另一组研究人员 Yang 等展示了一种电磁能量收集器的实验结果，该收集器能够在系统的 3 个第 1 固有频率（$f_1$ = 369Hz，$f_2$ = 938Hz 及 $f_3$ = 1184Hz）下进行能量收集[7]。已报道的 $f_1$ 最大输出电压和功率分别为 1.38mV 和 $0.6\mu W$；而对于第 2 模式，最大输出电压和功率分别为 3.2mV 和 $3.2\mu W$，激励振幅为 $14\mu m$，磁铁和线圈之间的间隙为 0.4mm。Liu 等的团队采用了类似的方法，利用一种 3D 激励新型电磁能量采集器来收集能量[32]。结果显示 1285Hz 的第 1 振动模式是平面外运动，而 1470Hz 和 1550Hz 的第 2 和第 3 振动模式分别与水平（$x-$）轴成 60°（240°）和 150°（330°）角的平面内角度。施加 $1gm/s^2$ 的激励加速度，在第 1、第 2 和第 3 频率的振动模式下，获得的最大功率密度分别为 $0.444\mu W/cm^3$、$0.242\mu W/cm^3$ 和 $0.125\mu W/cm^3$。Illy 等展示了另一项先进研究，通过利用摆动和冲击激励从人体运动中采集能量[33]。能量的采集是由两种类型的感应式能量采集器来完成的：一种是利用脚的摆动运动的多线圈拓扑收集器；另一种是在脚跟撞击时被激发共振的冲击式收集器。这两种设备的建模和设计都考虑了设备高度的关键约束，以便于集成到鞋底中。对这两台收集器在不同的运动速度下进行了表征测试，两名受试者在一台跑步机上使用摆动式收集器（设备总体积+外壳共 $21cm^3$），产生高达 0.84mW 的平均输出功率，而冲击式收集器（设备总体积为 $48cm^3$）的输出功率则高达 4.13mW。据报道，第 1 台设备的功率密度为 $40\mu W/cm^3$，而第 2 台设备为 $86\mu W/cm^3$，比文献 [32] 中展示的结果更好。

---

① 静电发电机利用电隔离的带电电容器板之间的相对运动来产生能量，对板间静电力所做的功提供了能量收集。

图 8.15 展示了英国 Perpetuum 有限公司[①]制造的典型机电收集器的实物图。

图 8.15　PMG17 收集器，其 PCB 和使用的超级电容器（引自英国 Perpetuum 有限公司），用 1/4 美元来比照物品的尺寸

## 8.3　振动激励下的双晶片电源

如图 8.16 所示，压电双晶片由一对完全相同的双压电层构成，其沿厚度方向极化，在中间由金属承载梁隔开。压电双晶片是一个悬臂梁，夹紧边以已知振幅 $A$ 和给定频率 $\omega$ 在垂直方向($z$)谐波激励。在悬臂梁的自由侧，连接一个质量块 $M_0$，压电电极与外部阻抗 $Z_L$ 并联。

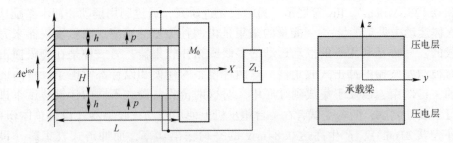

图 8.16　压电双晶片模型示意图

假设双晶片是一根细长的梁，该工况下的压电方程（式中，$x$、$y$、$z$ 系统通常为 1、2 和 3）可表示为

---

[①] www.perpetuum.com。

$$\begin{cases} S_x = s_{11}T_x + d_{31}E_z \\ D_z = d_{31}T_x + \varepsilon_{33}E_z \end{cases} \tag{8.45}$$

式中：$S_x$ 和 $T_x$ 分别为 $x$ 方向（沿图 8.16 所示梁的方向）上的应变和应力；$E_z$ 是振动引起的电场；$s_{11}$、$d_{31}$ 和 $\varepsilon_{33}$ 分别为在恒定电场下测得的轴向弹性模量、恒定应力下的横向压电常数和横向介电常数。对于目前的情况，有以下的应变和电场表达式：

$$\begin{cases} S_x = -zW,_{xx} \\ E_z = -V/h \end{cases} \tag{8.46}$$

式中：$W$ 为横向位移；$V$ 为由于诱导振动而在压电电极上累积的电压；$h$ 为压电层的厚度。

由式（8.45）中的第 1 个方程式求解 $T_x$，然后代入第 2 个方程式得出 $D_z$ 的表达式：

$$\begin{cases} T_x = S_{11}^{-1}S_x - S_{11}^{-1}d_{31}E_z \\ D_z = S_{11}^{-1}d_{31}S_x + \overline{\varepsilon}_{33}E_z \end{cases} \tag{8.47}$$

式中：

$$\begin{cases} \overline{\varepsilon}_{33} = \varepsilon_{33}(1-k_{31}^2) \\ k_{31}^2 = \dfrac{d_{31}^2}{\varepsilon_{33} \cdot s_{11}} \end{cases} \tag{8.48}$$

应注意，中间层（金属层）内的应力由下式得

$$T_x = S_{11}^{-1}S_x = E_{\text{substrate}}S_x = -E_{\text{substrate}}ZW,_{xx} \tag{8.49}$$

式中：$E_{\text{substrate}}$ 为金属层的杨氏模量。

弯曲力矩由下式得

$$\begin{aligned} M &= \int zT_x \mathrm{d}y\mathrm{d}z = -Dw,_{xx} + s_{11}^{-1} \cdot d_{31} \cdot \frac{V}{h} \cdot (H+h) \cdot h \cdot b = \\ &= -Dw,_{xx} + E_{\text{piezo}} \cdot d_{31} \cdot V \cdot (H+h) \cdot b \end{aligned} \tag{8.50}$$

式中：

$$D = \left\{ \frac{2}{3}E_{\text{substrate}} \cdot \left(\frac{H}{2}\right)^3 + \frac{2}{3}E_{\text{piezo}}\left[\left(\frac{H}{2}+h\right)^3 - \left(\frac{H}{2}\right)^3\right] \right\} \cdot b \tag{8.51}$$

梁中的剪切力由弯矩的微分给出，即

$$\check{V} = -M,_x = +D \cdot w,_{xxx} \tag{8.52}$$

然后将细长双晶片梁的弯曲振动表示为

$$M,_{xx} = m\ddot{w}(x,t) \quad \Rightarrow \quad -D \cdot w(x,t),_{xxxx} = m\ddot{w}(x,t) \tag{8.53}$$

式中：单位长度的质量为

$$m = H \cdot b \cdot \rho_{Al} + 2h \cdot b \cdot \rho_{piezo} \tag{8.54}$$

计算在 $z=(h+H/2)$ 处顶部电极上累积的电荷，将相关表面上的电位移积分，得

$$\begin{aligned}Q_{top} &= -\int_0^L \int_0^b D_{z@\left(z=\frac{H}{2}+h\right)} \mathrm{d}x \cdot \mathrm{d}y \\ &= -\left\{ b \cdot E_{piezo} \cdot d_{31}\left(\frac{H}{2}+h\right)\left[w_{,x}(L,t) - w_{,x}(L,t)\right] + b \cdot \bar{\varepsilon}_{33}\frac{V}{h}L \right\}\end{aligned} \tag{8.55}$$

电荷和电流之间的关系已知为

$$I = \frac{\mathrm{d}Q}{\mathrm{d}t} \tag{8.56}$$

这就引出了电压与两个压电层的函数关系：

$$2I = \frac{\hat{V}}{z_L} \Rightarrow \hat{V} = 2I \cdot z_L \tag{8.57}$$

具有压电双晶片的悬臂梁的边界条件可表示为

$$@\, x = 0, \quad w(0,t) = A\mathrm{e}^{\mathrm{i}\omega t}, \quad w_{,x}(0,t) = 0 \tag{8.58}$$

$$@\, x = L, \quad M(L,t) = 0, \quad \check{V}(L,t) = M_0 \frac{\partial^2 w(L,t)}{\partial t^2} \tag{8.59}$$

根据式（8.53）以及式（8.58）和式（8.59）给出的边界条件，可以提出以下形式的解决方案：

$$\begin{Bmatrix} w(x,t) \\ \hat{V}(t) \\ Q(t) \\ I(t) \end{Bmatrix} = \begin{Bmatrix} W(x) \\ \widetilde{V} \\ \widetilde{Q} \\ \widetilde{I} \end{Bmatrix} \mathrm{e}^{\mathrm{i}\omega t} \tag{8.60}$$

将提出的解代入式（8.53）和相应的边界条件，可得

$$D\frac{\mathrm{d}^4 W}{\mathrm{d}x^4} - m \cdot \omega^2 W = 0 \tag{8.61}$$

式（8.61）的解具有以下形式：

$$W = A_1 \cosh(\beta x) + A_2 \sinh(\beta x) + A_3 \cos(\beta x) + A_4 \sin(\beta x) \tag{8.62}$$

式中：系数 $A_1 \sim A_4$ 将由边界条件确定，并且

$$\beta^4 \equiv \frac{m\omega^2}{D} \tag{8.63}$$

由式（8.58）和式（8.59）给出的相关边界条件可以转化为以下表达式：

$$@\, x = 0, \quad W(0) = A, \quad W_{,x}(0) = 0 \tag{8.64}$$

$$\begin{cases} @\ x=L, & -D\cdot W,_{xx}(L)+E_{piezo}d_{31}\dfrac{\widetilde{V}}{h}\cdot(H+h)\cdot h\cdot b=0 \\ & D\cdot W,_{xxx}(L)=-M_0\cdot\omega^2\cdot W(L) \end{cases} \quad (8.65)$$

电流的表达式为

$$\widetilde{I}=i\cdot\omega\cdot b\cdot$$
$$\left\{E_{piezo}\cdot d_{31}\left(\dfrac{H}{2}+h\right)\cdot\beta\cdot[-A_1\sinh(\beta L)-A_2\cosh(\beta L)+A_3\sin(\beta L)-A_4\cos(\beta L)]-\bar{\varepsilon}_{33}\dfrac{\widetilde{V}}{h}L\right\} \quad (8.66)$$

应用边界条件（式（8.64）和式（8.65）），得到 4 个常数 $A_1 \sim A_4$ 的 4 个方程以及所产生电压的表达式为

$$A_1+A_3=A \quad (8.67)$$
$$A_2\beta+A_4\beta=0 \quad (8.68)$$

$$-D\cdot[A_1\cdot\beta^2\cdot\cosh(\beta L)+A_2\cdot\beta^2\cdot\sinh(\beta L)-A_3\cdot\beta^2\cdot\cos(\beta L)-A_4\cdot\beta^2\cdot\sin(\beta L)]+$$
$$E_{piezo}d_{31}\cdot\widetilde{V}\cdot(H+h)\cdot b=0 \quad (8.69)$$

$$-D\cdot[A_1\cdot\beta^2\cdot\cosh(\beta L)+A_2\cdot\beta^2\cdot\sinh(\beta L)-A_3\cdot\beta^2\cdot\cos(\beta L)-A_4\cdot\beta^2\cdot\sin(\beta L)]+$$
$$E_{piezo}d_{31}\cdot\widetilde{V}\cdot(H+h)\cdot b=0 \quad (8.70)$$

$$\dfrac{\widetilde{V}}{2Z_L\cdot i\cdot\omega\cdot b}=E_{piezo}\cdot d_{31}\left(\dfrac{H}{2}+h\right)\cdot\beta\cdot[-A_1\sinh(\beta L)-A_2\cosh(\beta L)+A_3\sin(\beta L)-$$
$$A_4\cos(\beta L)]-\bar{\varepsilon}_{33}\dfrac{\widetilde{V}}{h}L \quad (8.71)$$

或

$$\widetilde{V}\cdot\left(\dfrac{1}{2Z_L}+\dfrac{1}{Z_0}\right)=E_{piezo}\cdot d_{31}\left(\dfrac{H}{2}+h\right)\cdot\beta\cdot[-A_1\sinh(\beta L)-A_2\cosh(\beta L)+A_3\sin(\beta L)-$$
$$A_4\cos(\beta L)] \quad (8.72)$$

式中：

$$C_0=\dfrac{\bar{\varepsilon}_{33}\cdot b\cdot L}{h},\quad Z_0=\dfrac{1}{i\cdot\omega\cdot C_0} \quad (8.73)$$

注意，$Z_L$ 和 $Z_0$ 的阻抗取决于输入激励频率 $\omega$，其具体形式取决于输出电路的具体结构。下面给出 4 个常数和电压的显式表达式：

$$A_1 = \frac{A}{\Delta} [-2M_0 \cdot \omega^2 \cdot D \cdot \beta \cdot \cos(\beta \cdot L) \cdot \sinh(\beta \cdot L) - D \cdot \beta^4 \cdot \cos^2(\beta \cdot L)$$
$$M_0 \cdot \omega^2 \cdot \delta \cdot [1 - \cos(\beta \cdot L) \cdot \cosh(\beta \cdot L) - \sin(\beta \cdot L) \cdot \sinh(\beta \cdot L)]$$
$$-D^2 \cdot \beta^4 \cdot [\sin^2(\beta \cdot L) + \cos(\beta \cdot L) \cdot \cosh(\beta \cdot L) + \sin(\beta \cdot L) \cdot \sinh(\beta \cdot L)]$$
$$-D \cdot \beta^3 \cdot \delta \cdot \sin(\beta \cdot L) \cdot \cosh(\beta \cdot L) - \beta^3 \cdot \delta \cdot \sin(\beta \cdot L) \cdot \cos(\beta \cdot L)]$$
(8.74)

$$A_2 = \frac{A}{\Delta} [2M_0 \cdot \omega^2 \cdot D \cdot \beta \cdot \cos(\beta \cdot L) \cdot \sinh(\beta \cdot L) - \beta^4 \cdot \sin^2(\beta \cdot L) +$$
$$M_0 \cdot \omega^2 \cdot \delta \cdot [\sin(\beta \cdot L) \cdot \cosh(\beta \cdot L) + \cos(\beta \cdot L) \cdot \sinh(\beta \cdot L)] +$$
$$D^2 \cdot \beta^4 \cdot [\cos(\beta \cdot L) \cdot \sinh(\beta \cdot L) + \cosh(\beta \cdot L) \cdot \sin(\beta \cdot L) + \cos(\beta \cdot L) \cdot \sin(\beta \cdot L)] +$$
$$D \cdot \beta^3 \cdot \delta \cdot [2\sin(\beta \cdot L) \cdot \sinh(\beta \cdot L) + \sin^2(\beta \cdot L)] - D \cdot \beta^4 \cdot \delta \cdot \sin(\beta \cdot L) \cdot \cos(\beta \cdot L)]$$
(8.75)

$$A_3 = A - A_1 = \frac{A}{\Delta} [-2M_0 \cdot \omega^2 \cdot D \cdot \beta \cdot \cos(\beta \cdot L) \cdot \sinh(\beta \cdot L) - D \cdot \beta^4 \cdot \cos^2(\beta \cdot L)$$
$$+ M_0 \cdot \omega^2 \cdot \delta \cdot [1 - \cos(\beta \cdot L) \cdot \cosh(\beta \cdot L) - \sin(\beta \cdot L) \cdot \sinh(\beta \cdot L)]$$
$$-D^2 \cdot \beta^4 \cdot [\sin^2(\beta \cdot L) + \cos(\beta \cdot L) \cdot \cosh(\beta \cdot L) + \sin(\beta \cdot L) \cdot \sinh(\beta \cdot L)]$$
$$-D \cdot \beta^3 \cdot \delta \cdot \sin(\beta \cdot L) \cdot \cosh(\beta \cdot L) - \beta^3 \cdot \delta \cdot \sin(\beta \cdot L) \cdot \cos(\beta \cdot L)]$$
(8.76)

$$A_4 = -A_2 = -\frac{A}{\Delta} [2M_0 \cdot \omega^2 \cdot D \cdot \beta \cdot \cos(\beta \cdot L) \cdot \sinh(\beta \cdot L) - \beta^4 \cdot \sin^2(\beta \cdot L) +$$
$$M_0 \cdot \omega^2 \cdot \delta \cdot [\sin(\beta \cdot L) \cdot \cosh(\beta \cdot L) + \cos(\beta \cdot L) \cdot \sinh(\beta \cdot L)] +$$
$$D^2 \cdot \beta^4 \cdot [\cos(\beta \cdot L) \cdot \sinh(\beta \cdot L) + \cosh(\beta \cdot L) \cdot \sin(\beta \cdot L) + \cos(\beta \cdot L) \cdot \sin(\beta \cdot L)] +$$
$$D \cdot \beta^3 \cdot \delta \cdot [2\sin(\beta \cdot L) \cdot \sinh(\beta \cdot L) + \sin^2(\beta \cdot L)] - D \cdot \beta^4 \cdot \delta \cdot \sin(\beta \cdot L) \cdot \cos(\beta \cdot L)]$$
(8.77)

以及

$$\widetilde{V} = \frac{\gamma}{\alpha} \left(\frac{H}{2} + h\right) \cdot \beta \cdot [-A_1 \sinh(\beta L) - A_2 \cosh(\beta L) + (A - A_1)\sin(\beta L) + A_2 \cos(\beta L)]$$
(8.78)

其中，常数 $A_1 \sim A_4$ 在式 (8.74)~式 (8.77) 中定义，且 $\Delta$ 的定义如下：

$$\Delta = -2(D^2 \cdot \beta^4 - M_0 \cdot \omega^2 \cdot \delta)\cosh(\beta \cdot L) \cdot \cos(\beta \cdot L) - 2D^2 \cdot \beta^4 + M_0 \cdot \omega^2 \cdot \delta -$$
$$(D^2 \cdot \beta^3 \cdot \delta - \beta^3 \cdot \delta - 2M_0 \cdot \omega^2 \cdot \beta \cdot D) \cdot \sinh(\beta \cdot L) \cdot \sin(\beta \cdot L) -$$
$$(D^2 \cdot \beta^3 \cdot \delta - \beta^3 \cdot \delta) \cdot \sin(\beta \cdot L) \cdot \cosh(\beta \cdot L) + 2M_0 \cdot \omega^2 \cdot \delta \cdot \beta \cdot D$$
(8.79)

式中：

$$\alpha = \left(\frac{1}{2Z_L} + \frac{1}{Z_0}\right), \quad \gamma = E_{piezo} \cdot d_{31}, \quad \delta = \frac{\gamma^2}{\alpha} \cdot (H+h) \cdot \left(\frac{H}{2}+h\right) \cdot b \quad (8.80)$$

根据上述显式表达式，可计算得到外接阻抗 $Z_L$ 参数，在不同输入振幅 $A$ 和激励频率 $\omega$ 下压电双晶片的性能。

理解双压电晶片的不同类型的电气连接是具有指导意义的。图 8.17 展示了并联和串联两种类型的连接。应该注意，这些电气连接假定压电材料和金属基底之间存在一个非导电层。

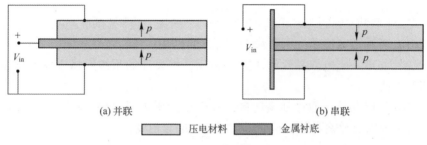

图 8.17 并联或串联的压电双晶片

## 8.4 具有增强频率带宽的压电收集器

### 8.4.1 引言

本节主要介绍 Idan Har-nes 硕士期间的研究工作[34-35]。

压电振动能量收集是一种从机械振动中收集电能的技术。将周围环境的机械振动转化为电能，能够达到远程操作无线传感器或低占空比无线电发射机等小型用电设备的目的。

环境振动能量收集的主要问题是频率及其相关振幅的随机特性。一个基本的收集器具有一个带有末端质量块的单一压电双晶片。由于需要将获取器的固有频率调整为平台激励频率，致使其固有频率带宽狭窄，所以使用单个悬臂收集器作为谐振子来收集振动能量是无效的。参阅文献［34-73］，可以了解关于宽带振动压电收集器的更多有趣应用。

利用文献［58］中提出的基本概念，本书的重点是先进能量收集系统，该系统基于 3 个相同的双晶片连接在 3 个不同端质量的悬臂梁上，这些悬臂梁通过不同的弹簧连接。图 8.18 是所提出的能量收集器的模型示意图。

从电学角度来看，传感器采用两个压电片串联的方式，而能量收集采用并联

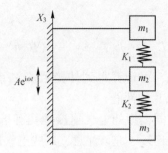

图 8.18 能量收集器模型示意图

的方式。串联和并联的对比可参阅文献［50］。所提出的概念的优点包括：

（1）频率带宽比单一压电双晶片设计的更宽。

（2）与单一压电双晶片解决方案相反，本设计的系统对输入频率的变化敏感度更低。

（3）压电双晶片通过弹簧连接，允许选择压电双晶片和基底梁尺寸、端质量块尺寸以及弹簧常数，从而增加了设计的自由度。

## 8.4.2 运动方程及其解的推导

为了推导图 8.18 中所示系统的运动方程，假设可以使用伯努利-欧拉理论对梁进行建模，该理论的压电层和基底梁之间的连接为理想状态，端质量块之间有线性弹簧。

图 8.19 展示了 3 个长度为 $L$ 的压电双晶片，每个压电双晶片都有一个端质量块和两个线性弹簧 $K_1$ 和 $K_2$，连接 3 个质量块 $M_1$、$M_2$ 和 $M_3$。根据牛顿第二定律，梁右端 3 个附加质量块的运动方程可以表示为以下形式：

$$\check{V}_1(L,t) - K_1[w_1(L,t) - w_2(L,t)] = M_1 \cdot \ddot{w}_1(L,t) \tag{8.81a}$$

$$\check{V}_2(L,t) + K_1[w_1(L,t) - w_2(L,t)] - K_2[w_2(L,t) - w_3(L,t)] = M_2 \cdot \ddot{w}_2(L,t) \tag{8.81b}$$

$$\check{V}_3(L,t) + K_2[w_2(L,t) - w_3(L,t)] = M_3 \cdot \ddot{w}_3(L,t) \tag{8.81c}$$

式中：$\check{V}_1$、$\check{V}_2$ 和 $\check{V}_3$ 分别表示 3 片双压电晶片各个自由端处的剪切力。

假设式（8.81a）~式（8.81c）中的变量可以写成

$$\begin{Bmatrix} w_i(L,t) \\ \check{V}_i(L,t) \end{Bmatrix} = \begin{Bmatrix} W_i(L) \\ \hat{V}_i(L) \end{Bmatrix} e^{i\omega t}, \quad i = 1, 2, 3 \tag{8.82}$$

式中：$\omega^2$ 为系统角频率的平方。将式（8.82）代入式（8.81a）~式（8.81c）

中，得到 $x=L$ 时的剪切力：

$$\check{V}_1(L) = K_1[W_1(L)-W_2(L)] - \omega^2 \cdot M_1 \cdot W_1(L) \tag{8.83a}$$

$$\check{V}_2(L) = -K_1[W_1(L)-W_2(L)] + K_2[W_2(L)-W_3(L)] - \omega^2 \cdot M_2 \cdot W_2(L) \tag{8.83b}$$

$$\check{V}_3(L) = -K_2[W_2(L,t)-W_3(L,t)] - \omega^2 \cdot M_3 \cdot W_3(L) \tag{8.83c}$$

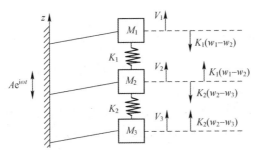

图 8.19　解析解的模型示意图

基于 8.3 节的推导，式（8.62）可以计算系统的每个梁的屈服强度，得

$$W_i(x) = A_{1i}\cosh(\beta_i x) + A_{2i}\sinh(\beta_i x) + A_{3i}\cos(\beta_i x) + A_{4i}\sin(\beta_i x), \quad i=1,2,3 \tag{8.84a}$$

式中：

$$\beta_i^4 \equiv \left(\frac{m\omega^2}{D}\right)_i, \quad i=1,2,3 \tag{8.84b}$$

式中：$D_i$ 和 $m_i$ 的定义分别见式（8.51）和式（8.54）（8.3 节），即

$$D_i = \left\{\frac{2}{3}E_{\text{substrate}} \cdot \left(\frac{H}{2}\right)^3 + \frac{2}{3}E_{\text{piezo}}\left[\left(\frac{H}{2}+h\right)^3 - \left(\frac{H}{2}\right)^3\right]\right\}_i \cdot b_i, \quad i=1,2,3 \tag{8.84c}$$

式中：$H$ 和 $h$ 分别表示基底厚度和压电层厚度：

$$m_i = (H \cdot b \cdot \rho_{\text{substrate}} + 2h \cdot b \cdot \rho_{\text{piezo}})_i, \quad i=1,2,3 \tag{8.84d}$$

式中：$\rho_{\text{substrate}}$ 和 $\rho_{\text{piezo}}$ 分别为基底层密度和压电层密度；$b$ 为双晶片宽度。

式（8.84a）给出了 3 个方程的相关边界条件可以写为

@ $x=0$

$$W_1(0)=A, \quad W_{1,x}(0)=0 \tag{8.85a}$$

$$W_2(0)=A, \quad W_{2,x}(0)=0 \tag{8.85b}$$

$$W_3(0)=A, \quad W_{3,x}(0)=0 \tag{8.85c}$$

@ $x=L$

$$-D_1 \cdot W_{1,xx}(L) + \left[E_{piezo} d_{31} \frac{\widetilde{V}}{h} \cdot (H+h) \cdot h \cdot b\right]_1 = 0 \quad (8.85d)$$

$$D_1 \cdot W_{1,xxx}(L) = K_1[W_1(L) - W_2(L)] - \omega^2 \cdot M_1 \cdot W_1(L)$$

$$-D_2 \cdot W_{2,xx}(L) + \left[E_{piezo} d_{31} \frac{\widetilde{V}}{h} \cdot (H+h) \cdot h \cdot b\right]_2 = 0$$

$$D_2 \cdot W_{2,xxx}(L) = -K_1[W_1(L) - W_2(L)] + K_2[W_2(L) - W_3(L)] - \omega^2 \cdot M_2 \cdot W_2(L) \quad (8.85e)$$

$$\begin{cases} -D_3 \cdot W_{3,xx}(L) + \left[E_{piezo} d_{31} \frac{\widetilde{V}}{h} \cdot (H+h) \cdot h \cdot b\right]_3 = 0 \\ D_3 \cdot W_{3,xxx}(L) = -K_2[W_2(L,t) - W_3(L,t)] - \omega^2 \cdot M_3 \cdot W_3(L) \end{cases} \quad (8.85f)$$

式中：$\widetilde{V}$ 为双晶片的电压，见 8.3 节。

由于壁面激励，每个双晶片上的感应电压写为（8.3 节中式（8.72））

$$\widetilde{V}_i = \Gamma_i \cdot \beta_i \cdot [-A_{i1}\sinh(\beta_i L) - A_{i2}\cosh(\beta_i L) + A_{i3}\sin(\beta_i L) - A_{i4}\cos(\beta_i L)], \quad i=1,2,3 \quad (8.86a)$$

式中：

$$\Gamma_i = \frac{\left[E_{piezo} \cdot d_{31}\left(\frac{H}{2} + h\right)\right]_i}{\left(\frac{1}{2z_L} + \frac{1}{z_0}\right)_i}, \quad i=1,2,3 \quad (8.86b)$$

式中：$z_L$、$z_0$ 分别为双晶片系统的外阻抗和内阻抗。

应用上述边界条件，我们得到了一组含有 12 个未知数的正则矩阵，写为

$$\begin{bmatrix} 1 & 0 & 1 & 0 & 0 & 0 & 0 & 0 & 0 & 0 & 0 & 0 \\ 0 & \beta_1 & 0 & \beta_1 & 0 & 0 & 0 & 0 & 0 & 0 & 0 & 0 \\ a_{31} & a_{32} & a_{33} & a_{34} & 0 & 0 & 0 & 0 & a & 0 & 0 & 0 \\ a_{41} & a_{42} & a_{43} & a_{44} & a_{45} & a_{46} & a_{47} & a_{48} & 0 & 0 & 0 & 0 \\ 0 & 0 & 0 & 0 & 1 & 0 & 1 & 0 & 0 & 0 & 0 & 0 \\ 0 & 0 & 0 & 0 & 0 & \beta_2 & 0 & \beta_2 & 0 & 0 & 0 & 0 \\ 0 & 0 & 0 & 0 & a_{75} & a_{76} & a_{77} & a_{78} & 0 & 0 & 0 & 0 \\ a_{81} & a_{82} & a_{83} & a_{84} & a_{85} & a_{86} & a_{87} & a_{88} & a_{89} & a_{810} & a_{811} & a_{812} \\ 0 & 0 & 0 & 0 & 0 & 0 & 0 & 0 & 1 & 0 & 1 & 0 \\ 0 & 0 & 0 & 0 & 0 & 0 & 0 & 0 & 0 & \beta_3 & 0 & \beta_3 \\ 0 & 0 & 0 & 0 & 0 & 0 & 0 & 0 & a_{119} & a_{1110} & a_{1111} & a_{1112} \\ 0 & 0 & 0 & 0 & a_{125} & a_{126} & a_{127} & a_{128} & a_{129} & a_{1210} & a_{1211} & a_{1212} \end{bmatrix} \begin{Bmatrix} A_{11} \\ A_{12} \\ A_{13} \\ A_{14} \\ A_{21} \\ A_{22} \\ A_{23} \\ A_{24} \\ A_{31} \\ A_{32} \\ A_{33} \\ A_{34} \end{Bmatrix} = \begin{Bmatrix} A \\ 0 \\ 0 \\ 0 \\ A \\ 0 \\ 0 \\ 0 \\ A \\ 0 \\ 0 \\ 0 \end{Bmatrix} \quad (8.87)$$

式 (8.87) 给出的矩阵中出现的各项如下：

$$\begin{cases} a_{31} = -D_1 \cdot \beta_1^2 \cdot \cosh(\beta_1 L) - x_1 \cdot \sinh(\beta_1 L), & a_{32} = -D_1 \cdot \beta_1^2 \cdot \sinh(\beta_1 L) - x_1 \cdot \cosh(\beta_1 L) \\ a_{33} = +D_1 \cdot \beta_1^2 \cdot \cos(\beta_1 L) + x_1 \cdot \sin(\beta_1 L), & a_{34} = +D_1 \cdot \beta_1^2 \cdot \sin(\beta_1 L) - x_1 \cdot \cos(\beta_1 L) \\ x_1 = [E_{\text{piezo}} \times d_{31} \cdot (H+h) \cdot b]_1 \cdot \Gamma_1 \end{cases}$$

(8.88a)

$$\begin{cases} a_{41} = +\alpha_1, \quad a_{42} = +\alpha_2, \quad a_{43} = +\alpha_3, \quad a_{44} = +\alpha_4, \quad a_{45} = +K_1 \cdot \cosh(\beta_2 L) \\ a_{46} = +K_1 \cdot \sinh(\beta_2 L), \quad a_{47} = +K_1 \cdot \cos(\beta_2 L), \quad a_{48} = +K_1 \cdot \sin(\beta_2 L) \\ \alpha_1 = D_1 \cdot \beta_1^3 \cdot \sinh(\beta_1 L) - (K_1 - \omega^2 \cdot M_1) \cdot \cosh(\beta_1 L) \\ \alpha_2 = D_1 \cdot \beta_1^3 \cdot \cosh(\beta_1 L) - (K_1 - \omega^2 \cdot M_1) \cdot \sinh(\beta_1 L) \\ \alpha_3 = D_1 \cdot \beta_1^3 \cdot \sin(\beta_1 L) - (K_1 - \omega^2 \cdot M_1) \cdot \cos(\beta_1 L) \\ \alpha_4 = -D_1 \cdot \beta_1^3 \cdot \cos(\beta_1 L) - (K_1 - \omega^2 \cdot M_1) \cdot \sin(\beta_1 L) \end{cases}$$

(8.88b)

$$\begin{cases} a_{75} = -D_2 \cdot \beta_2^2 \cdot \cosh(\beta_2 L) - x_2 \cdot \sinh(\beta_2 L), & a_{76} = -D_2 \cdot \beta_2^2 \cdot \sinh(\beta_2 L) - x_2 \cdot \cosh(\beta_2 L) \\ a_{77} = +D_2 \cdot \beta_2^2 \cdot \cos(\beta_2 L) + x_2 \cdot \sin(\beta_2 L), & a_{78} = +D_2 \cdot \beta_2^2 \cdot \sin(\beta_2 L) - x_2 \cdot \cos(\beta_2 L) \\ x_2 = [E_{\text{piezo}} \cdot d_{31} \cdot (H+h) \cdot b]_2 \cdot \Gamma_2 \end{cases}$$

(8.88c)

$$\begin{cases} a_{81} = +K_1 \cdot \cosh(\beta_1 L), \quad a_{82} = +K_1 \cdot \sinh(\beta_1 L), \quad a_{83} = +K_1 \cdot \cos(\beta_1 L) \\ a_{84} = +K_1 \cdot \sin(\beta_1 L), \quad a_{85} = \alpha_5, \quad a_{86} = \alpha_6, \quad a_{87} = \alpha_7, \quad a_{88} = \alpha_8 \\ a_{89} = -K_2 \cdot \cosh(\beta_3 L), \quad a_{810} = -K_2 \cdot \sinh(\beta_3 L), \quad a_{811} = -K_2 \cdot \cos(\beta_3 L) \\ a_{812} = -K_2 \cdot \sin(\beta_3 L) \\ \alpha_5 = D_2 \cdot \beta_2^3 \cdot \sinh(\beta_2 L) - [(K_1 + K_2) - \omega^2 \cdot M_2] \cdot \cosh(\beta_2 L) \\ \alpha_6 = D_2 \cdot \beta_2^3 \cdot \cosh(\beta_2 L) - [(K_1 + K_2) - \omega^2 \cdot M_2] \cdot \sinh(\beta_2 L) \\ \alpha_7 = D_2 \cdot \beta_2^3 \cdot \sin(\beta_2 L) - [(K_1 + K_2) - \omega^2 \cdot M_2] \cdot \cos(\beta_2 L) \\ \alpha_8 = -D_2 \cdot \beta_2^3 \cdot \cos(\beta_2 L) - [(K_1 + K_2) - \omega^2 \cdot M_2] \cdot \sin(\beta_2 L) \end{cases}$$

(8.88d)

$$\begin{cases} a_{119} = -D_3 \cdot \beta_3^2 \cdot \cosh(\beta_3 L) - x_3 \cdot \sinh(\beta_3 L), & a_{1110} = -D_3 \cdot \beta_3^2 \cdot \sinh(\beta_3 L) - x_3 \cdot \cosh(\beta_3 L) \\ a_{1111} = +D_3 \cdot \beta_3^2 \cdot \cos(\beta_3 L) + x_3 \cdot \sin(\beta_3 L), & a_{1112} = +D_3 \cdot \beta_3^2 \cdot \sin(\beta_3 L) - x_3 \cdot \cos(\beta_3 L) \\ x_3 = [E_{\text{piezo}} \cdot d_{31} \cdot (H+h) \cdot b]_3 \cdot \Gamma_3 \end{cases}$$

(8.88e)

$$\begin{cases} a_{125} = +K_2 \cdot \cosh(\beta_2 L), \quad a_{126} = +K_2 \cdot \sinh(\beta_2 L) \\ a_{127} = +K_2 \cdot \cos(\beta_2 L), \quad a_{128} = +K_2 \cdot \sin(\beta_2 L) \\ a_{129} = +\alpha_9, \quad a_{1210} = +\alpha_{10}, \quad a_{1211} = +\alpha_{11}, \quad a_{1212} = +\alpha_{12} \\ \alpha_9 = D_3 \cdot \beta_3^3 \cdot \sinh(\beta_3 L) - (K_2 - \omega^2 \cdot M_3) \cdot \cosh(\beta_3 L) \\ \alpha_{10} = D_3 \cdot \beta_3^3 \cdot \cosh(\beta_3 L) - (K_2 - \omega^2 \cdot M_3) \cdot \sinh(\beta_3 L) \\ \alpha_{11} = D_3 \cdot \beta_3^3 \cdot \sin(\beta_3 L) - (K_2 + \omega^2 \cdot M_3) \cdot \cos(\beta_3 L) \\ \alpha_{12} = -D_3 \cdot \beta_3^3 \cdot \cos(\beta_3 L) - (K_2 + \omega^2 \cdot M_3) \cdot \sin(\beta_3 L) \end{cases} \quad (8.88f)$$

只要找到式（8.84a）中3个方程式出现的系数 $A_{1i}$、$A_{2i}$、$A_{3i}$ 和 $A_{4i}$（$i=1$、2、3~12项），就可以估算每个双晶片（式（8.86a））上产生的电压，并可以计算下式给定激励频率下的收集功率 $P_i$：

$$P_i = I_i \cdot V_i \quad (8.89)$$

系统的固有频率可以通过消去系数矩阵的行列式来求解。相应地，在 MATLAB[①] 软件程序中编写了一个代码，用于计算给定激励频率下系统的固有频率、各系数和收集功率。

通过允许弹性柔度具有复数值，将阻尼包括在分析模型中。因此，$S_{11}$（式（8.15））将由 $S_{11}(1-iQ^{-1})$ 来替代（文献［58］和［63］中的讨论），式中 $Q$ 是双晶片的品质因数（假设 $Q=100$，常用于初步计算）。

由于品质因数的复杂性，功率计算将被更新为

$$P_i = \frac{1}{2}(\bar{I}_i \cdot V_i + I_i \cdot \bar{V}_i) \quad (8.90)$$

式中：$\bar{I}$ 和 $\bar{V}$ 分别为 $I_i$ 和 $V_i$ 的共轭数。

为了验证目前的概念，使用了改进的两个双晶片系统，如图8.20所示。

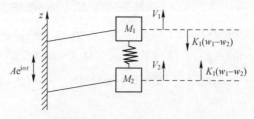

图8.20 两个双晶片弹簧连接的解析模型示意图

对于该系统，边界条件和产生的电压将具有以下形式：

---

① www.mathworks.com/products/matlab/。

$$@\ x = 0$$
$$W_1(0) = A, \quad W_{1,x}(0) = 0 \tag{8.91a}$$

$$W_2(0) = A, \quad W_{2,x}(0) = 0 \tag{8.91b}$$

$$\begin{cases} @\ x = L \\ -D_1 \cdot W_{1,xx}(L) + \left[ E_{\text{piezo}} d_{31} \dfrac{\widetilde{V}}{h} \cdot (H+h) \cdot h \cdot b \right]_1 = 0 \\ D_1 \cdot W_{1,xxx}(L) = K_1 [W_1(L) - W_2(L)] - \omega^2 \cdot M_1 \cdot W_1(L) \end{cases} \tag{8.91c}$$

$$\begin{cases} -D_2 \cdot W_{2,xx}(L) + \left[ E_{\text{piezo}} d_{31} \dfrac{\widetilde{V}}{h} \cdot (H+h) \cdot h \cdot b \right]_2 = 0 \\ D_2 \cdot W_{2,xxx}(L) = -K_1 [W_1(L) - W_2(L)] - \omega^2 \cdot M_2 \cdot W_1(L) \end{cases} \tag{8.91d}$$

$$\widetilde{V}_i = \varGamma_i \cdot \beta_i \cdot [-A_{i1}\sinh(\beta_i L) - A_{i2}\cosh(\beta_i L) + A_{i3}\sin(\beta_i L) - A_{i4}\cos(\beta_i L)], \quad i = 1,2 \tag{8.91e}$$

### 8.4.3 数值验证

为了将现有结果与文献 [58] 进行比较，使用了以下数据：

弹簧系数：$K_0 = \dfrac{3EI}{L^2}$，阻抗：$Z_0 = \dfrac{1}{\mathrm{i}\omega C_0}$，$Z_L = i \cdot Z_0$

式中：$I = \dfrac{b(\pi h + 2c)^3}{12}$，$C_0 = \dfrac{\varepsilon_{33} \cdot b \cdot L}{h}$。

双晶片由 PZT 5H 制成：$S_{11} = 16.5 \times 10^{-12} \mathrm{m}^2/\mathrm{N}$

$$d_{31} = -274 \times 10^{-12} \mathrm{C/N}, \varepsilon_{33} = 3400\varepsilon_0, \rho = 7500 \mathrm{kg/m}^3$$

其品质因数取作 $Q = 10^2$，而 $L = 25\mathrm{mm}$、$b = 8\mathrm{mm}$、$h = 2\mathrm{mm}$、$c = 2\mathrm{mm}$，加速度振幅：$\omega^2 A = I(\mathrm{m/s}^2)$，而对于弹性基底：$E = 70\mathrm{GPa}$，$\rho = 2700\mathrm{kg/m}^3$。

从图 8.21（a）~（d）可以看出，虽然文献 [5] 中的模型在 $X$ 轴值上有一个错误：定义为 $\omega(\mathrm{Hz})$ 的值实际上是 $\omega(\mathrm{rad/s})$（图 8.21（b）和（d）），但是两个模型之间存在很好的相关性。另一个重要的问题是两个双晶片之间的电气连接：文献 [58] 中的两个双晶片是电气连接在一起，为采集器提供单一输出电压。如本节所述，这实际上降低了输出功率，正确的解决方案应该是将每个双晶片连接到存储设备，并使用智能设计的电子卡来控制电源。

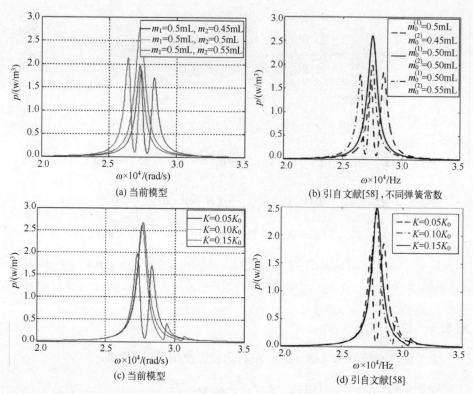

图8.21 不同端质量块条件下通过弹簧连接的两个双晶片的解析功率密度模型结果（见彩插）

### 8.4.4 实验验证

本节旨在对分析模型进行实验验证。共有3个测试案例：①3个双晶片系统的测试；②两个双晶片排列系统的测试；③没有弹簧连接的3个双晶片系统的测试。设计并搭建实验装置，测试了双晶片的各种配置，并记录了其结果供进一步处理。

实验装置如图8.22和图8.23（a）和（b）所示。其主要部件包括：压电收集双晶片系统、1kΩ电阻负载（对每个双晶片）、激光传感器（LG5A65PU，BANNER）、示波器、激光传感器电源和振动台。

每个双晶片分别连接到示波器，测量并记录其输出电压。示波器的另一个通道用于测量激光传感器的输出。激光传感器只能测量两个外部质量块的响应（由于没有与之直接相连的线路，所以可以测量中间质量块的响应）。3个双晶片弹簧连接系统的典型配置见图8.23（a）和（b）。

# 第 8 章 基于智能材料的能量收集

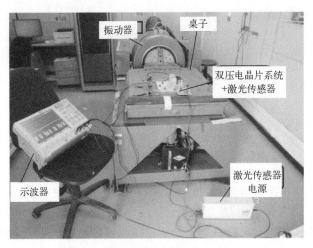

图 8.22 实验测试装置

(a) 3 个双晶片实验系统

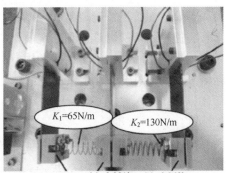

(b) 系统俯视图

图 8.23 3 个双晶片实验系统及其俯视图

每个双晶片由两个压电层和一个 301 不锈钢[①]带构成。压电层为 PI 陶瓷[②] P-876.A11，其具有 0.1mm 压电厚度（PIC255）以及保护压电材料并预加载该层的电子绝缘层（Kapton[③] 胶带）。压电贴片的尺寸为 61mm×35mm，而压电层的尺寸仅为 50mm×30mm，如图 8.24 所示。

---

① 301 型是一种奥氏体铬镍不锈钢，在冷加工时具有高强度和良好的延展性。它是 304 型的改进，降低了铬和镍的含量，以增加冷加工硬化范围。
② www.piceramic.com/en/。
③ Kapton 是杜邦公司在 20 世纪 60 年代末开发的一种聚酰亚胺薄膜，在从 -269~+400℃的广泛温度范围内性能保持稳定。

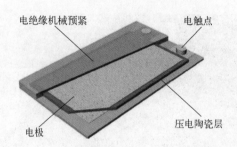

图 8.24　压电贴片和 PI 陶瓷 P-876.A11[35]

端质量块由 303 不锈钢①制成,并用螺钉和螺母固定在双晶片上的计数器板上,见图 8.23(b)。

在组装到双晶片上之前,对末端质量块、计数器板、螺钉和螺母进行了称重,两个外部双晶片的总质量分别为 58g 和 69g,而中间双晶片的总质量为 91g。

弹簧由 302 不锈钢②制成。在弹簧的第 1 圈和最后 1 圈之间添加一个小矩形板,通过两个螺钉连接到末端质量块上(图 8.23(b))。

解析模型和测试模型之间的固有差异可能会导致不同的结果。由于压电层与基底之间的黏合(胶水的位置和数量)不同、切割不准确导致基底尺寸的变化以及压电层之间的差异,每个双晶片(甚至设计成类似的)都有其自身的刚度。

由于双晶片的夹紧方式以及端质量块和弹簧的不同装配,整个系统的装配也会导致模型之间有一些差异。

为了能够以可靠的方式使用解析模型、利用实验结果,通过将每个双晶片的假设刚度乘以刚度因子(实验发现的),对模型进行了调整和修正。此外,品质因数 $Q$ 也被调整至 $10\sim20$ 的范围内。应注意,$Q$ 主要影响频率响应图的高度和宽度。

图 8.25 分别展示了每个双晶片($M_1 = 0.058\text{kg}$、$M_2 = 0.091\text{kg}$ 和 $M_3 = 0.069\text{kg}$)在没有弹簧连接的情况下($K_1 = K_2 = 0$)产生的电压与输入频率的关系。

---

① 303 合金是专门设计的,以表现出改进的可加工性,同时保持良好的力学性能和耐腐蚀性能。合金 303 是最容易加工的奥氏体不锈钢,然而硫的加入降低了 303 合金的性能。
② 302 合金是一种 18%铬/8%镍奥氏体合金的变种,这是不锈钢家族中最常用的。302 合金是 304 合金碳含量略高的版本,经常在带材和金属丝材中应用。

# 第 8 章 基于智能材料的能量收集

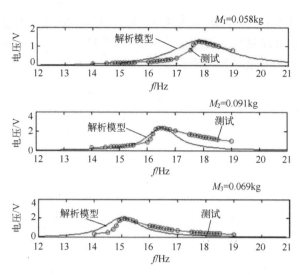

图 8.25 无弹簧连接的 3 个双晶片系统($K_1 = K_2 = 0$)：解析模型与实验结果的对比

注意，图 8.25 中给出的分析曲线是在校准了 $Q$ 因数和刚度因子后得出的。表 8.3 给出了从校准过程中获得的刚度和 $Q$ 因数，解析调谐模型与实验结果之间的比较如表 8.4 所列。

表 8.3　3 个没有弹簧连接的双压电晶片系统($K_1 = K_2 = 0$)的刚度系数和质量系数

| 双压电晶片 | 刚 度 系 数 | 质 量 系 数 |
| --- | --- | --- |
| 1 | 1.13 | 5.85 |
| 2 | 1.50 | 10.00 |
| 3 | 0.95 | 7.90 |

表 8.4　3 个没有弹簧连接的压电双晶片系统（$K_1 = K_2 = 0$）：分析预测与实验结果对比

| 质量/kg | 0.058 | 0.091 | 0.069 |
| --- | --- | --- | --- |
| 固有频率/Hz | 17.8 | 16.5 | 15.1 |
| 预测电压/V（解析模型） | 1.304 | 2.368 | 2.042 |
| 实验电压/V | 1.260 | 2.320 | 1.965 |
| 解析模型预测与实验结果的比值 | 1.035 | 1.020 | 1.039 |

由表 8.4 可知，分析模型（在其调整过程后）与实验结果吻合良好。分析模型和实验结果之间的差异在 4% 以内，这对于工程应用来说是足够小的。

图 8.26 分别展示了每个双晶片的产生电压与输入激励频率的关系。该系统有 3 个端质量块和两个弹簧。两个外部双晶片末端质量块的重量分别为

0.058kg 和 0.069kg，而中间双晶片的重量为 0.091kg。连接外部质量块（0.058kg）和中间质量块（0.091kg）的弹簧系数为 65N/m。连接外部质量块（0.069kg）和中间质量块的弹簧系数为 130N/m。

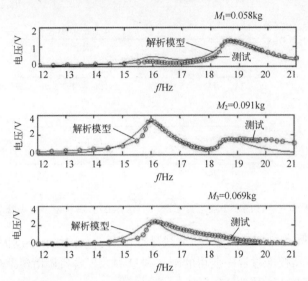

图 8.26　有弹簧连接的 3 个双晶片系统（$K_1 = K_2 = 0$）的解析模型与实验结果对比

注意，每个双晶片的电路连接（连接夹在基底梁中间的两个压电层）是并联的，如 8.3 节末尾所讨论和展示一样。结果如表 8.5 所列。

表 8.5　3 个双晶片系统并联的解析模型预测与实验结果对比

| 质量/kg | 0.058 | | 0.091 | | 0.069 | |
|---|---|---|---|---|---|---|
| 固有频率/Hz | 第 1 个峰 16 | 第 2 个峰 18.8 | 第 1 个峰 16 | 第 2 个峰 18.8 | 第 1 个峰 16 | 第 2 个峰 18.8 |
| 解析模型预测电压/V | 0.4942 | 1.303 | 1.661 | 0.7619 | 2.326 | 0.1895 |
| 实验电压/V | 0.2500 | 1.280 | 1.710 | 0.7650 | 2.335 | 0.6050 |
| 解析模型预测与实验结果的比值 | 1.9768 | 1.018 | 0.971 | 0.996 | 0.996 | 0.313[①] |

① 原书为 0.110，有误。——译者

表 8.5 中，当峰值固有频率等于双晶片匹配的固有频率时，每种双晶片都可以实现最佳拟合。以 0.058kg 的双晶片为例，显示出最佳相关性是在 18.8Hz 处的第 2 个峰值，该峰值接近其固有频率（17.8Hz）。在第 1 个峰值（16Hz）处，相关性不太好，但是它出现在最小电压振幅处，因此这一相关性的影响对总电压相关性的影响很小。

图 8.27 展示了每个双晶片产生的调谐电压与激励输入频率的关系。使用串联连接再次测试上述用于并联电路的相同系统。

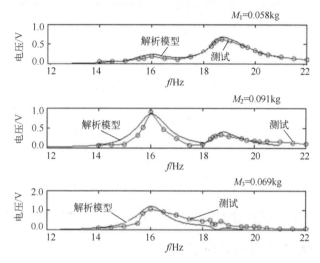

图 8.27 有弹簧连接（$K_1 = 65\text{N/m}$、$K_2 = 130\text{N/m}$）的 3 个双晶片（电气串联）系统的解析模型与实验结果对比

表 8.6 显示了解析模型的预测与实验结果之间相对良好的关联。当每个双晶片串联时，每个双晶片产生的电压以及系统的总电压都小于并联时的电压：串联情况下为 1V，并联情况下为 2V，固有频率保持不变。在本书展示的所有计算中，采用并联方式的主要原因是其输出产生的电压更高。

表 8.6 3 个双晶片系统串联的解析模型预测与实验结果对比

| 质量/kg | 58 | | 91 | | 69 | |
| --- | --- | --- | --- | --- | --- | --- |
| 固有频率/Hz | 第 1 个峰 16 | 第 2 个峰 18.8 | 第 1 个峰 16 | 第 2 个峰 18.8 | 第 1 个峰 16 | 第 2 个峰 18.8 |
| 解析模型预测电压/V | 0.2459 | 0.6812 | 0.856 | 0.3931 | 1.202 | 0.06639 |
| 实验电压/V | 0.195 | 0.625 | 0.900 | 0.325 | 1.055 | 0.450 |
| 解析模型预测与实验结果的比值 | 1.260 | 1.090 | 0.951 | 1.117 | 1.139 | 0.148 |

图 8.28 比较了 3 个双晶片和两个双晶片系统在系列测试中产生的总电压。可以观察到，与两个双晶片系统相比，3 个双晶片系统具有更低的固有频率，这是由于在 3 个双晶片系统中间质量块的质量更高且刚度因子更大。与预期的一样，3 个双晶片系统的电压和带宽比两个双晶片系统的更大。

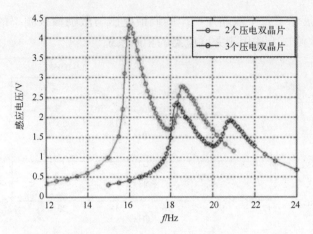

图 8.28 3个/2个双晶片系统实验结果(电压)对比

文献 [34-35] 正在对2个和3个压电双晶片系统参数进行研究。

### 8.4.5 3个与2个双晶片系统

如 8.4 节开头所述，本研究的主要目标是扩展功率输出带宽。具有更多双晶片的系统预期有更宽的带宽。图 8.29 展示了 3 个和 2 个双晶片系统之间的典型性能比较。数值比较如下：对于 3 个压电双晶片的系统，$M_1=0.03\text{kg}$、$M_2=0.06\text{kg}$、$M_3=50\text{kg}$、$K_1=40\text{N/m}$ 以及 $K_2=80\text{N/m}$，而对于 2 个双晶片的系统，其数据是 $M_1=0.03\text{kg}$、$M_2=0.05\text{kg}$ 以及 $K=50\text{N/m}$。

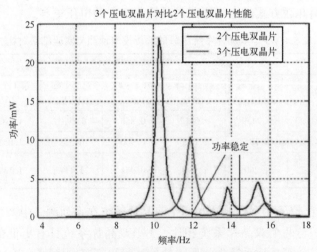

图 8.29 3个/2个压电双晶片系统性能对比(功率)

# 第8章 基于智能材料的能量收集

可以看出，与2个双晶片的系统相比，3个双晶片的系统具有更宽的带宽。此外，3个双晶片的系统具有更大的输出功率，并且通过添加额外的双晶片（中间的一个）可以减小"功率缺失"的宽度。类似的结果发表在文献[60]中。

另一个重要问题是双晶片之间的电路连接。双晶片可以通过电路连接在一起，或者每个双晶片可以单独连接到电路，并将所有双晶片的输出连接到存储系统。

连接的系统中，假设所有的正极彼此连接，所有的负极彼此连接。单个系统假设每个双晶片都是独立的，而电路输出的是电压、电流和功率总和。

如图8.30所示，通过3个双晶片的"整体"电气连接获得的功率输出小于单独的分离型连接（第1和第3固有频率除外）。虽然这些固有频率的输出功率更高，但带宽非常窄，它们之间的功率峰更宽，且第2个固有频率的功率几乎可以忽略不计。因此，为了达到更大的频率带宽和更高的总收集功率，建议使用单独的电路连接。

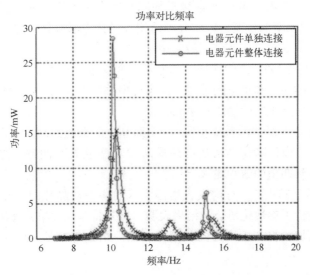

图8.30 单独和整体连接的3个双晶片系统与频率的数值结果对比（功率单位：mW）

应注意，图8.30中给出结果所用的双晶片长度、宽度和高度分别为 $L = 0.05\text{m}$，$b = 0.03\text{m}$ 和 $h = 0.0001\text{m}$。质量为 $M_1 = 0.03\text{kg}$，$M_2 = 0.06\text{kg}$，$M_3 = 0.05\text{kg}$，而弹簧系数为 $K_1 = 40\text{N/m}$ 和 $K_2 = 60\text{N/m}$。

### 8.4.6 结论和建议

根据上述分析和实验研究，可以得出以下结论和建议（另见文献 [34-35]）。

（1）事实证明，连接弹簧系数是至关重要的。如果弹簧系数与双晶片的刚度相比太小，系统表现与没有弹簧一样（$K=0$），其影响可以忽略不计。另外，弹簧系数与双晶片的刚度相比太大，弹簧则表现出类似刚性杆工作的特性。因此，对于两个双晶片系统，第 2 个固有频率将消失。首选的弹簧系数约为双晶片刚度的 15%（0.15（$EI/L^2$））。

（2）随着两个双晶片之间的质量比增加，"功率缺失"①（8.5.5 节）增加，从而降低系统效率。降低质量比会降低"功率缺失"，但会缩小频率带宽。因此，质量比的选择至关重要。根据系统几何形状和弹簧刚度的不同，理想的端部质量比应该是在 10~20 范围内。同样的结论可以在文献 [60，71] 中找到。

（3）注意，只有在两个非常接近的固有频率情况下，才能避免"功率缺失"，从而产生狭窄的带宽，因此并不适用。设计者必须使"功率缺失"最小化，从而使收集器输出功率带宽最大化。

（4）正如预期和预测的那样，相比于 2 个双晶片的系统，3 个双晶片的系统可产生更高的功率和更宽的带宽。向系统中添加更多的双晶片和弹簧有望进一步提高系统所产生的的输出功率和带宽。

（5）与 2 个双晶片的系统相比，3 个双晶片的系统为设计者提供了更大的自由度。因此，设计者将能够更好地应对"功率缺失"问题，并能将固有频率正确地调整到期望的激励输入。

（6）文献 [34-35] 对构成收集器系统的各种双晶片的"整体"和单独的电气连接进行了参数化研究，结果表明，首先将双晶片单独连接，然后将其产生的功率相加，将产生更好的收集器。这一结论与文献 [69] 中提出的说法类似。

## 参 考 文 献

[1] Caliò, R., Rongala, U. B., Camboni, D., Milazzo, M., Stefanini, C., de Petris, G. and Oddo, C. M., Piezoelectric energy harvesting solutions, Sensors 14, 2014, 4755-4790. doi: 10.3390/s140304755.

---

① "功率缺失"被定义为两个相邻固有频率之间产生的零功率。

[2] Roundy, S., On effectiveness of vibration-based energy harvesting, Journal of Intelligent Materials Systems and Structures 16, 2005, 809-823.

[3] Lopes, C. M. A. and Gallo, C. A., A review of piezoelectric energy harvesting and applications, 2014 IEEE 23rd International Symposium on Industrial Electronics (ISIE), 1-4 June, 2014, 1285-1288.

[4] Priya, S. and Inman, D. J. (eds.), Energy harvesting technologies, Appendix A: First draft of standard on vibration energy harvesting, 507-513, © Springer Science + Business media, LLC 2009, 517 pp.

[5] Priya, S. and Inman, D. J. (eds.), Energy Harvesting Technologies, © Springer Science + Business media, LLC, 2009, 517.

[6] Ertuk, A. and Inman, D. J., An experimentally validated bimorph cantilever model for piezoelectric energy harvesting from base excitations, smart Materials and Structures 18 (1), 2009, article ID 025009.

[7] Ertuk, A. and Inman, D. J., Issues in mathematical modeling of piezoelectric energy harvesters, smart Materials and Structures 17 (6), 2008, article ID 065016.

[8] Ertuk, A. and Inman, D. J., A distributed parameter electromechanical model for cantilevered piezoelectric energy harvesters, Journal of Vibration and Acoustics, Transactions of the ASME 130 (4), 2008, article ID 041002.

[9] Sebald, G., Kuwano, H., Guyomar, D. and Ducharne, B., Experimental Duffing oscillator for broadband piezoelectric energy harvesting, smart Materials and Structures 20, 2011, article ID 102001, 10.

[10] Vatansever, D., Hadimani, R. L., Shah, T. and Siores, E., An investigation of energy harvesting from renewable sources with PVDF and PZT, smart Materials and Structures 20, 2011, article ID 055019, 6.

[11] Pozzi, M. and Zhu, M., Plucked piezoelectric bimorph for knee-joint energy harvesting: modelling and experimental validation, smart Materials and Structures 20, 2011, article ID 055007, 10.

[12] Ling, B. K., Li, T., Hng, H. H., Boey, F., Zhang, T. and Li, S., Waste Energy Harvesting Mechanical and Thermal Energies, Chapter 2 - Waste Mechanical Energy Harvesting (I): Piezoelectric Effect, Springer, 2014, 25-133. http://www.springer.com/978-3-642-54633-4.

[13] Wekin, A. B. E., Characterization and comparison of piezoelectric materials for transducing power from a thermoacoustic engine, A master thesis submitted to the School of mechanical and Materials Engineering, Washington State University, August 2008, 130 pp.

[14] Bowen, C. R., Kim, H. A., Weaver, P. M. and Dunn, S., Piezoelectric and ferroelectric materials and structures for energy harvesting applications, Energy Environmental Science 7, 2014, 25-44.

[15] Sodano, H. A., Magliula, E. A., Park, G. and Inman, D. J., Electric power generation using piezoelectric devices, 13th International Conference on Adaptive Structures and Technologies (ICAST13), Breitbach, E. J., Campanile, L. F. and Monner, H. P. (eds.), CRC press, October 7-9, 2002, Potsdam, Germany, 153-161.

[16] Wang, X., Liang, X., Shu, G. and Watkins, S., Coupling analysis of linear vibration energy harvesting systems, Mechanical Systems an Signal Processing, 2015. doi: http://dx.doi.org/10.1016/j.ymssp.2015.09.006i.

[17] Peigney, M. and Siegert, D., Piezoelectric energy harvesting from traffic-induced bridge vibrations, smart Materials and Structures 22, 2013, article ID 095019.

[18] Dhingra, P., Biswas, J., Prasad, A. and Meher, S. S., Energy harvesting using piezoelectric materials, Special Issue of International Journal of Computer Applications (0975-8887), International conference on Electronic Design and Signal Processing (ICEDSP), 2012, 38-42.

[19] Mohamad, S. H., Thalas, M. F., Noordin, A., Yahya, M. S., Hassan, M. H. C. and Ibrahim, Z., A potential study of piezoelectric energy harvesting in car vibration, Journal of Engineering and Applied Sciences 10 (19), October 2015, 8642-8647.

[20] Sodano, H. A. and Inman, D. J., Comparison of piezoelectric energy harvesting devices for charging batteries, Journal of Intelligent Material Systems and Structures 16 (10), 2005, 799-807.

[21] Rödig, T. and Schönecker, A., A survey on piezoelectric ceramics for generator applications, Journal of American Ceramic Society 93 (4), 2010, 901-912.

[22] Jung, W.-S., Kang, M.-G., Moon, H. G., Baek, S.-H., Yoon, S.-J., Wang, Z.-L., Kim, S.-W. and Kang, C.-Y., High output piezo/triboelectric hybrid generator, www.nature.com/scientificre ports, Scientific Reports 5, March 2015. doi: 10.1038/srep 09309. 9309.

[23] De Jong, P. H., Power harvesting using piezoelectric materials-applications in helicopter rotors, Ph.D. Thesis, 160 pp., University of Twente, Enschede, The Netherlands, February 2013.

[24] Kim, H. W., Batra, A., Priya, S., Uchino, K. and Markley, D., Energy harvesting using a piezoelectric "cymbal" transducer in dynamic environment, Japanese Journal of Applied Physics 43 (9), 2004, 6178-6183.

[25] Abramovich, H., Tsikhotsky, E. and Klein, G., Experimental determination of the maximal allowable stresses for high-power piezoelectric generators, Journal of Ceramic Science and Technology 4 (3), 2013, 131-136. doi: 10.4416/JCST 2013-00006.

[26] Abramovich, H., Tsikhotsky, E. and Klein, G., An experimental investigation on PZT behavior under mechanical and cycling loading, Journal of the Mechanical Behavior of Materials, 22 (3-4), 2013, 129-136.

[27] Williams, C. B., Shearwood, C., Harradine, M. A., Birch, T. S. and Yates, R. B., Development of an electromagnetic micro-generator, IEE Proceedings on Circuits Devices Systems 148 (6), December 2001, 337-342.

[28] Beeby, S. P., Tudor, M. J. and White, N. M., Energy harvesting vibration sources for microsystems applications, Measurement Science and Technology 17, 2006, 175-195.

[29] Arnold, D. P., Review of microscale magnetic power generation, IEEE Transactions on Magnetics 43 (11), 2007, 3940-3951.

[30] Beeby, S. P., Torah, R. N., Tudor, M. J., Glynne-Jones, P., O'Donnell, T., Saha, C. R. and Roy, S., A micro electromagnetic generator for vibration energy harvesting, Journal of Micromechanics and Microengineering 17, 2007, 1257-1265.

[31] Yang, B., Lee, C., Xiang, W., Xie, J., He, J. H., Kotlanka, R. K., Low, S. P. and Feng, H., Electromagnetic energy harvesting from vibrations of multiple frequencies, Journal of Micromechanics and Microengineering 19, 2009, article ID 035001, 8.

[32] Liu, H., Soon, B. W., Wang, N., Tay, C. J., Quan, C. and Lee, C., Feasibility study of a 3D vibration-driven electromagnetic MEMS energy harvester with multiple vibration modes, Journal of Micromechanics and Microengineering 22, 2012, article ID 125020, 8.

[33] Ylli, K., Hoffmann, D., Willman, A., Becker, P., Folkmer, B. and Manoli, Y., Energy harvesting from human motion: exploiting swing and shock excitations, smart Materials and Structures 24 (2), 2015, article ID 025029, 12.

[34] Har-nes, I., Bandwidth expansion for piezoelectric energy harvesting, Master thesis, Faculty of Mechanical Engineering, Technion, I. I. T., 32000, Haifa, Israel, August 2016, 170.

[35] Abramovich, H. and Har-nes, I., Analysis and experimental validation of a piezoelectric harvester with enhanced frequency bandwidth, Materials 2018, 11 (1243), 41. doi: 10.3390/ma11071243 www.MDPI.COM/JOURNAL/MATERIALS.

[36] Arms, S. W., Townsend, C. P., Churchill, D. L., Galbreath, J. H. and Mundell, S. W., Power management for energy harvesting wireless sensors, Proceedings of the SPIE International Symposium on smart Structures and smart Materials, San Diego, CA, USA, 7-10 March, 2005.

[37] Murimi, E. and Neubauer, M., Piezoelectric energy harvesting: an overview, Proceedings of Sustainable Research and Innovation Conference, 2014, 4, 117-121.

[38] Tang, L., Yang, Y. C. and So, K., Broadband vibration energy harvesting techniques. In: Advances in Energy Harvesting Methods, New York, NY, USA, Springer Science Business Media, 2013, 17-61.

[39] Kong, L. B., Li, T., Hong, H. H., Boey, F., Zhang, T. and Li, S., Waste Mechanical Energy Harvesting (I): Piezoelectric Effect, Waste Energy Harvesting - Mechanical and Thermal Engines, Lecture Notes in Energy, Berlin/Heidelberg, Germany, Springer, 2014, 19-133.

[40] Challa, V. R., Prasad, M. G., Shi, Y. A. and Fisher, F. T., A vibration energy harvesting device with bidirectional resonance frequency tenability, smart Materials and Structures 2008, 17 (015035), 10. doi: 10.1088/0964-1726/01/015035.

[41] Challa, V. R., Prasad, M. G. and Fisher, F. T., Towards an autonomous self-tuning vibration energy harvesting device for wireless sensor network applications, smart Materials and Structures 2011, 20 (025004), 11. doi: 10.10.88/0964-1726/20/21025004.

[42] Challa, V. R., Prasad, M. G. and Fisher, F. T., High efficiency energy harvesting device with magnetic coupling for resonance frequency tuning, Proceedings of SPIE Sensors and smart Structures Technologies for Civil, Mechanical, and Aerospace Systems 6932, 2008, 69323Q. doi: 10.1117/12.776385.

[43] Ferrar, M., Ferrari, M., Guizetti, M., Ando, B., Baglio, S. A. and Trigona, C., Improved energy harvesting from wideband vibrations by nonlinear piezoelectric converters, Sensors and Actuators A, Physical 162 (2), 2010, 425-431.

[44] Cottone, F., Gammaitoni, L., Vocca, H. A. and Ferrari, V., Piezoelectric buckled beams for random vibration energy harvesting, smart Materials and Structures 21 (3), 2012, 035021, 11.

[45] Eichhorn, C., Goldschmidtboeing, F. A. and Woias, P. A frequency tunable piezoelectric energy converter based on a cantilever beam, Proceedings of the PowerMEMS2008+microEMS2008, Sendai, Japan, 9-12 November 2008, 309-312.

[46] Eichhorn, C., Goldschmidtboeing, F., Porro, Y. and Woias, P., A piezoelectric harvester with an integrated frequency tuning mechanism. In Proceedings of the PowerMEMS, Washington, DC, USA, 1-4 December 2009, 45-48.

[47] Eichhorn, C., Tchagsim, R., Wilhelm, N., Biancuzzi, G. and Woias, P., An energy-autonomous

self-tunable piezoelectric vibration energy harvesting system. In Proceedings of the 2011 IEEE 24th International Conference Micro Electro Mechanical Systems (MEMS), Cancun, Mexico, 23 – 27 January 2011, 1293-1296.

[48] Wu, X., Lin, J., Kato, S., Zhang, K., Ren, T. and Liu, L., Frequency adjustable vibration energy harvester, Proceedings of the PowerMEMS 2008 + microEMS2008, Sendai, Japan, 9 – 12 November 2008, 245-248.

[49] Ko, S. C., Je, C. H. A. and Jun, C. H., Mini piezoelectric power generator with multi-frequency response, Procedia Engineering 5, 2010, 770-773.

[50] Aridogan, U., Basdogan, I. and Erturk, A., Multiple patch-based broadband piezoelectric energy harvesting on plate-based structures, Journal of Intelligent Material Systems and Structures 25 (4), 2014, 1664-1680. doi: 10.1177/1045389x14544152.

[51] Berdy, D. F., Srisungsitthisunti, P., Jung, B., Xu, X., Rhoads, J. F. and Peroulis, D., Low-frequency meandering piezoelectric vibration energy harvester, IEEE Transactions on Ultrasonics, Ferroelectrics, and Frequency Control 59 (5), 2012, 846-858.

[52] Miah, H., Heum, D. A. and Park, J. Y., Low frequency vibration energy harvester using stopperengaged dynamic magnifier for increased power and wide bandwidth, Journal of Electrical Engineering and Technology 11 (3), 2016, 707-714.

[53] Erturk, A., Renno, J. M. and Inman, D. J., Modeling of piezoelectric energy harvesting from an L-shaped beam-mass structure with an application to UAVs, Journal of Intelligent Material Systems and Structures 20, 2009, 529-544.

[54] Wang, H. Y., Tang, L. H., Guo, Y., Shan, X. B. and Xie, T., A 2DOF hybrid energy harvester based on combined piezoelectric and electromagnetic conversion mechanisms, Journal of Zhejiang University-Science A (Applied Physics & Engineering) 15 (9), 2014, 711-722.

[55] Alghisi, D., Dalola, S., Ferrari, M. and Ferrari, V., Ball-impact piezoelectric converter for multi degree-of-freedom energy harvesting from broadband low-frequency vibrations in autonomous sensors, Procedia Engineering 87, 2014, 1529-1532.

[56] Ferrari, M., Bau, M., Cerini, F. and Ferrari, V., Impact-enhanced multi-beam piezoelectric converter for energy harvesting in autonomous sensors, Procedia Engineering 47, 2012, 418-421.

[57] Aryanpur, R. M. and White, R. D., Multi-link piezoelectric structure for vibration energy harvesting. Proc. SPIE 8341, Active and Passive smart Structures and Integrated Systems, 2012, 83411Y, April 26, 2012, doi: 10.1117/12.915438.

[58] Yang, Z. and Yang, J., Connected vibrating piezoelectric bimorph beams as a wide-band piezoelectric power harvester, Journal of Intelligent Material Systems and Structures 20, March 2009, 569-574, doi: 10.1177/104389x0800042.

[59] Lee, P., A wide band frequency-adjustable piezoelectric energy harvester, Master's Thesis, University of North Texas, Denton, TX, USA, 2012, 27.

[60] Zhang, H. and Afzalul, K., Design and analysis of a connected broadband multi-piezoelectric bimorph beam energy harvester, IEEE Transactions on Ultrasonics, Ferroelectrics, and Frequency Control 61 (6), 2014, 1016-1023.

[61] Meruane, V. and Pichara, K., A broadband vibration-based energy harvester using an array of piezoe-

lectric beams connected by springs, Shock and Vibration, 2016, 9614842, 13, doi: 10.1155/2016/9614842.

[62] Shahruz, S. M., Limits of performance of mechanical band-pass filters used in energy scavenging, Journal of Sound and Vibration 293 (1-2), 2006, 449-461.

[63] Shahruz, S. M., Design of mechanical band-pass filters for energy scavenging, Journal of Sound and Vibration 292 (3-5), 2006, 987-998.

[64] Xue, H., Hu, Y. and Wang, Q.-M., Broadband piezoelectric energy harvesting devices using multiple bimorphs with different operating frequencies, IEEE Transactions on Ultrasonics, Ferroelectrics, and Frequency Control 55, 2008, 2104-2108.

[65] Qi, S., Investigation of a novel multi-resonant beam energy harvester and a complex conjugate matching circuit, Ph.D. Thesis, University of Manchester, Manchester, UK, 2011.

[66] Lien, I. C. and Shu, Y. C., Array of piezoelectric energy harvesting by the equivalent impedance approach, smart Materials and Structures 21, 2012, 082001.

[67] Lin, H. C., Wu, P. H., Lien, I. C. and Shu, Y. C., Analysis of an array of piezoelectric energy harvesters connected in series, smart Materials and Structures 22, 2013, 094026. doi: 10.1088/10964-1726/22/9/094026.

[68] Lien, I. C. and Shu, Y. C. Piezoelectric array of oscillators with respective electrical rectification, Proc. Vol. 8688, Active and Passive smart Structures and Integrated Systems, Proceedings of the 2013 SPIE smart Structures and Materials+Nondestructive Evaluation and Health Monitoring, San Diego, CA, USA, 10-14 March 2013.

[69] Al-Ashtari, W., Hunsting, M., Hemsel, T. and Sextro, W., Enhanced energy harvesting using multiple piezoelectric elements: theory and experiments, Sensors and Actuators A: Physics 200, 2013, 138-146.

[70] Wu, P. H. and Shu, Y. C., Finite element modeling of electrically rectified piezoelectric energy harvesters, smart Materials and Structures 24, 2015, 094008. doi: 10.1088/0964-1726/24/9/094008.

[71] Dechant, E., Fedulov, F., Fetisov, L. Y. and Shamonin, M., Bandwidth widening of piezoelectric cantilever beam arrays by mass-tip tuning for low-frequency vibration energy harvesting, Applied Science 7, 2017, 1324, doi: 10.33901 app7121324.

[72] Yang, Y., Wu, H. and Soh, C. K., Experiment and modeling of a two-dimensional piezoelectric energy harvester, smart Materials and Structures 24, 2015, 125011. doi: 10.10.88/0964-1726/24/12/125011.

[73] Miller, L. M., Elliot, A. D. T., Mitcheson, P. D., Halvorsen, E., Paprotny, I. and Wright, P. K., Maximum performance of piezoelectric energy harvesters when coupled to interface circuits, IEEE Sensors Journal 16, 2016, 4803-4815.

# 第9章 光纤简介

## 9.1 几何光学：基本概念

几何光学是一种以光线来描述光传播的光学模型。几何光学中的光线有助于粗略估计光的传播路径。读者想必记得，真空中的光速及其相关频率为

$$c = 2.99792458 \times 10^8 \text{m/s}$$

$$f = c/\lambda \approx 10^{15} \text{Hz}$$

当光在一种介质（如水）中传播时，其速度会下降，由此产生了折射现象。折射率的计算公式为 $n = c/v$，其中 $v$ 是光通过该介质的速度（不同介质的 $n$ 值如表9.1所列）。

表9.1 常见材料的折射率（摘自《CRC化学物理手册》）

| 材料 | 空气 | 水 | 熔融石英 | 鲸油 | 冕玻璃 | 盐 | 沥青 | 金刚石 | 铅 |
|---|---|---|---|---|---|---|---|---|---|
| $n$ | 1.003 | 1.33 | 1.4585 | 1.46 | 1.52 | 1.54 | 1.635 | 2.42 | 2.6 |

波速（$v$）与频率（$f$）和波长（$\lambda$）之间的关系为 $v = f/\lambda$，且光的频率不随传播的介质变化而变化，折射率与光波长之间有如下关系：

$$\begin{cases} \lambda_0 = c/f (\text{真空}) \\ \lambda = v/f (\text{其他介质}) \\ \Rightarrow \dfrac{\lambda_0}{\lambda} = \dfrac{c}{v} = n \end{cases} \tag{9.1}$$

### 9.1.1 光的反射

当光线照射到光滑表面（如镜子）时，会被反射回来，如图9.1所示。反射定律规定：入射光线、法线和反射光线位于同一平面，并且 $\theta_i = \theta_r$。

# 第 9 章 光纤简介

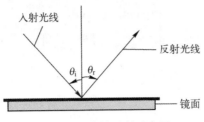

图 9.1 光线的反射示意图

## 9.1.2 斯涅耳定律-折射定律

当光从一个折射率为 $n_1$ 的介质传播到另一个折射率为 $n_2$ 的介质时,光的传播方向会改变,如图 9.2 所示。斯涅耳定律①可以表示为

$$n_1 \sin\theta_1 = n_2 \sin\theta_2$$
$$\Rightarrow v_1 \sin\theta_2 = v_2 \sin\theta_1 \tag{9.2}$$

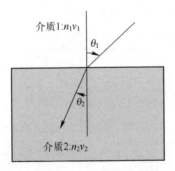

图 9.2 光线折射示意图——斯涅耳定律

## 9.1.3 临界角和全内反射

当光从一种 $n$ 值较高的介质传播到另一种 $n$ 值较低的介质时,应用斯涅耳定律可能会发现折射角正弦值将大于 1,这显然不会发生。因此,在这种情况下,光会被完全反射,这在文献中称为全内反射。临界角被定义为这种情况下允许发生折射的最大入射角,可表示为

---

① Willebrord Snellius(1580.6.13—1626.10.30),荷兰天文学家和数学家。

$$\begin{cases} n_1\sin\theta_1 = n_2\sin\theta_2 \\ \theta_{\text{crit}} = \arcsin\left(\dfrac{n_2}{n_1}\sin\theta_2\right) \\ \text{然而 } \sin\theta_2 = 1 \\ \text{所以 } \theta_{\text{crit}} = \arcsin\left(\dfrac{n_2}{n_1}\right) \end{cases} \quad (9.3)$$

### 9.1.4　光纤的数值孔径和接收角

数值孔径（Numerical Aperture，NA）是一个与给定角度的接收角相关的无量纲量，可表示为

$$\text{NA} = n_0\sin\theta_a \quad (9.4)$$

式中：$\theta_a$ 为光纤的最大半接收角；$n_0$ 为光纤外部材料的折射率。通常，外部材料是空气，因此 $n_0 = 1$，于是得到 $\text{NA} = \sin\theta_a$。定义数值孔径的另一种方法如图 9.3 所示。光线以最大半接收角传播到光纤中，会抵达光纤包层与纤芯之间的界面，并被反射。根据斯涅尔定律可以得

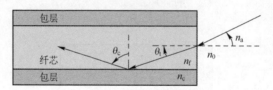

图 9.3　光线以半接收角入射光纤示意图

$$\begin{cases} n_0\sin\theta_a = n_f\sin\theta_t \quad (\text{空气-纤芯}) \\ n_f\sin\theta_c = n_c\sin\theta_r \quad (\text{纤芯-包层}) \\ \text{而 } \sin\theta_{\text{crit},c} = \left(\dfrac{n_c}{n_f}\sin\theta_r\right) \text{ 且 } \sin\theta_r = 1 \\ \Rightarrow \sin\theta_{\text{crit},c} = \left(\dfrac{n_c}{n_f}\right) \\ \text{又 } \theta_{\text{crit},c} + \theta_t = 90° \Rightarrow \sin\theta_{\text{crit},c} = \cos\theta_t \\ \Rightarrow \sin\theta_{\text{crit},c} = \left(\dfrac{n_c}{n_f}\right) = \cos\theta_t = \sqrt{1-\sin^2\theta_t} \\ \text{或 } \sin\theta_t = \sqrt{1-\dfrac{n_c^2}{n_f^2}} \end{cases} \quad (9.5)$$

以式 (9.5) 中的第 1 行替换式 (9.5) 中最后一行的等式，得

$$\begin{cases} n_0 \sin\theta_a = n_f \sin\theta_t \quad \text{（空气-纤芯）} \\ \text{而} \quad \sin\theta_t = \sqrt{1 - \dfrac{n_c^2}{n_f^2}} \\ \Rightarrow n_0 \sin\theta_a = n_f \sqrt{1 - \dfrac{n_c^2}{n_f^2}} = \sqrt{n_f^2 - n_c^2} = \mathrm{NA} \\ \text{或} \quad \sin\theta_{a,\max} = \sqrt{\dfrac{n_f^2 - n_c^2}{n_0^2}} \Rightarrow \theta_{a,\max} = \arcsin\left(\sqrt{\dfrac{n_f^2 - n_c^2}{n_0^2}}\right) \end{cases} \quad (9.6)$$

光锥的总接收角是式 (9.6) 中最后一个表达式所表示的角度的两倍。

纤芯的折射率从其中心向包层方向不断变化，折射率的分布情况如下：

$$\begin{cases} n^2(r) = n_f^2 \left[ 1 - 2\Delta \left( \dfrac{r}{0.5d} \right)^\alpha \right], \quad r < 0.5d \quad \text{（纤芯）} \\ n^2(r) = n_c^2 = \text{常数}, \quad r > 0.5d \quad \text{（包层）} \end{cases} \quad (9.7)$$

式中：

$$\Delta = \dfrac{n_f^2 - n_c^2}{2n_f^2}$$

且特殊情况下 $\alpha$ 为

$\alpha = 1$，三角折射率分布光纤

$\alpha = 2$，抛物线分布光纤

$\alpha \to \infty$，阶跃折射率光纤

另一个参数是归一化频率 $V$，表示为

$$V = 2\pi \dfrac{0.5d}{\lambda} \mathrm{NA} = \pi \dfrac{d}{\lambda} \mathrm{NA} \quad (9.8)$$

式中：$\mathrm{NA} = n_f \sqrt{2\Delta}$；$d$ 为纤芯直径；$\lambda$ 为光波长。参数 $V$ 决定光纤的工作状态，即单模传输或多模传输。

纤芯导模数量由上述公式决定，即

$$N \approx \dfrac{V^2}{2} \dfrac{\alpha}{\alpha + 2} \quad (9.9)$$

对于折射率阶跃型光纤，若 $V < 2.405$，则只有单模传导。这也称为标量模 $\mathrm{LP}_{01}$，其截止频率为 $V_c = 0$。

对于渐变折射率分布（$\alpha = 2$），可以得

$$V_c \approx 2.405 \times \sqrt{2} \approx 3.4 \quad (9.10)$$

## 9.1.5 干涉

当两束光线具有相同波长且相位相同时,它们将发生叠加,产生一个振幅为输入波的振幅之和的波,从而导致相长干涉,如图9.4所示。

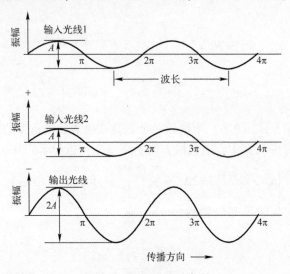

图 9.4 相长干涉示意图

另一种相反的情况即为相消干涉,其原因是一束光线与另一束光线相比处于"异相"状态。若相位差为 π,则输出光的振幅将消失,如图9.5所示。

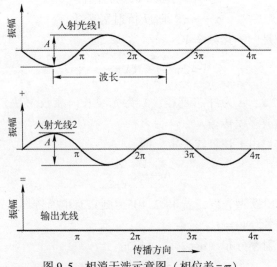

图 9.5 相消干涉示意图(相位差 = π)

### 9.1.6 光的衍射光栅

衍射光栅（简称光栅）是在玻璃甚至塑料材料制成的薄膜上，每毫米宽度刻蚀大量线条形成的。光栅的密度可从 250 线/mm 到 3000 线/mm，甚至更高。光源射出的光通过光栅时，将产生大量独立光源（图9.6），这些光源是同相源或相干源，每个光源都会向各个方向发射光线。将屏幕放置在距离光栅 $D$ 处，屏幕上会出现明亮的条纹（图9.6）。主极大条纹将会出现在中央明纹的两侧。请注意，由于所有的波源在最亮的条纹 $P_0$ 处同相，而并非所有的波在 $P_3$ 处同相，因此 $P_3$ 比 $P_0$ 要暗得多。

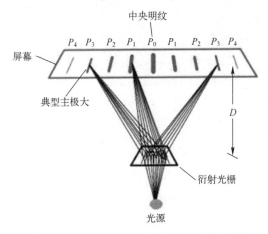

图 9.6 衍射光栅及屏幕示意图（未按比例）

假设两条相邻狭缝之间的距离为 $d$，$N$ 为每毫米光栅上可以刻写的狭缝数量，即

$$d = \frac{1}{N} \quad \text{或} \quad N = \frac{1}{d} \tag{9.11}$$

那么，求衍射光栅主极大的公式为

$$d\sin\theta_n = n\lambda, \quad n = 1, 2, 3, \cdots$$

式中：$\lambda$ 为波长。

## 9.2 光纤的基本特性

光纤被定义为通过非常细的玻璃纤维或塑料纤维进行光传输的科学（图9.7）。由于光纤相较于铜导体有很多优势，目前已被越来越广泛地应用。光纤的优势

包括带宽大（单模光纤的带宽大于 10GHz/km，多模光纤的带宽在 200～600MHz/km，而导电体的带宽为 10～25MHz/km）、抗电磁/射频干扰、易于安装、所需空间小、重量轻（比铜轻 10～15 倍）以及安装成本低。此外，光纤是电绝缘的，不会产生火花或短路，因此没有引起电击的风险。

图 9.7　各种光纤

光纤链路（图 9.8）由 4 个主要部分组成：

（1）发射机：将电信号转换为光信号。它包含一个光源，如发光二极管（Light Emitting Diode, LED）、激光器（通过受激辐射进行光放大）或垂直腔面发射激光器（Vertical Cavity Emitting Laser, VCSEL）。这些光源为红外波段（约 850nm[①]）。

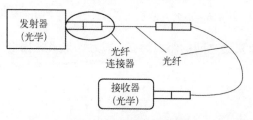

图 9.8　光纤链路示意图

（2）光纤：由 5 层材料组成的特殊设计的细管来传输光（参见图 9.9 中的示意图）。

① 纤芯是光纤的中心管，光线在其中传播。纤芯通常由熔融石英[②]（无定形 $SiO_2$）或掺杂二氧化硅（如 $GeO_2$ 或 $P_2O_5$ 作为掺杂剂）制成。

---

[①]　$1nm = 10^{-9}m$。
[②]　石英：二氧化硅（$SiO_2$），在自然界中称为石英，沙子中含有大量的二氧化硅。

② 包层是包围纤芯的第 2 层。它也由掺杂二氧化硅（用硼或氟作为掺杂剂）制成，但包层的折射率低于纤芯的折射率，这将使得光通过纤芯-包层界面处的全反射而限制在纤芯中。

③ 缓冲涂层的作用是在安装过程中保护光纤，厚度为 900μm。

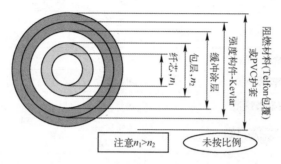

图 9.9　光纤线缆的典型横截面示意图

④ 第 4 层通常用 Kevlar① 制成，为光缆提供所需的强度。

⑤ 外层是由 PVC 制成的护套或涂有 Teflon② 的阻燃材料，以防止起火。

（3）连接器：为确保平稳运行，采用连接器将光缆连接到发射器和接收器，有以下两种连接方法。

① 直接连接：通过电弧或熔接机直接将两根光纤焊接在一起。该方法的优势在于，当光从一根光纤传输到另一根光纤时，几乎没有光损失；但这种方法具有永久性连接、连接处较脆弱和机器（熔接机）成本高等缺点，从而降低了该方法对用户的吸引力。

② 物理连接器：使用两个接头将两根光纤进行物理连接。该连接可以配对或不配对，支持数千次连接和断开。这种方法的缺点是：与熔接机方法相比光损失更高，需要专用工具，且工序更长。

当前有两种类型的连接器（表 9.2）：

a. 物理接触式连接器：使用环氧树脂将一个盖子连接到抛光的光纤上，确保两根光纤对齐。这种连接方式较为稳固，具有低光损耗（约 0.3dB）、易于清洁和成本效益高的特点。

---

① Kevlar（凯夫拉）：由 Stephanie Kwolek 女士在杜邦公司创造，是一种由芳纶（耐热性非常强）合成纤维，以其高比强度而闻名。
② Teflon（特氟龙）：四氟乙烯或聚三氟乙烯合成的含氟聚合物，熔点高（327℃），导热系数为 0.25W/（moK），化学式为 $(C_2F_4)_n$。

b. 扩光连接器：在每根光纤的出口处放置一个透镜，以加宽光线并准直穿过光纤。由于其光损耗更高（0.8~2.5dB），实现过程比前一种类型的连接器更复杂，市场上很少见到。

表9.2 光纤连接器（改编自 L-com® Global Connectivity）

| 连接器 | 类型 | 耦合类型 | 光线数量 | 应用 |
|---|---|---|---|---|
|  | LC | 直接固定 RJ45型 | 1 | 千兆以太网[1]，视频多媒体 |
|  | SC | 直接固定 | 1 | CATV[2]，测试设备 |
|   | ST | 扭转固定 | 1 | LANs[3]，军事 |
|   | FC | 螺丝固定 | 1 | 数据通信，电信 |
|   | MT-RJ | 直接固定 RJ45型 | 2 | 千兆以太网 ATM[4] |
|   | MPO（MTP） | 推/拉 | 6 或 12 | 有源设备收发器[5]，O/E[6]模块互联，QSFP[7]收发器 |

[1] 以太网是局域网中常用的一种计算机网络技术。
[2] CATV（Community Antenna Television）：社区公共电视天线系统（国内一般指广电有线电视系统），一种有线电视。
[3] LANs 代表局域网，在特定的位置由一系列计算机连接在一起形成网络。
[4] ATM 代表异步传输模式。
[5] 收发器是一种既能发送又能接收的设备。
[6] O/E 代表光电。
[7] QSFP 表示四通道 SFP 接口，用于硬件与光缆的接口。

（4）接收器：光纤链路的最后一部分，使用光电二极管将光信号转换为电信号。光电二极管可以是 PIN 型（Positive Intrinsic Negative，正-本征-负）或 APD 型（Avalanche Photo Diode，雪崩光电二极管）。

光纤另一个重要特性体现在用于制造光纤的玻璃类型。在当今的高速网络中，渐变折射率多模光纤（纤芯的折射率从纤芯向包层逐渐降低，通常利用抛物线定律产生光沿光纤传播时的"弯曲"）或阶跃折射率多模、单模光纤（由于材料折射率 $n$ 的阶跃，使得光在包层-纤芯界面发生反射）的使用能够改善光的长距离传输。多模光纤具有更大的纤芯，通常用于建筑物内的短距离光信号传输。单模光纤的纤芯较小，通常用于建筑物之间的远距离光信号传输。如图 9.10 所示，阶跃折射率单模（图 9.10（a））光纤中光信号保持不变，而阶跃折射率多模光纤（图 9.10（b））的输出信号与输入信号相比有相当大的衰减和较大的色散。使用渐变折射率多模光纤（图 9.10（c））能够降低色散，同时改善输出信号（减少衰减）。

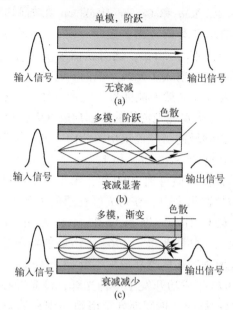

图 9.10　玻璃类型对光纤性能的影响示意图（改编自 L-com® Global Connectivity）

## 9.3　光纤-历史视角

从历史的角度来了解光纤的发展是很有启发性的。人们一致认为，由 Alexander Graham Bell 及其共同发明人 Charles Summer Tainter[1] 于 1880 年发明的一种能够通过光线传输声音的设备——光电话，这是光纤通信的起点。虽然 Bell 和 Tainter 于 1880 年 6 月 3 日进行的测试因大气介质而失败，但它可被视为光纤应用于通信的起点。

之后，英国伦敦帝国理工学院的 Harold Hopkins 和 Narinder Singh Kapany① （后者也被称为"光纤之父"）证明卷起来的玻璃纤维也可以传输光，这与之前"光只能在直介质中传播"的假设大相径庭。继荷兰科学家 Van Heel 开发了光学包层之后②，Kapany 于 1960 年 11 月在《科学美国人》发表的一篇名为

---

① Hopkins H, Kapany, N. S. A flexible fiberscope, using static scanning [J]. Nature, 1954, 173: 39–41.
② Simons, c. a. j, A. C. S. van Heel: teacher and inspirator of technical optics, Proc. SPIE 3190, Fifth International Topical Meeting on Education and Training in Optics, (8 December 1997). https://doi.org/10.1117/12.294379.

《光纤》的文章中首次提出了"光纤"的概念[2]。

1966 年，Charles K. Kao 和 George Hockham 在英国哈洛的 STC 实验室 (STC Laboratories, STL) 工作时提出了光纤一词。Kao 得出的结论是，玻璃透明度的基本极限是每千米低于 20dB，这对通信来说是可行的。Hockham 经过计算得出，包层光纤不应辐射太多光线。他们一起撰写了一篇论文，首次提出光纤通信的概念[3]。Kao 在文献中被称为"宽带教父""光纤之父"和"光纤通信之父"。他因在光纤中实现光传输的开创性成就而获得 2009 年诺贝尔物理学奖，并因在光纤通信领域的贡献而被伊丽莎白二世授予爵位。

应该注意的是，1958 年 Schawlow 和 Townes[4]发表过一篇论文，这被认为是激光运行的第 1 个证据。但直到 1960 年，第 1 个发光脉泽（更广为人知的名称是激光器）才被建造出来[5]，使得光纤在通信领域的广泛使用成为可能。

值得一提的是，C. H. Townes 与 N. G. Basov 和 A. M. Prokhorov 凭借在量子电子学领域的基础性工作共同获得了 1964 年诺贝尔物理学奖（这些基础工作为开发基于脉泽激光原理的振荡器和放大器奠定了基础）。

康宁玻璃厂于 1970 年成功开发了一种光纤，该光纤在通信中的衰减较低（约为 20dB/km）。与此同时，俄罗斯列宁格勒（现在的圣彼得堡）Ioffe 研究所的研究人员（Zhores Alferov 团队）开发了 GaAs 紧凑型半导体激光器，美国新泽西州贝尔实验室的 Panish 和 Hayashi 也独立地开发了 GaAs 紧凑型半导体激光器，适用于通过光纤线缆远距离传输光。

自 1970 年至今，光纤的传输能力有了大幅提高，如表 9.3 和图 9.11 所示。请注意，图 10.5 中的直线表示比特率-距离积（$BL$）将每年翻倍。

表9.3　不同年代的传输能力

| 代 | 实现年份 | 比 特 率 | 中继器间距① | 工作波长度 |
| --- | --- | --- | --- | --- |
| 第 1 代（渐变光纤） | 1980 | 45Mb/s | 10km | 0.8μm |
| 第 2 代（单模纤维） | 1985 | 100Mb/s~1.7Gb/s | 50km | 1.3μm |
| 第 3 代（单模激光） | 1990 | 10Gb/s | 100km | 1.55μm |
| 第 4 代（光学放大器） | 1996 | 10Tb/s | >10000km | 1.45~1.62μm |
| 第 5 代（拉曼放大） | 2002 | 40~160Gb/s | 24000~35000km | 1.53~1.57μm |

① 中继器：一种无线网络设备，重复无线信号以扩大范围，它并不通过电缆连接到路由器/调制解调器或者用户。

# 第9章 光纤简介

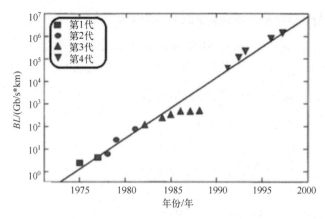

图9.11 前4代比特率-距离积（Bit①×长度）与年份的关系[1]

值得一提的是，无线网络的能力多年来取得了巨大进步，共经历了5个时代：20世纪80年代的第1代（1G）、20世纪90年代的第2代（2G）、21世纪初的第3代（3G）、21世纪10年代的第4代（4G）和21世纪20年代的第5代（5G）。5G预计将显示出增强型移动宽带（Enhanced Mobile Broadband，EMBB）、大规模机器型通信（Massive Maehine Type Communication，MMTC）和超可靠低延迟通信（Ultra-Reliable and Low-Lateney Communication，URLLC）等特性[1]。

当然，时至如今，光纤已经在众多书籍和论文中得到了很好的研究[6-16]。下文中，讨论的范围将仅限于布拉格光栅及其应用。

## 9.4 光纤布拉格光栅

光纤布拉格光栅（Fiber Bragg Grating，FBG）是一种能够精确测量机械应变和温度诱导应变的光纤传感器。该装置的名称来源于布拉格定律②。光纤布拉格光栅的开发归功于Kenneth Hill和他的合作研究者，他们于1978年在加拿大通信研究中心工作时开发了这种光栅[17-18]。除了能感应应变和温度，光纤布拉格光栅还能探测玻璃的压力、应力和折射率。光纤布拉格光栅应用于许多行业中，如航空航天、民用和海事，以及生化、健康和生物医学设备。

---

① Bit：在计算和数字通信中使用的值为0或1的基本信息单位。
② Lawrence Sir William Bragg 于1912年提出了X射线衍射的布拉格定律。该定律用于晶体和薄膜的测定与研究。Lawrence Sir William Bragg 和他的父亲 William Henry Bragg 在1915年获得诺贝尔物理学奖，以表彰他们在用X射线分析晶体结构方面的贡献。

典型的光纤布拉格光栅结构如图9.12所示。

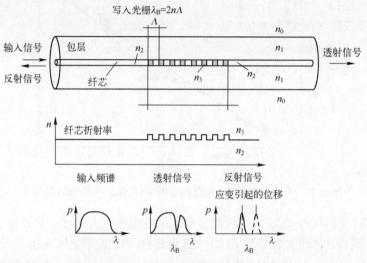

图9.12　FBG透射和反射光谱示意图

将光栅引入纤芯的制造过程称为刻制或写入，制作光栅的方法有以下几种[19]：

（1）干涉法：将纤芯的感光区域暴露在干涉图中，以产生整个光栅。这是通过照射具有光纤布拉格光栅周期的适当掩膜来实现的。请注意，这个流程也可以在无掩膜的情况下完成。

（2）直接逐点刻蚀法：通过控制激光的参数和激光脉冲的非线性吸收，从而使用激光写入光栅。这种方法需要将光纤固定在一个平移平台上，让平台移动，操纵激光器逐点刻蚀以形成光栅。

（3）连续纤芯扫描法：在这种方法中，唯一需要控制的参数与光纤平移台的移动相关。

如文献［20］所述，除了上述均匀光栅，还有啁啾光栅、长周期光栅、倾斜光栅、相移光栅和超结构FBG光栅。

对于具有周期性等间距扰动的简单FBG（图9.12），调制强度仅为正[19]。沿光栅的折射率将以方波的形式变化，其占空比为光栅的半周期$\Lambda$，可表示为

$$\begin{aligned} n_{\text{cora}} &= n_{\text{cora0}}, & 0 < x < \Lambda/2 \\ n_{\text{cora}} &= n_{\text{cora0}} + \Delta n, & \Lambda/2 < x < \Lambda \end{aligned} \quad (9.12)$$

式中：$\Delta n$为调制强度；$n_{\text{core0}}$为未接触纤芯的折射率。

通过光纤纤芯向前传输的部分光被光栅反射回来，其余部分将通过。这种

情况下,光束由两个组成部分,它们以两个相反的方向通过。当由光栅参数决定的特定波长满足布拉格条件时,将产生反射光的相长干涉。反射光的光谱将在布拉格波长 $\lambda_B$ 处呈现峰值,而透射光将呈现陷波滤波器的图案(如图9.12所示的传递函数)。那些波长为非 $\lambda_B$ 的光分量将不受干扰地传播。

布拉格定律由下式表达:

$$2d\sin\theta = n\lambda, \quad n=1,2,3,4,\cdots \quad (9.13)$$

式中:$d$ 为两个相邻晶面之间的距离;$\theta$ 为入射角;$\lambda$ 为波长,据此可以得到 $\lambda_B$ 的表达式;角度 $\theta$ 为90°;$d$ 为相邻峰值之间的距离(9.1节)。对于 $n=1$,得到 $\lambda = 2d$,并且必须考虑光纤的折射率(原始布拉格定律是在真空条件下推导的),于是可得

$$\lambda_B = 2n_{\text{eff}}\Lambda \quad (9.14)$$

因此,布拉格波长是光纤有效折射率 $n_{\text{eff}}$ 和光栅间距 $\Lambda$ 的函数。

如文献[19]所述,运用耦合模理论[21]可以计算反射率和透射率谱,并定义以下参数:

$$\alpha \equiv \frac{2\pi}{\lambda}\overline{\Delta n_{\text{eff}}}, \quad \kappa \equiv \frac{\pi}{\lambda}v\overline{\Delta n_{\text{eff}}}, \quad L \equiv N\Lambda \quad (9.15)$$

式中:$\alpha$ 为直流自耦合系数;$\overline{\Delta n_{\text{eff}}}$ 表示在光栅周期内空间平均的直流折射率变化;$\kappa$ 为交流耦合系数;$v$ 为FBG阶数(均匀FBG的单位);$N$ 为导致反射振幅的周期数。

$$R_{\text{am.}} = -\frac{\kappa\sinh(\sqrt{\kappa^2-\alpha^2}L)}{\alpha^2\sinh(\sqrt{\kappa^2-\alpha^2}L)+\mathrm{i}(\sqrt{\kappa^2-\alpha^2})\cosh(\sqrt{\kappa^2-\alpha^2}L)} \quad (9.16)$$

$$功率 = |R_{\text{am.}}^2| = -\frac{\sinh^2(\sqrt{\kappa^2-\alpha^2}L)}{\cosh^2(\sqrt{\kappa^2-\alpha^2}L)-\frac{\alpha^2}{\kappa^2}} \quad (9.17)$$

半峰宽(Full-Width-Half-Maximum,FWHM)或带宽(图9.13)写为

$$\text{FWHM}_\lambda = \lambda_B\beta\sqrt{\left(\frac{\Delta n}{2n_{c0}}\right)^2 + \left(\frac{1}{N}\right)^2} \quad (9.18)$$

式中:对于高反射率光栅,$\beta \approx 1$;对于弱反射率光栅,$\beta \approx 0.5$。如图9.13所示,光谱有一个中心峰,该中心峰的带宽可根据FBG的给定参数而改变,其两侧被旁瓣簇拥。

根据不同的用途[20-21],可考虑使用能够减少旁瓣的其他光栅面型,如啁啾或相移光栅。另一个有趣的面型称为变迹光栅,它具有沿光栅的非均匀调制强度(如升余弦或高斯面型),该面型被证实能够大幅减小旁瓣的振幅。

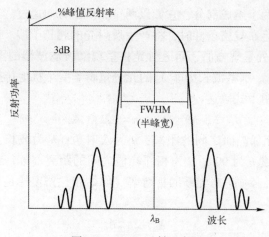

图 9.13　FBG 反射示意图

## 9.5　用作应变传感器的光纤布拉格光栅

见图 9.12，应变或温度变化将导致布拉格波长发生位移，这种位移从偏微分方程（9.15）可得

$$\Delta \lambda_B = 2\left(\Lambda \frac{\partial n_{\text{eff}}}{\partial \varepsilon} + n_{\text{eff}} \frac{\partial \Lambda}{\partial \varepsilon}\right)\Delta \varepsilon + \\ 2\left(\Lambda \frac{\partial n_{\text{eff}}}{\partial T} + n_{\text{eff}} \frac{\partial \Lambda}{\partial T}\right)\Delta T \qquad (9.19) \\ = (\Delta \lambda_B)_{\text{strain}} + (\Delta \lambda_B)_{\text{temp}}$$

式（9.19）左侧的第 1 部分表示由于应变导致的布拉格波长位移，而第 2 部分是由温度变化引起的。

式（9.18）可以用另一种表达式为[19]

$$\frac{\Delta \lambda_B}{\lambda_B} = (1-\rho_E)\varepsilon + (\beta + \chi)\Delta T \qquad (9.20)$$

式中：$\rho_E$ 为源自普克尔应力光学张量的光弹性系数[22]；$\varepsilon$ 为光纤方向上的应变；$\beta$ 为光纤材料（如二氧化硅）的热膨胀；$\chi$ 为材料的热光系数[23]（折射率随温度的变化率，$dn_{\text{eff}}/dT$）。文献［23］给出了与式（10.9）中的常数相关数据（掺锗石英光纤芯）：$\rho_E = 0.22$、$\beta = 0.55 \times 10^{-6}/℃$ 以及 $\chi = 8.6 \times 10^{-6}/℃$，于是有 $\Delta \lambda_B = \Delta \varepsilon = 1.2\text{pm}/\mu\text{s}$ 以及 $\Delta \lambda_B/\Delta T = 14.18\text{pm}/℃$。注意，在文献［24］

中，作者为应变-布拉格位移关系给出了一个不同的值：$\Delta\lambda_B/\lambda_B = 0.79\varepsilon$。

需要注意的是，当使用光纤布拉格光栅作为应变传感器时，应变输出通常包括附加的温度应变，需要进行补偿以消除温度的影响，使得输出仅包含机械应变。文献［23］中介绍了一种有趣的方法：将两个光纤布拉格光栅平行安装（一个黏合，另一个未黏合），使之提供温度诱导应变。该应变从第1个光纤布拉格光栅的读数中减去，以产生在结构特定部分（FBG附着在其上）的净应变。文献［25-32］是关于使用FBG感测应变的研究论文。

## 9.6　光纤布拉格光栅询问器

对FBG收集的信息进行解码、使用的过程称为解调或询问。基于FBG的传感器有几种不同的技术[19-20]，如体光学、滤波（具有被动边缘或主动带通滤波），以及干涉测量、激光传感等。据文献［19］所述，最常用的解调技术是滤波法和干涉测量法。滤波法采用收集FBG反射光谱光信号的光电二极管，将FBG输出的波长位移转换为电信号的变化（不使用传统的光谱分析仪OSA，因为该分析仪运行速度缓慢且使用成本高[19]）。干涉测量法将光学相位差分配给布拉格波长位移。通常使用非平衡不对称马赫-曾德尔干涉仪接收传感光栅的反射光谱，并将其传输至位于干涉仪一个臂上的相位调制器[24]。只要知道干涉仪相位及其相关强度输出，通过测量强度的变化就可以转化为相位差。然后，通过对光电二极管电压进行适当的信号处理，获得所需的波长位移，从而得到在$\text{pm}/\sqrt{\text{Hz}}$范围内的分辨率[19]。

更多关于解调仪的信息可在文献［19-20］或有关光纤的书籍（如文献［6-16］）中找到。

## 参 考 文 献

［1］　Liu, X., Evolution of fiber-optic transmission and networking toward the 5G era, iScience 22, December 20, 2019, 489-506.

［2］　Kapany, N. S., Fiber Optics Principles and Applications, Academic Press, 1967, 429.

［3］　Kao, K. C. and Hockham, G. A., Dielectric fibre surface waveguides for optical frequencies, Proceedings of the IEEE 113, 1966, 1151-1158.

［4］　Schawlow, A. L. and Townes, C. H., Infrared and optical masers, Physical Review 112 (6), 1958, 1940-1949.

［5］　Maiman, T., Stimulated optical radiation in ruby, Nature 187, 1960, 493-494.

[6] Saleh, B. E. A. and Teich, M. C., Fundamentals of Photonics, Chapter 8: Fiber Optics, John Wiley & Sons, Inc., 1991, 947.

[7] Ansari, F., (ed.), Applications of Fiber Optic Sensors in Engineering Mechanics, American Society of Civil Engineers, 1993, 230.

[8] Reese, R. T. and Kawahara, W. A., editors, Handbook on Structural Testing, Society for Experimental Mechanics (US), Fairmont Press, 1993, 402.

[9] Udd, E., editor, Fiber Optic Smart Structures, John Wiley & Sons Inc., 1995, 688.

[10] Culshaw, B., Smart Structures and Materials, Artech House, 1996, 207.

[11] Van Steenkiste, R. J. and Springer, G. S., Strain and Temperature Measurement with Fiber Optic Sensors, Technomic Publishing Co, 1997, 294.

[12] Ansari, F., editor, Fiber Optic Sensors for Construction Materials and Bridges, Technomic Publishing Co., 1998, 267.

[13] Ghatak, A. and Thyagarajan, K., Introduction to Fiber Optics, Cambridge University Press, 1998, 581.

[14] Measures, R. M., Structural Monitoring with Fiber Optic Technology, Academic Press, 2001, 716.

[15] Lopez-Higuera, J. M. editor, Handbook of Optical Fibre Sensing Technology, John Wiley & Sons Inc., 2002, 828.

[16] Al-Amri, M., El-Gomati, M. and Zubairy, M. S., Optics in our Time, Chapter 4: Applications, Springer Open, 2016, 504.

[17] Hill, K. O., Fujii, Y., Johnson, D. C. A. and Kawasaki, B. S., Photosensitivity in optical fiber waveguides: application to reflection fiber fabrication, Applied Physics Letters 32 (10), 1978, 647–649.

[18] Hill, K. O. and Meltz, G., Bragg grating technology fundamentals and overview, Journal of Lightwave Technology 15 (8), 1997, 1263–1276.

[19] Campanella, C. E., Cuccovillo, A., Campanella, C., Yurt, A. and Passaro, V. M. N., Fibre Bragg grating based strain sensors: review of technology and applications, Sensors 18, 2018, Id paper 3115, 27. doi: 10.3390/s18093115.

[20] Sahota, J. K., Gupta, N. and Dhawan, D., Fiber Bragg grating sensors for monitoring of physical parameters: a comprehensive review, Optical Engineering 59 (6), June 2020, Id paper 060901.

[21] Pierce, J. R., Coupling modes of propagation, Journal of Applied Physics 25, 1954, 179–183.

[22] Nye, J. F., Physical properties of Crystals: Their Representation by Tensors and Matrices, Oxford University Press, 1957, 324.

[23] Werneck, M. M., Allil, R. C. S. B., Ribeiro, B. A. and De Nazaré, F. V. B., A guide to fiber Bragg grating sensors, INTECH Open Science book, Chapter 1, 2013, 25. http://dx.doi.org/10.5772/54682.

[24] Gagliardi, G., Salza, M., Avino, S., Ferraro, P. and De Natale, P., Probing the ultimate limit of fiber-optic strain sensing, Science 330, 19th November. 2010, 1081–1084.

[25] Fidanboylo, K. and Efendioğlu, H. S., Fiber optic sensors and their applications, 5th International Advanced Technologies Symposium (IATS'09), May 13–15, 2009, Karabuk, Turkey.

[26] Pevec, S. and Donlagić, D., Multiparameter fiber-optic sensors: a review, Optical Engineering 58

## 第 9 章 光纤简介

(7), Id paper 072009, 2019, 26. doi: 10.1117/1.0E.58.7.072009.

[27] Guemes, A., Fernandez-Lopez, A. and Soller, B. J., Optical fiber distributed sensing-physical principles and applications, Structural Health Monitoring 9 (3), 2010, 233-245. doi: 10.1177/1475921710365263.

[28] Sánchez, D. M., Gresil, M. and Soutis, C., Distributed internal strain measurement during composite manufacturing using optical fibre sensors, Composite Science and Technology 120, 2015, 49-57. doi: 10.1016/j.compscitech.2015.09.023.

[29] Chandarana, N., Sánchez, D. M., Soutis, C. and Gresil, M., Early damage detection in composites by distributed strain and acoustic event monitoring, Procedia Engineering 188, 2017, 88-95. doi: 10.1016/j.proeng.2017.04.515.

[30] Caucheteur, C., Guo, T. and Albert, J., Polarization-assisted fiber Bragg grating sensors, Tutorial and review, Journal of Lightwave Technology 35 (16), 2027, 3311-3322.

[31] Sano, Y. and Yoshino, T., Fast optical wavelength interrogator employing arrayed waveguide grating for distributed fiber Bragg grating sensors, Journal of Lightwave Technology 21 (1), 2003, 132-139.

[32] Freydin, M., Ratner, M. K. and Raveh, E. D., Fiber-optics-based aeroelastic shape sensing, AIAA Journal 57 (12), 2019, 5094-5103.

# 第10章 其他主题

## 10.1 形状记忆合金丝增强板材的抗弯性能

### 10.1.1 形状记忆合金特性用于驱动技术：受限恢复现象

如第4章所述，形状记忆效应是由形状记忆合金的载荷引起的，从而产生塑性残余应变 $S_{res}$，如图10.1所示。残余应变是卸载后材料马氏体分数 ($\xi$) 的函数。当 $\xi=1$ 时，残余应变等于最大可恢复应变 $S_L$（图10.1（a））。注意，当加热温度超过奥氏体起始温度 ($A_s$) 时，由于应力诱发马氏体转变为奥氏体，将导致应变恢复。参考图10.1（a），如果应力足够高，足以诱发完全的马氏体相变，那么 $S_{res}$ 将等于最大可恢复应变 ($S_L$)；若应力较小，则马氏体分数为

$$\xi = \frac{S_{res}}{S_L} \tag{10.1}$$

当加热材料以恢复塑性（诱导）应变时，可以检查3个基本的力学边界条件。第1个是自由应变恢复，当其中一个力学边界不/受约束时，将发生自由应变恢复（图10.1（b））。在这种情况下，形状记忆合金将在加热过程中随着温度的变化而收缩。当材料加热时，若两个边界条件都被限制为轴向运动，将产生零应变，则会发生受限恢复（图10.1（c）），阻止材料收缩而产生回复应力 $\sigma^r$，$\sigma^r$ 在形状记忆合金中为加热过程中温度的函数。然而，如图10.1（d）所示，形状记忆合金丝的端部不可移动且连接有一个弹簧，当加热到 $M \to A$（马氏体到奥氏体）转变温度以上时，形状记忆合金丝恢复应变并趋于收缩，这会导致线性弹簧的伸长以及形状记忆合金丝中的应力相应增加。为了量化产生的应力，第4章中介绍的 Liang 和 Rogers 模型将用于3个基本边界条件。如第4章所述，假设温度诱发马氏体相变和应力诱发马氏体相变之间没有区别，本构关系为

$$\sigma - \sigma_0 = Y(\xi)(S - S_L \xi) - Y(\xi)(S_0 - S_L \xi_0) \tag{10.2}$$

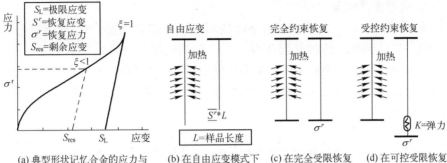

图 10.1 形状记忆效应

**1. 自由应变恢复**

对于自由恢复的情况,假设形状记忆合金已经过加载和卸载,从而导致残余应变 $S_{res}$,见图 10.1(a)。

假设应力为零(初始应力和应变也为零),从式(10.2)可得

$$S = S_L \xi \tag{10.3}$$

若温度高于马氏体起始温度 $M_s$,初始马氏体分数 $\xi_0 = S_{res} = S_L$,并将式(4.9)代入式(10.3)得出 Liang 和 Rogers 模型,则第 4 章中将有以下表达式和其余变量的定义:

$$S^r = \frac{S_{res}}{2}\{\cos[\alpha_A(T-A_s)]+1\} \tag{10.4}$$

式中:

$$\alpha_A = \frac{\pi}{A_f - A_s}$$

**2. 受限恢复**

假设 $\sigma_0 = 0$、$S_0 = 0$、$\xi_0 = 1$,忽略热膨胀,则式(10.2)如下:

$$\sigma = Y(\xi)(S - S_L\xi) + Y(\xi)(S_L) = Y(\xi)[S - S_L(\xi-1)] \tag{10.5}$$

受限恢复应力 $\sigma^r$ 可以写成

$$\sigma^r = \frac{Y(\xi)S^r}{2}\left\{1 - \cos\left[\alpha_A(T-A_s) - \frac{\alpha_A}{C_A}\sigma^r\right]\right\} \tag{10.6}$$

式中:$Y(\xi)$ 为式(10.3)中定义的马氏体分数的线性函数。应注意,恢复应力 $\sigma^r$ 出现在方程的两侧,因此式(10.6)是一个迭代方程。求解式(10.6)的最简单方法是选择一个温度并迭代,以确定与假设温度相对应的应力。

### 3. 可控受限恢复

见图10.1（d），形状记忆合金丝的端部不可移动且连接有一个弹簧，当加热到马氏体到奥氏体转变温度以上时，形状记忆合金丝恢复应变并趋于收缩，这会导致线性弹簧的伸长以及形状记忆合金丝中的应力相应增加。形状记忆合金丝与线性弹簧之间的相对位移关系与合金丝的应力及应变有关。因此，应变表达式如下：

$$S = -\frac{\Delta L}{L} = -\frac{F^r}{kL} = -\frac{\sigma^r A}{kL} \tag{10.7}$$

式中：$k$ 为弹簧系数；$L$ 为丝的长度；$A$ 为其横截面积。

受控可恢复应力可表示为

$$\sigma^r = S_L Y(\xi) \frac{1-\xi}{\left[1+\frac{AY(\xi)}{kL}\right]} = \frac{Y(\xi)S^r}{2\left[1+\frac{AY(\xi)}{kL}\right]} \left\{1-\cos\left[\alpha_A(T-A_S) - \frac{\alpha_A}{C_A}\sigma^r\right]\right\} \tag{10.8}$$

对于无限刚度（$k \to \infty$），式（10.8）将收敛于式（10.6）。应注意，为了快速计算预应力，可以假设应力与应变的函数是线性的，如图10.2所示。

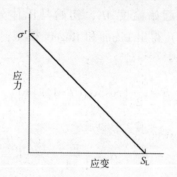

图10.2 可恢复应力与应变线性关系示意图

## 10.1.2 实验结果

在本书作者 Haim Abramovich 教授的带领下，以色列理工学院航空航天工程学院结构实验室（以色列，海法）开展了大量的试验，旨在通过实验证明镍钛诺合金丝增强受载板弯曲性能。

图10.3展示了用于试验的各种类型镍钛诺合金结构，如棒、弹簧和金属丝。形状记忆合金的表征分两个阶段进行。

# 第10章 其他主题

(a) 镍钛诺合金棒　　　　　(b) 镍钛诺合金弹簧　(c) 镍钛诺合金线

图 10.3　用于试验的 3 种镍钛诺合金结构：棒、弹簧和金属丝

第 1 阶段（图 10.4）包括形状记忆合金丝（棒或弹簧）在其塑性区域轴向拉伸，产生非常高的塑性应变（图 10.4（a）和表 10.1）。图 10.5（a）~（d）中展示了监测行为的典型曲线图（注意，这些曲线图没有显示返回零负载的情况）。该阶段结束于将轴向载荷减至零，并在试样中保留塑性应变。

第 2 阶段见图 10.4（b），其中，具有塑性变形的镍钛诺合金试样在两端夹紧，连接两根电线（图 10.4（b））使其加热温度超过转变温度。由于两端阻止了金属丝的膨胀，金属丝内产生了相当大的应力（图 10.4（c））以恢复塑性应变，如表 10.1 所列。典型结果如图 10.6（a）~（d）所示。应注意，在载荷达到峰值后，停止加热以降低载荷。如图 10.6（a）所示，再次加热可恢复载荷峰值。

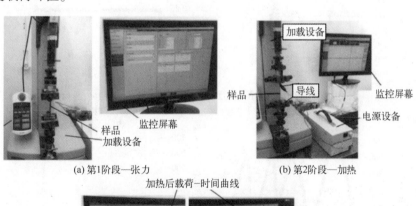

(a) 第1阶段—张力　　　　　　　　(b) 第2阶段—加热

(c) 加热后载荷与时间的典型关系

图 10.4　试验装置

407

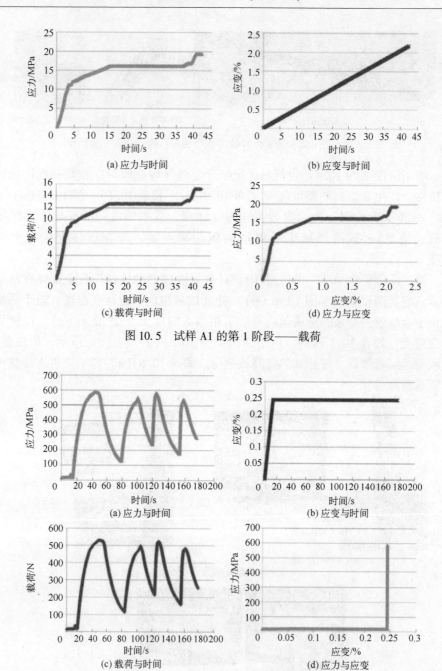

图 10.5 试样 A1 的第 1 阶段——载荷

图 10.6 试样 A1@第 2 阶段——载荷

## 第 10 章 其他主题

表 10.1 各类镍钛诺合金结构的实验测试数据

| 试样# | 横截面积/mm² | 转变温度/℃ | 最大应变/% | 生成的力/N |
|---|---|---|---|---|
| A1（丝线） | 0.900 | 44 | 2.1583 | 525.0000 |
| B1（丝线） | 0.785 | 75 | 0.7130 | 616.9300 |
| B2（丝线） | 0.785 | 75 | 0.5500 | 611.7856 |
| C1（丝线） | 0.900 | 37 | 7.7430 | 591.0000 |
| C2（丝线） | 0.900 | 37 | 8.4680 | 706.2384 |
| C3（丝线） | 0.900 | 37 | 7.6130 | 651.5474 |
| D1（棒） | 0.196 | 37 | 13.000 | 12.3000 |
| E1（弹簧） | — | 37 | 3.8500 | 15.8000 |

为了展示镍钛诺合金丝增强受载板的抗弯性能，设计了一个演示器，如图 10.7 和图 10.8 所示。它由 9 根直径为 0.5mm 的镍钛诺合金丝组成，能够在试验前使用扳手将其拉伸至预定载荷。金属丝连接着两块板，下板由其自重加载。加热方式有两种：使用电流或通过鼓风机加热周围空气，提供所需的平衡力来抵消自重，从而使底板保持平整。在试验过程中，使用粘在底板上的两个应变计监测底板的面外位移。图 10.9 展示了演示器的其他细节。

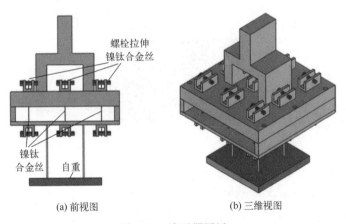

(a) 前视图　　　　　　(b) 三维视图

图 10.7 演示器图纸

在验收试验之前，使用 ANSYS 有限元分析软件对其进行有限元模拟。图 10.10（a）显示了由于自重载荷引起的底板偏转，图 10.10（b）展示了由于镍钛诺合金丝加热引起的合力。图 10.10（a）中的最大挠度为 0.023188mm，图 10.10（b）中的最大挠度为 0.022573mm。这两个数值非常接近，即镍钛诺合金丝的激活能够平衡由自重引起的挠度。

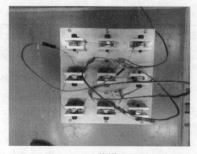 

(a) 俯视图　　　　　　　　　　　　(b) 前视图

图 10.8　实现的演示器

 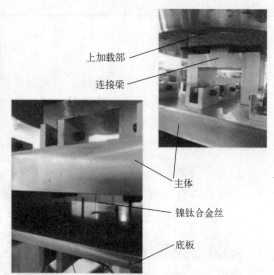

底板　加载电缆　自重

上加载部
连接梁
主体
镍钛合金丝
底板

图 10.9　演示器的其他细节

样机试验主要采用鼓风机进行，平均温度为 45℃，镍钛诺丝的转化温度为 37℃。将合金丝加热到 45℃，便可以转变为奥氏体相。典型的测试结果如图 10.11 所示，由于镍钛诺合金丝的激活，最大应变+20 微应变（由 1 号应变计读取）被消除，导致获得-30 微应变（平均值）。由于镍钛诺合金丝的激活恢复，停止加热会再次产生+40~80 微应变的机械峰值，这些峰值再次被消除也将导致获得-30 微应变。2 号应变计中也可以看到相同的趋势。

根据上述结果，可以总结出以下结论：

（1）使用各种镍钛诺合金丝和棒拉伸试样，利用约束状态下的形状记忆效应可获得相当大的载荷。

## 第 10 章　其他主题

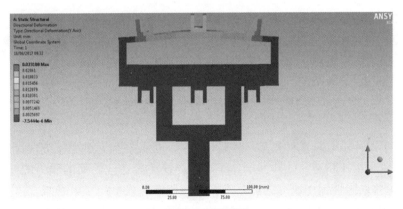

(a) 静态自重载荷下

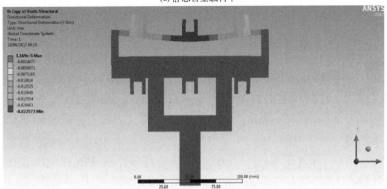

(b) 镍钛诺合金丝的激活

图 10.10　ANSYS 有限元模拟

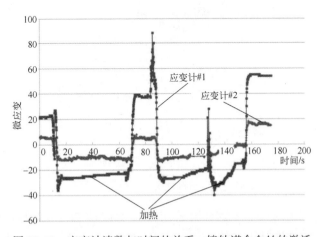

图 10.11　应变计读数与时间的关系：镍钛诺合金丝的激活

(2) 塑性变形越大，引起的载荷越高。

(3) 在演示器上进行的测试展示了较好的结果。镍钛诺合金丝的激活成功地平衡了由于机械自重引起的挠度，使底板的净挠度非常小。

## 10.2　压电纤维复合材料

由于单片压电陶瓷材料是一种脆性材料，因此在搬运和黏接或在使用过程中容易发生意外断裂。此外，单片压电陶瓷换能器适应曲面的能力较差（1.1节）。这些局限性促使研究人员开发新的制造方法以生产压电纤维。需要注意的是，除了陶瓷纤维，市场上也有聚偏二氟乙烯（PVDF）共聚物纤维和超细纤维。文献［1-19］是关于各种类型纤维为传感器或致动器的典型研究。

### 10.2.1　PZT 压电纤维

将直径小于 250μm 的 PZT 压电纤维插入聚合物基体中并配备叉指形电极，在文献中称为活性纤维复合材料（Active Fiber Composites，AFC）或长纤维复合材料（Macrofiber Composites，MFC），如图 10.12 所示。聚合物基体中 PZT 压电纤维的另一种结构称为 1-3 压电复合材料，如图 10.13 所示。注意，2000 年左右，NASA[7]开发了 MFC，其中 PZT 压电纤维是从常规压电陶瓷板上切下的矩形纤维。

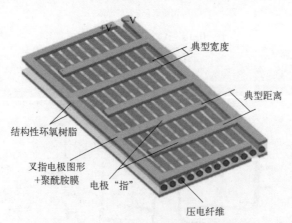

图 10.12　活性纤维复合材料示意图

目前，德国 Dresden 智能材料公司[7]生产制造 MFC。Schönecker 介绍了 PZT 压电纤维的其他制造方法[6]。

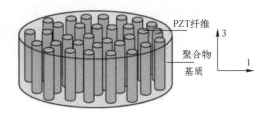

图 10.13　1-3 压电复合材料示意图

MFC 和 1-3 压电复合结构的极化是有挑战性的。由于压电纤维的几何尺寸（其中一个尺寸在纵向上较大，另两个尺寸在纤维截面上较小），使得叉指电极（Inter digitated Electrodes，IDE）极化并产生纵向电场（图 10.14），虽然电场沿纤维方向不是恒定的，但它为传感器提供 $d_{33}$ 常数。注意，在电极之间会产生"死区"，但通过精细设计电极的宽度和两个相邻电极之间的距离，可以将其最小化[3]。

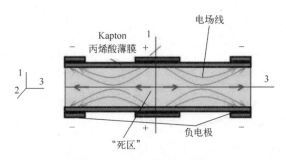

图 10.14　MFC 模块示意图

由于纤维嵌入聚合物基体中，MFC 的解析混合规则应根据成分（基体和 PZT 压电纤维）的材料特性和纤维的体积分数 $V_f$ 提供活性层的均质性。表 10.2 所示为 $d_{31}$ 和 $d_{33}$ MFC 单位体积的等效特性。

表 10.2　$d_{31}$ 和 $d_{33}$ MFC 单位体积的混合规则

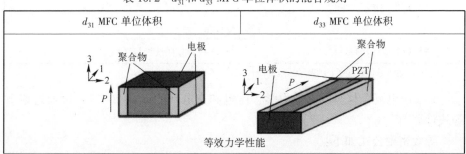

续表

| $d_{31}$ MFC 单位体积 | $d_{33}$ MFC 单位体积 |
|---|---|
| $E_1 = V_f E_1^p + (1-V_f) E_1^m$ | $E_3 = V_f E_3^p + (1-V_f) E_3^m$ |
| $E_2 = \dfrac{E_2^m}{V_f \dfrac{E_2^m}{E_2^p} + (1-V_f)}$ | $E_2 = \dfrac{E_2^m}{V_f \dfrac{E_2^m}{E_2^p} + (1-V_f)}$ |
| $U_{12} = V_f u_{12}^p + (1-V_f) u_{12}^m$ | $U_{32} = V_f U_{32}^p + (1-V_f) U_{32}^m$ |
| $U_{21} = U_{12} \dfrac{E_2}{E_1}$ | $U_{23} = U_{32} \dfrac{E_2}{E_3}$ |
| $G_{12} \equiv G_{21} = \dfrac{G_{12}^m}{V_f \dfrac{G_{12}^m}{G_{12}^p} + (1-V_f)}$ | $G_{32} \equiv G_{23} = \dfrac{G_{32}^m}{V_f \dfrac{G_{32}^m}{G_{32}^p} + (1-V_f)}$ |
| $= \dfrac{G_{21}^m}{V_f \dfrac{G_{21}^m}{G_{21}^p} + (1-V_f)}$ | $= \dfrac{G_{23}^m}{V_f \dfrac{G_{23}^m}{G_{23}^p} + (1-V_f)}$ |
| $G_{13} \equiv G_{31} = V_f G_{13}^p + (1-V_f) G_{13}^m$ $= V_f G_{31}^p + (1-V_f) G_{31}^m$ | $G_{31} \equiv G_{13} = V_f G_{31}^p + (1-V_f) G_{31}^m$ $= V_f G_{13}^p + (1-V_f) G_{13}^m$ |
| $G_{23} \equiv G_{32} = \dfrac{G_{23}^m}{V_f \dfrac{G_{23}^m}{G_{23}^p} + (1-V_f)}$ | $G_{21} \equiv G_{12} = \dfrac{G_{21}^m}{V_f \dfrac{G_{21}^m}{G_{21}^p} + (1-V_f)}$ |
| $= \dfrac{G_{32}^m}{V_f \dfrac{G_{32}^m}{G_{32}^p} + (1-V_f)}$ | $= \dfrac{G_{12}^m}{V_f \dfrac{G_{12}^m}{G_{12}^p} + (1-V_f)}$ |
| 等效压电性能 ||
| $d_{31} = \dfrac{V_f d_{31}^p E_1^p}{E_1}$ | $d_{33} = \dfrac{V_f d_{33}^p E_3^p}{E_3}$ |
| $d_{32} = -d_{31} U_{12} + V_f d_{31}^p (1 + u_{12}^p)$ | $d_{32} = -d_{33} U_{32} + V_f (d_{32}^p + d_{33}^p u_{32}^p)$ |
| 等效介电性能 ||
| $\bar{\varepsilon}_{33}^2 = V_f (\bar{\varepsilon}_{33}^2)^p + (1-V_f)(\bar{\varepsilon}_{33}^2)^m$ | $\bar{\varepsilon}_{33}^2 = V_f (\bar{\varepsilon}_{33}^2)^p + (1-V_f)(\bar{\varepsilon}_{33}^2)^m$ |

注：( )$^m$ 表示基体性能，( )$^p$ 表示压电纤维性能。

1-3 压电复合材料的等效力学、压电和介电性能（图 10.13）可通过以下等式计算[6,8]。

等效密度公式如下：

$$\rho_{\text{eq.}} = V_{\text{f}}\rho^{\text{f}} + (1-V_{\text{f}})\rho^{\text{m}} \qquad (10.9)$$

式中:$V_{\text{f}}$ 为压电体积分数。等效径向介电常数具有以下表达式:

$$(\bar{\varepsilon}_{\text{r}})_{\text{eq.}} = V_{\text{f}}(\bar{\varepsilon}_{\text{r}})^{\text{f}} + (1-V_{\text{f}})(\bar{\varepsilon}_{\text{r}})^{\text{m}} \qquad (10.10)$$

极化后恒定应力下等效介电常数的表达式如下:

$$(\bar{\varepsilon}_{33}^{\sigma})_{\text{eq.}} = V_{\text{f}}\left[(\bar{\varepsilon}_{33}^{\sigma})^{\text{f}} - \frac{(d_{33}^{\text{f}})^2}{(1-V_{\text{f}})(s_{33}^{E})^{\text{f}} + V_{\text{f}}(s_{33})^{\text{m}}}\right] + (1-V_{\text{f}})(\bar{\varepsilon}_{11})^{\text{m}} \qquad (10.11)$$

式中:$(s_{33}^{E})^{\text{f}}$ 为压电纤维在恒定电场($E$)下的弹性柔度;$(s_{33})^{\text{m}}$ 为基体(压电纤维周围的聚合物)的弹性柔度。

等效的压电电荷常数 $d_{33}$ 可以写为

$$(d_{33})_{\text{eq.}} = \frac{(d_{33})^{\text{f}}}{1 + \frac{(1-V_{\text{f}})(s_{33}^{E})^{\text{f}}}{V_{\text{f}}(s_{33})^{\text{m}}}} \qquad (10.12)$$

厚度模式下的等效机电耦合系数如下:

$$(k_{\text{t}})_{\text{eq.}} = \frac{(e_{33})^{\text{f}}}{\sqrt{(c_{33}^{D})_{\text{eq.}}(\bar{\varepsilon}_{33}^{S})_{\text{eq.}}}} \qquad (10.13)$$

等效声阻抗有以下表达式:

$$(Z_{\text{acoustic}})_{\text{eq.}} = \sqrt{(c_{33}^{D})_{\text{eq.}}\rho_{\text{eq.}}} \qquad (10.14)$$

式中:

$$\begin{cases} (e_{33})_{\text{eq.}} = V_{\text{f}}\left\{(e_{33})^{\text{f}} - \frac{2(1-V_{\text{f}})(e_{31})^{\text{f}}[(c_{13}^{E})^{\text{f}} - (c_{12})^{\text{m}}]}{V_{\text{f}}[(c_{11})^{\text{m}} + (c_{12})^{\text{m}}] + (1-V_{\text{f}})[(c_{11}^{E})^{\text{f}} + (c_{12}^{E})^{\text{f}}]}\right\} \\ (c_{33}^{E})_{\text{eq.}} = V_{\text{f}}\left\{(c_{33}^{E})^{\text{f}} - \frac{2(1-V_{\text{f}})[(c_{13}^{E})^{\text{f}} - (c_{12})^{\text{m}}]^2}{V_{\text{f}}[(c_{11})^{\text{m}} + (c_{12})^{\text{m}}] + (1-V_{\text{f}})[(c_{11}^{E})^{\text{f}} + (c_{12}^{E})^{\text{f}}]}\right\} + \\ \qquad (1-V_{\text{f}})(c_{11})^{\text{m}} \\ (\bar{\varepsilon}_{33}^{S})_{\text{eq.}} = V_{\text{f}}\left\{(\bar{\varepsilon}_{33}^{S})^{\text{f}} - \frac{2(1-V_{\text{f}})[(e_{31})^{\text{f}}]^2}{V_{\text{f}}[(c_{11})^{\text{m}} + (c_{12})^{\text{m}}] + (1-V_{\text{f}})[(c_{11}^{E})^{\text{f}} + (c_{12}^{E})^{\text{f}}]}\right\} + \\ \qquad (1-V_{\text{f}})(\bar{\varepsilon}_{11})^{\text{m}} \\ (c_{33}^{D})_{\text{eq.}} = (c_{33}^{E})_{\text{eq.}} + \frac{[(e_{33})_{\text{eq.}}]^2}{(\bar{\varepsilon}_{33}^{S})_{\text{eq.}}} \end{cases}$$

$$(10.15)$$

式中:$e_{ij}$、$s_{ij}$ 和 $c_{ij}$ 分别为压电应变常数、弹性柔度和弹性刚度。注意,表达式 $\bar{\varepsilon}_{33}^{S}$ 为恒应变极化后的介电常数。

## 10.2.2 PVDF 共聚物纤维

一般来说，压电纤维有熔体纺丝法和静电纺丝法两种制造方法。纺丝工艺是一种制造纤维聚合物（如 PVDF）的方法。它使用一种特殊的挤压形式，通过喷丝头生产多条连续的长丝。注意，为了实现聚合物纺丝，必须通过加热或化学反应将其溶解在溶剂中，从而转化为流体状态。

熔体纺丝工艺使用熔融和拉伸的 PVDF，将其加热以产生合适的黏度来生产纤维。熔化的聚合物穿过有小孔的喷丝头，每个小孔将产生一根单独的纤维，而喷丝头的孔数决定了纱线中纤维的数量。要将 PVDF 纱线极化，需要高电压（高达 20kV）。极化的典型条件为 80~90℃，快辊和慢辊之间的拉伸比为 5∶1，从而导致纱线拉伸[14]。PVDF 熔体纺丝工艺示意图如图 10.15 所示。

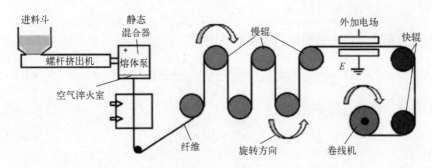

图 10.15　PVDF 纤维连续熔体纺丝工艺示意图[14]

静电纺丝工艺是一种利用电场将 PVDF 溶液或聚合物熔体的带电丝拉至直径达 100nm 的纤维生产方法。

众所周知，当向液滴施加高电压时，带电液滴导致静电排斥，该排斥作用于液滴表面张力，从而产生拉伸体。将电压增加到临界点，液体流将从表面喷发（喷发点在文献中称为泰勒锥，如图 10.16 所示）。

对于具有足够高分子内聚力的液体，流体不会被破坏，将形成带电液体射流。液体射流在飞行中会变干，随着电荷迁移到纤维表面，电流的类型将从欧姆电流变为对流电流。纤维弯曲处的静电斥力引起的鞭动过程（图 10.16）导致射流伸长，直至其沉积在接地收集器上。

PVDF 纤维可应用于各个方面，如可穿戴纺织品、传感器或采集器。应通过实验测量 PVDF 纤维的压电性能，以计算其作为传感器或致动器的性能[13-19]。

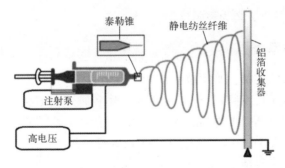

图 10.16　PVDF 纤维静电纺丝生产示意图

## 10.3　声能采集

声能是噪声产生的能量，它是一种寄生的环境能源，可以转化为有用的电能。常见的噪声源包括车辆（卡车、轿车和摩托车）、飞机、发电厂、机器、火车和扬声器。有趣的是，声能最终会以热能的形式传播和消散。由于人类可听到的频率范围为 20Hz~20kHz，应将该范围内的噪声转换为电能，以减少噪声造成的环境污染。很多文献已经对这一课题进行了讨论，典型研究见文献 [20-29]。

### 10.3.1　声学基础

声音在空气中以纵向波形传播，对于平面波情况，可以将其写成

$$\frac{\partial^2 p}{\partial x^2}=\frac{1}{c^2}\frac{\partial^2 p}{\partial t^2} \tag{10.16}$$

式中：$p$ 为声压；$c$ 为空气中的传播速度；$x$ 和 $t$ 分别表示一维特殊坐标和时间。在 20℃下，$c=343\text{m/s}$，也可以表示为 $c=f\cdot\lambda$，其中 $f$ 是频率，$\lambda$ 是波长。式（10.16）假设声音传播是绝热和非黏性的。然而，在某些情况下，忽略黏度会产生错误的结论[29]。由于可听声压具有很大的可变性，因此可以用以分贝（dB）为单位的对数指数声压级（Sound Pressure Level，SPL）表示：

$$\text{SPL}=20\log_{10}\left(\frac{p}{p_r}\right) \tag{10.17}$$

式中：$p$ 为均方根声压，$p_r=20\mu\text{Pa}$，为空气中的参考声压。另一个重要的表达式是声强（$I_s$）乘以入射表面积（$A_s$）得到的声功率：

$$功率=I_s\cdot A_s=(p\cdot v)\cdot A=\left(\frac{p^2}{Z}\right)\cdot A_s \tag{10.18}$$

式中：$v$ 为粒子速度；$Z$ 为声阻抗率。

### 10.3.2 声功率增强

为了能够收集声能，必须放大低声压。当 SPL=114dB 时，相应的声压为 10Pa[29]。最有效的一种声学谐振器是由空腔和腔颈组成的亥姆霍兹谐振器。声音在谐振器内传播，压力随之逐渐增加，进而产生足够的压力来刺激空腔底部。在空腔底部放置一个柔性壁和一个压电盘（图 10.17）可以使换能器弯曲，将声能转化为电能。

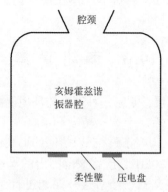

图 10.17 声学采集器示意图

更多示例可参见文献［20-22，27，29］。Monthéard 等介绍的如何使用采集到的空气噪声来为商业数据记录器供电是一份非常有前景的报告[23]，而 Noh 则研究了在铁路环境中采集声能的方法[26]。研究发现，高速列车在车厢内以及相邻两节车厢之间产生的噪声频率为 50~200Hz。他们针对 174Hz 的目标噪声设计了亥姆霍兹谐振器。实验结果表明，声压级为 100dB 时，产生的电压为 0.7V。最后，Monroe 和 Mir 分别在其硕士论文中论述了声能采集（Acoustic Energy Harvestigated，AEH）[24,28]。Monroe 研究了 AEH 设备的尺寸和带宽，并设计了一种基于 PVDF 压电薄膜的大型声能采集器，面积为 100cm²[24]。Mir 提出了一种新的超材料，它可以提供增强的隔声效果，同时将隔离的噪声转化为可用的电能[28]。数值计算预测 100Hz 的声音频率在 10kΩ 上产生大约 2mW 的电力。

### 10.4 使用形状记忆合金转化能量

第 4 章讨论了形状记忆合金，介绍了其性能和应用。结果表明，它们的动态性能无法与压电材料相比，因为形状记忆合金需要加热和冷却循环，这本身就需要时间。还有文献介绍了利用压电采集器或磁性形状记忆合金（Magnetic Shape Memory Alloys，MSMA）[1] 传感器消除附加能量的形状记忆合金研究。

### 10.4.1 磁性形状记忆合金

磁性形状记忆合金（Kari Ullakko 博士和他的同事于 1996 年在麻省理工学院首次发现）是一种铁磁性材料，可以在中等磁场下产生位移和力。通常，磁性形状记忆合金是镍、锰和镓（Ni-Mn-Ga）的合金（表 10.3）。磁性形状记忆效应发生在合金的低温马氏体相，其中构成合金的基本单元具有四方晶格。如果温度超过马氏体-奥氏体转变温度，合金将进入奥氏体相，其中基本单元具有立方晶格。这两种相之间的切换通常由磁场触发，但也可像传统形状记忆合金一样由温度变化或机械变形驱动。磁性形状记忆合金技术由于磁感应应变大、响应时间短，使得其对于创新型致动器的设计非常有吸引力，可应用于气动、机器人、医疗设备和机电一体化领域。磁性形状记忆合金的磁性能随形变而变化，这种伴随效应对于能量手机器或位移、速度和力的设计非常有用。

表 10.3　磁性形状记忆合金 Ni-Mn-Ga 单晶的性能研究[2]

| 特　性 | 数　值 |
| --- | --- |
| 磁场延伸率 | 通常为 3~5%，高达 6% |
| 响应 | 高达 1~2kHz |
| 力密度 | 约 2MPa |
| 输出功率（力×行程） | 最大 100kJ/m$^3$ |
| 疲劳寿命 | 几亿周期 |
| 磁场 | 小于 0.8T |
| 温度上限 | 马氏体转变为奥氏体的温度为 70℃ |
| 居里温度 | 95~105℃ |

### 10.4.2　形状记忆合金和磁性形状记忆合金的能量转化

形状记忆合金或磁性形状记忆合金可用于将寄生温度波动转化为电能。文

---

[1] Ullakko, K.,（1996 年）. Magnetically controlled shape memory alloys: A new class of actuator materials, Journal of Materials Engineering and Performance, Vol. 5, No. 3, 1996, pp. 405-409.

[2] Goodfellow——研发材料供应商，https://www.goodfellow.com。

献［30-42］介绍了此类收集器的典型研究。

Avirovik 等提出了一种基于压电双晶片悬臂梁结构的收集器[30]。悬臂的自由端与直径为 100μm 的预载荷 SMA 丝相连，SMA 丝被激光加热至 110℃ 并发生形变，产生弯曲悬臂的力。收集功率为预加载 SMA 及其沿悬臂位置的函数，测量范围为 0.003~0.006μW（RMS）。

Avirovik 等提出了另一种有趣的设备，以小型 SMA 热机的形式为无线传感器节点供电[31]。该装置包括热水器、滑轮系统和 SMA 丝。由于 SMA 丝在加热时从马氏体转变为奥氏体，在微型发电机上产生扭矩，从而使其旋转并产生电力，在 80Ω 电阻上测量到值为 1.8mW。文献［32-36，38-39，41］中介绍了 SMA 和压电贴片用于热能收集，而 Davidson 和 Mo 则介绍了用于结构健康监测的能量收集技术，其中包括基于 MSMA 的热能收集装置[37]。

以 MSMA 为材料构建共振收集器可以根据磁通梯度将振动能量直接转化为电能。因此，如果在恒定磁场中将铜线圈包裹在 MSMA 元件上，那么在 MSMA 器件上施加应变（或应力），就会在线圈中产生电流[39-40,42]。最后，Zacharov 等主张通过使用混合层压复合材料收集准静态温度变化，将热电、压电和形状记忆效应结合起来收集热能[34]。收集器由 MFC 压电片和 Ti-Ni-Cu 形状记忆合金组成，同时考虑了 MFC 压电片的热电效应。

## 10.5 利用压电传感器进行道路交通能量收集

如上所述，可利用在 $d_{33}$ 模式下工作的压电材料收集寄生机械能并将其转换为可用电能，文献［43-44］描述了嵌入在驾驶道路上的压电致动器，以从过往机动车中收集能量。该想法是建立一个能够感应经过车辆重量而变形的压电发电机（Piezoelectric Generator，PEG），并通过正压电效应产生电能。Abramovich 等的两篇早期论文[43-44]介绍了这些发电机在实验室条件下的试验（图 10.18），并讨论了由 PZT 陶瓷制成的大功率多元压电发电机在长期循环外部机械载荷作用下的性能。结果表明，在安全侧应力为 30MPa 的情况下，发电功率略有下降，如图 10.19 所示。由于施加在 PZT 盘上的机械应力相对较低，所产生的电能也相对较低，因此可以通过将压电材料层逐层堆叠来增加材料的体积，每个压电材料层承受相同的机械应力，这将从安全机械应力下得到所需的电力，而不会减少其输出。后续文献［45-51］主要回顾了压电道路发电机的最新技术，讨论了其实现前景，同时介绍了最新成果。Kim 等总结了压电式道路发电机的研究成果，并打算在美国乔治亚州的高速公路进行试验[45]。

Duarte 和 Ferreira 概述了道路路面能源收集技术的发展，试图将其划分为不同的类别，同时利用现有成果对这些技术进行了讨论、分析和比较[46]。Kour 和 Charif 旨在评估道路中的压电性能，以利用移动车辆产生的能量[47]。利用压电技术将能量转换为电能，以取代化石燃料在路灯中的应用。这项新技术宣称，道路压电是一种新的能源革命，可提供环境、经济和社会需求方面的可持续解决方案。

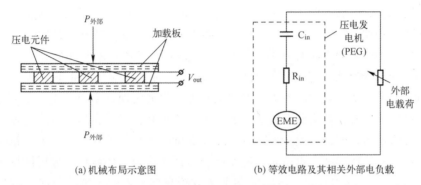

(a) 机械布局示意图　　(b) 等效电路及其相关外部电负载

图 10.18　压电发电机[43]

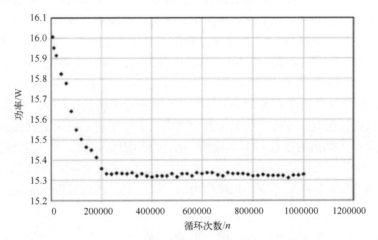

图 10.19　在 5Hz 的频率和 30MPa 载荷下压电道路发电机的输出功率[44]

Papagiannakis 等介绍了几种压电样机的开发，这些样机能够从高速公路上的交通活动中收集能量[48]。通过对移动轮胎载荷下路面应变能的有限元模拟，探索了可收集的能量。开发的样机包括各种圆柱形和棱柱形压电元件。研究分析表明，应将叠层压电元件的极性进行交替，且并联连接，以避免产生难以控制的高电压。他们对样机进行了频率为 10Hz 的正弦加载下的单轴压缩试验，

421

并对电功率与应力的实验室数据曲线进行了拟合。对于单程 44.48kN 载重汽车轮胎载荷,预计产生的电功率在 1.0~1.8W,与文献［43-44］中给出的实验结果相差甚远。Yang 等在最近的文献（2017 年）中介绍了使用堆叠模式的道路发电机的实验结果[49]。他们的模型能够承受高达 150kN 的重复载荷,压电换能器可经过 100000 次循环而不降低输出功率。作者采用了 280V 的开路电压,但未提及所获得的功率。值得一提的是,压电道路发电机所用的压电材料数量较少,期望输出功率较小。另一篇由 Qabur 等发表的关于压电道路发电机的文献（2018 年）,再次回顾了从道路收集能源的现状,并认为 PZT 因为其独特的特性而被认为是最有效的材料[50]。根据大量参考文献和实际应用,多种因素（如几何形状、厚度和结构）会影响压电过程的输出,而能量收集器可以通过最大限度地减少收集阶段的机械和电力损失实现效率增强或改良。他们总结指出压电材料在道路上的应用需要进一步研究,根据实际生活或实际应用中获得的不同数据进行更全面地分析。Walubita 等在其文章中回顾了道路能量收集技术的最新进展,重点是压电系统,包括从社会和环境角度分析该技术造成的影响[51]。总的来说,他们的参考文献研究结果表明,将道路能量收集技术扩展到大规模的实际运用是可行的,但应该从更广泛的角度明智地权衡这一任务。最后,文献对道路能源收集技术对人类社会能源持续和环境挑战的潜在贡献提出了积极的展望。该文章重点介绍了美国得克萨斯州农工运输研究所（Texas A&M Transportation Institute,TTI）进行的一项实验研究,在该研究中,一个名为公路传感和能量转换（Highway Sensing and Energy Conversion,HiSEC）的压电模块样机在 TTI 的一个实验室进行了实验。HiSEC 模块由 3 对压电片组成,压电片堆叠在一起,并与两个二极管相连。模块随后被封装在一个金属盒中,顶部覆盖着一个防撞帽。在 0.827MPa（120psi）的机械应力和 1Hz 的频率下,该模块获得的最大功率约为 13.5mW。请注意,他们的实验中由于使用的压电材料数量少、机械应力低、频率低,从而导致性能较差。

# 参 考 文 献

［1］Williams, R. B., Park, G., Inman, D. J. and Wilkie, W. K., An overview of composite actuators with piezoceramic fibers, International modal analysis conference, Proceedings of IMAC-XX, the Westin Los Angeles Airport, Los Angeles, CA, USA, 47, February 2002, 421-427.

［2］Mallik, N. and Ray, M. C., Effective coefficients of piezoelectric fiber-reinforced composites, AIAA Journal 41 (4), 2003, 704-710.

［3］Nelson, L. J., Bowen, C. R., Stevens, R., Cain, M. and Stewart, M., Modelling and measurement of piezoelectric fibres and interdigitated electrodes for the optimization of piezofibre composites, Proceeding Vol. 5053, Smart Structures and Materials, 2003: Active Materials: Behavior and Mechanics, 2003, San Diego, CA, US., 12. doi: 10.1117/12.484738.

[4] Berger, H., Kari, S., Gabbert, U., Ramos, R. R., Guinovart, R., Otero, J. A. and Catillero, J. B., An analytical and numerical approach for calculating effective material coefficients of piezoelectric fiber composites, International Journal of Solids and Structures 42, 2005, 5692-5714.

[5] Ralf, S., Modelling and characterization of piezoelectric 1-3 fibre composites., In: Chapter A8.4 from Piezoelectric and acoustic materials for transducer applications, Safari, A. and Akdoğan, E. K. (eds.), Springer Science+Business Media, LLC, 2008, 483.

[6] Schönecker, A., Piezoelectric fiber composite fabrication, Chapter 13., From Piezoelectric and acoustic materials for transducer applications, Safari, A. and Akdoğan, E. K. (eds.), Springer Science+Business Media, LLC, 2008, 483.

[7] Deraemaeker, A., Nasser, H., Benjeddou, A. and Preumont, A., Mixing rules for the piezoelectric properties of Macro Fiber Composites (MFC), Journal of Intelligent Material Systems and Structures 20 (2), 2009, 1475-1482. doi: 10.1177/1045389x09335615.

[8] Smith, W. A., Modeling 1-3 composite piezoelectrics: hydrostatic response, IEEE Transactions on Ultrasonics, Ferroelectrics, and Frequency Control 40 (1), 1993, 41-49.

[9] Kerur, S. B. and Ghosh, A., Active vibration control of composite plate using AFC actuator and PVDF sensor, International Journal of Structural Stability and Dynamics 11 (2), 2011, 237-255. doi: 10.1142/so219455411004075.

[10] Lin, X.-J., Zhou, K.-C., Zhang, X.-Y. and Zhang, D., Development, modeling and application of piezoelectric fiber composites, Transactions of Nonferrous Metals Society of China 23, 2013, 98-107.

[11] Nilson, E., Mateu, L., Spies, P. and Hagström, B., Energy harvesting from piezoelectric textile fibers, Procedia Engineering 87, 2014, 1569-1572.

[12] Jemai, A., Najar, F., Chafra, M. and Ounaies, Z., Mathematical modeling of an active-fiber composite energy harvester with interdigitated electrodes, Shock and Vibration 2014, 2014, 9, Paper Id 971597.

[13] Ji, S. H., Cho, J. H., Jeong, Y. H., Paik, J.-H., Yun, J. D. and Yun, J. S., Flexible lead-free piezoelectric nanofiber composites based on BNT-ST and PVDF for frequency sensor applications, Sensors and Actuators. A, Physicals 247, 2016, 316-322.

[14] Matsouka, D. and Vassiliadis, S., Piezoelectric melt-spun textile fibers: technological overview, Chapter 4 in Piezoelectricity-organic and inorganic materials and application, Intech-Open 2016, 66-82. doi: http://dx.doi.org/10.5772/intechopen.78389.

[15] Kumar, R. S., Sarathi, T., Venkataraman, K. K. and Bhattacharyya, A., Enhanced piezoelectric properties of polyvinylidene fluoride nanofibers using carbon nanofiber and electrical poling, Material Letters 255, 2019, 4, Paper Id 126515.

[16] Park, S., Kwon, Y., Sung, M., Lee, B.-S., Bae, J. and Yu, W.-R., Poling-free spinning process of manufacturing piezoelectric yarns for textile applications, Materials and Design 179, 2019, 10, Paper Id. 107889.

[17] Ghafari, E. and Lu, N., Self-polarized electrospun polyvinylidene-fluoride (PVDF) nanofiber for sens-

423

ing applications, Composites Part B 160, 2019, 1-9.

[18] Lam, T. -N., Wang, -C. -C., Ko, W. -C., Wu, J. -M., Lai, S. -N., Chuang, W. -T., Su, C. -J., Ma, C. -Y., Luo, M. -Y., Wang, Y. -J. and Huang, E. -W., Tuning mechanical properties of electrospun piezoelectric nanofibers by heat treatment, Materialia 8, 2019, 8, Paper Id 100461.

[19] Cui, N., Jia, X., Lin, A., Liu, J., Bai, S., Zhang, L., Qin, Y., Yang, R., Zhou, F. and Li, Y., Piezoelectric nanofiber/polymer composite membrane for noise harvesting and active acoustic detection, Nanoscale Advances 1, 2019, 4909-4914. doi: 10.1039/cqna00484y.

[20] Sherrit, S., The physical acoustics of energy harvesting, 2008 IEEE International Ultrasonics Symposium Proceedings, Beijing International Convention Center (BICC) Beijing, China, November 2-5, 2008, 1046-1055. doi: 10.1109/ULTSYM.2008.0253.

[21] Lin, J. -T., Lee, B. and Alphenaar, W., Non-linear energy harvesting with random noise and multiple harmonics, Chapter 12, Small-scale energy harvesting, Intech-Open 2012, 283-302.

[22] Rahman, A. and Hoque, M. E., Harvesting energy from sound and vibration, Paper ID: Am-22, International Conference on Mechanical, Industrial and Materials Engineering 2013 (ICMIME2013), 1-3 Nov. 2013, RUET, Rajshahi, Bangladesh, 6.

[23] Monthéard, R., Airiau, C., Bafleur, M., Boities, V., Dilhac, -J. -J., Dollat, X., Nolhier, N. and Piot, E., Powering a commercial datalogger by energy harvesting from generated aeroacoustic noise, Journal of Physics: Conference Series 557, 2014, 5, Paper Id: 012025. doi: 10.1088/1742-6596/557/1/012025.

[24] Monroe, N. M., Broadband Acoustic Energy Harvesting via Synthesized Electrical Loading, Master of Engineering Thesis, Department of Electrical Engineering and Computer Science, Massachusetts, USA, MIT, Cambridge, 2017, 149.

[25] Zhao, N., Zhang, S., Yu, F. R., Chen, Y., Nallanathan, A. and Leung, V. C. M., Exploiting interference for energy harvesting: a survey, research issues, and challenges, IEEE Access 5, 2017, 10403-10421. doi: 10.1109/ACCESS.2017.2705638.

[26] Noh, H. -M., Acoustic energy harvesting using piezoelectric generator for railway environmental noise, Advances in Mechanical Engineering 10 (7), 2018, 1-9. doi: 10.1177/1687814018785058.

[27] Choi, J., Jung, I. and Kang, C. -Y., A brief review of sound energy harvesting, Nano Energy 56, 2019, 169-183.

[28] Mir, F., Acoustoelastic metamaterial with simultaneous noise filtering and energy harvesting capability from ambient vibrations, Master of engineering thesis, Mechanical Engineering, College of Engineering and Computing, University of South Carolina, Columbia, South Carolina, USA, 2019, 62.

[29] Yuan, M., Cao, Z., Luo, J. and Chou, X., Recent developments of acoustic energy harvesting: a review, Micromachines 10 (1), 2019, 48, 21. doi: 10.3390/mi10010048.

[30] Avirovik, D., Kumar, A., Bodnar, R. J. and Priy a, S., Remote light energy harvesting and actuation using shape memory alloy-piezoelectric hybrid transducer, Smart Materials and Structures 22, 2013, 6, Paper Id: 052001.

[31] Avirovik, D., Kishore, R. A., Vuckovic, D. and Priya, S., Miniature shape memory alloy heat engine for powering wireless sensor nodes, Energy Harvesting and Systems 1 (1-2), 2014, 13-18.

[32] Gosliga, J. S. and Ganilova, O. A., Energy harvesting based on the hybridization of two smart materi-

als, Paper #170, EACS 2016-6th European Conference on Structural Control, Sheffield, England, 11 -13 July 2016, 12.

[33] Zakharov, D., Lebedev, G., Cugat, O., Delamare, J., Viala, B., Lafont, T., Gimeno, L. and Shelyakov, A., Thermal energy conversion by coupled shape memory and piezoelectric effects, Journal of Micromechanics and Microengineering 22, 2012, 7, Paper Id: 094005. doi: 10.1088/0960-1317/22/9/094005.

[34] Zakharov, D., Gusarov, B., Gusarova, E., Viala, B., Cugat, O., Delamare, J. and Gimeno, L., Combined pyroelectric, piezoelectric and shape memory effects for thermal energy harvesting, Journal of Physics: Conference Series 476, 2013, 5, Paper Id: 012012. doi: 10.1088/1742-6596/476/1/012021.

[35] Namli, O. C. and Taya, M., Design of piezo-SMA composite for thermal energy harvester under fluctuating temperature, Journal of Applied Mechanics 70, 2011, 8, paper Id: 03101.

[36] Namli, O. C., Jae-Kon, L. and Taya, M., Modeling of piezo-SMA composites for thermal energy harvester, Proc. SPIE 6526, Behavior and Mechanics of Multifunctional and Composite Materials 2007, 65261L, 12 April 2007, San Diego, California, USA., 12. doi: 10.1117/12.715786.

[37] Davidson, J. and Mo, C., Recent advances in energy harvesting technologies for structural health monitoring applications, Smart Materials Research 2014, 2014, 14, Paper Id: 410316. doi: 10.1155/2014/410316.

[38] Todorov, T., Nikolov, N., Todorov, G. and Ralev, Y., Modelling and investigation of a hybrid thermal energy harvester, MATEC Web of Conferences, Vol. 148, 2018, International Conference of Engineering Vibration (ICoEV), Paper Id. 12002, 6. doi: 10.1051/matecconf/2018148/2002.

[39] Viet, N. V., Zaki, W. and Umer, R., Analytical investigation of an energy harvesting shape memory alloy-piezoelectric beam, Archive of Applied Mechanics 90 (12), 2020, 2715-2738. doi: 10.1007/s0049-020-01745-9.

[40] Fasangi, M. A. A., Cottone, F., Sayyaadi, H., Zakerzadeh, M. R., Orfei, F. and Gammaitoni, L., Energy harvesting from structural vibrations of magnetic shape memory alloys, Applied Physical Letters 110, 2017, Paper Id: 103905, 4.

[41] Adeodato, A., Duarte, B. T., Monteiro, L. L. S., Pacheco, P. M. C. L. and Savi, M. A., Synergistic use of piezoelectric and shape memory alloy elements for vibration-based energy harvesting, International Journal of Mechanical Sciences 194, 2021, 10, Paper Id: 106206.

[42] Safari, O., Zakerzadeh, M. R. and Baghani, M., Study of a magnetic SMA-based energy harvester using a corrugated structure, Journal of Intelligent Material Systems and Structures 32, 2021, 12. doi: 10.1177/1045389x20983903.

[43] Abramovich, H., Tsikhotsky, E. and Klein, G., An experimental determination of the maximal allowable stresses for high power piezoelectric generators, Journal of Ceramic Science and Technology 4 (3), 2013, 131-136. doi: 10.4416/JCST2013-00006.

[44] Abramovich, H., Tsikhotsky, E. and Klein, G., An experimental investigation on PZT behavior under mechanical and cycling loading, Journal of the Mechanical Behavior of Materials 22 (3-4), 2013, 129-136.

[45] Kim, S., Shen, J. and Ahad, M., Piezoelectric-based energy harvesting technology for road sustain-

ability, International Journal of Applied Science and Technology 5 (1), 2015, 20-25.

[46] Duarte, F. and Ferreira, A., Energy harvesting on road pavements: state of the art, Proceedings of the Institution of Civil Engineers, Energy 169 (EN2), 2016, 79-90, Paper Id: 1500005.

[47] Kour, R. and Charif, A., Piezoelectric roads: energy harvesting method using piezoelectric technology, Innovative Energy and Research 5 (1), 2016, 6, Paper Id: 10000132. doi: 10.4172/2576-1463.1000132.

[48] Papagiannakis, A. T., Montoya, A., Dessouky, S. and Helffrich, J., Development and evaluation of piezoelectric prototypes for roadway energy harvesting, Journal of Energy Engineering, ASCE 143 (5), 2017, 7, Paper Id: 04017034. doi: 10.1061/ (ASCE) EY.1943-7897.0000467.

[49] Yang, H., Wang, L., Hou, Y., Guo, M., Ye, Z., Tong, X. and Wang, D., Development in stackedarray-type piezoelectric energy harvester in asphalt pavement, Journal of Materials in Civil Engineering, ASCE 29 (11), 2017, 9, Paper Id: 04017224.

[50] Qabur, A., Alshammari, K. and Systematic, A., Review of energy harvesting from roadways by using piezoelectric materials technology, Innovative Energy & Research 7 (1), 2018, 6, Paper Id: 10000191. doi: 10.4172/2576-1463.1000191.

[51] Walubita, L. F., Djebou, D. C. S., Faruk, A. N. M., Lee, S. I., Dessouky, S. and Hu, X., Prospective of societal and environmental benefits of piezoelectric technology in road energy harvesting, Sustainability 10, 2018, 383, 13. doi: 10.3390/su10020383.

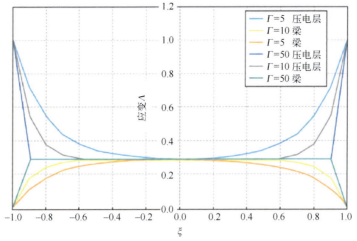

图 3.23 当 $\alpha=6$、$\psi=15$ 时,对于 $\Gamma=5$、10、50 情况下,压电层和沿梁长度承载结构中的无量纲应变

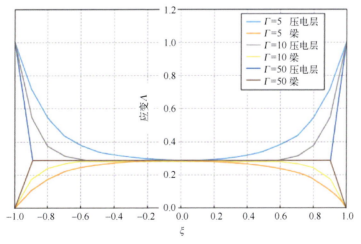

图 3.24 当 $\alpha=2$、$\psi=15$ 时,对于 $\Gamma=5$、10、50 情况下,压电层和沿梁长度承载结构中的无量纲应变

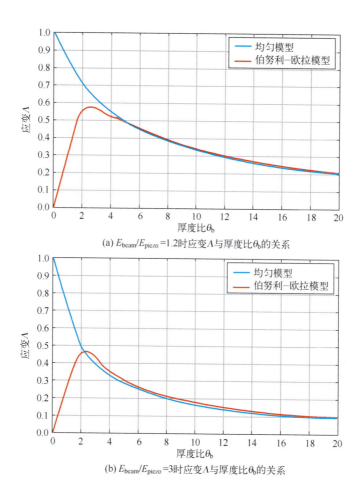

(a) $E_{beam}/E_{piezo}=1.2$ 时应变 $\Lambda$ 与厚度比 $\theta_b$ 的关系

(b) $E_{beam}/E_{piezo}=3$ 时应变 $\Lambda$ 与厚度比 $\theta_b$ 的关系

图 3.29 弯曲应变对比

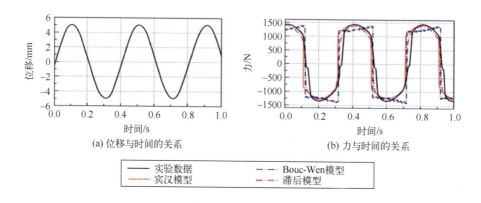

(a) 位移与时间的关系

(b) 力与时间的关系

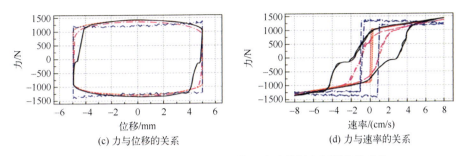

(c) 力与位移的关系  (d) 力与速率的关系

图 5.30　磁流变体减震器-多个模型试验数据和预测值的比较，
激励信号采用振幅 5mm、电流 1.5A 的 2.5Hz 正弦激励[48]

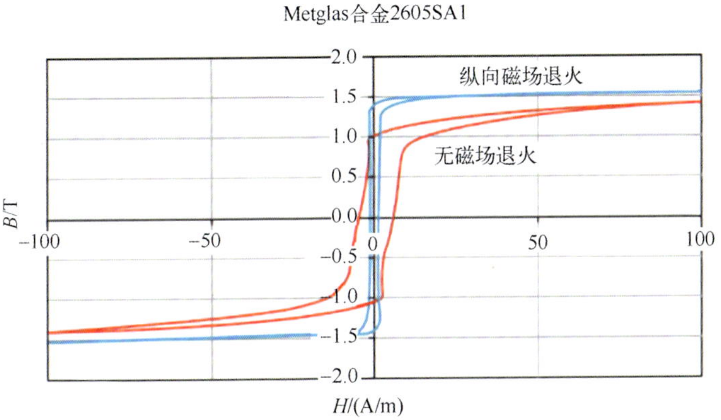

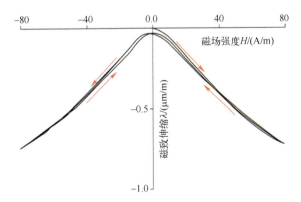

图 6.5　Metglas 合金 2605SA1——一种典型的磁滞回线

（资料来源：www.metglas.com/products/magnetic_materials/2605sa1.asp）

图 6.6　Mn-Zn 铁氧体的典型磁致伸缩迟滞回线

（资料来源：https://commons.wikimedia.org/wiki/File:Magnetostrictive_hysteresis_loop_of_Mn-Zn_ferrite.png）

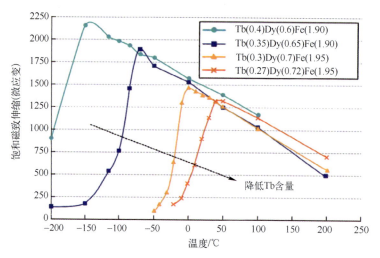

图 6. c 典型 Terfenol-D 合金：温度影响
（资料来源：ETREMA 专有资料）

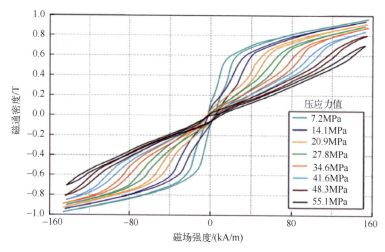

图 6. d 典型 Terfenol-D 合金：不同压应力下磁通密度与磁场强度之间的关系
（资料来源：ETREMA 专有资料）

彩 4

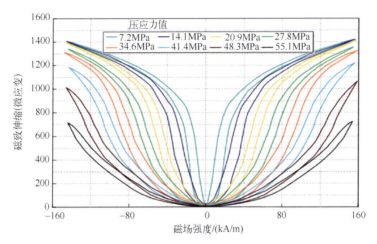

图6.e 典型 Terfenol-D 合金：不同压应力下磁致伸缩与磁场之间的关系
(资料来源：ETREMA 专有资料)

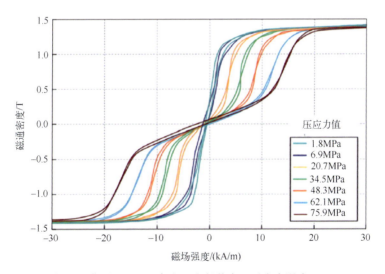

图6.f 典型 Galfenol 合金（生长状态，无应力退火 BH）：
不同压应力下磁通量密度与磁场强度之间的关系
(资料来源：ETREMA 专有资料)

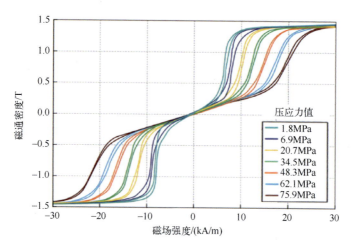

图 6.g 典型 Galfenol 合金（应力退火 BH）：
不同压应力下磁通密度与磁场强度之间的关系
（资料来源：ETREMA 专有资料）

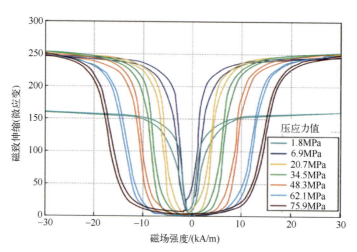

图 6.h 典型 Galfenol 合金（生长状态，无应力退火 BH）：
不同压应力下磁致伸缩与磁场强度之间的关系
（资料来源：ETREMA 专有资料）

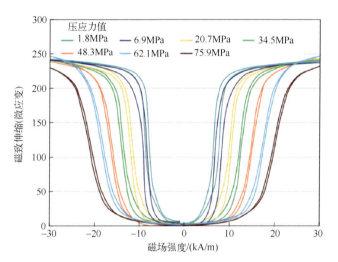

图6.i 典型 Galfenol 合金（应力退火 BH）：
不同压应力下磁致伸缩与磁场强度之间的关系
（资料来源：ETREMA 专有资料）

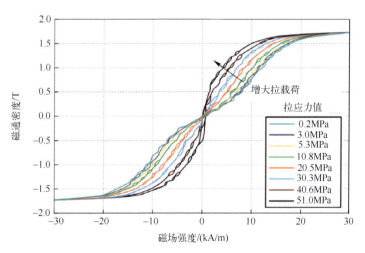

图6.j 典型 Galfenol 合金（生长状态，无应力退火 BH）：
不同拉应力下磁致伸缩与磁场强度之间的关系
（资料来源：ETREMA 专有资料）

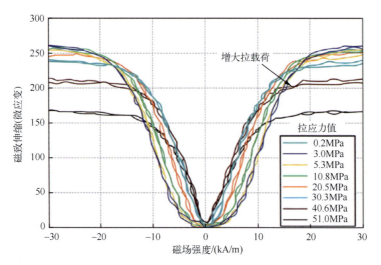

图 6.k 典型 Galfenol 合金（应力退火 BH）：
不同拉应力下磁致伸缩与磁场强度之间的关系
（资料来源：ETREMA 专有资料）

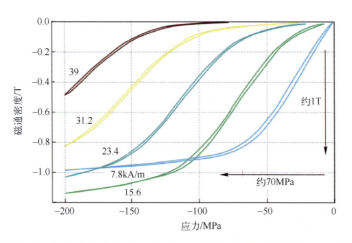

图 6.l 典型 Galfenol 合金：不同场强下通量密度相对于应力强度的降低
（资料来源：ETREMA 专有资料）

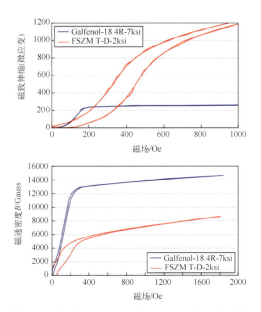

图6.m 典型 Terfenol-D 和 Galfenol 合金的比较

(资料来源：ETREMA 专有资料)

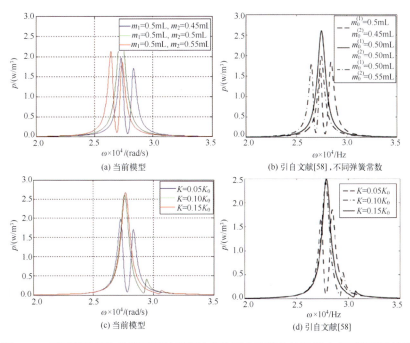

图8.21 不同端质量块条件下通过弹簧连接的两个双晶片的解析功率密度模型结果